Quantum Computing and Quantum Physics

Quantum Computing and Quantum Physics

A Linear Algebra Approach

Alice Flarend
Bellwood-Antis High School

Robert Hilborn
University of Maryland

OXFORD
UNIVERSITY PRESS

OXFORD
UNIVERSITY PRESS

Great Clarendon Street, Oxford, OX2 6DP,
United Kingdom

Oxford University Press is a department of the University of Oxford.
It furthers the University's objective of excellence in research, scholarship,
and education by publishing worldwide. Oxford is a registered trade mark of
Oxford University Press in the UK and in certain other countries.

Published in the United States of America by Oxford University Press
198 Madison Avenue, New York, NY 10016, United States of America

British Library Cataloguing in Publication Data
Data available

Library of Congress Control Number: 2025950476

ISBN 9780197904930
ISBN 9780197904923 (pbk.)

DOI: 10.1093/oso/9780197904930.001.0001

Printed and bound by
CPI Group (UK) Ltd., Croydon, CR0 4YY

MIX
Paper | Supporting responsible forestry
FSC® C013604
FSC www.fsc.org

The manufacturer's authorized representative in the EU for product safety is
Oxford University Press España S.A. of Parque Empresarial San Fernando de Henares,
Avenida de Castilla, 2 – 28830 Madrid (www.oup.es/en or product.safety@oup.com).
OUP España S.A. also acts as importer into Spain of products made by the manufacturer.

Contents

PART II Multiqubit States

PART III Quantum Algorithms

Preface

Dear Readers:

Let us introduce ourselves to let you know about the perspectives we bring to this book.

ALICE: A nuclear engineer and educator with a strong theoretical background in physics and pedagogy, she is very experienced with translating quantum computing and quantum cryptography for high school students and teachers.

BOB: An experimental physicist who likes to think about how quantum computers will actually be built and how you can use lasers and microwaves to prepare, manipulate, and analyze quantum states. But he also has a philosophical bent: What does quantum computing tell us about fundamental issues in quantum mechanics and information theory? What does it teach us about the fundamental nature of reality?

Many folks ask if we are the famous characters Alice and Bob who appear in both popular books and technical papers about quantum information, cryptography, and quantum computing. Unfortunately, we are not that Alice and Bob. In fact, "Alice and Bob" first appeared in print in 1978 in a cryptography article in the technical journal *Communications of the Association for Computing Machinery*. Nor are we the characters in the 1969 movie *Bob & Carol & Ted & Alice*. But the common mention of our names in quantum information and quantum computing books was part of the motivation for our putting together these materials. What better way to learn about quantum computing than directly from Alice and Bob!

You have probably picked up this book because you are intrigued or puzzled by what you have heard about quantum computing (QC) and quantum information science (QIS) and you want to learn more. Why should you keep reading? What is different in this book? We have tried to position this book between the highly technical books aimed at professional scientists and engineers with expertise in formal quantum mechanics and advanced mathematics, and those general audience books that use almost no math, though some are quite clever in finding pictorial substitutes for the math.

Our presentation is aimed at readers who want an introduction to quantum computing that gives them a strong basic understanding and prepares them to talk intelligently with the "experts." If readers are so inclined, they will be ready to dig into the more technical aspects of the field after working through this book. The material should be accessible to a typical undergraduate student whose mathematics background includes some linear algebra. We use a bit of calculus, but no prior experience with calculus is needed. No college-level background in physics is required, but if you are fortunate enough to have had an advanced placement physics course in high school or a college-level introductory physics course, what you learned there will provide a broader perspective on quantum computing. High school mathematics and physics teachers as well as college and university faculty who are not experts in QIS and QC should also enjoy, and benefit from, reading the book.

To distinguish our presentation from others, we

- Make the treatment accessible to those who do not have much background in physics by providing just enough information to enable you to understand the fundamental issues. We have avoided unnecessary physics and mathematical jargon, but we treat carefully what we do use with examples and explanatory comments.
- Spread out discussions to take time to build an understanding of the critical elements of quantum computing. The tone of the writing is more conversational than didactic.
- Develop, in some detail, applications of quantum computing to show where and how quantum computing has advantages over traditional computational methods. We use those applications to ground the mathematics in specific and accessible problems to be solved.
- Provide brief surveys of current QIS and QC devices, advice about what you should do next if you want to learn more about those fields, and an overview of career opportunities in those areas.

Although we have worked hard to make the book accessible to a wide audience, we also want to give you the concepts and vocabulary to read intelligently and knowledgeably about quantum computing. To accomplish that, we

- Introduce and thoroughly explore Dirac quantum state vector notation since it is the most widely used notation in quantum computing and, at the same time, the notation that you in our target audience have probably not seen before, even if you have taken introductory courses in physics, linear algebra, or computer science.
- Focus on multiple representations of quantum states and operations on quantum states, using abstract symbols, graphical (state space) diagrams, and mathematical column vectors and matrices. We emphasize multiple representations because quantum states are the bread and butter of quantum computing. Using multiple representations will deepen your understanding.
- Present clear ways of thinking about the nonintuitive role of quantum measurements in quantum computing and how you represent these measurements formally. We show you how to make a "quantum sandwich," which encapsulates everything we can learn from quantum measurements.
- Provide in-text exercises (*Try Its*) that you can use to build your fluency in writing descriptions of quantum states and to check your understanding of the concepts.
- Give a summary of quantum formulas and linear algebra terminology in the Appendix: Quantum Toolkit.

The QIS and QC revolutions are just getting underway. To give you a sense of the diversity in both people and their educational backgrounds in these rapidly growing fields, we have included at the end of the last chapter links to short biographical sketches of those involved who work in those fields in academia, in government labs, and in private business and industry.

Although the fundamental principles of QIS and QC are reasonably well established, the details of the applications of those principles are likely to change dramatically over the next few decades. In the last chapter of the book, we write about what we believe are likely to be the most exciting developments in these fields. We also discuss moral and ethical concerns raised by the possibility of dramatic changes in computational power that are likely to be the product of the second quantum revolution.

Perhaps we should also tell you about what is *not* in the book—things that you might have expected to see. We avoid

- Stern–Gerlach measurements using magnetic field gradients. To appreciate those measurements, which were important in the historical development of quantum mechanics, you need a deeper grounding in physics: atomic and molecular beams, magnetic dipole moments, and forces exerted on dipoles due to magnetic field gradients. More importantly, these kinds of measurements are unlikely to play a role in QC and QIS even though they show up in many books on quantum computing. We believe that polarized light provides a much more accessible physical system to illustrate the basics of quantum states. Furthermore, QIS and QC systems using polarized light are already underway. We do give an introduction to the so-called spin-1/2 quantum states but without the details of how physical measurements are carried out on those systems. We refer the reader who is interested in these details to many fine textbooks and online articles.
- Computational complexity. Many computer scientists and quantum computer people spend a lot of time building classifications of the difficulty of the problems addressed by computational algorithms. You may have read about P and NP problems. Those categories have proliferated into an alphabet soup of acronyms. Although these classifications help you to understand the kinds of problems quantum computers can advantageously solve (compared to traditional computers), we believe that knowledge of these categories is not necessary to understand the fundamentals of quantum computing and are best left aside until the basics are mastered. In a few specific examples, we do provide arguments that show that quantum computation often requires fewer (and, in some cases, far fewer) computational steps compared to the corresponding classical algorithms, but we avoid connecting these to the traditional computational complexity categories.

For those of you who are educators, this book provides more than enough content for a semester-long course. The core material is in Chapters 1–10. Chapters 11–16 provide detailed discussions of several quantum algorithms. The algorithms are relatively independent. Pick those that your students will find most interesting and discuss the details at a level appropriate for your students. In any of those algorithms they will see the key quantum ideas used in quantum algorithms including superposition, change in measurement basis, and possible entanglement. If you have time to do only one algorithm in detail, we suggest the Grover algorithm in Chapter 12. It requires the least mathematical background to understand the goal, and we have presented several different and complementary ways of understanding how the algorithm works.

Chapters 17–23 provide a deeper dive into quantum mechanics. These chapters are important for readers who want a more complete understanding of quantum mechanics in preparation for further coursework or reading. Instructors can choose between a more detailed approach to quantum algorithms and a broader look at the procedures of quantum mechanics, depending on the backgrounds and career interests of their students. Parts III and IV are relatively independent of one another to allow for more flexibility in course structures. Students interested in current technology and careers in QC and QIS can read the last chapter on their own.

Throughout the book, as we have mentioned, we make use of three ways of working with quantum states: symbolic state vectors, graphical representations (state space), and column vectors and matrices. Even though many students in our intended audience will have had linear algebra, some may be "rusty" in the application of matrix methods. We have included some exercises to help remove that rust. The book is also well suited to self-study. We encourage students to form collaborative groups to read the book together.

Our introduction to quantum computing is built around a fundamental theme: quantum states, quantum entanglement (we will explain what that means), and quantum measurements all provide "resources" for computation that in principle far exceed what can be done with traditional computers. We will play this leitmotif again and again throughout the book. However, quantum states are far from our everyday intuitions about how the world works. In fact, when it comes to quantum states and quantum computing, we are all "novices," and it takes many hours of work to move from being a novice to being an expert. So, it would be presumptuous for us to say that working through this book will make you an expert in quantum computing. Our goals are to provide you with the basic concepts and formalism of quantum computing, to show you how quantum states can be manipulated using quantum algorithms to carry out computationally useful processes, and to look to the future of where quantum computing is likely to go.

We treat many of the "classic" quantum algorithms as case studies to show how our fundamental theme has been applied to solve computational problems that are difficult if not impossible for traditional computers. We don't expect that most of you will be developing new quantum algorithms, just as the productive use of traditional computers does not require that you develop new numerical algorithms. In practice, programmers simply make use of what others have created and assemble the computer code to tackle the problem at hand. But we do want you to have the tools needed to be able to read and talk intelligently about quantum computing because there are many more opportunities in the field beyond writing algorithms.

As mentioned previously, we have, by design, limited the math background required to benefit from this book. That said, we do recognize that some experience with symbolic (algebraic) reasoning, with graphical and geometric representations, with the basics of linear algebra, and, more importantly, with persistence and perseverance will be important to grasp the fundamental concepts and ideas. Our advice to all readers: Give yourself plenty of time to work through and digest the material and to reread the arguments to deepen your understanding. Work and talk with others if possible. Keep in mind that it took physicists and mathematicians several decades and many conversations and arguments to complete the development of quantum mechanics, and another 50–60 years to recognize the revolutionary potential of quantum mechanics to rebuild completely the foundations of information science and computing. So, as another Bob (Bobby McFerrin) sang "don't worry, be happy" if some concepts seem rather puzzling at first. You are in good company. As dedicated educators, we need to remind you, as we remind our own students, if you are not confused, you are not learning. Or as a skiing instructor once told Bob, you are never going to learn to ski unless you fall down.

Let us close by mixing metaphors: We hope you will persevere through your confusion as you navigate the moguls on the slopes of quantum mechanics, quantum computing, and quantum information science.

The book's website QuantumComputingA2B.org will include solutions to the *Try Its*, additional problems for exams and group-problem solving, resources for teachers, links to other websites for concept assessments tools, and "breaking news" about quantum computing and quantum information science.

Your devoted authors,
Alice and Bob

Acknowledgments

We express our deep appreciation to Steve Arendt, John Donohue, and Diandra Leslie-Pelecky for providing feedback on the manuscript, encouraging us to clarify our thoughts and to focus on the central message. Many students and physics education colleagues over the years helped us to understand and learn to explain quantum mechanics more effectively, though that is a life-long journey. We express particular appreciation to the computer science, computer engineering, and math majors in the fall 2024 offering of Physics 360 (Quantum Mechanics: A Linear Algebra Approach) at the University of Maryland, whose comments and suggestions led to significant improvements in the book. We thank OUP Senior Editor Sonke Adlung for his enthusiasm and support for this project. OUP Senior Project Editor Guilia Lipparini and Project Manager Lavanya Sathiamurthy at Integra Software Services provided patient and practical advice in helping us navigate the manuscript preparation and refinement stages of the work. Our anonymous copy editor's efforts to enhance consistency and clarity throughout the manuscript are deeply appreciated.

PART I

Quantum State and Quantum Information Science Fundamentals

Introduction 1

In Nature's infinite book of secrecy, a little I can read.

Shakespeare, *Antony and Cleopatra*

1.1 What's the Big Deal about Quantum Computing?

Our goal in this book is to introduce people who aren't experts in physics and quantum mechanics to the key ideas and vocabulary of quantum computing (QC) and quantum information science (QIS). There are many books that take a fairly high-level approach, which is fine for math and physics majors, but leave most folks baffled. There are also several books that avoid math entirely and focus on general concepts, often with idiosyncratic models and weird terminology that don't allow contact with how most scientists, engineers, and mathematicians talk about quantum computers. We worked hard to find the "sweet spot" between insider jargon wrapped in unfamiliar mathematics and fluffy treatments that seem strange and inadequate to both experts and novices. Since both authors struggled to penetrate the many layers of jargon that expert authors have brought to the field, we felt we were in a good position to find that sweet spot. Jargon is fine of course for the experts, but many authors use it mainly to show that they are part of the "quantum tribe" and to separate the tribe from outsiders.

You may wonder about the difference between quantum information science and quantum computing. "Quantum information science" has become the general term for the use of quantum principles and technology for the transfer of information (what you might call quantum communications), the processing of information (quantum computing), and for the sensitive detection of various kinds of physical properties (quantum sensing). People are already talking about building a quantum internet to enhance the security of information exchange. In this book, we focus on the first two. To appreciate quantum sensing, you need substantial background in physics and how quantum systems interact with their environment.

Of course, quantum computing is built on the basic principles of quantum mechanics, and there are many parts of quantum mechanics that are conceptually quite challenging for people new to the field—and even to experts. The difficulty is no one has day-to-day experiences with the quantum realm like they do with the nonquantum realm. Of course,

Quantum Computing and Quantum Physics. Alice Flarend and Robert Hilborn, Oxford University Press.
 DOI: 10.1093/oso/9780197904930.003.0001

many scientists and mathematicians know how to do quantum calculations, but there is still a lot of controversy about what those calculations tell us about the nature of reality. We have tried to develop quantum computing ideas in a way that avoids, at least for a while, those conceptual challenges, but we will need to deal with them eventually. And, in their own way, they are rather cool challenges.

In selecting the mathematics level for this book, we have focused on a linear algebra approach to QC by choosing examples carefully and avoiding unnecessary mathematical jargon. The traditional physics approach to quantum mechanics leans heavily on differential equations. Both approaches are important, and we will show a connection between the linear algebra and differential equations. But for QC and QIS, we can do most of the work with just linear algebra, and in fact, with only a subset of linear algebra techniques. Linear algebra is also the mathematical language of classical computing.

We want you to learn to "read" the equations of QC; so, we will be using the symbols that are commonly deployed in the field. Those symbols were borrowed from the physics version of quantum mechanics, and they will look strange when you first meet them. But if you are patient and practice writing and reading those symbols, you will get used to them. Every community has its own language and for the community of physicists, engineers, and computer scientists, mathematics is the lingua franca.

Let us warn you that some experts may be annoyed by our approach because our examples do simplify the math. For example, ordinary quantum mechanics makes use of complex numbers, and we have found a way to avoid them at the beginning of this book. Our approach is to help you understand the basic ideas and then move into the more complex mathematics when we need to.

In Chapter 7, we show you *why* we need complex numbers in some aspects of QIS, but they are not essential for the main ideas. There is also some mathematical jargon we will mention only in passing in case you read about those ideas in the QC literature. We'll do our best to avoid excess jargon wherever possible.

1.2 A Brief Preview

Let us give you a quick preview of the story we tell in this book. To help you develop an understanding of quantum computing, we will need the key concept of a *quantum state*. Learning about quantum states, like learning about many concepts in science and mathematics, benefits from using multiple ways of representing quantum states. We will introduce you to (1) abstract state "vectors," (2) graphical pictures of quantum state "spaces," and (3) numerical forms of mathematical objects called column vectors and row vectors, which are straightforward generalizations of ordinary vectors. Using multiple representations deepens your understanding of the concept of quantum states; so, we encourage you to become fluent in using all these representations.

We don't expect you to be familiar with those concepts. We introduce them one by one and give you several ways of thinking about their role in QC and QIS. We have also found that many introductions to QC and QIS use confusing language about quantum measurements, collapse of quantum states, jumps between quantum states, and the results of measurements on entangled states. Keeping in mind how actual quantum measurements are carried out, we will show you a way of talking about those concepts that will help blow away the fog surrounding many of those discussions.

You may wonder about the differences among quantum theory, quantum physics, and quantum mechanics. They are essentially all the same. When quantum theory emerged in the first third of the 20th century, its way of thinking about how the world works was significantly different from what physicists had used before. To distinguish the new from the old, which physicists call "Newtonian mechanics" or "classical mechanics," they invented the term quantum mechanics. In physics, mechanics means theories of motion, forces, and so on. In this context, "classical" is different from *classical music*. You know Joseph Boulogne (Chevalier de Saint George), Wolfgang Mozart, Clara Schumann, Richard Wagner, Nadia Boulanger, William Grant Still, Marin Alsop, and the like. What "classical physics" means is physics that does not need quantum mechanics to explain what is going on. Classical physics can be just as mind-bending as quantum physics, particularly when it comes to things like chaos theory.

To give you more context about what we are going to do, let us explain a bit about the differences between the Newtonian worldview and that of quantum mechanics. In Newtonian mechanics—before quantum mechanics came along—we thought of the world as being made of objects whose positions and velocities we can track and predict if we know the interactions among those objects. With the addition of theories of electricity and magnetism (now unified as electromagnetism), physicists thought they had a reasonably comprehensive picture of how the world works. Interactions among material objects (such as atoms, protons, and electrons) and fields like the electromagnetic field and the gravitational field could explain everything. Of course, there were many details to be worked out, but the basic conceptual structure was in place, or so they thought. The world was essentially deterministic. What does that mean? Here is an example: If we knew the positions and velocities of all these objects that make up a system at some time, according to the Newtonian worldview, we could predict the future behavior of those objects. If you could apply that method to the bouncing numbered balls in a lottery machine, you could get rich very quickly.

But there is a catch. In real life, trying to carry out those predictions becomes practically impossible if there are more than a few objects in the system. To make matters worse, even for simple classical Newtonian systems with only a few objects, the precision with which we can make those predictions diminishes as the systems evolve in time because we can never really know the current state absolutely perfectly. Furthermore, with chaotic systems (like the weather), the precision of the predictions diminishes surprisingly rapidly because chaotic behavior changes dramatically with time. Such systems are hypersensitive to even small changes in their initial conditions. You may have heard of the butterfly effect: The flapping of a butterfly's wings in Brazil can cause a tornado in Texas. The bottom line is that there are many sources of uncertainty in predictions even before we consider quantum probabilities.

All of that said, science has moved beyond the Newtonian worldview. With the discovery of quantum mechanics in the early 20th century, we now recognize that the world at its most fundamental is not deterministic. As far as we know today, there is randomness and probability at the core of nature.

We will see that the strange aspects of quantum states—combinations of states called superposition states, a property of those states called entanglement, and the role of randomness and probability when we make observations on ("measure") a quantum system—will turn out to be of crucial importance in QIS and QC. In fact, sets of procedures—algorithms—in QIS and QC make use of all these strange aspects.

We will start off with a look at traditional computers, which are often called "classical computers" since they don't directly involve quantum mechanics. The concepts of classical computers will form a good platform from which to launch ourselves into quantum mechanics and ultimately quantum computing. We hope you will be patient because it will be a while before we get into the meat of quantum computing. We have found that if we don't spend some time on basic quantum concepts, then QC and QIS, will seem even stranger. Well armed with those concepts, you will find QC, though still strange, much more comprehensible.

Once we lay the foundation with the crucial quantum concepts, we will show you several algorithms that demonstrate the advantages of quantum computing over classical computing. We will also introduce some key issues in quantum cryptography and error correction. Those topics may sound rather specialized, but they are of great importance in any kind of computing and their quantum versions raise many intriguing issues about what is information and what is "noise."

We will end our tour of QC and QIS by visiting several quantum issues that are not directly part of quantum computing but, given the tools we will have learned about, we will be able to understand. Those topics reinforce the notion that the quantum world is conceptually and experimentally far removed from the Newtonian world. For example, there is some deep quantum weirdness in a famous result called Bell's theorem (which makes us question the nature of reality) and in the relationship between classical computing and quantum computing.

We'll conclude our voyage with a look at the future—not only your personal future, what you ought to do next if you want to learn more about QC, but also the future of QC itself.

Because the field of QC in some ways is still in its infancy, it is difficult to give an absolutely comprehensive overview of the field. But it is certain that QC allows us to carry out calculations that would be completely impractical with classical computers. More importantly, QC allows us to think about problems in a completely different way from the way we think about classical computation. As an analogy, we might point to the discovery of microbes (bacteria and viruses) as the carriers of diseases. After that discovery, we had entirely new ways of thinking about things we could not see with our naked eyes. This led to new ways of preventing and treating disease. Similarly, we believe that ultimately QC will open our eyes both to whole new ways of thinking and to issues that we don't even recognize today. QC is not just doing traditional computing more efficiently or faster, though that in itself would be worthwhile, but it is doing computation in entirely new ways. Even though we don't know exactly how that is going to work out, the possibilities are just mind-blowing. Already we know that QCs can find energies and configurations of molecules more efficiently than classical computers. That opens the door to new methods of drug discovery, for example. We also know that quantum computers can crack many of the encryption methods used to keep data secure. But quantum methods can be used to develop encryption systems that are yet more secure.

1.3 How to Use This Book: Encouragement and Coaching

To get you immersed in QC as smoothly as possible, we have broken up our story into chapters and the chapters into sections to provide a gradual introduction to new information and concepts. Along the way, we suggest various activities labeled "Try It" to encourage you to reflect on what you are learning and to examine the details that are important in understanding the concepts. By doing those Try Its, you will build your ability to think "quantumly," and you will increase your confidence that you can tackle cutting-edge concepts in QIS. Think of them as training sessions just like those used to learn a sport or any other skill. You need muscles and you need muscle memory. The Try Its build those cognitive muscles.

We have written this book to help direct your QC and QIS efforts in productive ways, to provide you with practice, and some coaching. QC and QIS, like all fields of human knowledge, have a new vocabulary unfamiliar to beginners, new concepts, and new ways of thinking about and reasoning with those concepts. We urge you to work with friends as you learn about QC and QIS to coach and to encourage each other when things seem difficult. No one learns alone. At least no one learns efficiently and easily alone. Find someone to talk these ideas over with. Humans use language and social interactions to learn.

When we teach physics classes, we often hear students say, "I understand the concepts but I'm not a math person." That statement does express students' feelings about math, but it is not grounded in the reality of how one learns. While learning about QC and QIS, it is important to realize that there are no "QC and QIS people." In fact, we all belong to the large group of people trying to understand how QC and QIS work, and what those fields are good for. The fields are brand new and are still in a state of growth and flux. Furthermore, the quantum world does not come naturally or intuitively to anyone—don't let the experts get away with saying "it is obvious that . . ." It only becomes obvious after you have spent hours, days, weeks, and longer figuring out the details and making connections among concepts.

We do acknowledge that people pick up some things more easily than other things. You might be great at drawing but have a tough time learning a foreign language. It all depends on what you already know and how big a stretch it is to understand and master the new activity. It also critically depends on how much effort you are willing to put in. We have written this book by breaking up an extremely rich meal into chewable and digestible bites and providing lots of encouragement and feedback.

If you don't have much experience applying linear algebra to physical situations, learning what is going on in QIS and QC, which is the goal of this book, will be challenging because the language of the quantum world is mathematics with a strong emphasis on linear algebra. However, your mathematical fluency should increase with practice, and we provide a gentle on-ramp for all the math that we need in this book. If you feel that the math in a particular section is too dense, feel free to skim through that section, if so inclined, and return to the details later. We will point out specific sections that are mathematically challenging and others that can be skimmed without impeding your overall understanding of the book.

Since your thoughts and attitudes about learning are critical to the benefits you will get from this book, we ask you to take a few minutes to reflect on what it takes to learn. This may seem like an unusual exercise for a book on QC and QIS, but education research and our experience have shown that this kind of reflection is important in your journey to become an effective learner. We, as learners, need to pay attention to what we do when we learn because there is more to it than just taking notes. Often, we are working hard but not noticing that reflection is an integral part of learning.

Try It 1.1 Write a few sentences about an experience you had learning something new (a sport, knitting, a game, cooking, wood working, drawing, playing a musical instrument . . .) and how you used persistence to become better over time at that activity. Did you do it all at once or did you get better after repeating it? Did you learn the activity just by reading about it or did you have to pick up the ball, for example, and practice throwing it? Did you find yourself becoming more comfortable with the language and the representations of the activity? Did talking about the activity with others help make more sense of what you were doing, and did you start noticing new details?

Can we learn QC and QIS? We urge you to join former President Barack Obama in shouting "Yes, we can!"

1.4 **Principles of Quantum Mechanics**

The following numbered statements are the commonly accepted fundamental principles of quantum mechanics. For each principle, we have included several comments, which provide some background and elaboration. In practice, "the Devil is in the details," and there are additional mathematical and conceptual facets that will be explained as we apply these principles to issues in QIS and QC.

We emphasize that you are not expected to understand this terminology and the set of principles upon this first reading; that understanding develops only with experience in applying the principles to specific situations. But it will be helpful in developing a comprehensive view of QIS and QC to refer back to these principles from time to time as we shall do throughout the book.

1. *The state of a quantum mechanical system is represented mathematically by a state vector, which is part of a complex-number, inner-product vector space.*
   a. Mathematicians call such a vector space a Hilbert space, after the early 20th century mathematician David Hilbert.

b. The state vector contains all possible information about the quantum state of a system. Since the state vectors "live" in a vector space, they may be written as linear combinations (superpositions) of other state vectors—a fundamental feature of linear algebra.

c. In classical (prequantum) physics and engineering, states are determined by the *properties* of the system such as energy, momentum, position, electrical current, electrical voltage, etc. In quantum mechanics, the quantum state contains the information that allows us to calculate the probability of measuring certain values of those properties. See Principle 4.

d. In the commonly used Dirac notation for state vectors, a state vector, written as $|\psi\rangle$ for a generic quantum state, is called a right vector. Each right vector has a "cousin" left vector represented by $\langle\psi|$. The symbol inside the Dirac "container" $|\rangle$ or $\langle|$ can be anything that is helpful in naming the state we are dealing with.

e. Left and right vectors can be multiplied to form an inner product $\langle\phi|\psi\rangle$, which is a complex number representing the overlap of the two state vectors.

2. *An observable quantity (property) is represented mathematically by a linear operator A that acts on quantum state vectors to produce other state vectors* $A|\psi\rangle = |\phi\rangle$.
   a. Observables are properties such as energy, momentum, position, angular momentum, number of particles, and so on.

   b. Linear operators may be viewed as linear transformations that act on state vectors.

3. *The possible results of a measurement of a property will always be one of the eigenvalues of the corresponding operator A. Each of the eigenvalues (say,* a_n*) is associated with an eigenstate* $|a_n\rangle$ *of the operator.*
   a. The techniques of linear algebra can be used to find the eigenvalues and eigenstates associated with an operator.

   b. The result of an operator acting on a state vector is another state vector, which can always be written as a linear combination of the operator's eigenstates: $A|\psi\rangle = |\phi\rangle = \sum_n c_n|a_n\rangle$, where the coefficients c_n are complex numbers.

4. *The Born rule: The probability of obtaining an eigenvalue* a_n *in a measurement of observable A when the system is described by the state* ψ *is given by* $P_{a_n}^{(\psi)} = |\langle a_n|\psi\rangle|^2$ *, the absolute value squared of the overlap between the state* ψ *and the eigenstate* $|a_n\rangle$.
   a. The probability depends on both the property and the quantum state.

5. *Measurement principle (Lüders rule): After a measurement of property A yields the eigenvalue* a_n*, the quantum system is described by a new state* $|a_n\rangle$*, the eigenstate corresponding to that eigenvalue.*
   a. This process is sometimes called the reduction or the collapse of the state vector. But the state vector is not a "thing"; so, "collapse" is just a metaphor.

6. *Quantum states may evolve with time.*

a. In the so-called Schrödinger picture, the time evolution is described by an equation relating a time-derivative operator to the action of another operator H, the Hamiltonian operator:

$$i\hbar\frac{d|\psi\rangle}{dt} = H|\psi\rangle\,.$$

where $i = \sqrt{-1}$ is the unit imaginary number and $\hbar$ is the reduced Planck constant, which sets the energy and timescale for quantum time evolution. H is often the energy operator for the system.

b. If the Schrödinger operator equation is applied to the position representation of the quantum state $\langle x|\psi\rangle$ (a complex number function, usually called the wave function), the result is a standard differential equation.

A technical aside: Principles 4 and 5 need to be expanded if it turns out that some eigenvectors in the system all have the same eigenvalue. That will not be an issue for most of the examples in this book.

1.5 Further Reading

A history of the characters "Alice and Bob" is available at https://en.wikipedia.org/wiki/Alice_and_Bob.

Traditional Computing

2

Logic imbues the world; the boundaries of the world are its boundaries.

Ludwig Wittgenstein, *Tractatus Logico-Philosophicus*, item 5.61 (1921)
Translation by this book's authors

2.1 Traditional Computing versus Quantum Computing

Before we get into quantum computing, we need to say a few words about traditional computing and traditional computers. Some physicists like to call this "classical" computing because, as we mentioned in Chapter 1, the history of physics is often divided into a pre-quantum era called "classical physics" and the contemporary era of "quantum physics." Classical physics is the realm of Isaac Newton and his study of motion and forces, James Clerk Maxwell, who worked on electricity and magnetism, and Albert Einstein, who extended those studies to objects traveling close to the speed of light. Although we have mentioned only three famous physicists, there were, of course, hundreds of others contributing to these studies, but scientists, like people in many fields, prefer focusing on a limited hagiology. Classical physics explains phenomena that were interesting to humans with the tools of the day. This means things that can (mostly) be seen with the naked eye and experienced by all people.

Quantum physics, not surprisingly, focuses on those physical phenomena where the concepts and theories of quantum mechanics are needed to understand what is going on. When talking about quantum physics, people mean the structure of atoms and molecules, nuclear physics, elementary particle physics, semiconductors, lasers, and so on. Many of these things are not accessible to the naked eye or everyday experience.

The important point is that the classical physics world was turned on its head in the 1920s with the development of quantum theory—the theory describing the world at the atomic and subatomic levels. That theory became known as quantum mechanics and you can think of QC and QIS more generally as applications of quantum mechanics to computational and information processing situations.

Classical (traditional) computing (prior to the development of quantum computers) is actually based on a mixture of classical and quantum physics. Quantum physics helped scientists and engineers develop the semiconductor devices that power almost all contemporary computers. Even though the understanding of the material properties of those

Quantum Computing and Quantum Physics. Alice Flarend and Robert Hilborn, Oxford University Press.
 DOI: 10.1093/oso/9780197904930.003.0002

semiconductors grew out of quantum theory, the devices, as used in classical computers, operate in regimes in which classical physics (mostly electricity and magnetism) is perfectly adequate to describe their behavior. The details of quantum physics are not needed.

Let's look under the hood of a classical computer. Most of us who use computers these days know that you have to tell computers what to do. Those instructions are computer programs. Writing them is called coding. An app for your smartphone is just a computer program that tells the computer inside the smartphone what to do. So, writing a best-selling app is "just" computer programming. Fortunately, decades of work in computer science have produced what are called high-level computer programming languages that are close enough to everyday language and everyday mathematics to make coding relatively straightforward. But a big part of programming a computer is first figuring out what you want the computer to do. That part is called developing an algorithm. And then you need to work out how to implement that algorithm in the programming language you decide to use.

We have to warn you that many aspects of quantum computing require that same level of attention to detail as standard computer programming and coding do. But we will deal with that when we have to. Right now, we want to focus on the basics that underpin all computation.

2.2 **Binary Digits**

To understand what quantum computing is about, we need to dig a bit deeper into the inner workings of a traditional computer. The code you write in a high-level language is translated by another set of instructions called a compiler into code that the machinery of the computer can understand directly: this "machine language" consists of strings of 1s and 0s known as bits ("bit" is an abbreviation of "binary digit"). The binary number system has just two symbols in it, traditionally labeled 0 and 1. Our everyday number system is a decimal system, with 10 symbols: 0, 1, 2, 3, 4, 5, 6, 7, 8, and 9. It turns out that using just 0s and 1s is sufficient for almost all computational work, and it greatly simplifies the design and construction of computers.

So, if you looked at the machine language version of a computer program, you would see something like 1001111000111001111100..., which is not very informative to the typical human. Fortunately, the ordinary computer user (that includes you when you use your smartphone) never has to worry about the binary digits though you often see them scrolling across a computer screen in movies with computer programmers. You only need the binary digits if you are trying to save the world from evil hackers or if you want to understand the fundamental systems that govern all traditional computers.

Figure 2.1 shows a photograph of the electrical circuitry of a traditional computer. The black rectangles are "integrated circuits" (ICs), which are connected to each other and to other electronic components by wires made from metals such as copper. Some of the wires are "ribbon" cables (such as the one labeled B00 in the upper left corner of photograph) with many wires connecting parts of the circuits. Others are built into the circuit boards and are

Fig. 2.1 A circuit board of a classical computer.
Source: Photo by R. Hilborn.

not visible in this view. The wire voltages can be either a "high" electrical voltage (usually 3–5 V, the equivalent of a few AA batteries stacked end to end) or a "low" voltage (usually 0). The ICs sense those voltages and carry out computational actions based on them.

In principle, you could be sensing other physical quantities (the presence or absence of light, the presence or absence of sound, a switch in an "on" or "off" position, etc.), so it is useful to use a more abstract representation of what is happening in the wires. We like to distinguish between the physical state of the wire (high voltage or low voltage) and the abstract representation, the "computational state," using the binary digits 0 and 1. The voltage state of each wire is associated with a bit, whose value can be either 0 or 1. Most often, a 0 is associated with the low-voltage state and a 1 is associated with the high-voltage state though there is no fundamental reason you couldn't make the opposite association.

If we measure the electrical voltage of a particular wire in an operating computer, we would record a series of low and high voltages, which could be translated into the abstract representation of a series of 0s and 1s.

The series of 0s and 1s might represent many things. For example, it could be a coded message: think of Morse code or a flashing light signal ("one if by land; two if by sea"). Several wires (bits) looked at together might represent a numerical value (the 8-bit sequence 00000011 could represent the regular (decimal) number 3) or an address within the computer where information is to be stored or retrieved. Or they could be a code for the instructions of what an IC in the computer is to do based on the input bits.

Going into the details of these representations would take us too far afield. What is more important for our understanding of quantum computing are some simple computer operations, which use one or two bits as input and produce an output of other bits.

2.3 Logic Gates and Truth Tables

These operations are similar to binary logic (sometimes called Boolean logic after George Boole, a mathematician active in the 1800s), which was originally developed to represent the logic of combinations of statements (sometimes called "propositions") that can be either True (T) or False (F). In the 1930s, Claude Shannon and others realized that binary logic could be used to describe the operations of digital computing devices.

Let's start with a few Boolean logic examples. A proposition A ("I am hungry") can be either True (I haven't had anything to eat since breakfast) or False (I just consumed an entire extra-large pizza). A Boolean logic operation called NOT can act on the proposition. For example, NOT-A is NOT-("I am hungry.") = "I am not hungry." A horizontal bar over the proposition symbol is used to label the result of NOT: for example, NOT A = $\bar{A}$. One of the quirky (but important) features of logic is that in most cases we don't care about the meaning of the proposition; we just focus on its logical structure. That drives most people nuts.

In the language of binary logic, we say that if proposition P is True, then NOT-P is False and vice versa. This assumes, of course, that the proposition is either True or False. In *binary* logic, there are only *two* possibilities. Other situations are not binary systems: for example, I could be hungry, ravenous, a little hungry, stuffed, not hungry, ... the list goes on. In general, hunger is not a binary (two-state) system.

For binary digits, if a bit has the value 0, then NOT produces a bit with the value 1 and vice versa. Whenever something acts on a bit or a set of bits, we call that an "operation." We pictorially represent the devices that carry out the operations as "gates" (not related to Bill Gates). An operation can be a simple mathematical function like adding or multiplying. It can also be less familiar functions, which we discuss later.

Figure 2.2 shows a pictorial representation of a NOT gate, indicated by the triangle with the small circle on the right. The input bit (or proposition in logic) is indicated by A on the left. The "output" that results from applying the gate operation to A is on the right. The horizontal line over the symbol A is used to indicate the bit "opposite" that of A (0 if A is 1 and 1 if A is 0) or the opposite "truth value" (False if A is True and True if A is False).

Let's think about the NOT operation in a slightly different way: a "truth table," as shown in Table 2.1 in two different representations, logic and binary. The left column represents

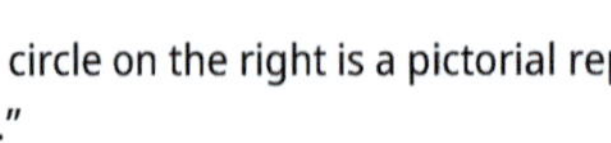

Fig. 2.2 The triangle with the small circle on the right is a pictorial representation of a NOT gate. The gate acts on the bit A to produce "NOT A."

Table 2.1 Truth table for a NOT gate.

A	NOT A	A	NOT A
False	True	0	1
True	False	1	0

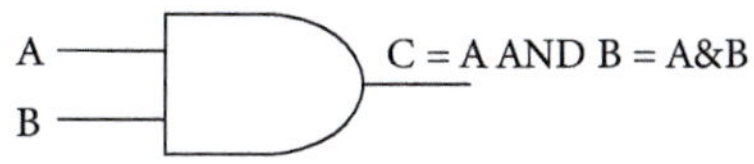

Fig. 2.3 The symbolic representation of an AND gate.

Table 2.2 Truth table for an AND gate.

A	B	C = A AND B	A	B	C = A AND B
False	False	False	0	0	0
False	True	False	0	1	0
True	False	False	1	0	0
True	True	True	1	1	1

the possible values of the input (True or False, 0 or 1), and the right column represents the corresponding output values. Truth tables (and their generalizations) will be helpful when we get to quantum computing operations.

In Table 2.1, the two columns on the left give the logical True/False values for A and NOT A. The two columns on the right give the binary 0 and 1 equivalents.

Of course, we can't do much with a gate with just one input. So, let's look at some gates that have two inputs (two bits). These are, not surprisingly, called two-bit gates. Our first example of a two-bit gate is an AND gate shown in Fig. 2.3. The AND operation has two inputs A and B and one output C.

Table 2.2 shows the output of the AND gate for the four possible "states" of the two input bits in both T/F and bits. Before reading further, we need to look at the table to find a pattern. In terms of binary logic, we say that the AND combination of two propositions is true if and only if both propositions are true. Otherwise, the combined proposition is false. Switching to bits, you should note a pattern that the output of an AND gate is 1 only if A and B are both 1s. Note that this table is larger than the one for the NOT gate. As we add bits to our repertoire, the number of possible combinations grows quickly. The larger repertoire enhances our capabilities for dealing with more complex algorithms.

Try It 2.1 Write two simple sentences in everyday language and convince yourself that the logic operation AND is close to what we mean by "and" in everyday language.

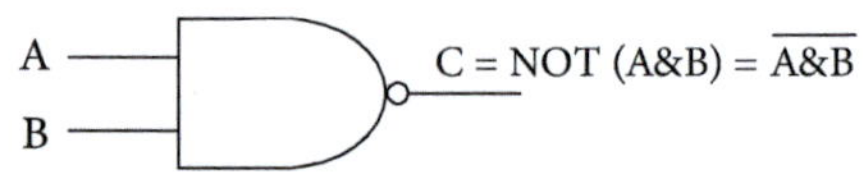

Fig. 2.4 A pictorial representation of a NAND gate—an AND gate followed by a NOT operation.

Table 2.3 Truth table for a NAND gate.

A	B	C = NOT A&B
0	0	1
0	1	1
1	0	1
1	1	0

Another important two-bit gate is called a NAND gate. Its output is the negative (opposite) of the AND operation, hence the N. You can think of the NAND gate as an AND gate, followed by a NOT gate acting on C, the output of the AND gate.

Figure 2.4 displays a symbolic representation of a NAND gate. Again, the horizontal bar over A&B means NOT. The output of the NAND gate is NOT applied to the output of the AND gate as shown in Table 2.3.

Note that the rightmost column in the NAND truth table has the 1s and 0s swapped compared to the AND truth table.

Try It 2.2 Using T and F (True and False) fill in the missing elements in the following NAND gate truth table.

A	B	$\overline{A\&B}$
F	F	
F		T
	F	T
T	T	

You may wonder how we build an AND gate, but we won't go into the details of the electronic circuitry that implements those gates. If you are interested, please take a look at Further Reading at the end of the chapter. We won't need to know anything about the actual devices in anything that follows. You can buy an IC that has four NAND gates built in for less than a dollar, and they do exactly what the table specifies without your knowing anything about the circuitry.

Here is something that is truly amazing: *any* binary logical operation can be implemented as a combination of NAND gates. These logic gates can be combined to add, subtract, multiply, and divide binary numbers and to carry out almost any computational task you can think of. How do you prove that? It is a theorem in binary logic. The theorem is

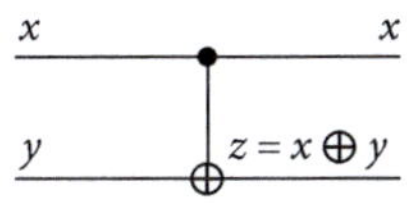

Fig. 2.5 The CNOT circuit. The symbol $\oplus$ is explained in the text. x is the "control bit," and y is the "target bit."

Table 2.4 Inputs and outputs of the CNOT gate.

Input		Output $z = x \oplus y$	
x	y	x	z
0	0	0	0
0	1	0	1
1	0	1	1
1	1	1	0

moderately complicated, and of course, no one in their right mind would build a computer that way. However, it is a statement that is very useful in theoretical aspects of computer science, and we bring it up here to model that sometimes in this book we will not go into all the details because it is not necessary for our purpose.

Let's look at one final example of a two-input gate. This one, called the controlled-NOT gate (CNOT), has two inputs and two outputs. Again, there are four possible states for the input bits. The pictorial representation of the CNOT gate is shown in Fig. 2.5. Note that we switched notation, replacing A, B, and C with x, y, and z, to match the notation often found in computer science and, as we shall see, in quantum computing.

The action of the CNOT gate is summarized in Table 2.4, which lists all possible two-bit inputs and the corresponding CNOT outputs. What is the value of the control bit that leads to a change in target bit?

The CNOT gate is called a controlled-NOT because what happens to y is controlled by the input x. If the control bit (x) is 1, the NOT operation changes $y = 0$ to 1 or 1 to 0. If the control bit is 0, then the output z is the same as the input y: the NOT operation is not invoked.

If you replace 0 with False and 1 with True, then $z = x \oplus y$ is the "Exclusive Or" (XOR) combination of x and y. Exclusive Or means that the combination is True (or 1) only if either x or y is True but not both. If both are True, the output of the gate is False (or 0). The standard Exclusive Or gate has two inputs but only one output (z). The CNOT gate, by design, has two inputs and two outputs.

The + in $\oplus$ reminds us that you can think of the CNOT gate as a two-bit "half adder." To see what that means, let's have a quick review of binary addition. The results we need are

$$\begin{array}{r}0\\+0\\\hline 0\end{array}\qquad\begin{array}{r}1\\+0\\\hline 1\end{array}\qquad\begin{array}{r}0\\+\ 1\\\hline 1\end{array}\qquad\begin{array}{r}1\\+\ 1\\\hline 10\end{array}\tag{2.1}$$

In other words, XOR implements binary addition of the two input bits but drops the "carry" 1 if the two inputs are 1 and 1.

You may remember the old math joke: There are 10 types of people in the world—those who understand binary and those who do not.

We won't need binary addition very much, but $\oplus$ and CNOT will show up fairly frequently in quantum algorithms. The XOR results are

$$0 \oplus 0 = 0 \quad 1 \oplus 0 = 1 \quad 0 \oplus 1 = 1 \quad 1 \oplus 1 = 0. \tag{2.2}$$

In the next chapter, we introduce the mathematical symbols used in quantum computing by applying those symbols to classical bits.

2.4 Chapter Summary

- Classical computers use binary digits (bits) to represent information, instructions, addresses, etc., which the computer circuitry manipulates to carry out computational actions, store and retrieve data, produce images, communicate with your printer, and to show cute kitten videos.
- Those bits can take on values represented by the symbols 0 and 1.
- The basic operations carried out by the computer can be represented by logic gates (or just gates, for short) that use one or more bits as "input" and produce one or more bits of "output."

2.5 Further Reading

Charles Petzold, *CODE: The Hidden Language of Computer Hardware and Software* (Microsoft Press, Redmond, WA, 2000). This book provides a nice introduction to the basics of classical computing.

3 Traditional Bits in New Clothing

Beware lest you lose the substance by grasping at the shadow.

Aesop. "The Dog and the Shadow"

3.1 Quick Review

In Chapter 2, we introduced bits as the representations of the states of circuit elements in classical computers. The states of those elements may be (and usually are) associated with the numerical binary digits 0 and 1.

Before we jump into quantum computing, let's look at several other ways of describing classical computing. In quantum computing, we use symbolic representations that are based on the way physicists describe quantum mechanics, blended with some notation and concepts from computer science. At first, those representations may look rather strange because you have not experienced them in your math, physics, or computer science courses, even though the underlying mathematics (mostly linear algebra) is not all that complicated. We thought it would be helpful to introduce this notation first within the framework of classical computing. This is not often done (one exception is Mermin (2007)) because the resulting representations are not useful in traditional computing. But in quantum computing, this notation is important and almost universally used. Thus, we think it is good to get used to it in a familiar context first. This approach will also help us understand what appears to be a deep gulf between classical and quantum computing. You should know that this way of talking about classical computers is hardly ever used by anyone else; so, if you mention this language to a computer science major, you will probably get a puzzled expression in return.

3.2 State Symbols

Let's start by thinking about two wires in a traditional computer. The voltage differences between wires in the computer are used for several purposes: to bring information to gates for processing and to indicate where data should come from and where it should go. Each wire has two states (or conditions): The voltage is "high" V_{high} or the voltage is "low" V_{low}. Because there are just two states, this is a *binary* system.

Quantum Computing and Quantum Physics. Alice Flarend and Robert Hilborn, Oxford University Press.
 DOI: 10.1093/oso/9780197904930.003.0003

We can represent those two voltage states by the symbols $|V_{\text{high}}\rangle$ and $|V_{\text{low}}\rangle$. Each symbol consists of a vertical line on the left and an angle bracket on the right. You should think of the combination $|\ \rangle$ as a container. The contents of the container are *names* that label the two possible states. So, we could have $|\text{On}\rangle$ and $|\text{Off}\rangle$ states or $|\text{True}\rangle$ and $|\text{False}\rangle$ states or $|\text{Awake}\rangle$ and $|\text{Asleep}\rangle$ states, and so on. These $|\ \rangle$ symbols are widely used in quantum computing to represent quantum states. You might think of state symbols as a form of emoji ☺. In and of themselves, the labels or names have no numerical or even direct physical meaning.

The $|\ \rangle$ symbol is part of what is called Dirac notation, invented by Paul Dirac, an English physicist who lived from 1902 to 1984—one of the pioneers in quantum mechanics. He developed this notation in 1939, about 13 years after the birth of quantum mechanics, to clarify the procedures of quantum mechanics. Quantum computing (QC) and quantum information science (QIS) make wide use of Dirac notation.

Try It 3.1 Write out by hand the state symbols given in the previous paragraph ("We can represent ..."). This exercise will get you familiar with writing the symbols and using them as containers. Make up some binary states of your own (for example, win–lose, dead–alive, or score–no-score) and write the corresponding state symbols.

Anticipating the terminology used in quantum computing, we might call these classical states Cbits (C for classical). Cbit is pronounced see-bit. The 0s and 1s are also called classical bits, or just bits; most people use the latter term.

In principle, we could develop computational systems with more than two states, but those are hardly ever used in classical computing or in QIS and QC. We have already seen that we can use clusters of bits to represent numbers, alphabetical symbols, and so on. So, binary systems and bits are all we will need to understand commonly used computers.

3.3 State Vectors

In quantum computing, these state symbols are called "state vectors" because they are given mathematical properties that are similar to the mathematics of vector quantities such as velocity, force, and acceleration, which are represented pictorially by arrows that point in a particular direction. The length of the vector indicates the "strength" of the quantity. You've probably seen vectors on weather maps, where the arrows (vectors) are used to indicate wind direction and speed. These state vectors are just like column vectors in linear algebra.

It is important to note that the "directions" for state vectors are in most cases just abstract ideas. In fact, the state vectors "live" in an abstract space called a "vector space" or "state space" whose "directions" have nothing to do with east or north or up or down in the physical world. The role of east or north is played by the vectors representing "basis states," which we discuss later in this chapter. Though this abstract space may sound like a crazy idea, it is crazy enough to be useful. In fact, the usual mathematics of quantum mechanics is entirely a "vector space" formalism.

Since this state space idea is weird to the uninitiated, we find it helpful to draw pictures. These pictures are widely used in quantum mechanics and some aspects of quantum computing.

Let's see how this works for the two states $|V_{\text{low}}\rangle$ and $|V_{\text{high}}\rangle$. In Fig. 3.1, we put the state vector $|V_{\text{low}}\rangle$ along the horizontal axis and $|V_{\text{high}}\rangle$ along the vertical axis. This is an arbitrary choice (like the usual choice of x along the horizontal direction and y along the vertical direction in graphs of mathematical functions). You don't have to do it this way, but most people in the field do. Reversing the roles of x and y in a graph will drive your friends nuts even though it is perfectly legitimate. All this is important, though, because the two state vectors are perpendicular to each other. Alternatively, we say the vectors are orthogonal. The two terms mean the same thing, but saying "orthogonal" gets you extra points in mathematics. In practical terms, orthogonal means that if the computer wire is in one Cbit state, then it can't be in the other state. They are mutually exclusive. If the wire is in the $|V_{\text{low}}\rangle$ state, it cannot be in the $|V_{\text{high}}\rangle$ state and vice versa. So, for a two-state classical computing system, the actual states of a wire could be represented by arrows either along the horizontal axis or the vertical axis in state space. For a classical computer, these are the only two options. There is no "in-between" state.

In Fig. 3.2, the red arrow in state space represents the state of a particular wire in a classical computer. When computations are being carried out, the wire will switch back and forth between those two states. If the system were described by a state with a "tilted" state vector pointing in some direction other than horizontal or vertical in Fig. 3.2, the vector would quickly move to be either horizontal or vertical in the state space diagram. Of course, wires need to switch from one state to another as the computer carries out operations, but the

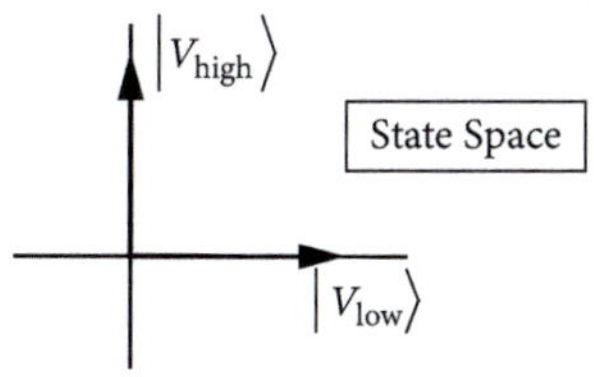

Fig. 3.1 A state space diagram for a two-state system. $|V_{\text{low}}\rangle$ and $|V_{\text{high}}\rangle$ are represented by the "vectors" along the two perpendicular axes in state space.

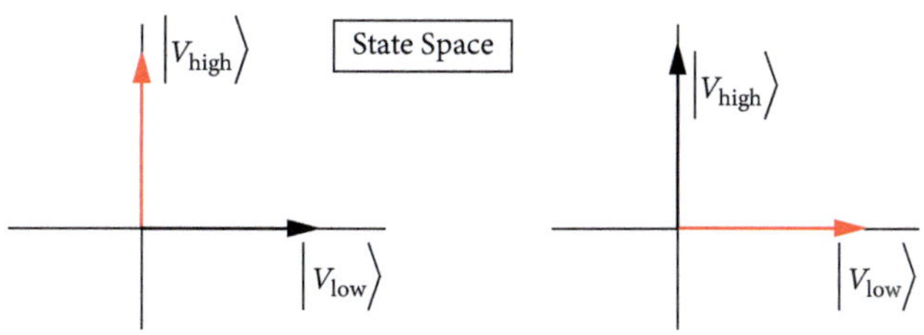

Fig. 3.2 On the left, the state of a traditional computer wire in the V_{high} state is represented by the vertical red arrow (vector) in state space. On the right, the wire is in the V_{low} state. For a classical computer, the two red arrows represent the only possible states.

circuitry is timed so the other parts of the computer "look" at the wires only after the wires have settled into the appropriate state. If the timing is not set correctly, all sorts of bad things can happen to the computer's calculations.

The length of a state vector plays no role in classical computing, but once we get into quantum computing, we'll see that it is important to associate a "length" with a state vector. State vector lengths lead to another helpful representation of computing states. Assume for the moment that the state vectors are one unit long. Let's also use the horizontal axis in the state space diagram as the x axis in a typical mathematical graph; the vertical axis corresponds to y.

We draw the state space vectors so that their tails are at $x = 0$ and $y = 0$, the point also known as the origin. We can then ask, what are the (x, y) coordinates of the tips of the arrows. For example, the state vector $|V_{\text{high}}\rangle$ has its tip at $x = 0$ and $y = 1$ in Fig. 3.2. For the $|V_{\text{low}}\rangle$ vector in Fig. 3.2, the tip is at $x = 1$ and $y = 0$.

We may group those tip coordinates in columns:

$$
\begin{aligned}
|V_{\text{high}}\rangle &\doteq \begin{pmatrix} x_{\text{tip}} \\ y_{\text{tip}} \end{pmatrix} = \begin{pmatrix} 0 \\ 1 \end{pmatrix} \\
|V_{\text{low}}\rangle &\doteq \begin{pmatrix} x_{\text{tip}} \\ y_{\text{tip}} \end{pmatrix} = \begin{pmatrix} 1 \\ 0 \end{pmatrix}.
\end{aligned} \tag{3.1}
$$

The x coordinate of the tip is listed at the top of the column and the y coordinate at the bottom. The large parentheses symbols are called "column vectors." They are akin to the way you may have labeled xy coordinates of a point in a graph as (x, y).

The symbol $\doteq$ in Eq. (3.1) is used to mean "is mathematically represented by." (Some authors use :=.) The terminology reminds us that a Dirac state symbol is not the same as a column vector. In fact, a state vector can be represented by many different column vectors depending on how you orient your x and y axes in the diagram.

By the way, some authors use square brackets rather than parentheses. In that case, Eq. (3.1) would look like

$$
\begin{aligned}
|V_{\text{high}}\rangle &\doteq \begin{bmatrix} 0 \\ 1 \end{bmatrix} \\
|V_{\text{low}}\rangle &\doteq \begin{bmatrix} 1 \\ 0 \end{bmatrix}.
\end{aligned} \tag{3.2}
$$

It is tempting to think that the 1s are bits equal to 1 and the 0s are bits equal to 0. But, we haven't yet made any connection to the bits 1 and 0. Let's see how that connection works out. As we talked about in Chapter 2, in classical computing we usually associate V_{high} with the bit value 1 and V_{low} with the bit value 0. So, following that choice, we associate the state vector $|V_{\text{high}}\rangle$ with the bit 1 and the state vector $|V_{\text{low}}\rangle$ with the bit 0. (V_{low} is often the zero-voltage level for the circuit, but that is not required.) It is customary to write this association as

$$|V_{\text{high}}\rangle \doteq 1 \text{ and } |V_{\text{low}}\rangle \doteq 0. \tag{3.3}$$

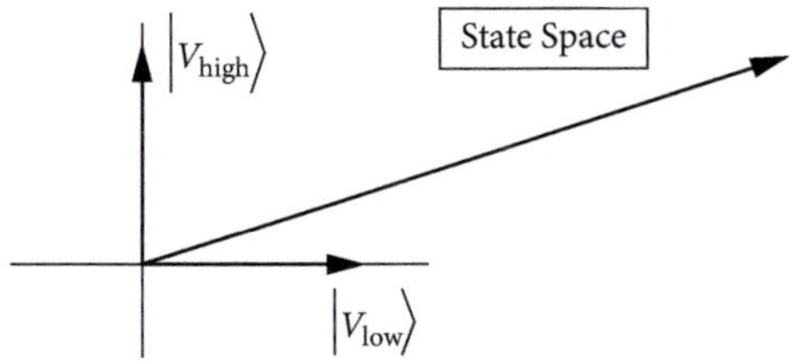

Fig. 3.3 A state space diagram with V_{low} and V_{high} basis states. The long arrow represents a state vector $3\,|V_{\text{low}}\rangle + |V_{\text{high}}\rangle$.

We agree that the notation is a *bit* confusing (Ha-ha!). The way we have worked this out is to assume that what is inside the state vector "container" is always a label—just a name.

We turn now to the concept of *basis states*. In traditional computing, the basis states are just $|V_{\text{low}}\rangle$ and $|V_{\text{high}}\rangle$ (the voltage basis) or $|\text{False}\rangle$ and $|\text{True}\rangle$ (the True/False basis) or $|0\rangle$ and $|1\rangle$ (the classical bit basis, sometimes called the computational basis). In all three cases, the basis state vectors point along the horizontal and vertical axes in Fig. 3.1.

The physical devices in traditional computing are constructed so that only the basis states actually occur. As we mentioned before, there are rapid switches from one to the other, but the computer is constructed to ignore those switching periods. In the next chapter, we find that the notion of basis states is more complicated for quantum states than for classical states.

For quantum states, we observe in Chapter 4 that the state need not lie along one of the basis state directions. It can lie in a linear combination (often called a superposition) of the states. Drawing such a state vector is just like drawing any other vector based on coordinates. Start with the tail at the origin ($x = 0$ and $y = 0$). The head will be at the coordinates given by the column vector entries. For example, Fig. 3.3 shows the state space vector for the state $\begin{pmatrix} 3 \\ 1 \end{pmatrix}$ with the V_{low} and V_{high} basis states. Even though the combination makes sense as a linear combination in linear algebra, it is *not* an allowed state for a classical computer.

Most of the time we want to be more general and not have to worry about the particular physical construction of the traditional computer. Then, we use the vectors $|0\rangle$ and $|1\rangle$, the computational basis states or standard basis states. In our way of talking about state vectors, the "0" and "1" inside the state vector containers are just the *names* of the states. But given the usual conventions of computer science, we may associate the state labels with the binary numbers 0 and 1. Generally, we don't need to worry about the distinction between labels and numbers. However, it would be a big error to write $0.5\,|1\rangle = |0.5\rangle$ or $|1\rangle + |1\rangle = |2\rangle$.

Let's pause and summarize what we have done. We've looked at three representations of the states of components (wires) in a classical computer:

1. *Abstract states*: For example, $|\text{True}\rangle$ or $|1\rangle$ or $|V_{\text{high}}\rangle$ in Dirac notation.
2. *State space vectors*: Arrow-like entities that "live" in the abstract state space.
3. *Column vectors*: A vertical array of numerical values giving the coordinates of the tips of the state vectors, assuming that the tails are at $x = 0$ and $y = 0$ in state space.

In almost every area of science, math, computer science, and engineering, a critical professional skill is being fluent in translating among different representations of concepts. We will get more fluent in those translations once we get into qubits and quantum state vectors in Chapter 4.

Try It 3.2 For the following column vectors $\begin{pmatrix}1\\1\end{pmatrix}$ $\begin{pmatrix}2\\1\end{pmatrix}$ $\begin{pmatrix}0.5\\0.5\end{pmatrix}$ $\begin{pmatrix}0.2\\0.5\end{pmatrix}$ $\begin{pmatrix}-1\\1\end{pmatrix}$, draw the corresponding state space vectors.

3.4 **Gates as Matrices**

We discussed in Chapter 2 that computation can be carried out by using logic gates to act on the Cbits carried by the wires in the computer's electrical circuits. We described the results of those operations using logic tables (or truth tables). There is a more quantum-esqe way of representing what gates do to classical computing bits. To illustrate this, let's consider the NOT gate. Remember that a NOT gate changes a bit 0 into a bit 1 and a bit 1 into a bit 0. In state vector language, these operations are written as

$$\begin{aligned} \text{NOT}|1\rangle &= |0\rangle \\ \text{NOT}|0\rangle &= |1\rangle . \end{aligned} \tag{3.4}$$

We read these left to right: In the first line of Eq. (3.4), the NOT operates on the symbolic state vector $|1\rangle$ to produce the symbolic state vector $|0\rangle$. In the second line, NOT acts on $|0\rangle$ to produce $|1\rangle$. These operations are called linear transformations in linear algebra.

In terms of column vectors, the representation in Eq. (3.4) looks like

$$\begin{aligned} \text{NOT}|1\rangle &\doteq (\text{NOT})\begin{pmatrix}0\\1\end{pmatrix} = \begin{pmatrix}1\\0\end{pmatrix} \\ \text{NOT}|0\rangle &\doteq (\text{NOT})\begin{pmatrix}1\\0\end{pmatrix} = \begin{pmatrix}0\\1\end{pmatrix}. \end{aligned} \tag{3.5}$$

NOT switches between the two column vectors, like it did between True and False or 0 and 1. So, the question is, what is the mathematical operation indicated by (NOT) that makes this work? The 1s change to 0s and vice versa? Well, the mathematics requires something more than just a label change. The answer is an array of 1s and 0s, which when multiplied times the column vectors gives the result stated in Eq. (3.5). Here is what the array looks like when it operates on the $|0\rangle$ state:

$$\begin{pmatrix}0 & 1\\1 & 0\end{pmatrix}\begin{pmatrix}1\\0\end{pmatrix} = \begin{pmatrix}0\\1\end{pmatrix}. \tag{3.6}$$

The array is called a *two-by-two matrix*: two-by-two because it has two rows and two columns, and "matrix" because it is an array of numbers (or symbols) that has certain mathematical properties. For now, we just need to know what happens when you multiply a column vector by a matrix. From linear algebra, the general two-by-two matrix times column vector multiplication rule is

$$\begin{pmatrix} a & b \\ c & d \end{pmatrix}\begin{pmatrix} e \\ f \end{pmatrix} = \begin{pmatrix} ae + bf \\ ce + df \end{pmatrix}. \tag{3.7}$$

One way to remember this rule is that multiplying the array times the column vector gives another column vector. The *top element* $ae + bf$ of the resulting column vector is the sum of the products of the elements in the *top row* of the array times the corresponding elements in the column vector. The *bottom element* $ce + df$ of the resulting column vector is the sum of the products of the *bottom row* elements of the array times the corresponding column vector elements. If you have taken a linear algebra course, you will recognize Eq. (3.7) as part of the way we may write two simultaneous linear equations

$$Ax = y \tag{3.8}$$

with

$$A \doteq \begin{pmatrix} A_{11} & A_{12} \\ A_{21} & A_{22} \end{pmatrix} \quad x \doteq \begin{pmatrix} x_1 \\ x_2 \end{pmatrix} \quad y \doteq \begin{pmatrix} y_1 \\ y_2 \end{pmatrix}. \tag{3.9}$$

Try It 3.3 Check that Eq. (3.7) is the correct way to write the product of a matrix and column vector that yields another column vector. Hint: Use the mnemonic $(m \times n) \times (n \times 1) = (m \times 1)$, indicating that the number of columns n of the first matrix must equal the number of rows n of the second so the middle two numbers are equal. The first m and last 1 give the number of rows and columns in the result. For Eq. (3.7), what is m and what is n?

We have learned that the gate operations introduced in Chapter 2 are equivalent to a matrix multiplying the corresponding column vector. We remind you that this is an unusual way of talking about classical computing, but it is at the heart of quantum computing, which we will take up in Chapter 4.

Try It 3.4 Use the multiplication rule in Eq. (3.7) to check the following results:

$$\begin{pmatrix} 0 & 1 \\ 1 & 0 \end{pmatrix}\begin{pmatrix} 1 \\ 0 \end{pmatrix} = \begin{pmatrix} 0 \\ 1 \end{pmatrix} \quad \begin{pmatrix} 1 & 0 \\ 0 & 0 \end{pmatrix}\begin{pmatrix} 0 \\ 1 \end{pmatrix} = \begin{pmatrix} 0 \\ 0 \end{pmatrix} \quad \begin{pmatrix} 1 & 1 \\ 1 & 0 \end{pmatrix}\begin{pmatrix} 1 \\ 0 \end{pmatrix} = \begin{pmatrix} 1 \\ 1 \end{pmatrix}$$

3.5 Chapter Summary

This chapter has introduced three state representations that we have used throughout the remainder of this book. This summary can serve as a quick reference guide to that notation.

- In classical computing, the "state" of a wire can be at a low or a high voltage and these states can be represented by state vectors $|V_{\text{low}}\rangle$ and $|V_{\text{high}}\rangle$. In quantum computing, you will see similar use of states.
- Basis state vectors are almost always orthogonal to each other. This has important mathematical ramifications for quantum computing algorithms.
- A state vector can be represented by
  (1) an arrow in state space,
  (2) a column vector of the x, y coordinates of the tip of the state arrow in state space, or
  (3) the Dirac state $|\,\rangle$ notation.
- These states live in an abstract state space, whose directions do not represent any physical space direction such as left and right or east and west. Instead, the basis state vectors represent the mutually exclusive orthogonal states, which could be labeled as True/False, high or low voltage, or other binary properties.
- Logic gates can be represented by matrix operations on column vectors or as symbolic operations acting on states in the $|\,\rangle$ notation.

3.6 Further Reading

N. David Mermin, *Quantum Computer Science, An Introduction* (Cambridge University Press, Cambridge, 2007). Chapter 1 uses Dirac notation to describe classical computing states.

"Matrix (mathematics)," Wikipedia. This provides more information about matrices and column vectors.

The Khan Academy has some nice short videos and articles about matrices. https://www.khanacademy.org.

Qubits and Quantum States

4

L'État, c'est moi

Attributed to King Louis XIV

4.1 What Is a Qubit?

This is a momentous occasion; we are now ready to tackle the formal description of quantum states—the essential ingredient of quantum computing. But, before we get into the details of quantum states, we want to point out that there are two ways to think about quantum computing: first, building computers out of quantum objects. Examples of such objects are polarized photons, trapped ions, electrons trapped in "vacancies" in materials (for example, in diamonds), trapped neutral atoms, or superconducting electrical circuits. Quantum mechanics is used to understand the physics of those objects.

The second way is developing quantum computing algorithms that make use of the mathematics of quantum mechanics. In either case, we will need to understand how the states of such objects are described in quantum mechanics, how we can manipulate those states to carry out computational tasks, and how using measurements of the objects tells us about the quantum states.

The term qubit (an abbreviation of "quantum bit") is used to denote the quantum equivalent of a classical bit, thinking of information in terms of just two values, in other words, a binary system. Sometimes the two values arise naturally. For example, the so-called spin of an electron, neutron, or proton has just two possible values: the spin component along a particular spatial direction can be either "up" or "down." In other cases, the quantum system is more complex: an atom or ion made up of protons, neutrons, and electrons, or a superconducting device built from millions or billions of atoms. In those cases, we manipulate the quantum system so that only two possible values of some property (such as energy) are relevant. The key element in both cases is that there are only two possible quantum states: one associated with one value of the property and the other state with the other value.

We need to emphasize that quantum computing is not just replacing classical bits with qubits. We will see that there are some similarities but also some fundamental differences.

Quantum Computing and Quantum Physics. Alice Flarend and Robert Hilborn, Oxford University Press.
 DOI: 10.1093/oso/9780197904930.003.0004

Those differences allow quantum computers to do things that are simply impossible for classical computers.

You may wonder why we limit ourselves to quantum objects with just two states. In fact, there has been some quantum information science (QIS) and quantum computing (QC) work with quantum systems with more than two states. But so far, most QIS and QC theories and experiments have focused on qubits.

We need to be careful with the word "qubit." Some authors use, as we have mentioned, the term qubit as a unit of quantum information—in analogy to a classical bit (a 0 or a 1), which itself can be viewed as a unit of information. Others use qubit to refer to the symbolic representation of the object's quantum state, which is introduced in the next section. In this book, we will use qubit to refer to a physical object. Sometimes we will write "physical qubit" when we need to re-enforce the idea of a physical object. Those multiple usages are a major cause of confusion for QC novices. Experts develop enough understanding to sort out which meaning is appropriate and when, but it is confusing to figure that out if you're just starting in QC and QIS.

The ambiguity becomes apparent in phrases such as "Alice sends Bob her qubit." Does that mean Alice sends Bob the physical object (a trapped ion, for example) or does she send information that specifies the quantum state of that object? Commonly, it means sending the physical object. So, we will use qubit to refer only to the physical object and employ the phrase "quantum state" (or "quantum state vector") to refer to the mathematical representation of the state of the qubit.

4.2 **What Is a Quantum State?**

Let's start with states in classical physics. There the state of a system is the collection of information (such as mass, position, velocity, angular momentum, electric charge . . .) sufficient to predict the behavior of the system when it is acted upon by various forces. For a classical particle, knowing its velocity and its position at some instant of time and knowing the forces acting on the particle are sufficient to predict its future position and velocity, using Newton's laws of motion. We say that the state of the particle at a particular time is given by its position and velocity at that time. Of course, for more complicated systems, we may need to specify and keep track of other properties as well.

In the world of quantum mechanics, the term "state" also means the collection of information that allows us to predict the future. But in quantum mechanics, the future is restricted to the *probabilities* of the outcomes of measurements made on that system (see Principle 4 in Chapter 1). The probability issue points to a critical philosophical difference between quantum states and classical states. Quantum states tell us only the probabilities of getting various results when we measure the quantum state, whereas in classical physics, the states allow us to make exact predictions, at least in principle, for the behavior of the system.

Although the meaning of quantum state is similar to the notion of state in other areas of physics, mathematics, and engineering, the mathematical structures used to describe quantum states and their behavior are significantly different (see Principles 1–3 in Chapter 1) from that used for classical states.

Most physicists believe that the classical states are just the large-scale limit of the quantum states. In other words, when dealing with large-size objects such as mites, grand pianos, or gazillions of quantum particles, the quantum states of those large objects are, for all practical purposes, the same as the classical states. In particular, those probabilities we mentioned become certainties. However, both the conceptual connections and the mathematical connections between quantum and classical states are less obvious. Using ideas from one realm in the other can lead to all sorts of confusion.

> Why are quantum states different from classical states? A wise guy answer is "They are just different. Get over it!" A better explanation is that quantum states allow us to describe two essential aspects of quantum phenomena: "interference" and probability. As we shall see, the interference part comes from wave effects, so we need a theory of states that accommodates those wave properties. Just like two waves on the surface of some water can overlap and enhance the height or amplitude of the resulting wave in some locations and diminish it in other locations, quantum systems have so-called amplitudes that can be summed. The square of that sum gives us the probability of getting certain measurement results, which we explore in Chapter 5.

In what follows, we introduce the concepts associated with quantum states in a way that gives you a good understanding of how they work and how they are different from the classical concepts. In Chapter 17, we discuss more about the connection between the quantum world and the classical world. Just a warning: thinking about that connection often gives us a feeling of vertigo, particularly when we try to imagine what is "actually" going on in the world as described by quantum mechanics.

4.3 Quantum States and Quantum Computers

Before we discuss the details of quantum states, it might be helpful to have an overview of how a quantum computer works. As shown in Fig. 4.1, reading from left to right, we start with a system of qubits (objects) described by a specified quantum state. The qubits then interact with a collection of quantum gates, and then finally, each qubit interacts

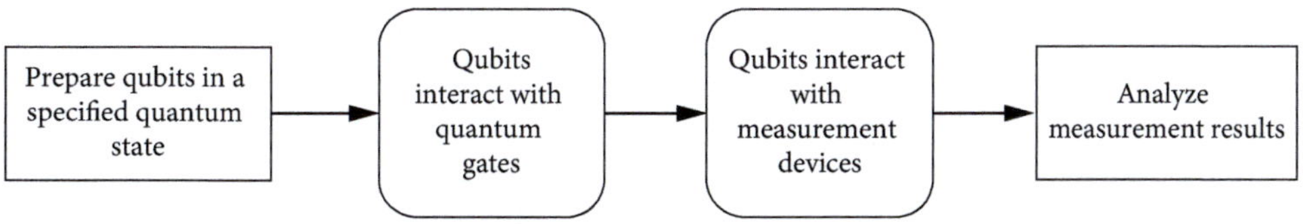

Fig. 4.1 Schematic of a quantum gate-array computer. The diagram is read from left to right.

with a measurement device. That is, in most QC and QIS situations, we "prepare" a system in some quantum state, act on it with quantum gates, and "measure" the resulting state. The quotation marks remind us that we will need to be more specific about what we mean by "preparing" a quantum state and "measuring" a quantum state. The measurement results give us the "answer" to the computation. We choose the appropriate quantum gates to implement a specific quantum algorithm, much like a classical computer program tells a classical computer the set of calculations to be carried out on the initial set of bits.

As we observed in classical computing, we almost always use objects (classical bits) that are constructed to be in one of two possible states: V_{high} or V_{low}, for example. These are then represented more abstractly as the binary digits 1 and 0. We say that a particular object (most often a wire in a circuit) can be in either a 1 state or a 0 state. Other circuit elements (gates) sense the object's state (usually with inputs of two or more bits) and produce outputs according to the design of the particular gate.

Classical computers are built from devices with large numbers of particles, and those large numbers allow the devices to be put into a state we can measure (with a voltmeter, for example) without significantly changing the state. It also means that the device (say a wire and its branches) can be connected simultaneously to many measuring or sensing devices (gates) without significantly affecting the state. It turns out that we can't do either of those things with quantum states.

In terms of symbolic representations, we know from Chapter 3 that the state of a classical bit (device) could be represented abstractly as a Dirac notation state space vector: $|0\rangle$ or $|1\rangle$, or equivalently, by column vectors $\begin{pmatrix}1\\0\end{pmatrix}$ or $\begin{pmatrix}0\\1\end{pmatrix}$, though for most traditional computing purposes these abstractions are not particularly useful.

But when we come to quantum computing and the description of quantum states, life gets not only more complicated but also much more interesting. The state space vectors become essential for describing quantum behavior and that's why those vectors are common throughout the scientific literature on quantum computing.

Scientists like to say that the physical system is "in a quantum state" as if the quantum state were a piece of clothing that the physical system puts on. That seems to imply that a quantum state is like a location: My electron is in a state in the northeast corner of the room. That metaphorical way of speaking is okay as long as you don't take the words literally. The quantum state is not a "thing" that the physical system has or does not have. We find it best to take the point of view that the quantum state is just a mathematical expression of the state describing the physical object.

There are a lot of disagreements among physicists, chemists, mathematicians, and philosophers of science about the status of the concept "quantum state." Is it something real (like mass) or is it "just" a mathematical construct that helps us understand quantum behavior? We will stick with the idea that the quantum state is simply an abstract notion that can be represented by symbols, which if manipulated with certain rules allow us to understand and predict the behavior of quantum objects (qubits). In particular, we can predict the probabilities of various measurement outcomes.

4.4 Quantum States and State Space

How do we describe the state of a qubit? One of the fundamental postulates of quantum mechanics is that quantum states are represented by vectors in an abstract state space. As we mentioned in Chapter 3, these vectors, like ordinary force or position vectors, have both a magnitude (length) and a direction. However, unlike ordinary vector quantities, quantum state vectors live in an abstract state space. For example, "up" in state space does not mean "toward the ceiling" in physical space. These kinds of vector spaces have been thoroughly studied by mathematicians in the field of linear algebra. However, in this book we will need only a small subset of the concepts and methods of linear algebra.

The abstract vector space is a collection of vectors. You can do two things to the vectors: (1) You can multiply a vector by a number to form another vector. That second vector will also be a member of the vector space. (2) You can add vectors together to form another vector, which is also part of the vector space. Adding state space vectors forms what is called a superposition state in quantum mechanics. If the numbers you use to multiply vectors are real numbers, the space is called a real vector space. If the numbers are complex numbers $a = u + iv$, then the space is a complex vector space. Quantum mechanics, as we shall see, uses a complex vector space. In Chapter 7, we explain why. Quantum state vectors are not themselves directly observable, but they allow us to predict the probability of the outcomes of measurements made on the qubits. It is the results of those measurements that can be observed, with some caveats described later.

Also, if the vector space has what is called an inner product (or sometimes, "scalar product") by which two vectors can be multiplied together to give a number (scalar), we say we have an inner-product vector space. The vector space used by quantum mechanics is, with all its appropriate adjectives, an inner-product complex vector space called Hilbert space, after the famous mathematician David Hilbert (1862–1943).

You can also act mathematically on a vector in the vector space with mathematical objects called operators (or, in QC language, quantum gates) that allow us to change one state vector into another. In linear algebra, such operators are called linear transformations because they transform one vector in the vector space into another vector in the vector space and the operation, in mathematical terms, is linear.

To emphasize that the state space is an abstract space and not a physical space, we quote a famous aphorism attributed to the quantum theorist Yakiv Aharonov in speaking about Hilbert space—the abstract quantum state vector space: "In Hilbert space, no one can hear you scream."

For a quantum system like those used in quantum computing devices, we write a general "state vector" as $|S\rangle$. Often, physicists like to use the notation $|\psi\rangle$ with the Greek letter ψ (psi, pronounced "sigh"). You can also remember the symbol as a version of Poseidon's trident (or Neptune's trident, if you prefer Roman to Greek gods): a three-pronged pitchfork.

The symbol ψ was a great choice of symbols because we always want to sigh when we are perplexed by the mysteries of quantum mechanics. In fact, there is a ψ sign on Erwin Schrödinger's grave in the village of Alpbach in Austria (Fig. 4.2). Schrödinger (1887–1961)

Fig. 4.2 Erwin Schrödinger's grave marker in Alpbach, Austria.
Source: Photo by R. Hilborn.

was one of the founders of quantum mechanics and is now most famous in popular culture for his Schrödinger's Cat paradox.

But in spite of Schrödinger's grave marker, we will use a variety of other labels to keep track of different kinds of quantum states. For example, for a trapped-ion qubit, the basis states are so-called energy eigenstates: states associated with a definite energy of the ion. Let's suppose that we focus on just two energy states: $|E_0\rangle$ and $|E_1\rangle$. Then, we can write a general quantum state of the trapped ion as

$$|S\rangle = \underbrace{\beta_0|E_0\rangle}_{\text{part of the state associate with } E_0} + \underbrace{\beta_1|E_1\rangle}_{\text{part of the state associate with } E_1}. \tag{4.1}$$

Eq. (4.1) expresses the key concept associated with the vector space postulate of quantum mechanics: A general quantum state can be written as a linear combination (*superposition*) of other quantum states. β_0 and β_1 are called the *state coefficients* or *state amplitudes*, which are just numbers, sometimes real and sometimes complex. The states $|E_0\rangle$ and $|E_1\rangle$ are called the *basis states*, and the overall state $|S\rangle$ is said to be a superposition of $|E_0\rangle$ and $|E_1\rangle$. In Chapter 5, we show you how the amplitudes are related to the probabilities of measurement outcomes. Another piece of terminology: The dimensionality of a vector space is the number of basis states needed to "span" the space. For our two-basis state example, it means that we can express any vector in the space as a sum of those two basis states and hence the dimensionality of the space is two.

We urge you to write all the equations in this chapter by hand. That will help you "read" the equations and give you practice identifying state vectors and state coefficients. Cognitive researchers have found you tend to understand things better when you write them down. The physical act of writing involves a lot of mental decisions that are helpful in reading the symbols and promoting understanding. In fact, it might be useful to write the vector "container" symbol $| \rangle$ in one color, the state vector label in a different color, and the amplitudes in a third color to increase your awareness of what you are writing.

In many situations, we're going to use the computational basis states $|0\rangle$ and $|1\rangle$ instead of $|E_0\rangle$ and $|E_1\rangle$; so, we don't have to worry about the detailed physics of the qubits. Let us remind you that the symbols in the state vector containers are just names. They have no immediate connections to the *numbers* 0 and 1. In fact, if it were not for the connection to classical computing and classical bits, we should probably label those states as $|\text{Zero}\rangle$ and $|\text{One}\rangle$. However, since almost everyone in QC and QIS uses the 0 and 1 forms, we just need to get used to that notation. But, as we mentioned in Chapter 3, don't be tempted to write something like $5|1\rangle = |5\rangle$ even if it seems correct.

Using the computational basis states $|0\rangle$ and $|1\rangle$, we may write any state vector $|S\rangle$ as a superposition of those two basis states:

$$|S\rangle = a_0|0\rangle + a_1|1\rangle, \tag{4.2}$$

where a_0 and a_1 are the state coefficients (amplitudes). These coefficients are just numbers, though in many cases in quantum mechanics they are complex numbers. For now, we use only real number coefficients.

You may wonder how we know what a_0 and a_1 are. As we mentioned before, the state amplitudes a_0 and a_1 carry information about the results of measurements on the qubit whose state is $|S\rangle$. So, you can usually think of the amplitudes as numbers determined by measurements or by theories that predict the results of measurements.

Figure 4.3 shows a graphical representation of the basis state vectors $|0\rangle$ and $|1\rangle$, and a general state vector $|S\rangle$. Geometrically, a_0 is found by drawing a line from the tip of $|S\rangle$ to the line along the $|0\rangle$ direction. We call a_0 the perpendicular projection of $|S\rangle$. Similarly, for a_1. As shown on the right of Fig. 4.3, we could also write the state in terms of column vectors in three ways:

$$\begin{aligned} |S\rangle &\doteq a_0\begin{pmatrix}1\\0\end{pmatrix} + a_1\begin{pmatrix}0\\1\end{pmatrix} \\ &= \begin{pmatrix}a_0\\0\end{pmatrix} + \begin{pmatrix}0\\a_1\end{pmatrix} \\ &= \begin{pmatrix}a_0\\a_1\end{pmatrix}. \end{aligned} \tag{4.3}$$

As we mentioned in Chapter 3, we use the symbol $\doteq$ to indicate mathematical equivalence between different representations.

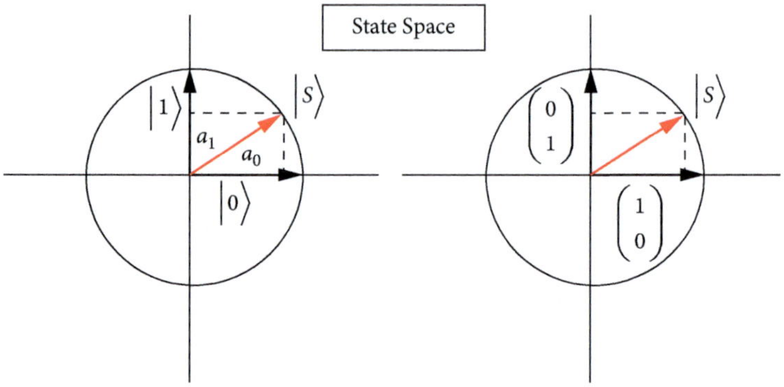

Fig. 4.3 A state space diagram showing the two computational basis states, $|0\rangle$ and $|1\rangle$ on the left, and a general state vector $|S\rangle$. a_0 and a_1 are the perpendicular projections of the vector $|S\rangle$ along the $|0\rangle$ and $|1\rangle$ directions, respectively. All three vectors in each diagram have length equal to 1. Consequently, their tips lie on a circle with radius 1. On the right, the column vector form of the basis states is used.

Let's pause to recall the different meanings of the 0s and 1s in these equations. In Eq. (4.2), the symbols inside the state vectors are merely names given to the two basis states. In Eq. (4.3), the 1s and 0s are the coordinates of the tips of the basis state vectors in state space, assuming the tails are at (0, 0).

In other words, Eqs. (4.2) and (4.3) assert that the state $|S\rangle$ of the quantum object is a combination (superposition) of the computational basis states $|0\rangle$ and $|1\rangle$. Superposition is a critical characteristic of quantum states that makes them different from the classical computing states described in Chapter 3. Classical computer states are always one or the other of the individual basis states—never a superposition. In fact, superposition is the "secret sauce" of QIS and QC. We will see the power of superposition in almost all the quantum algorithms we will discuss. Remember that the state of a classical computer's wire (for example) is either $|0\rangle$ or $|1\rangle$. The wire's state can't be a combination of these two states, so superposition makes no sense for a classical computer.

Another important quantum principle is that you have a choice of basis states. Describing the same quantum state in terms of different basis states is a powerful tool in QC. In Chapter 8, we take up the details of using different basis states.

One notational issue is important: We will always list the coefficient associated with the computational basis state $|0\rangle$ first (the top entry in a column vector) and the other, associated with $|1\rangle$, next (the bottom entry in a column vector). This order is illustrated in Eq. (4.3) in the last line. We are then using what is called an "ordered basis set." Using an ordered basis set helps in the bookkeeping of state vectors and amplitudes. We do this routinely in everyday life: We almost always say "(x, y) coordinates" or, if you are a baseball fan, "the count is 3 and 2," with the understanding that 3 refers to the number of balls and 2 is the number of strikes.

Back to the fundamental issues. You may (should) worry about what it means to have a state that is a mix of 0 and 1. That indeed is a crucial issue in quantum physics. It turns out to be a rather subtle notion and, in a way, it clashes with our everyday understanding of what it means for an object to "have" some property.

If we think back to the Boolean logic we used in Chapter 2, we see the conundrum: What would it mean for a proposition in a superposition of True and False? Is the proposition then partly True and partly False? Or can a wire in a classical computer be partially V_{high} and partially V_{low}? It certainly does not mean that the wire has a voltage between those two values (as a steady state).

All of this seems rather abstract, and it is. So, let's look at a specific example: so-called *polarization states* of light. We choose this example for two reasons: (1) Many QC and QIS experiments make use of light polarization and (2) with only a few dollars' worth of equipment, you can carry out experiments at home which illustrate many of the basic features of light polarization.

4.5 **Polarized Light**

For more than 150 years, we have known that a light beam (think of a laser-pointer beam or even a sunbeam) can be described as oscillating electric and magnetic fields that travel through space as illustrated in Fig. 4.4. These are the same fields that are responsible for "static electricity," for sparks that jump from your fingers to a doorknob after you rub your shoes on a carpet, and for refrigerator magnets. We say light is polarized if the oscillation pattern of the electric and magnetic fields that constitute the light wave is systematically organized. We are interested in the case when the electric field in the light beam oscillates along a fixed direction. In that case, the light beam is said to be "linearly polarized." By convention, the *polarization direction* of light is the direction along which the light wave's electric field oscillates. For example, the electric field might oscillate along a horizontal axis

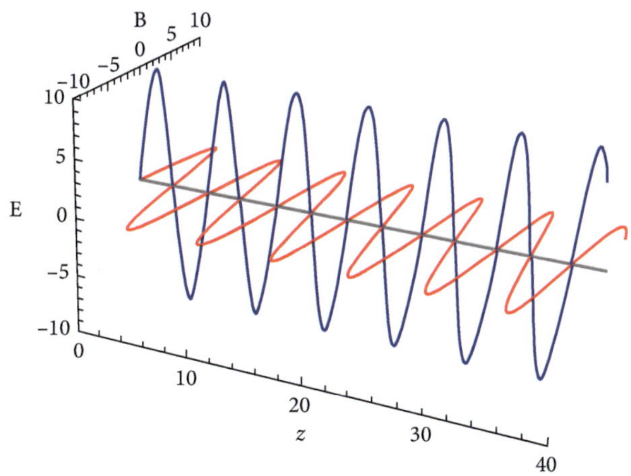

Fig. 4.4 A light beam, traveling to the right in the *z* direction, can be described by a combination of oscillating electric and magnetic fields. In this case, the electric field (blue) E is oscillating in the vertical direction. The magnetic field (red) B is oscillating in the horizontal direction. The units of E and B have been chosen to make the amplitudes equal. The figure shows the field strengths as a function of position at a specific time.

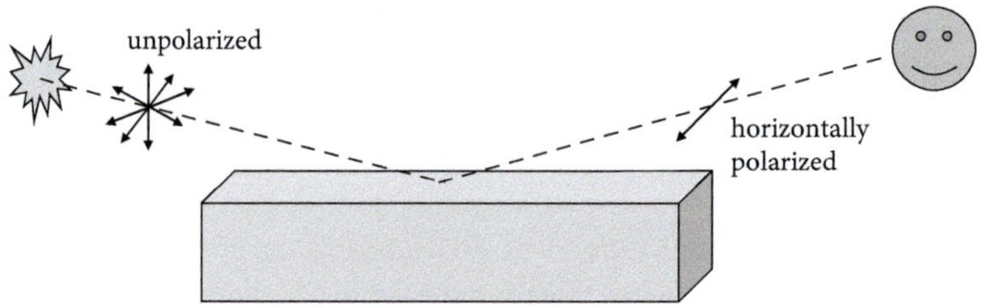

Fig. 4.5 A diagram showing how unpolarized light from the light source becomes horizontally polarized after reflecting off a horizontal surface such as a piece of dark polished granite.

or a vertical axis or along an axis tilted at 32.1° with respect to a horizontal axis, and so on. In Fig. 4.4, the light is linearly polarized along the vertical axis, labeled by E.

Linear polarization is relatively easy to detect using polarizing material readily available in "polarized" sunglasses. Inexpensive sheets of polarizing material are also available from science equipment suppliers (see Further Reading).

The polarizing material in sunglasses is oriented to block light with electric field vectors oscillating in a horizontal direction while transmitting light polarized in the vertical direction. Why? You would like polarized sunglasses to block "glare" from horizontal surfaces: a road, the hood of a car, or the surface of a lake or the ocean (see Figs. 4.5 and 4.6). It turns out that when unpolarized light, such as from the sun, reflects off a horizontal surface, the reflected light has most of its electric field oscillating in a horizontal direction, parallel to the horizontal surface from which it was reflected. We say that the light is horizontally polarized.

A polarizing sheet is a stretched polymer material that absorbs light polarized along the stretch direction and transmits (more or less) light polarized perpendicular to the stretch direction. By rotating the polarizing sheet (or your sunglasses), you can determine if a given light beam is "unpolarized." If the brightness of the light transmitted is unchanged when you rotate the filter, then the light is unpolarized (strictly speaking, it has no linear polarization). It usually means that there are equal amounts of light with all polarization directions.

You can use glare light reflected from a horizontal surface to find the transmission axis of your polarizing sheet or your sunglasses. Look at the glare light through your polarizing sheet and rotate the sheet so that the glare light transmitted through the sheet is *maximized.* Then, in that position the transmission axis of your sheet is horizontal. The difference between the maximum intensity and the minimum intensity is most dramatic if the angle between the surface and the ray of light coming toward your eyes is about 30° (see Fig. 4.6).

Once you know the polarization axis direction for transmission for your sunglasses (or your polarizing sheet), you can find another light beam's polarization direction by rotating the sunglasses until you find a maximum in the amount of light coming through. The orientation of the polarizing material's transmission axis is then the predominant electric field oscillation direction. If rotating the sunglasses 90° from the maximum intensity direction gives no light coming through, we say the light is linearly polarized in the axis direction that gives maximum intensity. If there is a small intensity of light at 90° from the maximum intensity direction, we say the light has at least partial linear polarization. Note that the two

Fig. 4.6 Photos of the surface of a shiny granite countertop. On the left, you can see light reflected from the top surface viewed through a polarizing filter with its transmission axis horizontal. On the right, the photo is taken through a polarizing filter whose transmission axis is vertical. The horizontally polarized glare light is absorbed by the filter and you can see the pattern in the granite more clearly, without the "haze" from the glare light.
Source: Photo by R. Hilborn.

polarization axes (maximum and minimum transmission) are perpendicular to each other. In the next section, we discuss this perpendicularity in the abstract state space.

Figure 4.6 illustrates how a properly oriented polarizing sheet can remove glare light. But we want to raise one detail. There is another kind of light polarization called circular polarization, which might appear to be unpolarized by the criterion we posed. But for our current purposes, all we need is linearly polarized light.

Try It 4.1 Look at glare light reflected from a shiny floor through a polarizing sheet or polarizing sunglasses. Do the statements in the previous few paragraphs describe what you see? Do you see a change in intensity of the light coming through the polarizing sheet when you rotate the sheet about the viewing direction? Now look at light from computer and tablet screens. Light from most computer and tablet screens is polarized. Light from smartphone screens is partially polarized. Can you tell the difference between fully and partially polarized? What about your TV screen?

Polarizing sheets can be used both to analyze the polarization state of the light beam and to produce polarized light with an electric field oscillation direction you want.

(Actually, to completely characterize the polarization state, you need to use a few other devices in addition to the polarizing sheet, but the basic idea still holds.)

In other words, a polarizing sheet can be used both to prepare a light beam with its polarization direction along the sheet's transmission axis and to analyze ("measure") the polarization state, usually by rotating the polarization sheet about a direction perpendicular to the sheet and detecting how the power of the transmitted light varies with the orientation of the transmission axis.

4.6 **Photons**

As we mentioned before, polarized light is often used in QIS and QC applications. The *quantum* nature of light manifests itself in the empirical observation that light is made up of small, discrete bundles of energy. Those bundles are called *photons.* The discovery of the photon nature of light was one of the most fundamental and important experiments leading to the development of quantum mechanics early in the 20th century. Photons came as a surprise because 19th century theories of light focused on the wave behavior of light. Waves are spread out in space while particles are localized in space. Quantum mechanics has forced us to accept wave–particle duality: Some light phenomena are best explained as wave effects, while others are best explained with particles (photons). We will return to these issues later. Ordinary light beams transmit billions and billions of photons per second, so in most everyday situations, we don't notice the individual photons.

As an analogy, think about rain. Rain is made up of individual rain drops. But if we are having a heavy rainstorm, all we hear is a whoosh of water hitting the roof of our house. However, during a light sprinkle, you can hear the individual raindrops hitting the roof.

Another analogy is the sound made by popping popcorn. When only a few corn kernels are popping, you hear the individual "pops," which seem to occur randomly. If a lot of kernels pop at about the same time, you hear a continuous "roar."

The detailed quantum properties of photons become apparent only when we have the sensitivity to detect individual photons. So, when we use photon qubits for QIS and QC, we will want to think about manipulating light photon by photon. This means we need to describe the quantum polarization state of an individual photon.

Here is how that works: In our preferred language, a photon is a physical qubit, a quantum object for which there are just two quantum basis states—one $|\text{hlp}\rangle$ for horizontal linearly polarized light and the other $|\text{vlp}\rangle$ for vertical linearly polarized light.[1]

Figure 4.7 shows a state space diagram for a linear polarization state (indicated by the red arrow) that is along neither the horizontal nor vertical directions. A general quantum polarization state $|S\rangle$ for the photon can be written as a superposition of the two basis states. We use $|\text{hlp}\rangle$ and $|\text{vlp}\rangle$ in place of $|0\rangle$ and $|1\rangle$ in Eq. (4.2) and write

$$|S\rangle = a_{\text{h}}|\text{hlp}\rangle + a_{\text{v}}|\text{vlp}\rangle\,. \tag{4.4}$$

[1] Note that many authors use $|H\rangle$ and $|V\rangle$ for the linear polarization basic state vectors, but H is used for several quantum gates and operators, as we shall see.

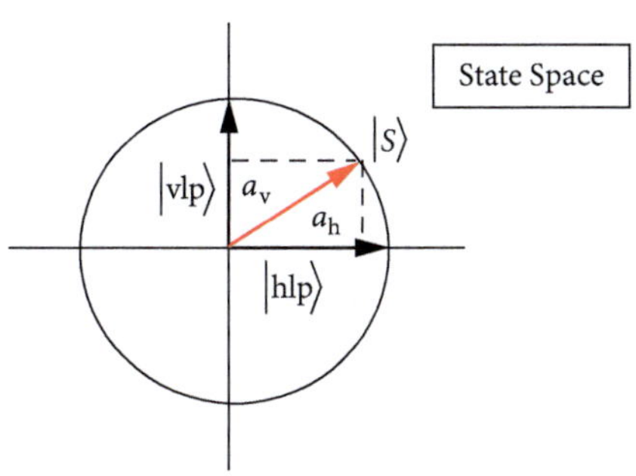

Fig. 4.7 A state space diagram for linearly polarized light (photons). The two basis state vectors are |hlp⟩ for horizontal linear polarization and |vlp⟩ for vertical linear polarization. The vectors all have unit length, so their tips fall on a circle of radius 1. a_h and a_v are the amplitudes of the state vector along the basis state directions.

In other words, the state is a linear combination (a superposition) of horizontal and vertical polarization.

Linearly polarized basis states constitute a special case in the sense that the directions associated with the basis states can be the same as the real (physical) space horizontal and vertical directions. Later, we will meet up with other quantum basis states that have similar state space diagrams but very different physical space representations.

It might be tempting to call the state $|S\rangle$ a *mixture* of the two basis states, but that would be misleading. When we say "mixture" you might be thinking about what happens if we grab a handful of candies from a bowl. Some of the candies are red and some are blue. A handful will have a mixture of red and blue candies. A superposition state is more subtle. In the current context, the superposition state refers to just one qubit (just one piece of candy). Furthermore, it does not mean the piece of candy is purple, a color mixture of red and blue. As discussed in the next chapter, the superposition state tells us that if we observe the candy's color, sometimes we will see red and sometimes we will see blue.

4.7 Ordinary Vectors

Let's pause for a few minutes and remind ourselves that this way of writing a quantum state vector is like the way we describe other vectors in mathematics, physics, engineering, and other sciences. For example, we use vectors to describe velocity, position, acceleration, forces, and electromagnetic fields in physics and engineering.

Let's review the connection between an ordinary vector and its components relative to the traditional x and y axes. To be concrete, suppose we have a vector represented by the symbol $\vec{B}$ and, in keeping with our restriction to two-state systems, we assume that the vector "lives" in a two-dimensional space. The vector can be described as a sum of two vectors, $b_x\hat{x}$ in the x direction and $b_y\hat{y}$ in the y direction. $\hat{x}$ is a vector of unit length pointing in the x direction with an analogous definition for $\hat{y}$. The unit vectors $\hat{x}$ and $\hat{y}$ play the role of basis vectors. We could also use two column vectors or a column vector with two entries. So, $\vec{B}$ can be written in three equivalent ways:

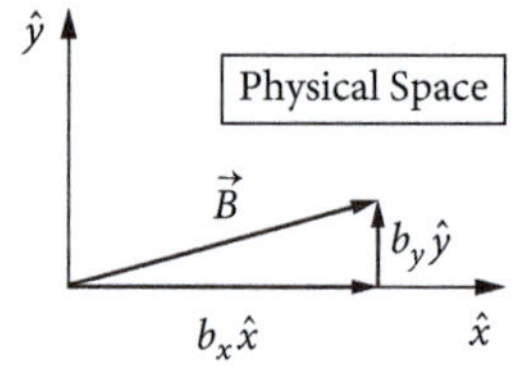

Fig. 4.8 A graphical representation of $\vec{B} = b_x\hat{x} + b_y\hat{y}$.

$$\vec{B} = b_x\hat{x} + b_y\hat{y} \doteq b_x\begin{pmatrix}1\\0\end{pmatrix} + b_y\begin{pmatrix}0\\1\end{pmatrix} = \begin{pmatrix}b_x\\b_y\end{pmatrix}. \tag{4.5}$$

Now comes a crucial question: How do we express the length of the vector $\vec{B}$ in terms of the coefficients b_x and b_y? The answer is given by the Pythagorean theorem: The square of the hypotenuse of a right triangle (the square of the length of the vector $\vec{B}$ in Fig. 4.8) is equal to the sum of the squares of the other two sides. For the right triangle in Fig. 4.8, we have

$$\left\|\vec{B}\right\|^2 = B^2 = b_x^2 + b_y^2, \tag{4.6}$$

where the first two terms are different ways of writing the square of the length of the vector $\vec{B}$. The double vertical lines are the most common notation for the length of the vector—what mathematicians call the "norm" of a vector. In linear algebra language, we are using a vector space that has an associated norm to define the length of a vector in that vector space. As mentioned previously, in quantum physics we will often need to use complex numbers for the state amplitudes, so we are using a complex vector space. The norm in such a case is defined as

$$\left\|\vec{B}\right\|^2 = B^2 = b_x^* b_x + b_y^* b_y, \tag{4.7}$$

where the superscript * indicates the complex conjugate of the complex number. We won't need the complex amplitudes until we get to Chapter 7, in which we explain why they are used. A brief review of complex numbers can be found in The Quantum Toolkit Appendix.

4.8 Back to Polarized Light

Figure 4.7 shows the analogous vector component diagram for polarized light. By comparing Figs. 4.7 and 4.8, we see that we can express the square of the "length" of our quantum state vector in Eq. (4.4) as

$$\|\,|S\rangle\,\|^2 = a_{\mathrm{h}}^2 + a_{\mathrm{v}}^2. \tag{4.8}$$

We used quote marks around length for a state vector because that length is not something you can measure with a ruler. For almost all our work in quantum computing, we use state vectors whose length is 1. (They are called "unit vectors" or "normalized vectors.") Following that convention, we insist that

$$a_{\text{h}}^2 + a_{\text{v}}^2 = 1. \tag{4.9}$$

We motivate that requirement in Chapter 5. From Eq. (4.9), we see that each of the amplitudes must be less than or equal to 1. This is a requirement that shows up in almost all aspects of QIS and QC.

In addition to polarizing sheets and polarizing sunglasses, another way to measure the polarization state of a light beam is to use devices called polarizing beam splitters (PBSs for short). Most PBSs are made from a transparent crystalline material that is sensitive to the polarization direction associated with the light traveling through the device. From a classical physics point of view, the light's electric field interacts with the PBS material in such a way as to have light with vlp travel in a direction different from light with hlp. When an unpolarized light beam interacts with a PBS, the beam is split into two parts. One of the parts is associated with vlp and the other with hlp. What is "vertical" and what is "horizontal" is determined by the orientation of the PBS.

Using photon language, we say that the PBS sorts a stream of incoming photons into two separated beams, one of which is described by $|\text{hlp}\rangle$ and the other by $|\text{vlp}\rangle$. In contrast, recall that a polarizing sheet transmits only one polarization state whose direction is parallel to the sheet's transmission axis.

In the Top View part of Fig. 4.9, the beam going to the right is described by a linear polarization state $|\text{hlp}\rangle$ with the polarization direction parallel to a horizontal edge of the prism while the other beam is described by a linear polarization state represented by $|\text{vlp}\rangle$, parallel to a vertical edge of the prism. The PBS "splits" the incoming beam into two beams: one with hlp and the other with vlp.

Try It 4.2 Write the state vector in Eq. (4.4) for each of the light beams coming out of the PBS in both state vector and column vector forms. Explain which coefficients (amplitudes) are zero and nonzero for each of the beams.

The PBS can be used two ways: First, the device can be used as what we like to call a quantum *state separator.* If you send in a light beam described by an arbitrary polarization state, the PBS splits up the light beam into sub-beams each of which is associated with one of the linear polarization basis states. The PBS has separated the light beam into its two linear polarization components. The intensity of the light in each beam (proportional to the number of photons in each beam) can be predicted from the quantum state sent into the PBS. The exact connection is discussed in Chapter 5. The state separator is different from a measurement device because it does not detect the photons.

The PBS can also be used as a quantum *state preparation* device. For example, the light that emerges in the right-going beam shown in the Top View part of Fig. 4.9 has horizontal linear polarization no matter what the polarization state of the incoming beam. So, by

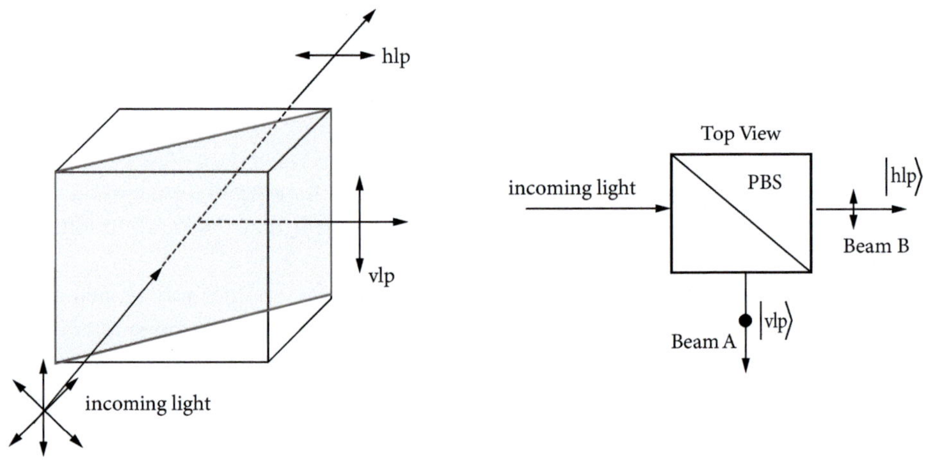

Fig. 4.9 A polarizing beam splitter (PBS). The diagonal plane (shaded gray) marks the boundary between the two parts of the splitter. The incoming light beam is "unpolarized," an equal mix of all polarization directions. The beam that comes out of the PBS to the right has its electric field oscillating in the vertical direction; that is, it is linearly polarized in the vertical direction. The other beam has its electric field oscillating in the horizontal direction. On the right is a top view of the polarizing beam splitter. The black filled circle indicates that the electric field vector for Beam A is oscillating perpendicular to the plane of the figure.

using just that right-going beam, we know we are working with photons whose polarization quantum state is $|\text{hlp}\rangle$.

We could carry out a similar state preparation using a polarizing sheet whose transmission axis is aligned horizontally. The photons that get through the sheet are described by the hlp state. As we said before, in the case of the polaroid sheet, the photons described by the vlp state are absorbed by the polaroid sheet.

Another interesting light polarization device is a calcite crystal, which behaves somewhat like a polarizing beam splitter except that the two light beams emerge from the crystal relatively close to each other. Calcite crystals can be purchased for a few dollars at gem stores, but unfortunately, calcite is difficult to polish; so, the surfaces on the inexpensive crystals are often scratched, making it hard to see the details of the two emerging beams. Calcite crystals with more highly polished surfaces are available online for $15–20. By way of contrast, a good polarizing beam splitter cube costs $200 or more from optical suppliers online. Inexpensive laser pointers (red or green) make good sources to see the polarization separation behavior of the calcite crystals or the polarizing beam splitter cube.

4.9 Orthogonality of Two-State Vectors

Now that we have developed a basic understanding of a quantum state vector, let's look at a few simple mathematical techniques that will be useful in understanding quantum computation. Many operations in QC involve multiplying two state vectors together to obtain a number (called a scalar) rather than another vector. A "scalar" is just a regular number with no direction associated with it. This kind of product is, as mentioned previously, often called a *scalar product* or an *inner product*, or a *dot product* since the dot product notation $\vec{C} \cdot \vec{B}$ is often used in standard vector algebra.

Let's see how the scalar product works with the notation we have introduced so far. Let's start with the column vector form for two state vectors $|B\rangle$ and $|C\rangle$:

$$|B\rangle \doteq \begin{pmatrix} b_0 \\ b_1 \end{pmatrix} \qquad |C\rangle \doteq \begin{pmatrix} c_0 \\ c_1 \end{pmatrix}, \tag{4.10}$$

where the state amplitudes $b_0, b_1, c_0,$ and c_1 are just regular numbers (scalars). To multiply $|B\rangle$ by $|C\rangle$ using the dot (inner) product $\vec{C} \cdot \vec{B}$, we first convert $|C\rangle$ to a *row vector* $\begin{pmatrix} c_0 & c_1 \end{pmatrix}$ keeping the order of the numbers the same (top to bottom in the column vector becomes left to right in the row vector).

As mentioned previously, in the most general version of state vectors (as used in quantum mechanics), state coefficients are complex numbers. Complex numbers have two parts: historically called the "real" part and the other the "imaginary" part. In that situation, c_0 and c_1 are replaced by their complex conjugates c_1^* and c_0^* when writing the row vector. The row vector is then $\begin{pmatrix} c_0^* & c_1^* \end{pmatrix}$. The complex conjugate of a complex number has the same real part as the original number, but its imaginary part changes sign. If the state coefficients are "real" numbers, then the complex conjugate is the same as the original number.

Math Aside In linear algebra, the complex conjugate is often represented by an over-bar symbol. In this book, we use the superscript * notation because that is the form most frequently used in physics. Just remember $c^* = \bar{c}$.

Now let's get back to the dot product. The row vector is represented by the "backward" state vector symbol $\langle C|$. To help remember which is which, we call the original vector $|C\rangle$ a "right vector" because the angle bracket points to the right, while the "backward" row vector $\langle C|$ is called a "left vector" because the angle bracket points to the left.

Jargon Aside Some authors use another pair of somewhat strange names for these two types of state vectors. Those names were introduced by Paul Dirac, the inventor of Dirac notation. He had realized that expressions involving angular brackets, such as $\langle C|A\rangle$, are critical in understanding transformation rules in quantum mechanics (as described in Chapters 7 and 8) and that those brackets could be "factored" into two parts, $\langle C|$ and $|A\rangle$; he apparently decided to split the word "bracket" to give names to those two parts. But instead of using "brac" and "ket," he used (whimsically? sophomorically?) "bra" and "ket," which, alas, continue to be widely used in QC and QIS. So, if you read or hear "ket" or "ket vector," think "right vector" or "column vector" and if you read or hear "bra" or "bra vector," think "left vector" or "row vector."

To form the scalar product of two state vectors $|B\rangle$ and $|C\rangle$, we use the column vector (right vector) $|B\rangle$ and the row vector (left vector) $\langle C|$ with the following rule:

$$\langle C||B\rangle = \langle C|B\rangle \doteq \begin{pmatrix} c_0^* & c_1^* \end{pmatrix} \begin{pmatrix} b_0 \\ b_1 \end{pmatrix} = c_0^* b_0 + c_1^* b_1. \tag{4.11}$$

The result, as promised, is just a number although, in general, it is a complex number. In the first two terms in Eq. (4.11), you can see that the right vector $|B\rangle$ appears on the right of the angular bracket term $\langle C|B\rangle$ while the left vector $\langle C|$ appears on the left. The last two terms in Eq. (4.11) show how to multiply the row vector of state amplitudes times a column vector of state amplitudes to obtain the scalar product result. Also note that the multiplication of the state vector symbols on the left in Eq. (4.11) involves two vertical line symbols, which are almost always replaced by a single vertical line as indicated in the second term in Eq. (4.11).

Aside Both column vectors and row vectors are common "characters" in linear algebra and often associated with separate (but related) vector spaces often called the column space and the row space. For the latter, you sometimes hear the phrase "dual space," and a row vector is called a "dual vector." For the most part, we will not need that distinction.

If we restrict ourselves to real number coefficients, it is easy to see that the quantum state inner-product multiplication rule mimics the way regular vectors are multiplied using the scalar product (dot product) rule: $\vec{D} \cdot \vec{E} = d_x e_x + d_y e_y$.

With the geometry defined in Fig. 4.10, we have a useful result for unit vectors:

$$\langle C|B\rangle = \cos\theta_{BC}. \tag{4.12}$$

Note that this result is valid only if the Dirac bracket is a real number and if both $|B\rangle$ and $|C\rangle$ are unit length vectors.

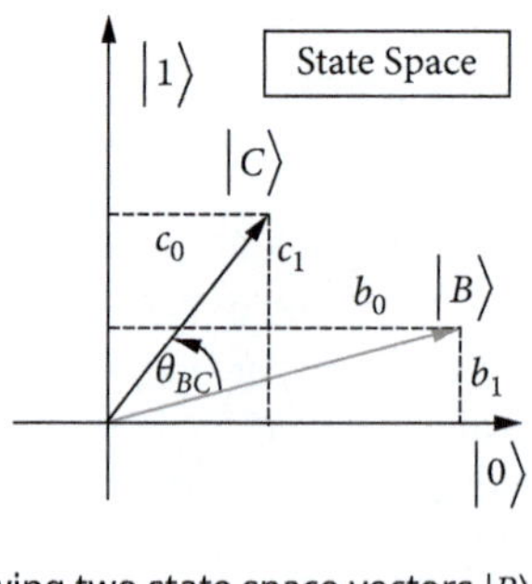

Fig. 4.10 A state space diagram showing two state space vectors $|B\rangle$ and $|C\rangle$ and the angle θ_{BC} between them. b_0 and b_1 are the components of $|B\rangle$ along the state space basis vectors. c_0 and c_1 are the corresponding components of $|C\rangle$. In this diagram, the components are assumed to be real numbers (see Eq. (4.10)).

If the state coefficients are complex numbers, our two-dimensional state space is no longer able to provide a visual representation of row vectors. In Chapter 7, we describe how to handle that situation by drawing vectors representing the quantum states in a three-dimensional state space rather than the two-dimensional version we have used here.

You may wonder if it is possible to multiply two right vectors or two left vectors. When we get to describing multiqubit systems, we'll see that we can make use of the product of two (or more) right vectors to form what are called product states for the qubits. More of that is discussed later.

Let's look at the scalar product $\langle C|B\rangle$ in more detail. If $\langle C|B\rangle = 0$ and if neither vector itself is a 0 vector—we are beginning to sound like mathematicians—then we say that the two state vectors $|B\rangle$ and $|C\rangle$ are *orthogonal*. In that case, the Dirac bracket (assuming real coefficients) is

$$\cos\theta_{BC} = \langle C|B\rangle = c_0 b_0 + c_1 b_1 = 0. \tag{4.13}$$

In a state space diagram, two orthogonal state vectors are perpendicular to each other. Orthogonality is an expression of the mutual exclusivity of the associated quantum states. For example, if your basis states are $|\text{vlp}\rangle$ and $|\text{hlp}\rangle$, then if you are in state $|\text{vlp}\rangle$ you cannot also be in state $|\text{hlp}\rangle$; the states are mutually exclusive. But you can be in a state that is a linear superposition of the two states. Figure 4.7 illustrates the orthogonality of $|\text{hlp}\rangle$ and $|\text{vlp}\rangle$.

Try It 4.3 Check orthogonality with our standard basis vectors expressed in column vector format

$$|0\rangle \doteq \begin{pmatrix} 1 \\ 0 \end{pmatrix} \quad |1\rangle \doteq \begin{pmatrix} 0 \\ 1 \end{pmatrix} \tag{4.14}$$

Hint: Use the multiplication rule expressed in Eq. (4.11).

Try It 4.4 Create two state vectors that do *not* lie along the basis state directions by giving the coefficients in Eq. (4.11) numerical values between -1 and 1 with $c_0 b_0 + c_1 b_1 = 0$. (The coefficients may be positive or negative.) Are those state vectors orthogonal? Draw state space diagrams for the vectors and explain how the state space diagrams indicate orthogonality when $c_0 b_0 + c_1 b_1 = 0$.

It is mathematically possible to use basis state vectors that are not orthogonal, but then many subsequent calculations become overly complicated. Both to satisfy the notion of "mutual exclusivity" and to make the math easier, we always use orthogonal basis states.

One more question: What is the scalar product (dot product) of a state vector with itself? Let's look at $|A\rangle = a_0|A_0\rangle + a_1|A_1\rangle$ and write the scalar product with itself as

$$\langle A|A\rangle \doteq \begin{pmatrix} a_0^* & a_1^* \end{pmatrix}\begin{pmatrix} a_0 \\ a_1 \end{pmatrix} = |a_0|^2 + |a_1|^2. \tag{4.15}$$

We see that the scalar product of a vector with itself is the *square* of the length of the vector. This is just the Pythagorean theorem with the length of the hypotenuse equal to 1. If the square of the length is 1, then the length itself is 1, and we say that the vector is "normalized." For quantum state vectors, we almost always use $|a_0|^2 + |a_1|^2 = 1$ as mentioned previously.

Similarly, if we have two state vectors, each with length 1, and if the two state vectors point in the same direction in state space, then the Dirac bracket will be

$$\langle C|B\rangle = c_0 b_0 + c_1 b_1 = 1. \tag{4.16}$$

Comparing Eqs. (4.13) and (4.16), we see that the Dirac bracket ranges from 0 (when the two normalized vectors are orthogonal) to 1 (when the two unit vectors are parallel). If the vectors lie along the same line, but point in opposite directions, then we obtain $\langle C|A\rangle = -1$. In general, if both $|B\rangle$ and $|C\rangle$ have unit length, then $\langle C|B\rangle = \cos\theta_{BC}$, where θ_{BC} is the angle between $|B\rangle$ and $|C\rangle$ in state space.

Try It 4.5 When the two state vectors are orthogonal, the angle between them in state space is $90°$, and when they are parallel, the angle between them is $0°$. Look up $\cos 90°$ and $\cos 0°$ to check the previous paragraph's statement about the numerical range for the Dirac bracket for two normalized state vectors.

4.10 Chapter Summary

- Qubits are actual physical objects whose properties we can measure. For qubits, there are two possible results for the measurement outcomes.
- The associated quantum states are represented by state vectors in an abstract two-dimensional state space analogous to a classical vector in physical space with x and y spatial components. A state vector can be represented by right vector symbols $|A\rangle$, by arrows in a state space diagram, and by column vectors.
- The quantum state carries all the possible information—the probabilities—about the outcomes when you measure that state.

- Quantum algorithms employ quantum states interacting with gates and measurement devices. The measurement devices make use of the qubit's physical properties.
- Unlike classical states, a general quantum state vector can be represented by a linear combination (superposition) of basis states. A state of superposition means that the qubit is described by both basis states weighted with numerical coefficients (amplitudes).
- For the computational basis states, the superposition is written as

$$|S\rangle = a_0|0\rangle + a_1|1\rangle \,. \tag{4.17}$$

- Linearly polarized photons are examples of qubits. Their basis states are vertical linear polarization $|\text{vlp}\rangle$ and horizontal linear polarization $|\text{hlp}\rangle$.
- Polarizing sheets and polarizing beam splitters can be used both to prepare and analyze linear polarization in a light beam. This dual use occurs often in quantum systems.
- Often a quantum algorithm will require state vectors to be multiplied together. The dot product (inner product or scalar product) of two state vectors is represented by the Dirac bracket $\langle A|B\rangle$ between a left vector and a right vector, which can be evaluated by finding the product of the row vector corresponding to $\langle A|$ and the column vector corresponding to $|B\rangle$. The result of this multiplication will be a single number, and if that number is zero, then the state vectors are orthogonal.
- The dot product of a state vector with itself, for example $\langle A|A\rangle$, is the square of the length of that state vector in state space. Normalized state vectors have a length equal to 1.
- The inner (dot) product of two normalized state vectors (each have length 1) is equal to the cosine of the state space angle between the two vectors: $\langle C|A\rangle = \cos\theta_{AC}$ when the state coefficients are real numbers.

4.11 Further Reading

Michael G. Raymer, *Quantum Physics: What Everyone Needs to Know* (Oxford University Press, Oxford, 2017). This book has a nice section on linearly polarized light.

Linear polarizing sheets are readily available online. Some suggested sources (the authors have no commercial connections to any of these sources) are Amazon www.amazon.com, American Polarizers, Inc. https://www.apioptics.com/product/ap38-006t/, Edmund Optics https://www.edmundoptics.com, and Arbor Scientific https://www.arborsci.com/products/polarizing-filters.

Schrödinger's grave marker has been updated since the photo in Fig. 4.2 was taken. The circular plaque now has Schrödinger's equation $i\hbar\,\partial\psi/\partial t = H\psi$, which is a differential equation for the wave function ψ. The name plate also lists his wife, who, in fact, is not buried there. To see the new marker, do a web search for "Schrödinger's grave."

5 Quantum Measurements

It is much easier to make measurements than to know exactly what you are measuring.

J. W. N. Sullivan (1928)

5.1 Quantum Measurements: What Are They?

We are now ready to tackle quantum measurements, an integral part of quantum information science (QIS) and quantum computing (QC). First, we provide a few words about classical measurements. For a classical system, we may measure its properties (such as high voltage or low voltage on a computer wire or its position in three-dimensional space) without significantly disturbing its state. Moreover, for identically prepared systems, you get the same measurement result each time you measure the property (except for experimental uncertainties, which, in principle, can be made as small as you want). However, we will see that none of that is true for a quantum measurement. Focusing on qubit (two-state) systems, we will see that the best quantum mechanics allows for, in general, is finding the probabilities of getting the two possible measurement outcomes. Moreover, the quantum state describing the qubit changes because of the measurement process and, in general, there is no way to reduce that change to zero.

Measurements always involve interactions between the qubit and a "measurement device." Ideally, the measurement device is equipped with signals such as light bulbs or sound generators or electrical outputs that tell you which output channel has detected the qubit. For most purposes in QIS and QC, we don't need to worry about the details of the measurement devices, but for those who want to build quantum computers, they are a big deal.

At a fundamental level, however, there are many unresolved conceptual issues concerning quantum measurement devices. By definition, a measurement device provides a stable, permanent indication of the results of the measurement. Consequently, most measurement devices involve large numbers of atoms and molecules to provide that indicator (a lit light bulb, a pointer needle, an electrical signal sent to your computer, etc.). But how many atoms and molecules are necessary to make a functioning measurement device that provides a permanent indication of the results of the measurement? Surprisingly, that is a question for which quantum mechanics does not yet have a direct answer.

Quantum Computing and Quantum Physics. Alice Flarend and Robert Hilborn, Oxford University Press.
 DOI: 10.1093/oso/9780197904930.003.0005

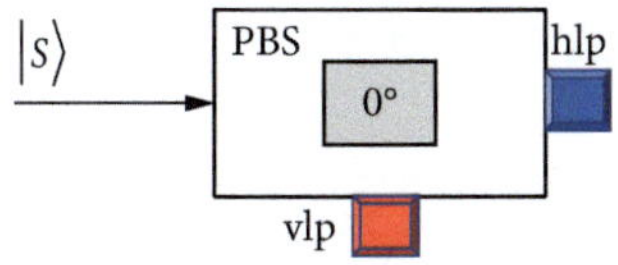

Fig. 5.1 A quantum measurement device for linearly polarized light. A photon qubit described by the state vector $|S\rangle$ interacts with a polarizing beam splitter (PBS), a linear polarization measurement device. The measurement device is set to sort out horizontal and vertical polarization relative to a "horizontal" axis at 0° indicated by the central display. If the photon is detected in the vlp beam, the red light will be lit. If it is detected in the hlp beam, the blue light will be lit.

Given that ambiguity about exactly what constitutes a measurement device, we simply assume that at a practical level we can build appropriate measurement devices (and indeed we can). A specific example will be helpful.

Let's start with linearly polarized light and a polarizing beam splitter (PBS) as shown in Fig. 5.1. We have added photon detectors to the PBS setup to produce a measurement device. If a detector senses a photon, it will activate a light (an irreversible process), indicating that the device has made a measurement. Just as in Chapter 4, we denote the linear polarization state associated with the incoming photon as the superposition of the two linear polarization basis states:

$$|S\rangle = a_{\text{h}} |\text{hlp}\rangle + a_{\text{v}} |\text{vlp}\rangle . \tag{5.1}$$

Let's unpack the measurement process illustrated in Fig. 5.1. We read the diagram from left to right. A photon described by a quantum state $|S\rangle$ interacts with a linear polarization measurement device, the PBS. The PBS orientation angle is indicated by the reading of a display on the device. We say that the two linear polarization states (horizontal and vertical relative to that orientation angle) form the "measurement basis states."

Suppose we have added the appropriate equipment to be able to detect single photons. If a single photon interacts with the measurement device, the output will indicate either red or blue but never both at the same time. Those are fundamental principles of quantum mechanics: The qubit is not divisible (it can't split up with part going one way and part the other way), and vlp and hlp are mutually exclusive outcomes. We know the physics of how the PBS interacts with light, so we know which output corresponds to vlp and which to hlp.

The important question is what do we learn about the quantum state in Eq. (5.1) after a single photon interacts with the measuring device. That issue was mentioned in Chapter 4. If the qubit interacts with the measurement device and the measurement device indicates vlp, we really can't say much about the qubit's state *before* it interacted with the measurement device. The state before might have been $|\text{vlp}\rangle$ or it could have been a superposition state $|S\rangle = a_{\text{h}} |\text{hlp}\rangle + a_{\text{v}} |\text{vlp}\rangle$ with $a_{\text{v}} \neq 0$. The only thing we know for sure is that the state wasn't purely $|\text{hlp}\rangle$ because if it were we would never have observed a vlp result.

In the next sections, we show how we can extract information about the probabilities of the two possible outcomes. Then, we take up the question of describing the state of the qubit after the measurement. Both of those results are important for QC and QIS applications.

5.2 Light Intensity from a Polarizer–Analyzer Setup

To tease out the physical meaning of the state amplitudes in Eq. (5.1), let's use a polarizing sheet to produce a beam of linearly polarized photons. We then employ a PBS as a measurement device and measure how much of the linearly polarized light beam emerges in each of the PBS's output beams. We then ask a critical question: What happens if we rotate the PBS relative to the polarization transmission axis of the polarizing sheet? We will first answer this question by assuming that the light beam can be described as a classical system.

Let's think about how the two intensities (the numbers of photons per second) will vary as we change the angle between the polarizing sheet and the PBS direction as shown in Fig. 5.2. A good way to start is to think about the extreme cases. If $\theta = 0$, the transmission directions of the polarizing sheet and the PBS are aligned horizontally. It means that whatever comes through the first polarizing sheet will also come out in the hlp beam I_{2h} from the PBS. When the angle is 90°, we expect nothing will be in the hlp beam from the PBS, so all the light must come out in the vlp beam I_{2v}. What happens for angles between 0° and 90°? We find from actual experiments with light beams (which have a lot of photons) that the intensity (power) of the light transmitted by the PBS in its hlp beam is given by

$$I_{2h} = I_1 \cos^2 \theta. \tag{5.2}$$

and that in the vlp output beam is given by

$$I_{2v} = I_1 \sin^2 \theta, \tag{5.3}$$

where θ is the angle between the transmission axis of the sheet and the hlp direction of the PBS. I_1 is the intensity of the light beam before it interacts with the PBS.

The graphs of Eqs. (5.2) and (5.3) are shown in Fig. 5.3. Let's check that the results given in Fig. 5.3 make sense for various cases. For $\theta = 0$, the two transmission axes are parallel, and

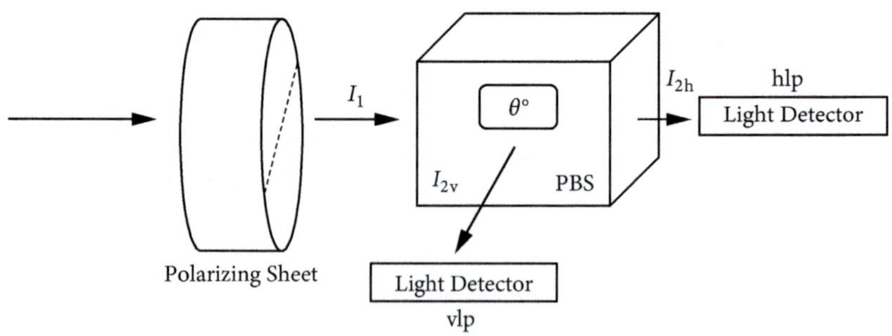

Fig. 5.2 A light beam on the left approaches a polarizing sheet with its transmission axis indicated by the dashed line. The beam with intensity I_1 then encounters a polarizing beam splitter (PBS) with its hlp transmission axis rotated by an angle θ relative to the transmission axis of the polaroid sheet. Light detectors produce signals proportional to the intensity (number of photons per second) of the light beams after they emerge from the PBS.

we would expect, as we mentioned before, that everything that gets through the first sheet is also transmitted by the PBS in its horizontal polarization output beam (I_{2h}) and none in the vertical polarization output beam (I_{2v}). That is consistent with $\cos(0) = 1$ and $\sin(0) = 0$ in Eqs. (5.2) and (5.3). If we rotate the PBS by $\theta = 90°$, the beam splitter's hlp output gets no light since the hlp's polarization direction is now perpendicular to the polarizing sheet's transmission axis. We expect that none of the light will get through in that beam, and all the light shows up in the vlp beam. For $\theta = 45°$, 50% of the light goes into one PBS output beam and 50% into the other, which is consistent with $\cos^2(45°) = 1/2$.

Try It 5.1 Explain in detail how those results are consistent with the graphs in Fig. 5.3. In particular, explain the cases $\theta = 0°$, $45°$, and $90°$.

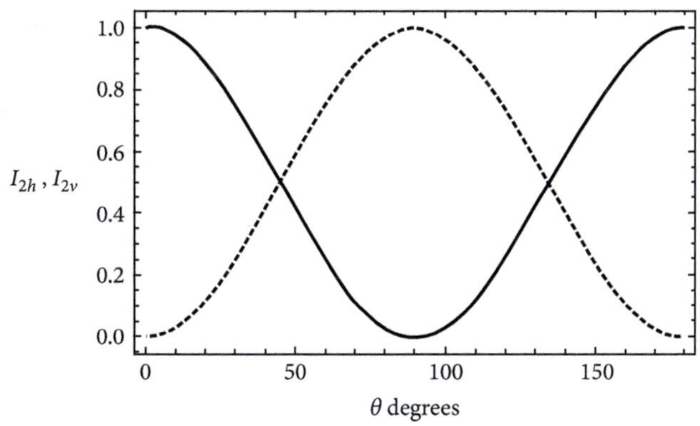

Fig. 5.3 A plot of the relative intensities I_{2h}/I_1 (solid curve) and I_{2v}/I_1 (dashed curve) in the two output beams for the PBS as a function of the angle between the polarizing sheet transmission axis and the PBS hlp axis. See Eqs. (5.2) and (5.3) for the definitions of I_{2h} and I_{2v}.

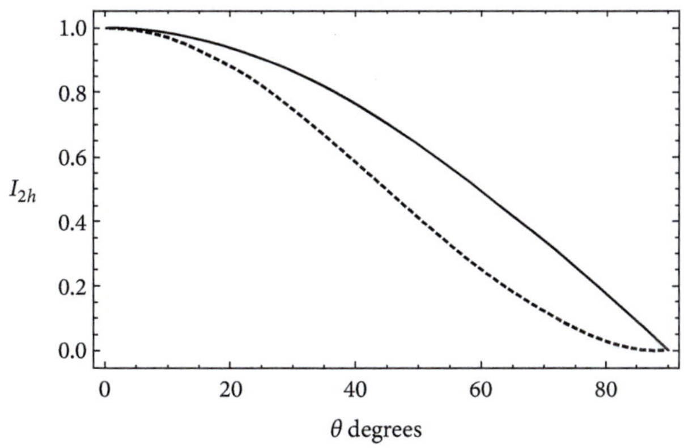

Fig. 5.4 A plot of $\cos^2\theta$ (dashed curve) and $\cos\theta$ (solid curve).

Try It 5.2 Suppose Prof. Blah claims that her measurements indicate that $\cos\theta$ and $\sin\theta$ should replace $\cos^2\theta$ and $\sin^2\theta$ in Eqs. (5.2) and (5.3). Draw graphs of both the original results and Prof. Blah's results. How might you distinguish between them (see Fig. 5.4)? Are there theoretical reasons that might rule out Prof. Blah's results? Hint: What happens to $\cos\theta$ for angles between 90° and 180°?

5.3 State Amplitudes and Probabilities

Let's now assume we have the detectors shown in Fig. 5.2 needed to detect and count the number of individual photons in each beam during a specified period of time, and the photons are described by the quantum state in Eq. (5.1). The equipment to detect individual photons is readily available in most college and university physics departments. Unfortunately, the equipment is too expensive to buy for yourself. (We give links to several videos about detecting individual photons in the Further Reading section.)

We will go through this process in some detail because it gives us a paradigm we can use for getting information about the quantum state of other qubits. Using the setup illustrated in Fig. 5.2, Alice prepares a beam of light in a linearly polarized state using the first polarizing sheet, but she doesn't tell us the transmission axis direction. Our task is to find the polarization state of Alice's beam using the output of the detectors in Fig. 5.2 for the PBS vertical and horizontal polarization directions. That is, we want to find the numerical values of the state amplitudes a_h and a_v in Eq. (5.1).

Let's suppose a photon is detected in the vlp output beam of the polarization measurement device (the PBS). What does that result tell us about its state $|S\rangle$ before it encountered the PBS? All we know is that its state $|S\rangle$ was not *purely* hlp. In terms of the amplitudes, we can say that $a_\mathrm{v} \neq 0$, but we know nothing about a_h. If the state $|S\rangle$ had been purely hlp, no photons would end up the in PBS vlp channel. But other than that, having detected just one photon, that is all we can say.

We can learn more about the state $|S\rangle$ by asking Alice to prepare more photons described with that same state and to send them one by one toward our polarization measurement device. If we detect a photon in the vlp channel, we write down a "V." If it is detected in the hlp channel, we write down "H."

After a long time (depending on how rapidly Alice can fire photons at us), we end up with a listing of measurement event outcomes; in this case, a long string of Vs and Hs: VHHVVVHVHHH... with no apparent pattern. The sequence looks completely random. Let's count n_v, the number of Vs, and n_h, the number of Hs, in the list. The total number of photons detected is $n_\mathrm{v} + n_\mathrm{h} = N$.

The fraction of Hs, that is, n_h/N, is the probability that one of Alice's photons ends up in the hlp beam. In other words, for Alice's photon state $|S\rangle$, the probability of detecting a horizontally polarized photon is

$$P_\mathrm{h} \approx n_\mathrm{h}/N. \tag{5.4}$$

Similarly, the probability for detecting a vertically polarized photon is

$$P_v \approx n_v/N. \tag{5.5}$$

We use approximate equality because with a finite number of photons received, the ratios are *estimates* of the probabilities (in the language statisticians like to use). For example, if we repeatedly flip a coin, we almost never exactly have the same number of heads and tails. But we expect to find $P_{\text{tails}} \approx 1/2$ and $P_{\text{heads}} \approx 1/2$.

Since there are only two possible outcomes for the photon polarization state (hlp and vlp), the two probabilities must add up to 1:

$$P_v + P_h = 1. \tag{5.6}$$

Eq. (5.6) is just like the normalization condition for state amplitudes $|a_h|^2 + |a_v|^2 = 1$ presented in Chapter 4. This suggests that we may interpret the square of each amplitude individually as the probability for that particular outcome

$$|a_h|^2 = P_h \qquad |a_v|^2 = P_v. \tag{5.7}$$

The square of the length of a state vector is given by the sum of the probabilities of the outcomes:

$$\langle S|S\rangle = |a_h|^2 + |a_v|^2 = 1. \tag{5.8}$$

Try It 5.3 Suppose the two amplitudes are $a_h = 1/2$ and $a_v = \sqrt{3}/2$, what are the probabilities for measuring the photon with horizontal and vertical polarization? Answer the same question for $a_h = 1/4$ and $a_v = \sqrt{15}/4$.

Figure 5.5 shows a column chart which is a common way of displaying the probabilities for a qubit quantum state. We use this kind of chart again in Chapter 12 when we work with multiqubit states in quantum algorithms and in Chapter 22 when we describe the density operator and mixed states.

Recognizing that the squares of the state amplitudes give us measurement probabilities is perhaps the most important principle in quantum mechanics. That statement is called the *probability principle of quantum mechanics*. This principle is also known as the *Born rule* (see Chapter 1, Principle 4). It was introduced by Max Born in 1926, early in the development of quantum mechanics.

Probability principle of quantum mechanics (Born rule): The absolute-value-squared of a basis state amplitude is the probability of the qubit's interaction with the measurement device yielding a result associated with that basis state.

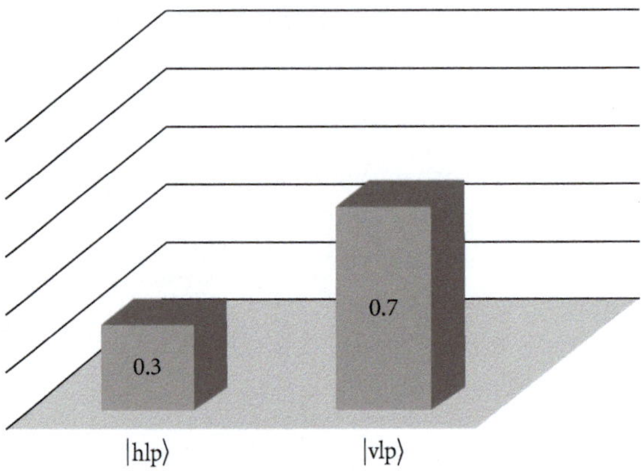

Fig. 5.5 A column chart for a single linearly polarized photon qubit. The height of the columns is proportional to the probability of finding the photon in the specified states. In this case, the probability of finding the photon in the hlp state is 0.3 and 0.7 for the vlp state. The labels below the columns indicate the two possible polarization states (see Eq. (5.1)).

There is a crucial conceptual issue here, which we alluded to before. Note that the probability interpretation does *not* say that the qubit *had* that value of the measured property before the interaction with the measurement device. As David Mermin has succinctly written in *Quantum Computer Science* (Mermin, 2007, p. 40), "[there is a] mistaken view that the quantum state encodes a property inherent in the Qbits [qubits]. The state encodes only the possibilities available for the extraction of information from those Qbits."

5.4 Quantum States after a Measurement

Now let's discuss the state describing the qubit after a measurement. If we don't detect the photon but let it travel on from the PBS, the state that should be used to describe the outgoing photon is the device's basis state associated with the output channel the photon takes. This assertion is called the *measurement principle* of quantum mechanics (see Chapter 1, Principle 5). According to the *measurement principle*, the outgoing qubit state vector is one or the other of the basis states of the device depending on which output channel the qubit takes. For linearly polarized light, after interacting with the PBS, the photon's state is either $|\text{hlp}\rangle$ or $|\text{vlp}\rangle$, but never both, as we mentioned before. The PBS has separated the stream of photons into two beams, but a single photon will be detected in only one of the output beams.

The previous sections show how to find the probability that the photon is in a particular output beam in terms of the absolute-value-squared of the appropriate quantum state amplitude (coefficient). Suppose we do not detect which beam a particular photon is in. In that case, we have not yet completed a quantum *measurement*. This may seem a bit fussy (and it is), but it highlights the importance of qubit detection in quantum measurements. If we don't detect the photon as it leaves the PBS, all we have done is to use the PBS as a quantum state selector or analyzer. The photons that end up in the vlp beam will, by design, have vlp polarization and similarly for hlp. This way of thinking about measurements is captured in the *measurement principle*.

Measurement principle: Immediately after a qubit interacts with a measurement device, the qubit's quantum state is the basis state associated with the device's output channel in which the qubit is found.

How do we check the *measurement principle*? There are two parts to the answer. First, we need to understand the physics of the measurement device. For polarized light, in particular, we need to know how the device interacts with light and how that interaction depends on the polarization state. For our purposes, we assume that work has been done and we know the physics. The second part of the answer is that you can check which state we have by letting the photon interact with a second linear polarization measurement device. (Of course, we need to be sure that the photon does not interact with anything that might change its state before it gets to the second device.) If the photon's quantum state after the first PBS is $|\text{vlp}\rangle$, for example, then the photon interacting with the second PBS will, assuming that the two PBSs have the same settings, always end up in the second's vlp output beam and never in its hlp beam. If it *always* ends up in the vlp beam of the second PBS, then we know that its state before interacting with the second measurement device must have been $|\text{vlp}\rangle$.

Another complication arises if the measurement process leads to the disappearance of the qubit from our quantum system. For example, if the measurement system includes a photon detector (see Fig. 5.2), then that detector "absorbs" the photon, which is no longer available for further quantum manipulations. In those cases, which in fact are the most common in QC and QIS applications, there is no state assigned to the absorbed photon. Similar results apply to electrons that might be detected by being absorbed into a metal wire or atoms that are ionized (one or more electrons are removed from the atoms), and the resulting electrical current is measured by standard electronic instruments. Such measurements are often called "destructive" measurements. The quantum states (and the physical qubit) are gone after a destructive measurement. When we get to multiqubit quantum states in Chapter 9, we introduce the *generalized Born rule* that tells us how to assign states for the multiqubit system when only one of the qubits is subject to a measurement interaction.

However, there is another kind of quantum measurement called a quantum nondemolition measurement (QND) that preserves the information about the qubit's state before the measurement. Such QND measurements are often difficult to implement, and not all observable properties can be handled by QNDs (see Further Reading).

In cases for which we do not detect the qubit immediately after the measurement device, for example, if we don't absorb the photon in a light detector after the PBS in Fig. 5.1, we may let the outgoing photon interact with further quantum gates and measurement devices, which lie along one path or the other before we finally detect the photon. In essence, we have "tagged" the photon by the path that it follows. We can infer that if the photon is eventually detected in the vlp path, it was described by vlp when leaving the first PBS and, conversely, hlp for the other path.

There are many subtleties associated with measurements in quantum mechanics. For now, we assume that we have a practical understanding of how a measurement device operates based on classical physics (which is all we need to describe how a PBS works).

Try It 5.4 Suppose that an unpolarized beam of photons interacts with a PBS with its "horizontal" direction set at $0°$. The hlp output beam from the PBS continues on and then interacts with another PBS whose "horizontal" direction is set at $90°$. What will be the output from the second PBS?

5.5 A "Minus 1" Mystery and a Change of Basis States

We have argued that making a long series of measurement events, resulting in the list of Hs and Vs for example, allows us to determine the *square* of the quantum state amplitudes. Can we find the "unsquared" amplitudes a_h and a_v? The state amplitudes are sometimes called "probability *amplitudes*" to distinguish them from the probabilities themselves. It seems easy enough. If the amplitudes are real numbers, all we have to do is take the square root of the probability. But the difficulty is that there are two possible square roots: $\sqrt{a_\mathrm{h}^2} = \pm a_\mathrm{h}$. How do we know which is correct? In other words, to get the complete quantum state (that is, to "measure" the quantum state), we also need to find out if the coefficients are positive or negative. Finding those signs (+ or –) is called finding the "relative phase" of the coefficients. Those relative phases turn out to be important in QC and QIS. We look at some simple examples of finding those relative phases. But first let's examine the effect of an *overall* sign change for a quantum state.

If the amplitude of a quantum state is negative, what does that mean? For ordinary vectors, $-\vec{F}$ is a vector that has the same length (magnitude) as the vector $\vec{F}$ but points in the opposite direction. The "obvious" generalization to quantum states is that in state space $-|\mathrm{vlp}\rangle$ points in the direction opposite to that of $|\mathrm{vlp}\rangle$. Note that $-|\mathrm{vlp}\rangle \neq |\mathrm{hlp}\rangle$.

Figure 5.6 shows how two vectors $|S\rangle$ and $-|S\rangle$ are related in state space. We now apply that idea to the linear polarization basis states. The distinction between $-|\mathrm{vlp}\rangle$ and $|\mathrm{hlp}\rangle$ is even more obvious if we write things out using column vectors:

$$-|\mathrm{vlp}\rangle \doteq -\begin{pmatrix}0\\1\end{pmatrix} = \begin{pmatrix}0\\-1\end{pmatrix} \neq \begin{pmatrix}1\\0\end{pmatrix} \doteq |\mathrm{hlp}\rangle\ . \tag{5.9}$$

Note that having the minus sign does not change the length of the state vector. It does, however, change the sign of the components of the state vector along the basis vector directions, but remember that it is the *square* of those components that give the probabilities that measurement results lie along the basis state directions. So, those are unchanged by an *overall* minus sign. Let's elaborate on that important result. Since quantum state vectors are meant to describe the quantum states of a qubit and those states are supposed to carry all the information about possible measurements, we need to understand how the

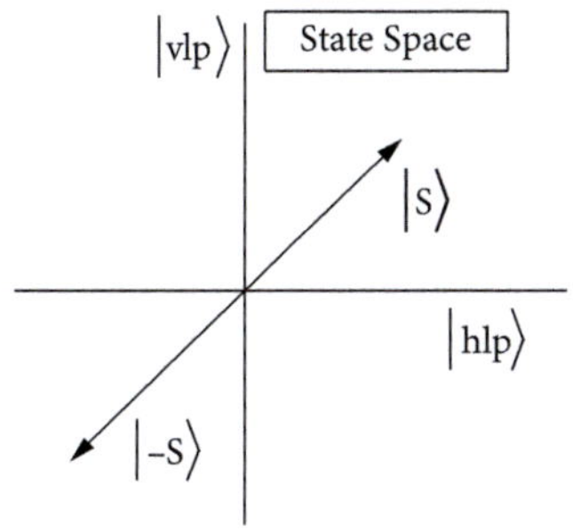

Fig. 5.6 A state space diagram showing a state vector $|S\rangle$ and its negative $-|S\rangle$.

probabilities associated with the quantum state change if the state vector is replaced by its negative.

Let's consider a quantum state that is just one of the polarization basis states, say, horizontal polarization. That statement means that every polarization measurement on a qubit described by that state will yield horizontal polarization. In other words, the probability for getting horizontal polarization is 1, and we know that the coefficients in Eq. (5.1) satisfy $|a_h|^2 = 1$ and $a_v = 0$. If we think of state vectors in state space, it means that $+|\text{hlp}\rangle$ and $-|\text{hlp}\rangle$ predict the same measurement outcomes. In our two-dimensional state space (with real coefficients), all unit-length state vectors pointing along the same line (see Fig. 5.6) give the same measurement predictions. So, in that sense, they all represent the same quantum state. If we have complex number coefficients, what happens? It is easy to see that $z\,|S\rangle$ with $|z| = 1$ will predict the same measurement outcomes as $|S\rangle$. The complex number z is called a phase factor. +1 and −1 are special cases of a phase factor. In Chapter 7, we treat states with complex coefficients in more detail.

Let's go back to photons and look at what happens when we do not limit ourselves to only horizontal and vertical orientations. Suppose we have prepared the qubit state to be one with linear polarization at 45° with respect to the axis of a PBS and then let the photon interact with the PBS. We recall that for any single photon observation, we find the photon in either the vlp beam or the hlp beam of the PBS. As far as we know, according to the rules of quantum mechanics, there is no way to predict which beam a particular photon ends up in except for the special case in which the photon happens to have been prepared with a state corresponding to vlp or hlp. All that quantum mechanics gives us is the probability for the photon's ending up in one or the other of the two measurement output states.

Figure 5.7 shows two new basis states $|+45°\rangle$ and $|-45°\rangle$. We now express the photon linear polarization state as a linear combination of the $|+45°\rangle$ and $|-45°\rangle$ basis states shown in Fig. 5.7:

$$|S\rangle = a_-\,|-45°\rangle + a_+\,|+45°\rangle\,. \tag{5.10}$$

We want to find the probabilities for getting hlp and vlp as the measurement outcomes from the PBS if the photon is described by Eq. (5.10). The Born rule tells us to find the amplitude $\langle \text{hlp}|S\rangle$ and then take the absolute-value-squared of that to get the probability of

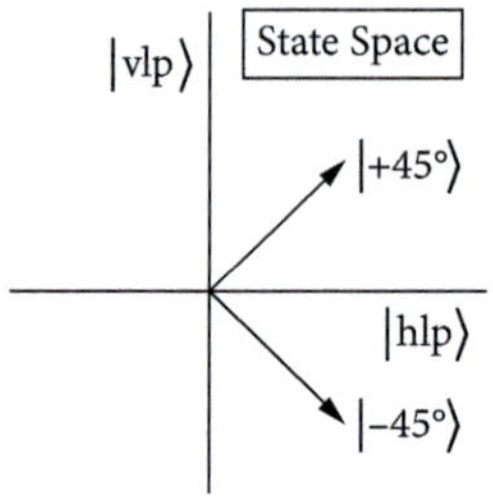

Fig. 5.7 A state space diagram showing basis states $|+45°\rangle$ and $|-45°\rangle$.

observing the detected photon in the hlp channel. There is an analogous expression for the vlp amplitude. The required amplitudes are

$$\langle \text{hlp}|S\rangle = a_- \langle \text{hlp}|-45^\circ\rangle + a_+ \langle \text{hlp}|+45^\circ\rangle \tag{5.11}$$

and

$$\langle \text{vlp}|S\rangle = a_- \langle \text{vlp}|-45^\circ\rangle + a_+ \langle \text{vlp}|+45^\circ\rangle . \tag{5.12}$$

How do we find the Dirac brackets $\langle \text{hlp}|-45^\circ\rangle$ and $\langle \text{hlp}|+45^\circ\rangle$ and so on? The geometry in Fig. 5.7 shows that

$$|+45^\circ\rangle = \cos(45^\circ)|\text{hlp}\rangle + \sin(45^\circ)|\text{vlp}\rangle = \frac{1}{\sqrt{2}}|\text{hlp}\rangle + \frac{1}{\sqrt{2}}|\text{vlp}\rangle . \tag{5.13}$$

And consequently, we obtain the Dirac bracket

$$\langle \text{hlp}|+45^\circ\rangle = \frac{1}{\sqrt{2}}\langle \text{hlp}|\text{hlp}\rangle + \frac{1}{\sqrt{2}}\langle \text{hlp}|\text{vlp}\rangle = \frac{1}{\sqrt{2}} \tag{5.14}$$

because$\langle \text{hlp}|\text{hlp}\rangle = 1$ and $\langle \text{hlp}|\text{vlp}\rangle = 0$.

Try It 5.5 Use similar arguments to show that $\langle \text{hlp}|-45^\circ\rangle = 1/\sqrt{2}$, $\langle \text{vlp}|+45^\circ\rangle = 1/\sqrt{2}$, and $\langle \text{vlp}|-45^\circ\rangle = -1/\sqrt{2}$.

Using the results in Eq. (5.14) and Try It 5.5, we find

$$\langle \text{hlp}|S\rangle = \frac{a_-}{\sqrt{2}} + \frac{a_+}{\sqrt{2}} = \frac{a_- + a_+}{\sqrt{2}} \tag{5.15}$$

and

$$\langle \text{vlp}|S\rangle = -\frac{a_-}{\sqrt{2}} + \frac{a_+}{\sqrt{2}} = \frac{-a_- + a_+}{\sqrt{2}}. \tag{5.16}$$

Finding the absolute-value-squared of those amplitudes gives us the probabilities. In particular, the probability of getting vlp is $|-a_- + a_+|^2/2$, and the probability for getting hlp is $|a_- + a_+|^2/2$. We see that those two probabilities depend on both the magnitudes of and the sign difference between a_- and a_+. In other words, if we measure those probabilities, we can reverse the process to find the magnitudes of those amplitudes and their sign difference. The takeaway message is that measuring the probabilities using measurement basis states different from the basis states in which the qubit was prepared gives us more information about the quantum state than using the same basis states for preparation and measurement. That result is widely used in QC and QIS applications. We present many applications of that result in subsequent chapters. In Chapter 8, we show you how to change basis states systematically.

5.6 Quantum Measurements and the No-Cloning Theorem

Let's pause for a moment to review what we have learned. We "measure" a quantum state by determining the numerical values of the state vector coefficients. This almost always requires repetitions of (1) preparing the qubit in a specified quantum state (though we may not know what the state is) and (2) observing the result of having the qubit interact with a measurement device and recording the channel into which the qubit goes (e.g., horizontal polarization versus vertical polarization for a photon, or spin-up versus spin-down for a spin system). The collection of individual observations then constitutes (at least a part of) a state measurement. Those repetitions give information that can be used to find the state vector coefficients, up to an overall minus sign or overall phase factor like $e^{i\alpha}$.

Here is one of the situations in which the careful use of terminology can help avoid confusion. We like to say that for each interaction of a qubit with the measurement device we have a measurement event (sometimes called an *observation*). Recognizing the limits on what we have learned from a single measurement event, we know we need to repeat the measurement many times with identically prepared qubits to acquire a series of measurement events from which we can find average values and other statistical properties of the overall measurement process.

An analogy might be helpful. If you flip an unknown coin just once and it comes up with a "head," you can't say much about the coin except that the probability of observing a head is not zero. You don't know if both sides of the coin are "heads" or if the coin is weighted in such a way to make observing a head much more likely than observing a tail or if the probabilities for heads and tails are the same. To *measure* the probability of getting heads for that coin, we need many *observations* of individual coin flips (measurement events). The overall series of events constitutes a measurement of the probabilities.

From a single measurement event, as we pointed out, you learn only a little about what the qubit's quantum state was before the observation. That doesn't seem too bad, but it has an implication that is extremely important for QIS and QC. The implication is that we cannot clone a quantum state because, after one observation, we have only limited information about the amplitudes of the basis states used to describe the qubit's state. More importantly, after the qubit interacts with the measurement device, the appropriate state to describe the qubit is the basis state associated with the measurement outcome. Further measurements on that postmeasurement qubit will not give us any additional information about the state describing the qubit before its interaction with the measurement device. All these considerations lead to the *no-cloning theorem* which states that a single measurement event does not give us enough information to clone a quantum state—that is, we can't find all the amplitudes and relative phases, which are what we need if we want to clone a state.

The *no-cloning theorem* does not prevent us from preparing several qubits with the same quantum state. What it does prohibit is having someone else learn all the details of a qubit's unknown quantum state from one observation on that qubit alone. In other words, what they do learn from a measurement event is insufficient to allow them to prepare another qubit in that same state or to restore the observed qubit to its state before the observation. By "unknown" we mean that we do not know the state amplitudes and relative signs (or phases). If we knew them, of course, we could prepare another qubit in that state by the preparation procedures we will learn about later, or we could restore the observed qubit's original state.

In Chapter 8, we see how to use the *no-cloning theorem* to detect if someone is eavesdropping on a quantum information channel. In Chapter 17, we show how the theorem is intimately tied to the other fundamental concepts of quantum states.

Some authors write about "measuring" the quantum state of a photon by simply putting a PBS in the light beam path (or analogous devices for other types of qubits). Interacting with the PBS is necessary, but it is not sufficient to constitute a quantum measurement as we have defined "measurement." We haven't *measured* anything until we actually know which path a particular photon takes and that requires we detect the presence of that photon. Usually, this means detecting the photon by absorbing it with a photodetector to produce an electrical signal we can record.

If we let the photon exit the measurement device without detection, all we can legitimately say is that *if* the photon is in the vlp beam, the appropriate quantum state vector is $|\text{vlp}\rangle$, and *if* it is in the hlp beam, the appropriate quantum state vector is $|\text{hlp}\rangle$, but without further detection, we don't know which has actually occurred.

There are some other terminological issues. You will often read that "a measurement collapses the quantum system's state vector." Or that "the measurement device causes the qubit to *jump* from the state $|S\rangle$ to the state $|\text{vlp}\rangle$," for example. At best, these statements try to give you a mental picture of what happens in a quantum measurement. At worst, they muddle thinking about quantum mechanics. We prefer to say that after the qubit interacts with a measurement device, we have new information about its state after the measurement event. Once the qubit leaves the device, its state is now one or the other of the device's basis states, so it is appropriate to use that information to describe the qubit's state. With the postmeasurement state, we are simply updating our knowledge about the qubit's state.

From this point of view, thinking about measurements re-enforces the notion that state vectors are not things. For those who insist that state vectors are real things, wouldn't all those collapsing state vectors make a terrible noise? And what a mess to clean up!

Many QIS and QC people avoid the detailed quantum description of what happens in measurement devices, and they simply say: We measure a qubit's state and get 0 or 1 (or vlp or hlp)—without worrying about the many technical difficulties of carrying out measurements on quantum systems and how those measurements are reflected in our formal description of quantum states. For designing quantum algorithms, that less detailed description is fine for most practical purposes.

There is another way of doing quantum measurements. Instead of repeatedly preparing a state and recording the outcomes of the measurement events, we could think of preparing identically at one time the states of a large number of qubits and then measuring them all at once with an array of measurement devices. The qubits are identically prepared in the sense that the quantum states of all the qubits are the same. If the qubits do not interact with each other, then we may assume that the results of measuring the system of many qubits with measurement devices are the same as those obtained from a sequence of qubits measured one at a time. Either way, from those observations we can calculate ("estimate") the probabilities of observing the qubit in each of the measurement device's basis states.

No matter what terminology you use, measurements are key to all applications of quantum mechanics, including QIS and QC. The theoretical physicist Julian Schwinger, who won the Nobel prize for his development of relativistic quantum electrodynamics, characterized quantum mechanics as "the symbolism of atomic measurements" (Schwinger, 2001). By that he meant that quantum state vectors carry information about quantum measurements, not about inherent properties of the quantum objects.

You might well ask: Can't we do better than just probabilities? Can't we know the "actual state" of the object? The answer is no. As far as we know, those probabilities are a manifestation of how nature works at the quantum level. Probabilities show up in almost all quantum situations. For example, if we shine a brief pulse of light from a laser on a collection of atoms, the atoms will make transitions from a lower energy state to a higher energy state as they absorb the light. Later each atom will emit a photon as it makes a transition from a higher energy state to a lower energy state. But some of the atoms will emit their photons shortly after the laser pulse and others will do so later. We can use quantum mechanics to calculate the probability per unit time that an atom will emit a photon. However, there is no way, as far as we know, to predict when any particular atom will emit its photon.

Try It 5.6 You can observe this spread-out-in-time emission of light directly. If you have a TV set turned on in a darkened room, hit the on/off button (or the equivalent remote control button) and watch the screen. It will continue to "glow" for a short period of time after the set turns off. The light-emitting pixels in the TV screen take some time to "decay" back to their "off" states. Some do that more quickly than others. This is an example of quantum randomness. Another example is the clicking of a Geiger counter when it is exposed to radioactive material. Some of the atomic nuclei in the radioactive material emit subatomic particles, which are registered by the Geiger counter. But the decays occur randomly, hence the erratic clicking of the counter. For a large sample of particles, half of them will decay in a time called the "half-life."

We have seen that a quantum state vector encodes information about the results of measurements made on the qubit (the physical object). But we need to remind ourselves that in the quantum world we can't legitimately say that the measurements tell us anything definite about the dynamical properties (properties that can change) associated with the qubit other than the probabilities of various measurement outcomes. For example, as we discussed before, if we prepare a beam of photons with linear polarization at 45° with respect to a PBS's axes, the photon state can be expressed as a superposition of the hlp and vlp states associated with the PBS directions. In that situation, half the photons come out in the vlp beam and half in the hlp beam. We are tempted to say that before the photons met the measurement device, those that come out in the vlp beam must have had some property different from those that come out in the hlp beam. That seems like common sense. But as far as we know, there are no such properties. The photons are randomly distributed between the two output beams. Pretty scary, huh?

But if randomness rules at the quantum level, doesn't that mean that the world is not deterministic, effect does not follow cause, and we don't know for sure if the Sun will rise tomorrow? Doesn't that show that the answers produced by a QC will be random? How do we know what the actual answer is? These are some of the most puzzling aspects of quantum mechanics. In the last half of the 20th century, a few scientists tried to "explain" this randomness by saying that there are "hidden variables," properties of the quantum system that are hidden from us. Atoms that emit photons at different times have different values of those hidden variables. That certainly sounds plausible. In the linearly polarized photon case, the photons that come out in the vlp beam must have had some property different from those that come out in the hlp beam. However, since the 1980s, experiments have shown that almost all plausible hidden variable theories are ruled out. Quantum randomness reigns supreme!

5.7 Eigenvectors, Eigenvalues, and Average Values of Properties

So far, we have focused on just the probabilities of getting particular results, for example, photons with vlp, from measurements. However, it is often useful to know the average value of a series of measurements of a particular property. In fact, that notion can be extended to find the complete probability distribution of measurement results. For now, we focus on just the average.

Let's begin by reviewing how one finds an average. Suppose we have a coin with a +1 on one side and −1 on the other. We flip the coin N times and record the results. Let's label this property Z. The average of those measurements is given by

$$\langle Z \rangle = (1 + 1 - 1 - 1 - 1 + 1 \cdots + 1)/N, \tag{5.17}$$

where we used the angle bracket expression $\langle Z \rangle$ to indicate the average value of the measurement outcomes. We add up all the individual results and then divide by the number of values in the series.

Try It 5.7 A total of 13 students take an exam. Their grades on the exam are 88, 90, 95, 90, 88, 40, 40, 55, 35, 92, 60, 64, and 59. What is the average grade for the class? Note that the average is not equal to any of the actual exam grades.

For many situations, it is helpful to write the average in a slightly different way by making use of the probability or relative frequency of each measurement outcome. Let's suppose we measure a property M, which has two possible values: m_α and m_β. Further, suppose we get m_α for n_α trials and m_β for n_β trials with $n_\alpha + n_\beta = N$. That means the measurement outcome is m_α for n_α/N fraction of the measurements and m_β for n_β/N fraction of the measurements. In other words, the relative frequency of getting m_α is $P_\alpha = n_\alpha/N$, and the relative frequency for m_β is $P_\beta = n_\beta/N$. If N is fairly large, we expect that those relative frequencies will be close to the probabilities. So, we can write the average in terms of the probabilities for each measurement outcome and the outcome values (+1 and − 1):

$$\langle M \rangle = \left(\text{probability of getting } m_\alpha\right) \times m_\alpha + \left(\text{probability of getting } m_\beta\right) \times m_\beta. \tag{5.18}$$

Try It 5.8 Redo the calculation of the average exam score from the previous Try It using the probability formulation of the average. Verify that you get the same result as you did for the direct calculation of the average.

By the Born rule in quantum mechanics, the probabilities are given by quantum state amplitudes $P_\alpha = |a_\alpha|^2$ and $P_\beta = \left|a_\beta\right|^2$, and allowing for the possibility that the amplitudes are complex numbers, we have

$$\langle M \rangle = m_\alpha |a_\alpha|^2 + m_\beta |a_\beta|^2. \tag{5.19}$$

The advantage of writing the average this way is that it is easy to generalize: You add up the products of the probabilities for each of the measurement outcomes and the value associated with M for that outcome.

There is a nice way of writing averages of measurements in quantum mechanics using the Dirac bracket notation. This method often shows up in discussions of QC. The basic idea is that the average can be written, as we have seen, as a sum of products of probabilities and measurement outcome values. We also know that the probabilities can be written as the squares of the amplitudes associated with that part of the state associated with that measurement outcome.

The final ingredient is the fact that in quantum mechanics every measurable property M can be represented by an operator (a linear transformation that acts on quantum states) and that the appropriate measurement basis states are the states that satisfy

$$\begin{aligned} M|\alpha\rangle &= m_\alpha|\alpha\rangle \\ M|\beta\rangle &= m_\beta|\beta\rangle, \end{aligned} \tag{5.20}$$

where m_α and m_β are the numerical values associated with the operator M. The symbols α and β are simply labels to distinguish the two states. States that satisfy these conditions are called eigenstates or eigenvectors associated with the operator M, and m_α and m_β are called the eigenvalues for M. "Eigen" is a German word meaning "characteristic." The procedures for finding the eigenvalues and eigenvectors of a linear transformation (operator) are standard fare in linear algebra. The Appendix Quantum Toolkit provides a brief review of those procedures.

How does this tie into average values? For a single qubit, we assume that any quantum state vector can be written as a sum of M's eigenvectors:

$$|\psi\rangle = a_\alpha|\alpha\rangle + a_\beta|\beta\rangle. \tag{5.21}$$

(As usual, there are some mathematical exceptions to that statement. We will take them up when we need to.) When the operator M acts on this state, we obtain, using Eq. (5.20),

$$M|\psi\rangle = a_\alpha M|\alpha\rangle + a_\beta M|\beta\rangle = a_\alpha m_\alpha|\alpha\rangle + a_\beta m_\beta|\beta\rangle. \tag{5.22}$$

In the matrix form, Eq. (5.22) becomes

$$(M)\begin{pmatrix} a_\alpha \\ a_\beta \end{pmatrix} = \begin{pmatrix} m_\alpha a_\alpha \\ m_\beta a_\beta \end{pmatrix} = \begin{pmatrix} M_{\alpha\alpha} & 0 \\ 0 & M_{\beta\beta} \end{pmatrix}\begin{pmatrix} a_\alpha \\ a_\beta \end{pmatrix}, \tag{5.23}$$

where $M_{\alpha\alpha}$ and $M_{\beta\beta}$ are called the matrix elements of the operator M in the α, β basis.

The next step might not be obvious, but it turns out to be just what we need to calculate the average of a sequence of measurements of M. We multiply Eq. (5.22) on the left by the left vector associated with the state $|\psi\rangle$:

$$\langle\psi|M|\psi\rangle = \langle M\rangle_\psi \doteq \begin{pmatrix} a_\alpha^* & a_\beta^* \end{pmatrix}\begin{pmatrix} a_\alpha m_\alpha \\ a_\beta m_\beta \end{pmatrix} = |a_\alpha|^2 m_\alpha + |a_\beta|^2 m_\beta. \tag{5.24}$$

As a reminder, a_α^* means the complex conjugate of a_α. Surprisingly, the result is just the average of the M measurements, a generalization of the average value expression in Eq. (5.18). The Dirac bracket $\langle\psi|M|\psi\rangle$ is a reminder that the left side of Eq. (5.24) is indeed an average of the measurements of the property represented by the operator M given that the qubit has been prepared in the state $|\psi\rangle$. Some authors like to call the bracket $\langle\psi|M|\psi\rangle$ a "quantum sandwich": The operator M is sandwiched between two slices of ψ bread. You can put any combination of operators in the bracket. Evaluating the bracket will give you the average of the measurement outcomes for that combination of operators when the system is prepared in the state $|\psi\rangle$.

5.8 State Preparation

At this point, you should be ready to ask about how we prepare qubits in quantum states. First, there is the fundamental issue: Are qubits "in" quantum states, do they "have" quantum states, or are they simply "described by" quantum states? As we will soon discover, there are conceptual traps if we claim that the qubit "has" or is "in" a quantum state. So, we will stick with the notion of "description."

Now, we start with state preparation itself. There are two parts of the process. We will see how this plays out in QIS and QC throughout the book. Almost always, we prepare a qubit so it is described by one of the basis states for the appropriate physical property (or collection of properties) we are concerned with. For example, for photon polarization, we might start with $|\text{hlp}\rangle$ by having the photon pass through a PBS and keeping only the photons that are in the hlp outgoing beam.

In many QIS and QC situations, we then manipulate that initial state to convert it into a desired superposition state. (We will see how to do this in Chapter 6.) The important point is that preparing an initial state is, in some cases, just like a measurement process. The physics of the state preparation device is often the same as the physics of the measurement device except that we are producing photons, say from a laser or a light-emitting diode or a flame, and using a polarizing device to keep only those that are described by the desired polarization beam. So, in quantum physics, state preparation and state measurement are two sides of the same coin. After a measurement, we know almost nothing about the state before the measurement. Similarly, preparing a qubit in a state wipes out information about the state before the preparation operation. Figure 5.8 illustrates the overall process.

This kind of conceptual analysis is often taken for granted in QIS and QC, where we assume we know how to do state preparation and state measurement, and we get on with the real work of state manipulation to build QCs and quantum communication devices and invent algorithms to carry out quantum information tasks. But, keep in mind that there are many subtleties in both state preparation and state measurement. We elaborate these subtleties in Chapter 17.

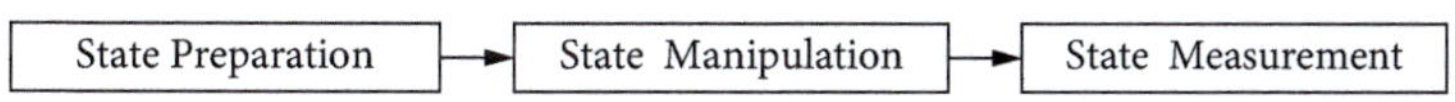

Fig. 5.8 A schematic representation of many QIS and QC processes.

5.9 Chapter Summary

- Quantum measurement is different from classical measurement because it does not generally yield a deterministic result. If you measure the velocity of identically prepared classical objects, you will get the same value each time, within experimental uncertainty. If you measure identically prepared quantum objects described by a superposition state, the outcomes vary randomly from one observation to the next. You will not get the same answer each time, regardless of experimental uncertainty.
- When a quantum measurement is made, the outcome will be the value of the physical property associated with one of the measurement device's basis states. If the qubit is not absorbed during the observation, the quantum state of the qubit after the interaction is the basis state associated with the observed outcome. All information about its prior state is lost.
- The numerical coefficients in a superposition quantum state are measurement outcome probability amplitudes—not probabilities.
- The probabilities of observation outcomes are given by the absolute-values-squared of the coefficients of the basis states in the superposition state. This result is known as the Born rule or the *probability interpretation of quantum mechanics.* For the state $|S\rangle = a_0|0\rangle + a_1|1\rangle$, the probability of getting the state $|0\rangle$ as a result of a measurement is $P_0 = |a_0|^2$ and $P_1 = |a_1|^2$ for the $|1\rangle$ state.
- Because information about the original state is lost upon observation, the observer cannot make a clone of the original quantum state with only one observation. This is the *no-cloning theorem.*
- Averages of a series of measurement results can be calculated using a quantum sandwich: $\langle M\rangle = \langle\psi| M |\psi\rangle$, where M is the operator representing the property and $|\psi\rangle$ is the quantum state. In general, the average will be different for different quantum states.
- The measurement outcomes for a property represented by an operator M will be one of the eigenvalues of that operator.

5.10 Further Reading

Michael G. Raymer *Quantum Mechanics: What Everyone Needs to Know* (Oxford University Press, Oxford, 2017). This book has clear and thoughtful discussions of quantum measurements, along with excellent introductions to other aspects of quantum physics—all done with almost no mathematics.

N. David Mermin, *Quantum Computer Science, An Introduction* (Cambridge University Press, Cambridge, 2007). A moderately sophisticated treatment of quantum computing aimed at computer scientists.

Julian Schwinger and B.-G. Englert (editor), *Quantum Mechanics: Symbolism of Atomic Measurements* (Springer, Berlin, 2001). The title indicates that Schwinger emphasizes that quantum mechanics is a theory of quantum measurements and their correlations. This text is written at the upper-level undergraduate or graduate-student level.

Nondemolition measurements are discussed in S. Haroche and J.-M. Raimond, *Exploring the Quantum: Atoms, Cavities, and Photons* (Oxford University Press, Oxford, 2006), pp. 313–326.

Online videos

The videos show experimental evidence that light is composed of individual quanta called photons.:
https://www.youtube.com/watch?v=GzbKb59my3U
https://www.youtube.com/watch?v=I9Ab8BLW3kA

6 Quantum Gates and Quantum Circuits

Probable impossibilities are to be preferred to impossible possibilities.

Aristotle, *Poetics*

6.1 What Is a Quantum Gate?

To describe quantum computing, we need devices that will operate on and change the quantum state associated with a qubit (the physical object). In quantum mechanics, those devices are represented by mathematical *operators*. In linear algebra language, an operator is a linear transformation that acts on a vector in a vector space and produces another vector in that vector space. However, in analogy to the language used to describe the manipulation of bits in classical computing, we call the corresponding quantum devices quantum *gates*. For example, we might think about a system with two qubits as input to some device, which then produces some output. A generic quantum gate is shown in Fig. 6.1.

We read Fig. 6.1 from left to right. We say that $|S_A\rangle$ is an input qubit state, prepared by Alice for example, while $|S_B\rangle$ is an input qubit state prepared by Bob. The gate operates on those states to produce output states $|O_1\rangle$ and $|O_2\rangle$. A combination of such gates is a quantum circuit. Remember, we talked about how to prepare a quantum state in Chapter 5, where we observed that state analyzer devices such as a polarizing beam splitter can also be used as state preparation devices. Although the details depend on the physical nature of the qubits being used, the state analyzers result in states that are the basis states associated with the device. In this chapter, we first discuss how to use quantum gates (operators) to manipulate those states to produce other states that are eventually used in quantum computing (QC) and quantum information science (QIS). We also describe the distinction between operators and state analyzers or measurement devices.

The thin horizontal lines in Fig. 6.1 are called "quantum wires" in analogy with the regular wires used to bring voltage signals to gates and integrated circuits in traditional computers. In most QC cases, they are just a visual indication of how a qubit interacts with a series of gates, not physical wires.

You should read the quantum circuit as a timeline, proceeding from left (earlier in time) to right (later in time). The quantum wires "carry" the quantum state vector information. The gate acts on those quantum states and produces output states shown on the quantum

Quantum Computing and Quantum Physics. Alice Flarend and Robert Hilborn, Oxford University Press.
 DOI: 10.1093/oso/9780197904930.003.0006

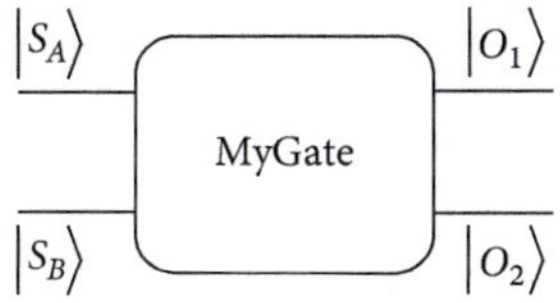

Fig. 6.1 A generic two-qubit quantum gate. There are two input qubits whose quantum states are shown on the left and two output quantum states shown on the right.

wires on the right side of the diagram. Those wires (that is, the associated quantum states) may then go on to other gates for further processing.

6.2 Single-Qubit Quantum Gates

Let's start with some quantum gates that manipulate single-qubit quantum state vectors. The gate shown in Fig. 6.1 is called a two-qubit gate, with two inputs and two outputs. We tackle those kinds of gates in Chapter 9. In this chapter, we focus on how gates manipulate the computational basis states $|0\rangle$ and $|1\rangle$. From there we can figure out how they process any other single-qubit state since any quantum state can be represented as a superposition of the computational basis states. As mentioned previously, in linear algebra language, a single-qubit gate is known as a linear transformation, which acts on a vector in the vector space to produce a new vector in that vector space.

Perhaps the simplest quantum gate is the NOT gate. Recall that in binary logic the NOT gate changes a 0 to a 1 and vice versa. A quantum NOT gate changes the basis state vector $|0\rangle$ into the other basis state vector $|1\rangle$ and vice versa. We represent that process graphically as a quantum circuit in Fig. 6.2.

We can write these results in a "logic (truth) table" just the way we did for a classical computing NOT gate (Table 6.1).

It turns out to be useful to represent the quantum gate by a matrix, which multiplies the column vectors representing the basis states. In other words, a quantum gate is just a physical implementation of a linear transformation, which acts on a column vector. For the NOT gate, the matrix in the computational basis is the following:

$$\text{NOT} \doteq \begin{pmatrix} 0 & 1 \\ 1 & 0 \end{pmatrix}, \tag{6.1}$$

and the action on the basis states is

$$\begin{aligned} \text{NOT}|0\rangle &\doteq \begin{pmatrix} 0 & 1 \\ 1 & 0 \end{pmatrix}\begin{pmatrix} 1 \\ 0 \end{pmatrix} = \begin{pmatrix} 0 \\ 1 \end{pmatrix} \\ \text{NOT}|1\rangle &\doteq \begin{pmatrix} 0 & 1 \\ 1 & 0 \end{pmatrix}\begin{pmatrix} 0 \\ 1 \end{pmatrix} = \begin{pmatrix} 1 \\ 0 \end{pmatrix}. \end{aligned} \tag{6.2}$$

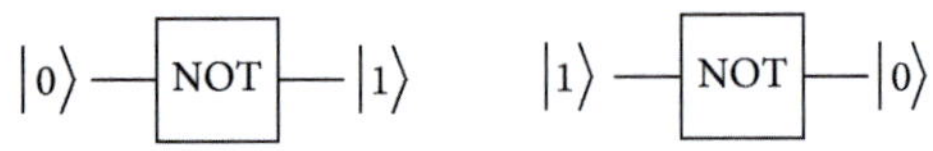

Fig. 6.2 Quantum NOT gate.

Table 6.1 The input/output table for a quantum NOT gate.

NOT Gate Input	NOT Gate Output
$\|0\rangle$	$\|1\rangle$
$\|1\rangle$	$\|0\rangle$

Reading Eq. (6.2) from left to right, we see that the 2×2 NOT matrix acts on the 2×1 column vector for a basis state vector to produce the other basis state vector. In QC, the NOT matrix is usually called the X matrix:

$$X \doteq \begin{pmatrix} 0 & 1 \\ 1 & 0 \end{pmatrix}. \tag{6.3}$$

Recall from Chapter 3 that a matrix multiplying a column vector gives a new column vector. Here is the multiplication rule:

$$\begin{pmatrix} a & b \\ c & d \end{pmatrix} \begin{pmatrix} e \\ f \end{pmatrix} = \begin{pmatrix} ae + bf \\ ce + df \end{pmatrix}. \tag{6.4}$$

As a mnemonic, you can think of the matrix on the left of Eq. (6.4) as a stack of two row vectors $(a\ b)$ and $(c\ d)$. The *top* entry in the column vector on the right after the equal sign of Eq. (6.4) is just the product of the matrix's *top* row vector multiplying the column vector $\begin{pmatrix} e \\ f \end{pmatrix}$. The *bottom* entry on the right is the product of the *bottom* row vector on the left times the column vector $\begin{pmatrix} e \\ f \end{pmatrix}$.

Try It 6.1 Check that multiplying a generic qubit quantum state vector $|S\rangle = \alpha_0|0\rangle + \alpha_1|1\rangle$ by the NOT matrix gives a state vector with the amplitudes swapped between the two basis states.

In addition to NOT, another simple, but surprisingly useful matrix is called the Identity Matrix. It is the matrix equivalent of multiplying a mathematical object by the number 1: you just get the same mathematical object as output. The Identity Matrix, often indicated by I, is the following:

$$I \doteq \begin{pmatrix} 1 & 0 \\ 0 & 1 \end{pmatrix}. \tag{6.5}$$

Try It 6.2 Check that multiplying the computational basis states (in column vector form) by the Identity Matrix gives the states back again.

Try It 6.3 Check that multiplying a generic qubit quantum state vector $|S\rangle = \alpha_0|0\rangle + \alpha_1|1\rangle$ by the Identity Matrix gives the same state vector back again.

The Identity Matrix is an example of what is called a diagonal matrix: The only nonzero entries in the matrix lie along the principal (upper-left to lower-right) diagonal of the matrix. As we shall see, diagonal matrices have properties that make them particularly simple to work with.

The X matrix and the Identity Matrix are part of a set of four matrices, called the Pauli matrices, commonly used in quantum mechanics. They are named after Wolfgang Pauli (1900–1958), one of the pioneers in quantum mechanics. The other two are called the Y and the Z matrices:

$$Y \doteq \begin{pmatrix} 0 & -i \\ i & 0 \end{pmatrix} \tag{6.6}$$

and

$$Z \doteq \begin{pmatrix} 1 & 0 \\ 0 & -1 \end{pmatrix}. \tag{6.7}$$

In Chapter 7, we explain why we need $i = \sqrt{-1}$ in Y. This is also a good point to note that physicists like to denote the Pauli matrices by the lowercase Greek σ (sigma):

$$\sigma_x = X \quad \sigma_y = Y \quad \sigma_z = Z. \tag{6.8}$$

Let's see what happens when Z operates on the computational basis states:

$$Z|1\rangle \Rightarrow \begin{pmatrix} 1 & 0 \\ 0 & -1 \end{pmatrix}\begin{pmatrix} 0 \\ 1 \end{pmatrix} = \begin{pmatrix} 0 \\ -1 \end{pmatrix} = -\begin{pmatrix} 0 \\ 1 \end{pmatrix} = -|1\rangle. \tag{6.9}$$

$$Z|0\rangle \Rightarrow \begin{pmatrix} 1 & 0 \\ 0 & -1 \end{pmatrix}\begin{pmatrix} 1 \\ 0 \end{pmatrix} = +1\begin{pmatrix} 1 \\ 0 \end{pmatrix} = +1|0\rangle. \tag{6.10}$$

Note that Z acting on the basis states gives you simply a number times the original state vector. The numbers -1 and $+1$ are the eigenvalues of the operator Z, and the states $|1\rangle$ and $|0\rangle$ are the corresponding eigenstates (or, equivalently, eigenvectors).

While introducing single-qubit gates, we should show you one more, the Hadamard gate, which turns out to be very useful in QIS and QC. The Hadamard gate H has the matrix representation

$$H \doteq \frac{1}{\sqrt{2}}\begin{pmatrix} 1 & 1 \\ 1 & -1 \end{pmatrix}. \tag{6.11}$$

Suppose H acts on $|0\rangle$, then we have

$$H\,|0\rangle = \frac{1}{\sqrt{2}}\{|0\rangle + |1\rangle\} = |+\rangle\,, \tag{6.12}$$

and for the other computational basis state, we obtain

$$H\,|1\rangle = \frac{1}{\sqrt{2}}\{|0\rangle - |1\rangle\} = |-\rangle\,. \tag{6.13}$$

We have introduced special names $|+\rangle$ and $|-\rangle$ for those states because they occur frequently in QIS and QC algorithms. The states $|+\rangle$ and $|-\rangle$ are called the Hadamard basis states. We discuss them again in Chapter 10. Figure 6.3 provides a state space representation of $|+\rangle$ and $|-\rangle$.

Eqs. (6.12) and (6.13) show that if we start with a quantum state that is just one of the basis states, then when a Hadamard gate acts on the state, the result is a superposition (sum) of the two basis states. Compare Eqs. (6.12) and (6.13) with Eq. (6.9) and you can see the presence of two state vectors versus a single state vector.

Try It 6.4 Why did we include the $1/\sqrt{2}$ in the definition of the Hadamard gate? Hint: What is the length of the superposition states produced by the Hadamard gate in Eqs. (6.12) and (6.13).

Those states look just like states for polarized light tilted at 45° which we can think of as a superposition of hlp and vlp states. The important point of the Hadamard gate is it acts on the individual basis states to produce a quantum state that is a superposition of the basis states.

What happens if H acts on a general quantum state that is already a superposition state represented by the column vector $\begin{pmatrix} \alpha_0 \\ \alpha_1 \end{pmatrix}$? Recall that the state can be written in two ways:

$$|\psi\rangle = \alpha_0|0\rangle + \alpha_1|1\rangle \doteq \begin{pmatrix} \alpha_0 \\ \alpha_1 \end{pmatrix}.$$

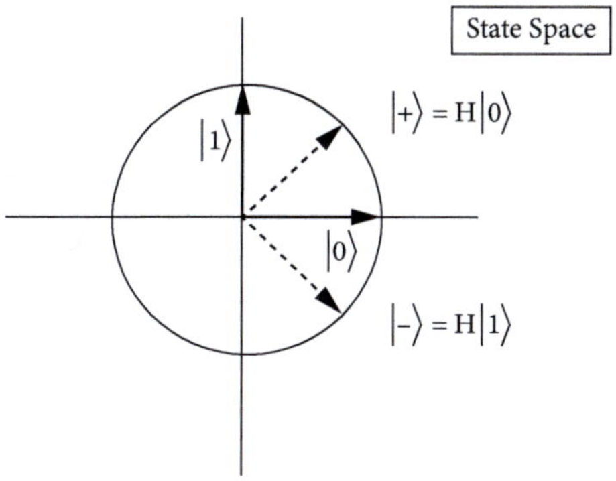

Fig. 6.3 A state space diagram illustrating the state vectors that result from having a Hadamard gate act on the computational basis states.

So, we may write the result of having the Hadamard gate H acting on $|\psi\rangle$ in two ways:

$$\begin{aligned} H\,|\psi\rangle &\doteq \frac{1}{\sqrt{2}}\begin{pmatrix}1 & 1\\ 1 & -1\end{pmatrix}\begin{pmatrix}\alpha_0\\ \alpha_1\end{pmatrix} \doteq \frac{1}{\sqrt{2}}\begin{pmatrix}\alpha_0+\alpha_1\\ \alpha_0-\alpha_1\end{pmatrix} \doteq \frac{1}{\sqrt{2}}(\alpha_0+\alpha_1)\,|0\rangle + \frac{1}{\sqrt{2}}(\alpha_0-\alpha_1)\,|1\rangle \\ &= \frac{1}{\sqrt{2}}\alpha_0\,\{|0\rangle + |1\rangle\} + \frac{1}{\sqrt{2}}\alpha_1\,\{|0\rangle - |1\rangle\} \\ &= \alpha_0\,|+\rangle + \alpha_1\,|-\rangle\,. \end{aligned} \tag{6.14}$$

Note that we end up with a superposition of the Hadamard basis states.

Try It 6.5 Write out the details of all the matrix multiplications in Eq. (6.14) and check the results.

Try It 6.6 Suppose $\alpha_0 = -\alpha_1$ in Eq. (6.14). What is the result for $H|\psi\rangle$?

We ought to point out that the equivalent of the Hadamard gate was developed back in the 1930s and 1940s in the field that became known as nuclear magnetic resonance. The quantum systems (what we would now call qubits) were atomic nuclei with spin angular momentum and magnetic dipole moments (magnetic North Pole–South Pole combinations). Putting those quantum systems into superposition states is a key component of magnetic resonance imaging, which, as you probably know, is a big deal in medical diagnostics. Analogous superposition states can be produced in atoms and molecules using bursts of laser light. So, the technology needed to produce superposition states has long been part of every quantum mechanic's toolbox. Magnetic resonance is discussed in detail in Chapter 19.

Try It 6.7 Can the Hadamard gate H be represented as a combination of Z and X gates $H = \left(1/\sqrt{2}\right)(Z + X)$? Hint: Remember that to add matrices, you add the elements that are in the corresponding positions, that is $\begin{pmatrix} a & b \\ c & d \end{pmatrix} + \begin{pmatrix} e & f \\ g & h \end{pmatrix} = \begin{pmatrix} a+e & b+f \\ c+g & d+h \end{pmatrix}$.

Try It 6.8 Draw quantum circuits that show how each of the X, Y, Z, and H gates processes the computational basis states $|0\rangle$ and $|1\rangle$ (see Fig. 6.2).

Try It 6.9 Show how each of the X, Y, and Z gates processes a general single-qubit quantum state $|\psi\rangle = \alpha_0|0\rangle + \alpha_1|1\rangle$ as we did with Eq. (6.14).

6.3 **Successive Quantum Gates**

A single quantum gate is fine, but to do something useful in QIS and QC, we need more than one gate. We can combine quantum gates in succession to do some interesting things. In linear algebra language, the successive gates form a "composite" linear transformation. As a first look at such a situation, consider the quantum circuits shown in Fig. 6.4.

In the left circuit of Fig. 6.4, an input quantum state $|\psi\rangle$ is first acted on by a Hadamard gate and then the output from the Hadamard gate is acted upon by an X gate. In the circuit on the right, the order in which the gates act is reversed. In abstract form, we write

$$\begin{aligned} |\psi_{HX}\rangle &= X(H|\psi\rangle) = XH|\psi\rangle \\ |\psi_{XH}\rangle &= H(X|\psi\rangle) = HX|\psi\rangle. \end{aligned} \tag{6.15}$$

In the first equality in Eq. (6.15), we group the operations with parentheses to remind ourselves that H acts first and then X. As shown on the far right of Eq. (6.15), the operations are usually written without the parentheses, which is cleaner typographically, but we need to remember the convention about the order of operations.

Try It 6.10 Draw the quantum circuit diagram for $XHXH|\psi\rangle = |\Phi\rangle$.

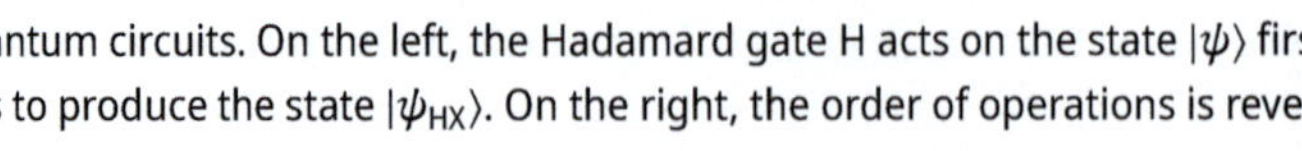

Fig. 6.4 Two quantum circuits. On the left, the Hadamard gate H acts on the state $|\psi\rangle$ first and then the NOT (X) gate acts to produce the state $|\psi_{HX}\rangle$. On the right, the order of operations is reversed.

Note that by convention, in the *equations*, we read the operations from right to left, the opposite of how we read the *quantum circuit diagram* (left to right). That seems to be inconvenient, but that is what almost all QIS and QC people do; so, we need to get used to that. It is worth making a note of those conventions because we tend to read text from left to right.

As we shall see, the Pauli operators play an important role in many quantum algorithms. Often, we have a succession of Pauli operators act on a state. For example, we might have XYY$|\psi\rangle$. The string of symbols "XXY" is called a Pauli string.

We now want to consider the matrix form of a composite linear transformation. In linear algebra, the "matrix product" of two matrices is given by

$$AB \doteq \begin{pmatrix} a_{11} & a_{12} \\ a_{21} & a_{22} \end{pmatrix} \begin{pmatrix} b_{11} & b_{12} \\ b_{21} & b_{22} \end{pmatrix} = \begin{pmatrix} a_{11}b_{11} + a_{12}b_{21} & a_{11}b_{12} + a_{12}b_{22} \\ a_{21}b_{11} + a_{22}b_{21} & a_{21}b_{12} + a_{22}b_{22} \end{pmatrix}. \tag{6.16}$$

Note that the top left entry of the result is the scalar product of the top row of the A matrix and the leftmost column of B with similar construction for the other entries. The adjective "matrix" in matrix product is important because there are several ways to multiply two matrices together (see Appendix: Quantum Toolkit).

Let's see if it actually matters what direction we read the matrix equations. The matrix representation of those operations is just the matrix product of the two matrices:

$$\text{XH} \Rightarrow \frac{1}{\sqrt{2}} \begin{pmatrix} 0 & 1 \\ 1 & 0 \end{pmatrix} \begin{pmatrix} 1 & 1 \\ 1 & -1 \end{pmatrix} = \frac{1}{\sqrt{2}} \begin{pmatrix} 1 & -1 \\ 1 & 1 \end{pmatrix} \tag{6.17}$$

$$\text{HX} \Rightarrow \frac{1}{\sqrt{2}} \begin{pmatrix} 1 & 1 \\ 1 & -1 \end{pmatrix} \begin{pmatrix} 0 & 1 \\ 1 & 0 \end{pmatrix} = \frac{1}{\sqrt{2}} \begin{pmatrix} 1 & 1 \\ -1 & 1 \end{pmatrix} \tag{6.18}$$

Surprise! The two products are not the same. With quantum gates, the order matters! We say that the two operators (matrices) X and H do not *commute*.

Try It 6.11 Apply XH to each of the standard computational basis states. Then do the same with HX. Do you get the same result?

6.4 Mathematical Interlude

It turns out that we can't use any combination of numbers in the 2×2 matrices representing quantum gates. To see why, we need to look at some of the math associated with matrices.

In QC and QIS, we most often deal with square matrices for which the number of rows equals the number of columns. Unless we state otherwise, we always refer to square matrices. The matrices allowed to represent quantum gates have to satisfy the following requirements. We explain the terminology in what follows.

1. The determinant of the matrix must be equal to $+1$ or -1. This assures that the matrix has an inverse. More generally, the determinant needs to be nonzero for the matrix to have

an inverse. The restriction to + 1 or –1 means that the matrix maintains the magnitude of the area (volume or hypervolume) defined by its row vectors.

The product of the matrix and its transpose $U^T U = I$ must equal I, the Identity Matrix. In this case, we say that the matrix is orthogonal. If we end up with matrix elements that are complex numbers, we require that the "adjoint" (or Hermitian conjugate) multiplying the original matrix gives I. The adjoint is a combination of taking the transpose and also the complex conjugate of all the matrix entries.

2. The columns in the matrix form a set of orthogonal unit vectors. So do the rows.

Let's check the Hadamard gate matrix to see if it satisfies those conditions and, along the way, we explain what determinant and transpose mean. For almost all this book, we restrict ourselves to gates that satisfy these conditions, and you won't need to worry about the mathematical details. However, it is good to know some of the mathematical terminology in case you come across these terms in talks about QIS and QC and in other books and articles.

The determinant of a 2 × 2 matrix is the product of upper-left and lower-right elements minus the product of the lower-left and the upper-right elements. In particular

$$\mathrm{Det}\begin{pmatrix} a & b \\ c & d \end{pmatrix} = ad - bc. \tag{6.19}$$

Here is an example with numerical entries in the matrix:

$$\mathrm{Det}\begin{pmatrix} -0.5 & 0 \\ 0 & 2 \end{pmatrix} = -1 - 0 = -1. \tag{6.20}$$

Try It 6.12 Evaluate the determinant of the Hadamard gate matrix.

The transpose of a matrix is a matrix with the rows and columns interchanged. In other words, in the transpose of a matrix the old rows become the new columns. The transpose will be designated by a superscript T attached to the symbol for the matrix:

$$\begin{aligned} A &= \begin{pmatrix} a_{11} & a_{12} \\ a_{21} & a_{22} \end{pmatrix} \\ A^T &= \begin{pmatrix} a_{11} & a_{21} \\ a_{12} & a_{22} \end{pmatrix}. \end{aligned} \tag{6.21}$$

If the transpose of a matrix is the same as the matrix, that is, if $A^{\mathrm{T}} = A$, then the matrix is said to be *symmetric*. Here is a numerical example for which the transpose is not equal to the original matrix:

$$\begin{pmatrix} 0.5 & -0.2 \\ 0.1 & 0.4 \end{pmatrix}^{\mathrm{T}} = \begin{pmatrix} 0.5 & 0.1 \\ -0.2 & 0.4 \end{pmatrix} \neq \begin{pmatrix} 0.5 & -0.2 \\ 0.1 & 0.4 \end{pmatrix}. \tag{6.22}$$

Math Aside Later, when we use matrices whose elements are complex numbers, we need to use the complex transpose: take the complex conjugate of all the matrix elements and swap rows and columns. The complex transpose $A^{\dagger}$ is indicated by the superscript † and is often called the Hermitian transpose. As an example, let's consider the Y gate matrix $\mathrm{Y} \doteq \begin{pmatrix} 0 & -i \\ i & 0 \end{pmatrix}$. Its Hermitian transpose is $\mathrm{Y}^{\dagger} \doteq \begin{pmatrix} 0 & -i \\ i & 0 \end{pmatrix}$, which is the same as Y. A matrix (and its corresponding gate) whose Hermitian conjugate is the same as the original matrix is said to be Hermitian. Hermitian matrices and gates play a major role in quantum mechanics and quantum computing.

Let's try this out with the Hadamard gate's matrix representation and calculate the product of the matrix and its transpose:

$$\mathrm{H}^{\mathrm{T}}\mathrm{H} \doteq \frac{1}{2}\begin{pmatrix} 1 & 1 \\ 1 & -1 \end{pmatrix}\begin{pmatrix} 1 & 1 \\ 1 & -1 \end{pmatrix} = \frac{1}{2}\begin{pmatrix} 2 & 0 \\ 0 & 2 \end{pmatrix} = \begin{pmatrix} 1 & 0 \\ 0 & 1 \end{pmatrix} \doteq \mathrm{I}. \tag{6.23}$$

Matrices that satisfy Eq. (6.23) are called *orthogonal* if the elements are real and *unitary* if the elements are complex numbers. Recall that unitary matrices are important in quantum mechanics because when they operate on a quantum state vector they produce a new state vector with the same length as the original state vector. The name "unitary" is used because we almost always use state vectors of unit length (the length = 1).

We see that the Hadamard gate matrix satisfies condition 2. Now consider condition 3. In analogy with our definition of orthogonal state vectors, the orthogonality condition for the Hadamard gate's two column vectors is

$$\begin{pmatrix} 1 & 1 \end{pmatrix}\begin{pmatrix} 1 \\ -1 \end{pmatrix} = 1 - 1 = 0 \tag{6.24}$$

We summarize these results: Quantum gates are represented by *unitary* matrices. A unitary matrix U satisfies $U^{\dagger}U = \mathrm{I}$, where $U^{\dagger}$ is the Hermitian adjoint matrix formed by taking the complex conjugates of all of U's elements and then applying the transpose operation. In other words, a unitary matrix satisfies $U^{\dagger} = U^{-1}$. That is, the Hermitian conjugate is equal to the inverse. When a unitary matrix U acts on a vector, the resulting vector has the same length as the original vector. As we explore more quantum gates, we discover other useful properties of unitary matrices. For a summary, see Appendix: Quantum Toolkit.

Try It 6.13 Show that all these conditions are met by the other Pauli matrices X, Y, and Z introduced in the previous section. Are any of the Pauli matrices symmetric?

Later, we find that the matrices representing quantum gates can be built out of combinations of the basis state vectors in their right vector and left vector forms. So, there is a close connection between these matrices and the column and row representations of the right and left state vectors.

6.5 Successive Application of the Same Quantum Gate

Many quantum gates have another interesting property. If you apply the gate to a quantum state and then apply it again to the result of the first application, you end up with the quantum state you started with. These gates are in a sense "reversible": You can get back to your starting state if you apply the gate twice (assuming you haven't done anything else to the state produced by the first application).

Here is another piece of terminology that occurs in QIS and QC. The inverse of a matrix A, written symbolically as A^{-1}, is defined to be that matrix which when multiplied by the original matrix gives the Identity Matrix I:

$$A^{-1}A = \mathrm{I} = AA^{-1}. \tag{6.25}$$

If a matrix is equal to its inverse, we say that the matrix is *involutory*. In other words, we have $AA = A^2 = \mathrm{I}$. We say that the matrix A is a square root of the Identity Matrix. If the involutory matrix is symmetric, it is an orthogonal matrix if the elements are real, and unitary if the elements are complex.

It is always a good idea to try out new ideas with specific examples. Let's start with the Hadamard gate acting on the computational basis states $|0\rangle$ and $|1\rangle$ as shown in Eq. (6.12). The first Hadamard operation inside the square brackets on the left produces a superposition state, and then the second H acts on that superposition state:

$$\begin{aligned} \mathrm{H}\,[\mathrm{H}\,|0\rangle] &\doteq \frac{1}{\sqrt{2}}\begin{pmatrix} 1 & 1 \\ 1 & -1 \end{pmatrix}\left[\frac{1}{\sqrt{2}}\left\{\begin{pmatrix} 1 \\ 0 \end{pmatrix} + \begin{pmatrix} 0 \\ 1 \end{pmatrix}\right\}\right] \\ &= \frac{1}{2}\left\{\begin{pmatrix} 1 & 1 \\ 1 & -1 \end{pmatrix}\begin{pmatrix} 1 \\ 0 \end{pmatrix} + \begin{pmatrix} 1 & 1 \\ 1 & -1 \end{pmatrix}\begin{pmatrix} 0 \\ 1 \end{pmatrix}\right\} \\ &= \frac{1}{2}\begin{pmatrix} 1 \\ 1 \end{pmatrix} + \frac{1}{2}\begin{pmatrix} 1 \\ -1 \end{pmatrix} = \frac{1}{2}\begin{pmatrix} 2 \\ 0 \end{pmatrix} = \begin{pmatrix} 1 \\ 0 \end{pmatrix} \doteq |0\rangle . \end{aligned} \tag{6.26}$$

In the first line of Eq. (6.26), we applied the first H to $|0\rangle$ and obtained the superposition state expressed in Eq. (6.12). In the second line, we applied the second H to each of the components of the superposition state. We then grouped terms and found that we end up with the quantum state we started from.

Try It 6.14 Work through Eq. (6.26) step by step to make sure you understand the various matrix and column vector operations.

Try It 6.15 Repeat the previous Try It but use an X gate rather than a Hadamard gate. Explain what you find. Repeat for a Y gate. Are the Pauli matrices involutory?

Try It 6.16 Go through the same repeat application of the Hadamard gate acting on a general superposition state as the starting state: $|S\rangle = a_0|0\rangle + a_1|1\rangle \doteq \begin{pmatrix} a_0 \\ a_1 \end{pmatrix}$. Do you get $\mathrm{H}\left[\mathrm{H}|S\rangle\right] = |S\rangle$?

6.6 Measurement Devices Are Not Reversible

In Chapter 5, we introduced measurement devices and noted that we avoided calling them gates because they are not (in almost all circumstances) represented by matrices that satisfy the conditions laid out in the previous two sections. In particular, they are not reversible. For example, after a photon interacts with a polarization state analyzer, its state might be $|\text{vlp}\rangle$. As we have emphasized before, from that information, we know very little about the state describing the photon before it arrived at the device. So, there is no way to apply the measurement operation to get back to the original state.

In quantum circuits in the QIS and QC literature, measurement devices are often represented by the symbol on the left of Fig. 6.5. We prefer the symbol on the right because nowadays you rarely see any measuring apparatus that uses a needle pointer (except for some speedometers in cars). The symbol on the right contains an M to remind us that we are representing a measurement device. It is also easier to draw, either by hand or by computer.

State analyzers can, however, be represented by "projection operators" constructed as products of right vectors and left vectors. For example, the operation of a state analyzer that puts the qubit in the device's $|0\rangle$ state is represented by

$$M_0 = |0\rangle\langle 0|\,. \tag{6.27}$$

Fig. 6.5 Symbols used to represent measurement devices in quantum circuits.

If a state $|S\rangle = a_0|0\rangle + a_1|1\rangle$ interacts with (is acted upon by) that device, we write

$$M_0 \underbrace{|S\rangle}_{\text{input state}} = \overbrace{\underbrace{|0\rangle}_{\text{output state}} \langle 0|}^{\text{projection operator}} S\rangle. \tag{6.28}$$

The output state (consistent with the *measurement principle*) is the basis state $|0\rangle$. The probability of getting that result is the square of $\langle 0| S\rangle$, which we recognize as the amplitude a_0 of $|S\rangle$ associated with the basis state $|0\rangle$. If the qubit is subsequently detected, we have completed a quantum observation.

As we discussed in Chapter 5, the probability of getting a particular measurement outcome is found experimentally by repeating the state preparation–measurement cycle many times. From a single observation, which is the important point here, we don't know anything about the amplitude $\langle 0| S\rangle$ except that it is not 0.

What happens if we apply M_0 to the state on the right side of Eq. (6.28)?

$$M_0 \underbrace{|0\rangle}_{\substack{\text{state after} \\ \text{1st } M_0}} \langle 0|S\rangle = |0\rangle \underbrace{\langle 0|0\rangle}_{=1} \langle 0|S\rangle = |0\rangle \langle 0|S\rangle. \tag{6.29}$$

For the last equality in Eq. (6.29), we used $\langle 0|0\rangle = 1$ since the basis state is a unit vector. We see that we don't get back to the original state $|S\rangle$. As we mentioned in Chapter 5, some authors like to say the measurement "collapses the state" and that collapse is irreversible. We prefer to say that the state analyzer device projects the state of the incoming qubit along one of the device's basis states. Since there are many (in fact infinitely many) different incoming states that get projected to the same output state, the measurement device projection is not reversible.

It is important to note that the projection operator simply gives a description of the results of interaction with a state analyzer device; it does not tell us the physical mechanism by which the projection occurs. As we mentioned in Chapter 5, even though quantum mechanics is nearly 100 years old, there is still no unique formulation of how a state analyzer measurement device works. See Chapter 17 for more discussion on these issues.

We certainly know how to carry out the measurements but translating their operation into the formal theory of quantum mechanics remains an open question. We should point out that similar interactions occur for qubits even outside the organized measurement devices found in laboratories and quantum computers around the world. In "real life," qubits such as photons and electrons are always interacting with billions and billions of other qubits, and those interactions are much like the interactions with our measurement devices. That means we don't have a widely accepted way of describing those kinds of interactions either. The projection operator and its extensions do give us a workable formal description, but that is different from having a detailed mechanistic theory.

6.7 Chapter Summary

- Quantum computers make use of quantum gates. The resulting gate arrays can be drawn as circuits just like classical computers. Gates can be represented by formal operators or matrices. The matrix representations of the X (or NOT), the Y, and the Z gates are

$$\mathrm{X} \doteq \begin{pmatrix} 0 & 1 \\ 1 & 0 \end{pmatrix} \quad \mathrm{Y} \doteq \begin{pmatrix} 0 & -i \\ i & 0 \end{pmatrix} \quad \mathrm{Z} \doteq \begin{pmatrix} 1 & 0 \\ 0 & -1 \end{pmatrix}.$$

- An important gate in QC is the Hadamard gate which acts on an input basis state and results in a superposition state. The matrix representation of this gate is

$$\mathrm{H} \doteq \frac{1}{\sqrt{2}} \begin{pmatrix} 1 & 1 \\ 1 & -1 \end{pmatrix}.$$

- Just like in classical computing, multiple gates are employed for most computational tasks. Learning to read these quantum circuits is like learning a different language. In English, the adjectives come before the noun while in Spanish the noun comes first. The order of the words is important in these languages, and the order of the gates is important in QC. For better or worse, the order in which the gates operate is left to right in a circuit diagram. Read the left side of the following diagram as "$|\psi\rangle$ meets up first with H and after that with X." On the right, "$|\psi\rangle$ meets up first with X and after that with H."

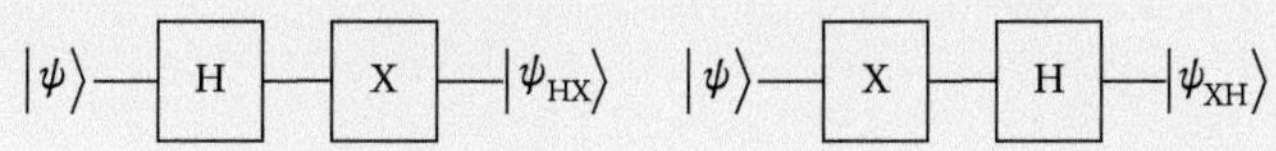

Fig. 6.4 (repeated)

In the state vector representation, we read from right to left:

$$|\psi_{\mathrm{HX}}\rangle = \underset{\substack{\text{X acts}\\ \text{second}}}{\underline{\mathrm{X}}} \left(\underset{\text{H acts first}}{\underline{\mathrm{H}|\psi\rangle}} \right) = \mathrm{X\,H}|\psi\rangle$$

$$|\psi_{\mathrm{XH}}\rangle = \underset{\substack{\text{H acts}\\ \text{second}}}{\underline{\mathrm{H}}} \left(\underset{\text{X acts first}}{\underline{\mathrm{X}|\psi\rangle}} \right) = \mathrm{H\,X}|\psi\rangle.$$

- Not all possible matrices represent quantum gates. For a gate, the matrix's columns must represent orthogonal unit vectors. The matrix determinate must equal ± 1 to ensure that the matrix has an inverse.
- One of the most important requirements is that the gate must be reversible, meaning that if the gate is applied to a state vector twice, the result is the original state vector.
- Measurement devices cannot be represented by reversible gates. Measuring a quantum state only tells us that the state's amplitude for a specific basis vector direction is not zero.

Another way of stating this is that the measurement device projects the state vector onto a particular device basis vector. All other information is lost.

- We introduced several matrices that are related to a matrix A:
  - A inverse = A^{-1} such that $A^{-1}A = I$
  - A transpose = A^{T}, which has the rows and columns of A interchanged.
  - A adjoint or complex transpose or Hermitian conjugate = $A^{\dagger}$, which has the rows and columns of A interchanged and the matrix elements replaced by their complex conjugates.
  - Det (A) = determinant of A. If the determinant of a matrix is not zero, the rows of the matrix are linearly independent, and the inverse of A exists. In that case, we say that A is invertible.
- More matrix terminology
  - Symmetric: $A^{T} = A$. The matrix elements across the primary diagonal are equal $A_{jk} = A_{kj}$ for all k and j. The eigenvalues of symmetric matrices are real numbers, and the eigenvectors can be chosen to be mutually orthogonal.
  - Orthogonal: A square matrix such that $A^{T}A = AA^{T} = I$. In other words, $A^{-1} = A^{T}$. Such a matrix is symmetric with real entries. The columns of the matrix form an orthonormal basis set for the vector space. Multiplication of a column vector by an orthogonal matrix preserves the length of the vector. The eigenvalues of an orthogonal matrix can always be written as $e^{i\theta}$, with θ a real number.
  - Unitary: $A^{\dagger}A = I$ (“orthogonal” generalized to matrices with complex entries).
  - Hermitian: $A^{\dagger} = A$.
  - Involutory: $A^{-1} = A$ or $A^{2} = I$.

6.8 Further Reading

Chris Bernhardt, *Quantum Computing for Everyone* (MIT Press, Cambridge, MA, 2019). This has a nice introduction to quantum gates and their matrix representations.

An extensive list of quantum gates along with their matrix and circuit representations can be found in “List of quantum logic gates” in Wikipedia. The url is https://en.wikipedia.org/wiki/List_of_quantum_logic_gates

Putting a Spin on Spin

7

Let the great world spin forever down the ringing grooves of change.

Tennyson, "Locksley Hall"

7.1 Quantum Spin Systems

The types of quantum spin systems usually talked about in quantum computing (QC) are two-state systems, whose formal description is just like the photon polarization and computational basis states we mentioned before. In other words, a quantum "spin" system is just another example of a qubit, another physical system. Physicists love spin systems because theoretically they are relatively simple but nevertheless have many practical applications. Physicists have studied such systems in detail from the early days of quantum mechanics in the 1920s and continue today. Most of the representations of the quantum states of those systems are exactly like those we used before for photon polarization and our abstract computational basis states, with only one or two minor modifications. In this chapter, we explore the quantum mechanics of spin systems and use them to deepen our understanding of qubit states.

We begin with a quick introduction to the basics of spin systems. Many quantum particles (such as electrons, protons, neutrons, and quarks) and systems built from them, such as nuclei, atoms, and molecules, behave as if they were spinning tops. Exactly how that comes about, particularly for "elementary particles" like electrons and quarks, is not entirely understood, but it is a well-established empirical fact.

We can associate a direction with spinning motion: For example, the axis about which the disk in Fig. 7.1 is spinning. The usual convention is to use a right-hand rule to choose a direction along the axis: wrap the fingers of your right hand around the top in the direction in which the edge of the top is moving. The thumb of your right hand will then point along the spin axis in the direction associated with the spin angular momentum vector $\vec{S}$.

Quantum mechanics tells us that if we measure the orientation of the spin vector of an electron along some specified direction (say north), we get only two possible results: The spin vector points along that direction (we call that "spin-up") or it points opposite to that direction (we call that "spin-down"). In Section 7.3, we more specifically discuss what "along that direction" means.

Quantum Computing and Quantum Physics. Alice Flarend and Robert Hilborn, Oxford University Press.
 DOI: 10.1093/oso/9780197904930.003.0007

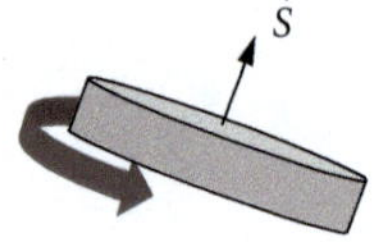

Fig. 7.1 A rotating disk has spin angular momentum represented by the vector $\vec{S}$. The direction of $\vec{S}$ is along the rotation axis and its orientation (up or down) is given by the right-hand rule.

For other quantum objects, there may be more than two possible results. The number of possible orientation values is specified by what historically is called a spin "quantum number" s. The number of possible orientation states is $2s + 1$. For electrons, protons, neutrons, and quarks, measurements show that $s = ½$, so the number of orientation states is $2 \times (1/2)+1 = 2$. For nuclei and atoms, for example, built up from protons, neutrons, and electrons, quantum mechanics tells us that the possible values for s are 0, 1/2, 1, 3/2, 2, and so on.

In QC with spin systems, we almost always assume that we are using qubits with spin quantum number $s = ½$. So, these are called "spin-1/2" systems (pronounced "spin-one-half"). As discussed later, having qubits with just two possible states is sufficient to do all sorts of interesting quantum information and quantum computing operations. Hence, we stick with spin-1/2 systems.

7.2 Operators and Representations of Spin-1/2 Systems

In the study of spin-1/2 qubits, the focus is on the spin angular momentum vector $\vec{S}$ and its three components in physical (three-dimensional) space S_x, S_y, and S_z. Like other physical properties, they are represented by operators (linear transformations) acting on state vectors. Of course, there are matrices associated with each of the operators. In fact, the spin angular moment operators are proportional to the Pauli X, Y, and Z operators we discussed in Chapter 6. The proportionality factor is $\hbar/2$, where, as before, $\hbar = 1.05 \ldots \times 10^{-34}$ joule-sec.

The connections between the spin angular momentum operators and the X, Y, and Z operators are

$$\begin{aligned}
S_z &= \frac{\hbar}{2} Z \doteq \frac{\hbar}{2}\begin{pmatrix} 1 & 0 \\ 0 & -1 \end{pmatrix} \\
S_x &= \frac{\hbar}{2} X \doteq \frac{\hbar}{2}\begin{pmatrix} 0 & 1 \\ 1 & 0 \end{pmatrix} \\
S_y &= \frac{\hbar}{2} Y \doteq \frac{\hbar}{2}\begin{pmatrix} 0 & -i \\ i & 0 \end{pmatrix}.
\end{aligned} \tag{7.1}$$

For each component of the spin angular momentum vector, there are two possible results for measurement of that component along some direction, say the z-direction. Because the

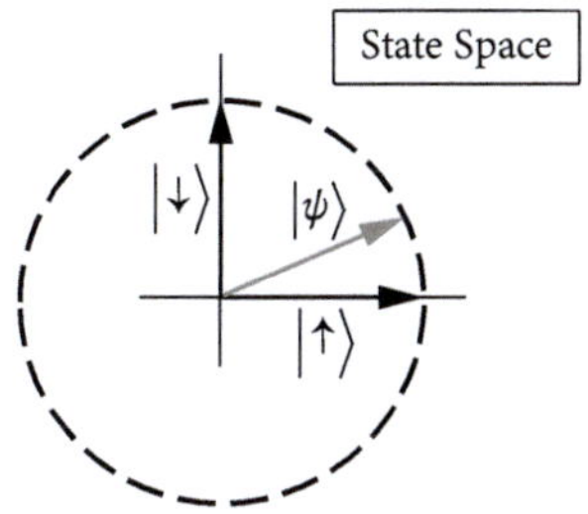

Fig. 7.2 The state space representation of the quantum state of a spin-1/2 qubit. $|\uparrow\rangle$ (spin-up) and $|\downarrow\rangle$ (spin-down) are the two basis states. $|\psi\rangle$ is a general state vector. The dashed circle has a radius = 1 and is often called the "unit circle."

two possible results are $+\hbar/2$ or $-\hbar/2$, we say that the angular momentum components are "quantized." By way of contrast, in classical physics the angular momentum components can take on any value between some maximum amount and some minimum amount. Those values are continuous within that range.

To describe the *states* of spin-1/2 qubits, we can use the three representations introduced for other quantum states: abstract quantum state vectors (right vectors), column vectors, and graphical representations in state space. The basis state vectors are often written as $|\uparrow\rangle$ for spin-up and $|\downarrow\rangle$ for spin-down for the z component of the spin angular momentum. A spin-1/2 state-space diagram is shown in Fig.7.2. The standard convention is to have the spin-up state vector be horizontal in the state space diagram. Since we are now using spin-up, spin-down basis states and not computational basis states, we have relabeled the axes in state space.

The two basis states can be represented by column vectors in the usual way:

$$|\uparrow\rangle \doteq \begin{pmatrix} 1 \\ 0 \end{pmatrix} \qquad |\downarrow\rangle \doteq \begin{pmatrix} 0 \\ 1 \end{pmatrix}. \tag{7.2}$$

Note that although spin-up and spin-down in ordinary physical space are in opposite directions (180° apart), the basis state vectors in state space are orthogonal (perpendicular) to one another. You need to remember that the basis state vectors live in a quantum state space and not in physical space. In the quantum state space, they are chosen to be orthogonal to each other. Recall that the orthogonality is a way of enforcing the fact that the properties associated with the basis states are mutually exclusive: If you observe the spin orientation of the qubit, you get either spin-up or spin-down but never both.

However, we often use the physical-space orientation of the spin to label the states as we have done in Eq. (7.2). Experiments show that for a spin-1/2 system, the spin vector orientation observations give us the two results—either along the observation direction or opposite the observation direction, or more succinctly, up or down.

Try It 7.1 Before looking at the following figures, draw an arrow representing a spin-up state and another arrow representing a spin-down state in a state-space diagram like Fig. 7.2.

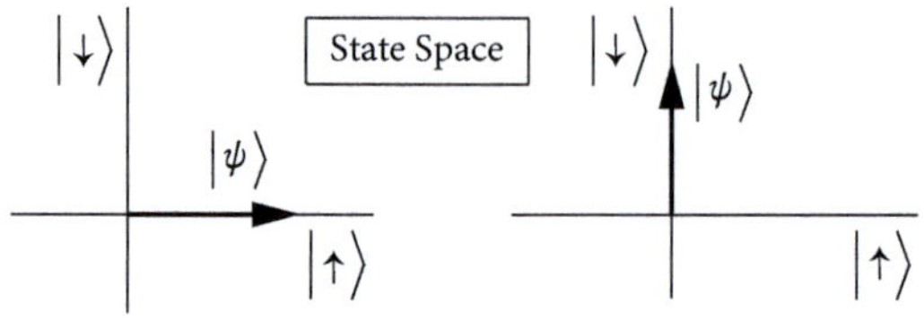

Fig. 7.3 On the left, the dark arrow represents a spin-up basis state vector for a spin-1/2 system. On the right, the dark arrow represents a spin-down basis state vector.

As Eq. (7.2) shows, the spin-up state $|\uparrow\rangle$ of a spin-1/2 qubit is customarily associated with the $|0\rangle$ computational basis state (which lies along the "horizontal" state-space axis as shown in Figs. 7.2 and 7.3) and the spin-down state $|\downarrow\rangle$ with the $|1\rangle$ computational basis state (which lies along the "vertical" state-space axis), although there is no fundamental reason you couldn't use the opposite association if you wanted to. The set of correspondences is

$$|\uparrow\rangle = |0\rangle \doteq \begin{pmatrix} 1 \\ 0 \end{pmatrix} \qquad |\downarrow\rangle = |1\rangle \doteq \begin{pmatrix} 0 \\ 1 \end{pmatrix}. \tag{7.3}$$

Of course, a general spin-1/2 state vector can be written as a linear combination (superposition) of the basis states. For the spin-up, spin-down basis set, we have

$$|\psi\rangle = c_\uparrow |\uparrow\rangle + c_\downarrow |\downarrow\rangle, \tag{7.4}$$

where, as usual, the state coefficients may be complex numbers. Figure 7.2 shows a state $|\psi\rangle$ that is a superposition of spin-up and spin-down with real coefficients.

These state vector orientations in Fig. 7.3 happen because of the collusion of two conventions: (1) associating spin-down with the $|1\rangle$ computational basis state and (2) associating the direction of $|1\rangle$ in state space with the vertical axis. This is just a reminder that the directions in state space need not be the same as the directions in physical space and that the labels in the state vectors are not necessarily related to directions in state space. For example, note that the spin-up basis state vector (in state space) is not pointing up. Similarly, the basis state arrow for the spin-down state is not pointing down in the state-space diagram. When we write "spin-up" or "spin-down," we are referring to the direction of the spin angular momentum in real space. You might want to keep a drawing of the spin-1/2 state-space diagram handy to consult because that mix of directions may be counterintuitive to you.

To measure the spin orientation, we make use of the fact that common elementary particles such as electrons, protons, and neutrons behave as if they are tiny bar magnets with a North Magnetic Pole (N) and a South Magnetic Pole (S). The critical point is that the direction associated with the spin lies along the particle's magnetic north–south direction. So, if we can carry out measurements to determine if the North Pole is up or down, we have determined whether the spin orientation is up or down. Furthermore, their N and S poles respond to external magnetic fields almost the way a compass needle responds to Earth's magnetic field. For a proton, the N pole is in the same direction as the spin angular momentum vector. For an electron (negative electrical charge), they are in opposite directions.

Note the cautious "as if" language we used. In some aspects, the quantum particle's North Pole–South Pole behavior is like a toy bar magnet. In other respects, it is quite different. We say "as if" because simple mechanical models of rotating spheres of electrical charge or other circulating charges cannot explain the magnetic properties of elementary particles. In fact, as we mentioned before, current science does not have completely satisfactory explanations of why elementary particles have spin. We can describe spin and do calculations involving spin, but we can't totally explain *why* particles have spin. Regardless of where spin comes from, we can measure those North Pole–South Pole orientations using the magnetic fields associated with them. You may have heard about "Stern–Gerlach" experiments, which do just that. The experiments were first carried out by Otto Stern and Walter Gerlach in the 1920s and provided some of the first measurements of atomic magnetic dipoles and confirmed their association with spin. Most importantly, they confirmed the "quantization" of the spatial components of spin angular momentum.

If the spin-1/2 state switches from up to down, we can think of the bar magnetic flipping 180° in our three-dimensional world, but in state space, the state vector flips only 90° from one orthogonal state to the other as we mentioned before. We have mentioned this observation several times because it is a critical one in quantum mechanics and its applications in atomic, molecular, optical, nuclear, and particle physics. By way of contrast, if you use a polarizing beam splitter (PBS) to measure the polarization state of a photon, the two mutually exclusive outcomes are vertical linear polarization and horizontal linear polarization. In that case, the physical-space directions associated with those outcomes are orthogonal and, of course, so are the state-space basis vectors. One of the reasons we decided to start our study of qubits using photon polarization states was to avoid just that added bit of complexity with spin-1/2 states.

Why are photons different from spin-1/2 systems? The standard textbook explanation is that photons have spin-1 states. How that is tied to horizontal and vertical linear polarization is a bit complicated; so, we won't go into the details, but the Further Reading section points you to books where you can learn about those details. Why do photons have spin-1? The standard explanation is that photons are the quantum particles associated with electric and magnetic fields. Those fields can be described as vectors, and the quantum theory of vector fields requires spin-1 particles. That may sound a bit circular, but it turns out that the spin of quantum particles is associated with the scalar, vector, or tensor nature of the underlying fields. However, for our QC work, we don't need those details.

7.3 Quantum Gates, Eigenvectors, and Eigenvalues for Spin-1/2 Systems

Let's take a quick look at how the quantum gates introduced in Chapters 5 and 6 apply to spin-1/2 systems. What do you expect to happen if a NOT (X) gate acts on a spin-up basis state? The X gate is supposed to flip the state to its orthogonal partner. So, we would expect that we get a spin-down state.

In more formal terms, we write

$$X|\uparrow\rangle = |\downarrow\rangle \qquad X|\downarrow\rangle = |\uparrow\rangle. \tag{7.5}$$

In the matrix form, Eq. (7.5) becomes

$$\begin{aligned} X|\uparrow\rangle &\doteq \begin{pmatrix} 0 & 1 \\ 1 & 0 \end{pmatrix}\begin{pmatrix} 0 \\ 1 \end{pmatrix} = \begin{pmatrix} 1 \\ 0 \end{pmatrix} \\ X|\downarrow\rangle &\doteq \begin{pmatrix} 0 & 1 \\ 1 & 0 \end{pmatrix}\begin{pmatrix} 1 \\ 0 \end{pmatrix} = \begin{pmatrix} 0 \\ 1 \end{pmatrix}. \end{aligned} \tag{7.6}$$

What about the Z gate? The Z gate action is easily found using the matrix multiplication form

$$Z|\uparrow\rangle = Z|0\rangle \doteq \begin{pmatrix} 1 & 0 \\ 0 & -1 \end{pmatrix}\begin{pmatrix} 1 \\ 0 \end{pmatrix} = \begin{pmatrix} 1 \\ 0 \end{pmatrix} \doteq |0\rangle = |\uparrow\rangle \tag{7.7}$$

$$Z|\downarrow\rangle = Z|1\rangle \doteq \begin{pmatrix} 1 & 0 \\ 0 & -1 \end{pmatrix}\begin{pmatrix} 0 \\ 1 \end{pmatrix} = \begin{pmatrix} 0 \\ -1 \end{pmatrix} = -\begin{pmatrix} 0 \\ 1 \end{pmatrix} \doteq -|1\rangle = -|\downarrow\rangle. \tag{7.8}$$

Let's remind ourselves about the quantum jargon, borrowed from linear algebra, that we introduced previously: If an operator acts on a state vector and returns the same state vector multiplied by a number, then that state vector is said to be an *eigenvector* of that operator, and the corresponding number is called an *eigenvalue*. That jargon is used a lot in quantum mechanics. From Eqs. (7.7) and (7.8), we see that the spin-up and spin-down basis states $|\uparrow\rangle$ and $|\downarrow\rangle$ are eigenvectors of the operator Z and the eigenvalues are +1 for $|\uparrow\rangle$ and −1 for $|\downarrow\rangle$.

The eigenvectors for the X operator can be written in terms of the Z eigenvectors:

$$\begin{aligned} |\uparrow\rangle_x &\doteq \frac{1}{\sqrt{2}}\begin{pmatrix} 1 \\ 1 \end{pmatrix}_z \doteq \frac{1}{\sqrt{2}}\left[|\uparrow\rangle_z + |\downarrow\rangle_z\right] \\ |\downarrow\rangle_x &\doteq \frac{1}{\sqrt{2}}\begin{pmatrix} 1 \\ -1 \end{pmatrix}_z \doteq \frac{1}{\sqrt{2}}\left[|\uparrow\rangle_z - |\downarrow\rangle_z\right], \end{aligned} \tag{7.9}$$

where $|\uparrow\rangle_x$ and $|\downarrow\rangle_x$ are the eigenvectors for spin-up and spin-down along the x-axis, respectively. We have used the eigenvectors in both their column vector and Dirac right-vector forms. Figure 7.4 shows those eigenvectors in a state-space diagram. Similarly, the eigenvectors of the Y operator in terms of the Z eigenvectors are

$$\begin{aligned} |\uparrow\rangle_y &\doteq \frac{1}{\sqrt{2}}\begin{pmatrix} 1 \\ i \end{pmatrix}_z \doteq \frac{1}{\sqrt{2}}\left[|\uparrow\rangle_z + i|\downarrow\rangle_z\right] \\ |\downarrow\rangle_y &\doteq \frac{1}{\sqrt{2}}\begin{pmatrix} 1 \\ -i \end{pmatrix}_z \doteq \frac{1}{\sqrt{2}}\left[|\uparrow\rangle_z - i|\downarrow\rangle_z\right]. \end{aligned} \tag{7.10}$$

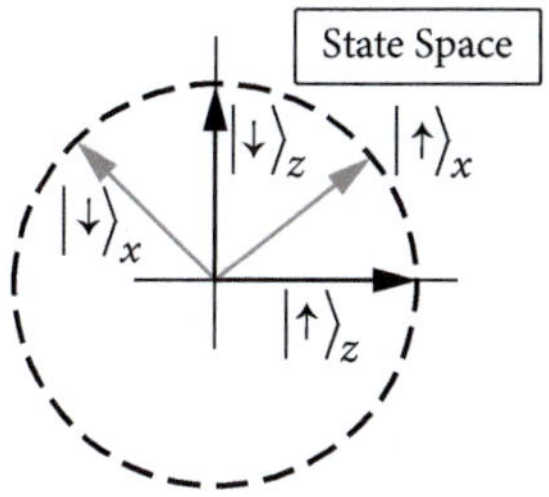

Fig. 7.4 State space diagram of the X (gray) and Z eigenvectors (black).

Try It 7.2 Using the column vector form of the X and Y operators and the eigenvectors as column vectors, show that the eigenvalues for both the X and Y operators are $+1$ for $|\uparrow\rangle_{x \text{ or } y}$ and -1 for $|\downarrow\rangle_{x \text{ or } y}$. Explain why it is reasonable that X, Y, and Z have identical eigenvalues.

Math Aside Here is a brief summary of eigenvalues and eigenvectors for 2×2 matrices. Suppose we have an operator A that acts on a state and reproduces the same state multiplied by a number λ: $A|\lambda\rangle = \lambda|\lambda\rangle$. In that case, λ is said to be an eigenvalue for that operator and $|\lambda\rangle$ is the corresponding eigenvector. Linear algebra tells us that for a 2×2 matrix for which the determinant is not zero, there will be two eigenvalues λ_1 and λ_2. If the matrix is in the diagonal form $\begin{pmatrix} a & 0 \\ 0 & b \end{pmatrix}$, you can read off the eigenvalues directly: $\lambda_1 = a$ and $\lambda_2 = b$. Even if the matrix is not in the diagonal form, the two eigenvalues satisfy the following two equations: $\lambda_1\lambda_2 = \det A$, where det is the determinant of A, and $\lambda_1 + \lambda_2 = \mathrm{Tr}\, A$, where $\mathrm{Tr}\, A$ is the trace of the matrix and given by the sum of the principal (upper-left and lower-right) matrix elements. This means that it is easy to find the eigenvalues of a 2×2 matrix because the determinant and the trace are the same for any matrix representation of the operator. Those facts are proved in most textbooks on linear algebra. If the determinant of the matrix is zero, then one of the eigenvalues is 0. The other can be found from $\lambda_1 + \lambda_2 = \mathrm{Tr}\, A$. Once you have the eigenvalues, you can find the eigenvectors (strictly speaking, the eigenvector coefficients in the representation that is used to construct the A matrix) by solving the two simultaneous equations for the coefficients that come from the matrix and column vector form of $A|\lambda\rangle = \lambda|\lambda\rangle$.

Try It 7.3 Use the trace-and-determinate method in the previous Math Aside to find the eigenvalues of the Pauli X, Y, and Z operators.

Try It 7.4 Figure out the results of the X operator acting on the z-direction spin-up and spin-down states. Draw the results in a state-space diagram. Hint: See Eq. (7.6).

Math Aside (for those who like mathematical proofs) In quantum mechanics, we generally assume that measured quantities have properties represented by real numbers. To ensure that the eigenvalues are real, we use the so-called Hermitian operators. Since this is such an important result, it is worth taking a few minutes to understand why that works. Recall that in Chapter 6 we defined the Hermitian adjoint $A^\dagger$ of an operator A. Let's refresh our memories: In matrix language, the Hermitian adjoint is a combination of two operations: (1) taking the complex conjugate of the matrix elements and (2) taking the transpose of the matrix. For example, for the matrix $A \doteq \begin{pmatrix} \alpha & \beta \\ \gamma & \delta \end{pmatrix}$, the Hermitian conjugate (often described as "A-dagger") is $A^\dagger = \begin{pmatrix} \alpha^* & \gamma^* \\ \beta^* & \delta^* \end{pmatrix}$. An operator is said to be Hermitian if it is equal to its Hermitian conjugate: $A = A^\dagger$. In abstract Dirac notation, we can define the Hermitian conjugate as the operator that satisfies $\langle\phi|A|\psi\rangle = \langle\psi|A^\dagger|\phi\rangle^*$. Swapping ϕ and ψ is equivalent to taking the transpose. We can now show that a Hermitian operator has real eigenvalues. We can do this in two ways: by using (1) Dirac notation and (2) standard linear algebra notation.

In the Dirac notation method, we introduce an eigenfunction of the operator A and its corresponding eigenvalue λ, that is, $A|\psi\rangle = \lambda|\psi\rangle$. We now form a quantum sandwich with $\langle\psi|$:

$$\langle\psi|A|\psi\rangle = \lambda\langle\psi|\psi\rangle. \tag{7.11}$$

By the definition of the Hermitian adjoint, we may write

$$\langle\psi|A|\psi\rangle^* = \langle\psi|A^\dagger|\psi\rangle = \langle\psi|A|\psi\rangle, \tag{7.12}$$

where the last equality holds if A is a Hermitian operator. Eq. (7.12) shows that the quantum sandwich $\langle\psi|A|\psi\rangle$ is a real number if A is a Hermitian operator. Note that the Dirac bracket $\langle\psi|\psi\rangle$ is a real number. Using those results, we see that Eq. (7.11) indicates that the eigenvalue λ must be a real number.

In the notation of linear algebra, we use x to represent the eigenvector. Then, the eigenvalue equation is $Ax = \lambda x$. Because we want to show that $\lambda^* = \lambda$, we take the complex conjugate of that expression to obtain $A^*x^* = \lambda^*x^*$. Left multiply the first of those equations by x^* to obtain

$$x^*Ax = \lambda x^*x \tag{7.13}$$

and the second by x to obtain $xA^*x^* = \lambda^*xx^*$. We now apply the transpose operation to both sides of that equation to obtain $x^*(A^*)^T x = \lambda^*x^*x$. For a Hermitian operator, we have $A = (A^*)^T$ by definition. So, we find

$$x^*Ax = \lambda^*x^*x. \tag{7.14}$$

Comparing Eqs. (7.13) and (7.14) gives $\lambda = \lambda^*$. Any number equal to its complex conjugate is a real number. In symbolic form, we see that if $z = u + iv$ equals its complex conjugate $z^* = u - iv$, then v (the imaginary part of z) must equal 0, leaving only the real part u End of proof.

Try It 7.5 Figure out the results of the Hadamard operator (Chapter 6) acting on the spin-up and spin-down states. Draw the results in a state-space diagram.

Math Aside Before returning to spin-1/2 systems, we want to point out another operator property. Once we have found the eigenvalues λ_j and eigenvectors $|\lambda_j\rangle$ associated with an operator, we may write the operator A as a linear superposition of the projection operators associated with each of the eigenvectors:

$$A = \sum_j \lambda_j |\lambda_j\rangle\langle\lambda_j|. \tag{7.15}$$

This form is called the spectral decomposition of the operator A because the set of eigenvalues $\{\lambda_j\}$ is the "spectrum" of the eigenvalues. The spectral decomposition is useful when we need to define a mathematical function of an operator:

$$f(A) = \sum_j f(\lambda_j) \underbrace{|\lambda_j\rangle\langle\lambda_j|}_{\text{projection operator}}. \tag{7.16}$$

We want to point out another peculiar feature of spin systems. It turns out that a "spin-up" state does not have the spin pointing exactly along the "up" direction in real space as illustrated in Fig. 7.5. Although the z component of the spin is certain for the spin-up state, its x and y components are unknown. This result is due to what is called the *uncertainty relation* in quantum mechanics, which we discuss later in this chapter. If the spin pointed exactly along the z-axis (for example), then its x and y components would both be exactly zero. The quantum theory of spins tells us that we can't prepare a state with both the x and y components of spin known precisely. That rules out having the spin pointing exactly along the z-axis. This peculiar feature is expressed by the mathematical fact that the operators (gates) X, Y, and Z do not commute with each other. That is, XY $\neq$ YX, XZ $\neq$ ZX, and ZY $\neq$ YZ.

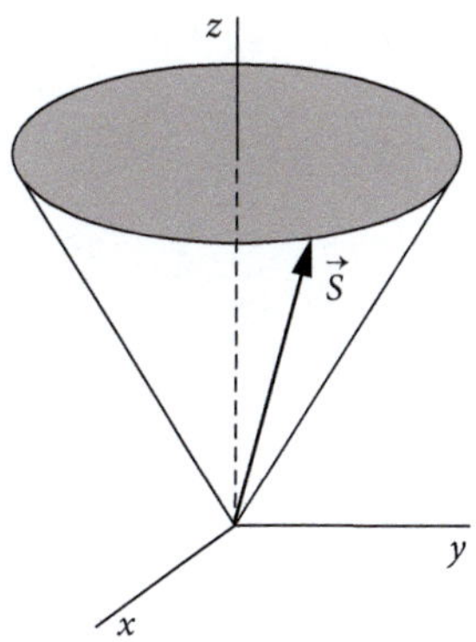

Fig. 7.5 A sketch of a quantum spin vector $\vec{S}$ with a definite component along the z-direction. The vector, with its tail anchored at the origin, should be thought of as smeared around the outside surface of the cone, so that its x and y components average to 0. The angle between $\vec{S}$ and the z-axis stays fixed for this state.

Now that we have the basics of spin-1/2 states in hand, we turn to three other issues. Relative phases with complex amplitudes are discussed in Section 7.4. Section 7.5 introduces a visual representation of spin-1/2 systems with complex coefficients. Section 7.6 explains why the imaginary i must be included in the Y spin-1/2 operator. In Section 7.7, we discuss the details of the *uncertainty relation.*

7.4 Relative Phases in Quantum States

In Chapter 5, we showed that the overall sign or phase of a quantum state does not change the measurement predictions calculated from that state. We also observed that the *relative* sign (a particular type of phase) between the amplitudes in a quantum superposition state leads to observable results, that is, to different measurement outcomes. In particular, if the superposition state is expressed in terms of one set of basis states, but measured with respect to a different set of basis states, the measurement outcomes can tell us the relative sign (and its generalization to complex phase terms) between the superposition amplitudes in the original superposition state. We will look at relative phases using complex exponentials to be completely general. See Appendix: Quantum Toolkit for a brief math review of complex numbers, including complex exponentials.

Let's start with a general phase difference (expressed as a complex exponential of a real number ϕ) between the amplitudes in a superposition state for a single qubit. For the sake of simplicity, we assume that the two amplitudes have a magnitude $1/\sqrt{2}$:

$$|\psi\rangle = \frac{1}{\sqrt{2}}\left[|0\rangle + e^{i\phi}\,|1\rangle\right]. \tag{7.17}$$

Putting the complex exponential in front of the $|1\rangle$ basis state and not the $|0\rangle$ basis state is just a convention but a commonly used one. In the more general case, we could have a complex exponential for both amplitudes $a_0 = |a_0|\,e^{i\phi_0}$ and $a_1 = |a_1|\,e^{i\phi_1}$ as shown in the following example:

$$\begin{aligned} |\psi\rangle &= \frac{1}{\sqrt{2}}\left[a_0|0\rangle + a_1|1\rangle\right] \\ &= \frac{1}{\sqrt{2}}\left[e^{i\phi_0}\,|a_0||0\rangle + e^{i\phi_1}\,|a_1||1\rangle\right] \\ &= \frac{e^{i\phi_0}}{\sqrt{2}}\left[|a_0||0\rangle + e^{i(\phi_1-\phi_0)}\,|a_1||1\rangle\right] \\ &= \frac{e^{i\phi_0}}{\sqrt{2}}\left[|a_0||0\rangle + e^{i\phi}\,|a_1||1\rangle\right], \end{aligned} \tag{7.18}$$

where $\phi = \phi_1 - \phi_0$. Note that the last line of Eq. (7.18) looks just like Eq. (7.17) except for an overall complex exponential factor. The complex exponential multiplying the overall state does not change the probabilities calculated from the state. But if the coefficient of $|1\rangle$ contains a factor that is the relative phase between $|0\rangle$ and $|1\rangle$, the measurement outcomes may depend on ϕ. The takeaway message is that only the relative phase between the two amplitudes matters, not the global value of the phase of the two amplitudes. As we mentioned, by convention we assign the complex exponential representing that phase to the $|1\rangle$ state.

Try It 7.6 Work through each line of Eq. (7.18) and explain both the manipulation of the complex exponentials and the Dirac state vector notation.

In Chapter 5 (Section 5), we gave an example of using different basis states with real number coefficients. Now we show that we can do the same with complex coefficients. In particular, we find the relative phase ϕ by making measurements with a different set of basis states. Let's use the state in Eq. (7.17) as an example. The method and notation match the ones those are used in Chapter 8, where we express the original state in terms of the basis states $|B_0\rangle$ and $|B_1\rangle$ that are used by the measurement device. For the present example, we use a simple transformation relationship

$$\begin{aligned} |0\rangle &= \frac{1}{\sqrt{2}}\left[|B_0\rangle + |B_1\rangle\right] \\ |1\rangle &= \frac{1}{\sqrt{2}}\left[|B_0\rangle - |B_1\rangle\right]. \end{aligned} \tag{7.19}$$

We then use Eq. (7.19) in Eq. (7.17), regrouping terms for the B basis states:

$$\begin{aligned} |\psi\rangle &= \frac{1}{\sqrt{2}}\left[\frac{1}{\sqrt{2}}\left(|B_o\rangle + |B_1\rangle\right) + \frac{e^{i\phi}}{\sqrt{2}}\left(|B_o\rangle - |B_1\rangle\right)\right] \\ &= \frac{1}{2}\left[\left(1 + e^{i\phi}\right)|B_o\rangle + \left(1 - e^{i\phi}\right)|B_1\rangle\right]. \end{aligned} \tag{7.20}$$

To find the probabilities for the measurement outcomes, we calculate, as usual, the absolute-value-squared of the state vector amplitudes. The probability for the measurement outcome B_0 is

$$\begin{aligned} \text{Prob}(B_0) &= \frac{1}{4}\left(1 + e^{i\phi}\right)^*\left(1 + e^{i\phi}\right) \\ &= \frac{1}{4}\left(1 + e^{-i\phi}\right)\left(1 + e^{i\phi}\right) = \frac{1}{4}\left(1 + e^{i\phi} + e^{-i\phi} + 1\right) \\ &= \frac{1}{2}\left(1 + \cos\phi\right), \end{aligned} \tag{7.21}$$

where we used the "reverse" Euler formula $\cos\phi = \left(e^{i\phi} + e^{-i\phi}\right)/2$.

Try It 7.7 Use the same type of calculation to show that $\text{Prob}(B_1) = 1/2(1 - \cos\phi)$. Do the measurement probabilities add up to 1? What are the probabilities if $\phi = \pi$ and $\phi = \pi/2$?

The important "takehome message" is that measurements carried out with basis states different from those in which the original state was prepared allow us to find the relative phase ϕ between the two parts of the qubit state. Or conversely, different relative phases lead to different measurement outcomes. We learn more about the state by making measurements relative to a different basis.

For quantum information science and quantum computing applications, we would like to have a quantum gate that can change the relative phases in a quantum state because that change will be observable in the measurement outcomes. In fact, there are many phase operators of various kinds in QC, but we will stick to a relatively simple but powerful version for single qubit systems. Our version is developed using the Z operator, which we covered in Chapter 4, and the complex exponential we have just become familiar with. The phase operator is

$$P(\phi) = e^{-i\phi Z/2}. \tag{7.22}$$

What does it mean to have an operator Z in an exponential function? This concept is treated in most linear algebra courses. We work out a special case, so you can see the basic ideas. The operator we consider is $\exp[aZ]$, where a is a number. We make use of the general power series expansion of an exponential function:

$$e^x = 1 + x + \frac{x^2}{2} + \frac{x^3}{3!} + \cdots \tag{7.23}$$

Because the Z operator has the nice (involutory) property that $ZZ = \text{I}$ (the identity operator), Eq. (7.23) becomes

$$\begin{aligned} \exp[aZ] &= \text{I} + aZ + \frac{a^2Z^2}{2} + \frac{a^3Z^3}{3!} + \cdots . \\ &= \text{I} + aZ + \frac{a^2}{2}\text{I} + \frac{a^3Z}{3!} + \cdots \end{aligned} \tag{7.24}$$

We now use the matrix form of Z:

$$\begin{aligned} \exp[aZ] &\doteq \begin{pmatrix} 1 & 0 \\ 0 & 1 \end{pmatrix} + \begin{pmatrix} a & 0 \\ 0 & -a \end{pmatrix} + \begin{pmatrix} a^2/2 & 0 \\ 0 & a^2/2 \end{pmatrix} + \begin{pmatrix} a^3/3! & 0 \\ 0 & -a^3/3! \end{pmatrix} + \cdots \\ &= \begin{pmatrix} 1 + a + a^2/2 + a^3/3! + \cdots & 0 \\ 0 & 1 - a + a^2/2 - a^3/3! + \cdots \end{pmatrix}. \end{aligned} \tag{7.25}$$

We see that the upper-left matrix entry is just the power series expansion of exp$[a]$ while the lower-right matrix entry is the power series expansion of exp$[-a]$. The final result is

$$\exp[aZ] \doteq \begin{pmatrix} e^{a} & 0 \\ 0 & e^{-a} \end{pmatrix}. \tag{7.26}$$

Alternatively, we could use the spectral decomposition of the Z operator as mentioned in the previous Math Aside to express exp$[aZ]$. Later in this book, we encounter several examples of exponential functions of operators (matrices).

Applying what we just learned about the Z operator in Eq. (7.26) to the general phase operator in Eq. (7.22), we obtain

$$P(\phi) \doteq \begin{pmatrix} e^{-i\phi/2} & 0 \\ 0 & e^{i\phi/2} \end{pmatrix}. \tag{7.27}$$

Applying Eq. (7.27) to a general qubit state, we obtain

$$\begin{aligned} P(\phi)(a_0\,|0\rangle + a_1\,|1\rangle) &\Rightarrow e^{-i\phi/2}\begin{pmatrix} 1 & 0 \\ 0 & e^{i\phi} \end{pmatrix}\begin{pmatrix} a_0 \\ a_1 \end{pmatrix} \\ &= e^{-i\phi/2}\begin{pmatrix} a_0 \\ a_1 e^{i\phi} \end{pmatrix} \doteq e^{-i\phi/2}\left(a_0\,|0\rangle + a_1 e^{i\phi}\,|1\rangle\right). \end{aligned} \tag{7.28}$$

So, we see that the operator creates a relative phase between the two state amplitudes and produces an overall phase factor in front of the matrix, which we can ignore because it disappears when we calculate probabilities. This is just what we need. The ability to adjust phase differences as well as amplitudes in superposition states gives us powerful tools to build quantum algorithms. Finding the relative phase is often called "phase estimation" and is important in the quantum Fourier transform (Chapter 15) and in quantum algorithms for solving systems of linear equations.

In linear algebra terminology, $P(\phi)$ is a unitary operator because its Hermitian conjugate is equal to its inverse:

$$P_Z^{\dagger} \doteq \begin{pmatrix} e^{i\phi/2} & 0 \\ 0 & e^{-i\phi/2} \end{pmatrix} \doteq P^{-1}. \tag{7.29}$$

Try It 7.8 Check that $P^{\dagger}\,P = P^{-1}\,P = \mathrm{I}$.

Since P is a diagonal matrix, we can see from Eq. (7.24) that its eigenvalues are $e^{-i\phi/2}$ and $e^{i\phi/2}$. Those eigenvalues follow the rule that the eigenvalues of a unitary matrix satisfy $|\lambda|^2 = 1$. The phase estimation comes in by noting that the eigenvalues of any unitary matrix U can be expressed by

$$U\,|\lambda\rangle = e^{2\pi i\alpha}\,|\lambda\rangle, \tag{7.30}$$

where α is a real number parameter and $|\lambda\rangle$ is the eigenstate associated with the eigenvalue $\lambda = e^{2\pi i\alpha}$ of the matrix U.

Try It 7.9 What are the eigenstates of P in the computational basis?

For more details on how quantum information science (QIS) makes use of quantum phase estimation, see Further Reading (Qiskit, Sections 3.6 and 4.1.1). We elaborate phase estimation when we discuss the quantum Fourier transform in Chapter 15.

7.5 Bloch Sphere and Bloch Circle

Since the amplitudes in a general quantum state are complex numbers, we need a visual representation different from our customary two-dimensional state-space diagram, which in its usual form can handle only real number amplitudes. The angle of the state vector in two-dimensional state-space diagram shows only the magnitude (with plus or minus signs) of the state amplitudes of the superposition state; it does not show information about the relative phases like $e^{i\phi}$.

The most common visual representation that includes the relative phase is called the Bloch sphere, named after Felix Bloch, a Nobel Laureate in physics and a pioneer in the field of nuclear magnetic resonance. Those techniques eventually led to magnetic resonance imaging, one of the most important medical imaging techniques available to us today. Bloch was concerned with the nuclei of atoms and their magnetic and spin properties. In fact, the nucleus of a hydrogen atom is a proton and therefore is an example of a qubit because it has spin-1/2. If you read more about qubits and their quantum states, you will eventually come across this pictorial representation: the Bloch sphere.

The Bloch sphere (Fig. 7.6) may look familiar if you recall how we specify geographical positions on Earth. To describe a place on the surface of the planet, we usually specify a latitude and longitude. For example, New York City is at 40.7128°N (latitude) and 74.0060°W (longitude). For the Bloch sphere, longitude corresponds to ϕ. The angle θ_B is called the "colatitude" because it is the complement of the angle used to describe latitude on Earth. Note that Bloch sphere longitude ϕ by convention increases in a direction opposite to Earth's longitude.

Try It 7.10 On a baseball, volleyball, or basketball, draw two dots representing the $|0\rangle$ and $|1\rangle$ "poles" and curved arrows (compare Fig. 7.6) representing Bloch sphere colatitude and longitude.

A quantum state whose Bloch vector tip is at 90° colatitude on the Bloch sphere is a superposition state where the probabilities associated with the measurement outcomes $|0\rangle$

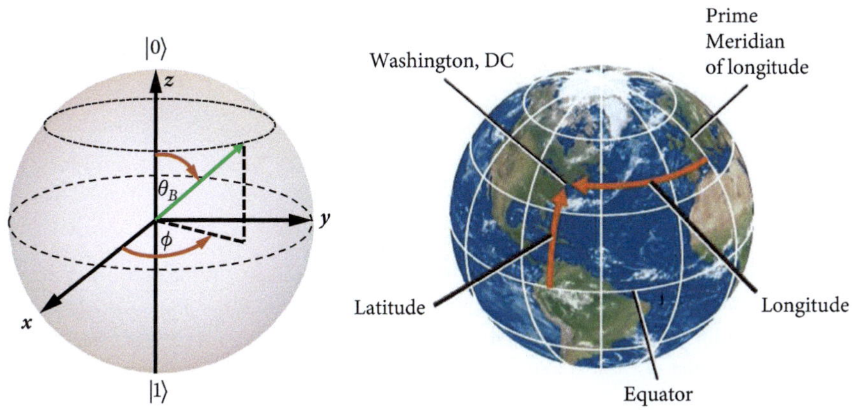

Fig. 7.6 On the left, the Bloch sphere for single-qubit states. The green vector represents the qubit state by a *Bloch vector*. The *colatitude* angle θ_B is related to the magnitudes of the state vector amplitudes while the *longitude* angle ϕ represents the phase difference between the amplitudes. The central dashed oval indicates the Bloch sphere's equator. The upper dashed oval is the constant colatitude curve (parallel to the equator) corresponding to the colatitude of the tip of the state vector. On the right, a picture of latitude and longitude lines on the surface of Earth.

and $|1\rangle$ are equal (50/50). Other angles between 0° and 90° and between 90° and 180° are superposition states with unequal probabilities. A Bloch vector pointing to the "North Pole" at 0° colatitude represents a state that is not a superposition but is simply $|0\rangle$.

The Bloch sphere representation twists the basis state vectors so the corresponding Bloch vectors point in opposite directions. In fact, state vectors that are perpendicular (orthogonal) in state-space point in opposite directions in the Bloch sphere picture. The Bloch sphere, in a sense, opens up the state-space diagram, so we can display information about the phase difference between the amplitudes. We should point out that in many cases we use spin-up and spin-down basis states instead of $|0\rangle$ and $|1\rangle$, respectively. By convention, the spin-up state $|\uparrow\rangle = |0\rangle_z$ points upward along the z-axis (toward the North Pole of the Bloch sphere) and the spin-down state $|\downarrow\rangle = |1\rangle_z$ points toward the South pole.

Longitude in the Bloch sphere diagram represents the relative phase of the complex exponentials, which we discussed in the previous section. In the Bloch sphere representation, the longitudinal angle ϕ is measured counterclockwise from the x-axis (looking down on the North Pole). So, a superposition state Bloch vector lying in the Bloch diagram xz plane has no phase difference between the amplitudes in the basis state expansion.

Let's work with the Bloch sphere by expressing a general qubit state in terms of the computational basis states and a complex exponential as part of the amplitude of the $|1\rangle$ state:

$$|\psi\rangle = \cos(\theta_B/2)|0\rangle + e^{i\phi}\sin(\theta_B/2)|1\rangle, \tag{7.31}$$

where we introduced $\cos(\theta_B/2) = a_0$ and $e^{i\phi}\sin(\theta_B/2) = a_1$ for the basis state amplitudes. This way of writing the state leads to the geometric picture depicted in Fig. 7.6. Note that

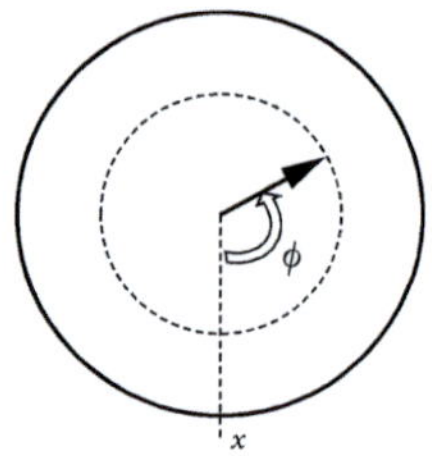

Fig. 7.7 Looking down on the North Pole of the Bloch sphere. The angle ϕ is the phase angle for the phase difference between the two basis state amplitudes. By convention, the angle is measured counterclockwise from the Bloch sphere x-axis. The dashed circle is a circle of constant colatitude. We call this diagram a phase dial.

$\theta_B/2$ is the standard *state-space* angle between the $|0\rangle$ basis state vector and the qubit state vector $|\psi\rangle$. When $\theta_B = \pi$, the Bloch vector points toward the South Pole and $|\psi\rangle = |1\rangle$ as we noted before.

Try It 7.11 Draw the Bloch vector on a Bloch sphere diagram with $\theta_B = 60°$ and $\phi = 30°$. Repeat with $\theta_B = 60°$ and $\phi = 90°$. Hint: $\cos 60° = 0.5$ and $\sin 30° = 0.5$. Then, use Eq. (7.31) to express the state as a superposition of the computational basis states.

In Eq. (7.31), the complex exponential phase term, as usual, is part of the coefficient of the $|1\rangle$ basis state. States that have the same angle θ_B have the same magnitudes of the amplitudes but different phases as specified by ϕ. For example, all state vectors that lie in the equatorial plane (for which $\theta_B = 90°$) are superposition states that give equal probabilities for spin-up and spin-down but differ in phase between the two basis states. Another way of saying this is that changing the phase difference moves the Bloch vector tip around a circle of constant colatitude as shown in Figs. 7.6 and 7.7.

In Fig. 7.7, we look down on the North Pole of the Bloch sphere. The angle ϕ is the longitude angle as shown in Fig. 7.6. This diagram is almost the same as the visualization that Richard Feynman used in his book *QED*: The Strange Theory of Light and Matter, which explains the interaction between photons and matter. Rather than using a Bloch sphere to keep track of the quantum state phase, Feynman used a "phase dial" (like a clock face), as shown in Fig. 7.8. You could also attach the phase dial to state vectors in our usual two-dimensional state space. (In such a case, the two-dimensional state-space unit circle might be called the Bloch circle.) The position of the clock hand indicates the relative phase between the amplitudes of the basis states in a superposition state.

The Bloch sphere diagram is a nice way of visualizing how quantum gates act on single-qubit quantum states. The gates can affect both the magnitudes of the amplitudes and their phase difference. In fact, there are some nice online animated visualizations of how various single-qubit gates affect the Bloch vector (see Further Reading). That said, the Bloch sphere representation is generally not helpful if we have systems of more than one qubit.

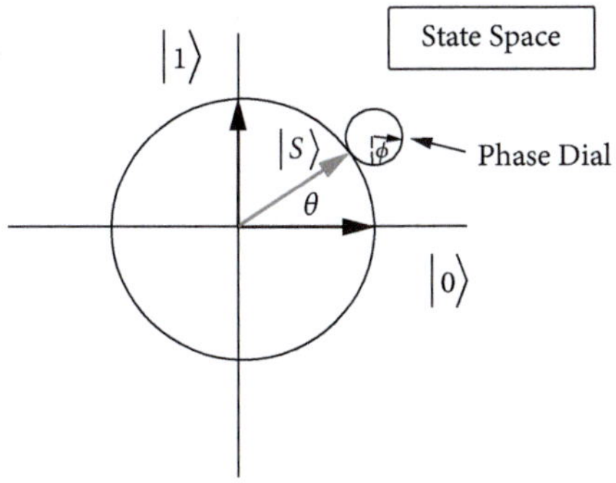

Fig. 7.8 A state space diagram showing a state space vector $|S\rangle$ and the associated phase dial, indicating the phase difference ϕ between the amplitudes of the components of the state space vector along the basis states. We might call this diagram a Bloch circle.

In addition to providing a nice visualization of single-qubit states, the Bloch sphere diagram helps us understand what is called the U_B (for "universal") single-qubit gate. (The subscript B reminds us that the gate operates on a Bloch sphere vector.) U_B is universal in the sense that it can operate on a Bloch vector to produce any other Bloch vector. In particular, $U_B(\alpha, \beta, \phi)$ operates on a Bloch sphere vector with phase ϕ and adds α to the phase and β to the colatitude. For a single qubit state, the U_B gate in the computational basis is represented by a matrix

$$U_B(\alpha, \beta, \varphi) \doteq \begin{pmatrix} \cos(\beta/2) & -e^{-i\phi}\sin(\beta/2) \\ e^{+i(\phi+\alpha)}\sin(\beta/2) & e^{i\alpha}\cos(\beta/2) \end{pmatrix}. \tag{7.32}$$

Note that we have assumed that ϕ appears in the complex exponential as shown in Eq. (7.31).

When U_B operates on the computational basis states' Bloch vectors (which have $\phi = 0$), it generates superposition states:

$$\begin{aligned} U_B(\alpha, \beta, 0)|0\rangle &= \cos(\beta/2)\,|0\rangle + e^{i\alpha}\sin(\beta/2)\,|1\rangle \\ U_B(\alpha, \beta, 0)|1\rangle &= -\sin(\beta/2)\,|0\rangle + e^{i\alpha}\cos(\beta/2)\,|1\rangle. \end{aligned} \tag{7.33}$$

Try It 7.12 Use the column vector forms of the basis states $|0\rangle$ and $|1\rangle$ and Eq. (7.32) to check the results in Eq. (7.33). Verify that the resulting states are normalized states.

The universal gate operator looks just like a generalization of the Hadamard gate we used before. In fact, the connection is

$$U_B(\pi, \pi/2, \pi) \doteq \frac{1}{\sqrt{2}}\begin{pmatrix} 1 & 1 \\ 1 & -1 \end{pmatrix} \doteq \mathrm{H}. \tag{7.34}$$

The U_B operator is widely used in quantum algorithms because of its ability to set both amplitudes and relative phases in a superposition state. There are also special cases of U_B, the S operator and the T operator, that appear in many QC and QIS applications:

$$S \doteq \begin{pmatrix} 1 & 0 \\ 0 & e^{i\pi/2} \end{pmatrix}$$
$$T \doteq \begin{pmatrix} 1 & 0 \\ 0 & e^{i\pi/4} \end{pmatrix}. \tag{7.35}$$

The operators in Eq. (7.35) are special cases of the general "phase shift" operator (see Eq. (7.27)):

$$P(\phi) \doteq \begin{pmatrix} 1 & 0 \\ 0 & e^{i\phi} \end{pmatrix}. \tag{7.36}$$

As shown in Eq. (7.28), when P acts on a qubit state, the coefficient of the $|1\rangle$ part of the superposition state is multiplied by $e^{i\phi}$:

$$P(\phi)|\psi\rangle \doteq \begin{pmatrix} 1 & 0 \\ 0 & e^{i\phi} \end{pmatrix}\begin{pmatrix} a \\ b \end{pmatrix} = \begin{pmatrix} a \\ e^{i\phi}b \end{pmatrix} \doteq a|0\rangle + e^{i\phi}b|1\rangle. \tag{7.37}$$

Note that these phase shift matrices are not Hermitian matrices and hence cannot represent directly measurable properties.

Try It 7.13 What values of α, β, and ϕ in Eq. (7.32) lead to the operator matrices in Eq. (7.35)?

We can derive Eq. (7.32) by using the z-axis rotation operator in Eq. (7.28), which changes the phase, and the operator for rotations about the Bloch y-axis. The operator for a rotation by an angle θ about the Bloch sphere y-axis is

$$R_y(\theta) \doteq \begin{pmatrix} \cos(\theta/2) & -\sin(\theta/2) \\ \sin(\theta/2) & \cos(\theta/2) \end{pmatrix}, \tag{7.38}$$

which is covered in Chapter 8. Note that this is the form only for Bloch vectors that lie in the Bloch xz plane.

Try It 7.14 Challenge: You can use Eqs. (7.28) and (7.38) to derive the U_B matrix in Eq. (7.32). First, operate on the Bloch sphere state in Eq. (7.31) with $R_z(-\phi)$. This brings the state longitude to 0°. Then, apply the matrix $R_y(\beta)$ to add β to the colatitude and finish by applying a longitude rotation $R_z(\alpha+\phi)$ to make the relative phase $\alpha+\phi$. This procedure will lead to $U_B(\alpha,\beta,\phi) = R_z(\alpha+\phi)R_y(\beta)R_z(-\phi)$.

The Bloch sphere also provides a nice visualization of different paths by which a state can evolve from one position to another, for example, from spin-up to spin-down. Paths can be chosen to minimize phase disruptions or amplitude disruptions to decrease errors, for example.

It is worth mentioning that there is a construction, similar to the Bloch sphere, for polarization states for light. This construction is called a Poincaré sphere. The mathematician and physicist Henri Poincaré developed this representation in 1892, long before the development of quantum mechanics. There is also a way to use matrices and column vectors to represent states of polarized light and how those states change when they interact with devices such as PBS. The method is called the Jones calculus after its inventor R. C. Jones, who developed the analysis in 1941 (see the Further Reading section for more information).

Math Aside The Hadamard operator, the S gate, and the CNOT gate generate what is called the set of Clifford gates. The set constitutes a mathematical "group" in that any of its members acting, as described in the next sentence, on a member of the group results in another group member. For example, if the Hadamard operator acts on the X operator, it produces a Z operator $HXH^{\dagger} = Z$ and $HZH^{\dagger} = X$. So, X and Y are also part of the Clifford group. The Clifford gates and their quantum circuits can be efficiently simulated using a classical computer. As an important corollary, quantum gates that are not Clifford gates cannot be efficiently simulated using a classical computer. For example, the T gate in Eq. (7.35) is not a Clifford gate. The Clifford group and the T gate together constitute a *universal set* of quantum gates in the sense that any quantum circuit can be expressed in terms of those gates alone. That is not usually an important practical issue in building quantum computers, but it is useful theoretically to decide on what can be done with a quantum computer and what cannot.

7.6 **Why *i*?**

In this section, we show why it is necessary to use complex numbers in some state coefficients (amplitudes). In the language of linear algebra, quantum state space is a complex vector space because the coefficients in the linear combinations of vectors may be complex numbers. As a reminder, quantum state space, as we have already shown, includes an inner product of two vectors. The inner product can be a complex number. A complex inner product vector space is called a Hilbert space. This section is optional in the sense that it is not needed for the material in subsequent chapters. If you are willing to accept complex numbers in quantum simply as a requirement imposed by nature, feel free to skip this section.

To provide a concrete example, we explore the spin-1/2 states and the Y spin-1/2 operator. We have already shown in Eq. (7.9) that states for spin-up and spin-down along the

x-direction can written as superpositions of states relative to the z-direction. We repeat those relations here:

$$|\uparrow\rangle_x = \frac{1}{\sqrt{2}}(|\uparrow\rangle_z + |\downarrow\rangle_z)$$
$$|\downarrow\rangle_x = \frac{1}{\sqrt{2}}(|\uparrow\rangle_z - |\downarrow\rangle_z). \tag{7.39}$$

Perhaps all we need to do to generate the y basis states is to swap signs on the $|\downarrow\rangle_z$ components. Here are what the suggested states look like:

$$|\uparrow\rangle_y \overset{?}{\rightarrow} \frac{1}{\sqrt{2}}(|\uparrow\rangle_z - |\downarrow\rangle_z)$$
$$|\downarrow\rangle_y \overset{?}{\rightarrow} \frac{1}{\sqrt{2}}(|\uparrow\rangle_z + |\downarrow\rangle_z). \tag{7.40}$$

How do we check if this works? There are two mathematical requirements for basis states: (1) Each of the states must be normalized and (2) the two states must be orthogonal. Both of those requirements are satisfied by the suggested states.

Try It 7.15 Show that the states in Eq. (7.40) are normalized and orthogonal. Hints: For normalization, the sum of the squares of the amplitudes should be 1. For orthogonality, ${}_y\langle\uparrow|\downarrow\rangle_y$ should be 0. Use the row vector and column vector forms.

But, in addition to the mathematical requirements, a condition in physics should also be satisfied. Remember that if we prepare a spin-1/2 qubit in the state $|\uparrow\rangle_x$ but measure the spin-orientation along z, the probability of getting spin-up along z is 0.5 and the probability of getting spin-down along z is 0.5. The same result should be true if we prepare the qubit state to be $|\uparrow\rangle_y$ and then measure along x. Quantum mechanical democracy means that x, y, and z should all be treated equally. Let's see if that works with the proposed y states? There are several ways to do this, but it is relatively easy to use the row and column vectors for the amplitudes in the z basis. We gave the results before, but let's follow our current line of reasoning. Here are the proposed relationships:

$$|\uparrow\rangle_x \doteq \frac{1}{\sqrt{2}}\begin{pmatrix} 1 \\ 1 \end{pmatrix}_z \quad |\downarrow\rangle_x \doteq \frac{1}{\sqrt{2}}\begin{pmatrix} 1 \\ -1 \end{pmatrix}_z$$
$$|\uparrow\rangle_y \doteq ? \frac{1}{\sqrt{2}}\begin{pmatrix} 1 \\ -1 \end{pmatrix}_z \quad |\downarrow\rangle_y \doteq ? \frac{1}{\sqrt{2}}\begin{pmatrix} 1 \\ 1 \end{pmatrix}_z. \tag{7.41}$$

For a specific example, let's calculate the probability amplitude ${}_x\langle\downarrow|\downarrow\rangle_y$, which by our previous reasoning ought to have its square equal to 0.5. Here is what we find for the suggested states:

$$ {}_x\langle\downarrow|\downarrow\rangle_y \doteq \frac{1}{2}\begin{pmatrix} 1 & -1 \end{pmatrix}\begin{pmatrix} 1 \\ 1 \end{pmatrix} = \frac{1-1}{2} = 0. \tag{7.42} $$

Try It 7.16 Calculate the other amplitudes with x states and the suggested states. Are the results what you expect?

Yes, we have a problem. It turns out, though we won't go through the details, that no combination of purely real number amplitudes in the y states expressed as superpositions of the z basis states will satisfy all the requirements. So, what do we do? Perhaps there is a fundamental flaw in quantum mechanics, and we need an entirely new theory. That would be a mess (and we would not have written this book building up this formalism). We can "fix" the problem by allowing the amplitudes to be complex numbers. For the y spin-1/2 states written in terms of the z basis states, it turns out that only two of the amplitudes need to be complex. We will work through the details, making some reasonable assumptions to simplify the argument.

First, we write the y spin states with general amplitudes in terms of the z basis states:

$$ \begin{aligned} |\uparrow\rangle_y &= \frac{1}{\sqrt{2}}\left(a|\uparrow\rangle_z + b|\downarrow\rangle_z\right) \\ |\downarrow\rangle_y &= \frac{1}{\sqrt{2}}\left(c|\uparrow\rangle_z + d|\downarrow\rangle_z\right). \end{aligned} \tag{7.43} $$

Note that we included the ubiquitous $1/\sqrt{2}$ needed for normalization with two-state systems. You may think we are overworking $\sqrt{2}$, but it is resilient and doesn't think much about being used over and over again. It's not very rational in that respect. Ha, ha! Sorry about that. We couldn't resist a good physics dad joke.

Getting back to showing why we need complex numbers, we note that only b and d need to be complex numbers to satisfy all our requirements. Consequently, we can use $a = 1 = c$ in Eq. (7.43) to speed up our argument. Normalization requires that

$$ \frac{1}{2} + \frac{b^*b}{2} = 1 \quad \text{and} \quad \frac{1}{2} + \frac{d^*d}{2} = 1. \tag{7.44} $$

So, we need to make sure that $b^*b = 1 = d^*d$. We also require that the two y basis states be orthogonal:

$$ {}_y\langle\uparrow|\downarrow\rangle_y \Rightarrow \begin{pmatrix} 1 & b^* \end{pmatrix}\begin{pmatrix} 1 \\ d \end{pmatrix} = 1 + b^*d = 0. \tag{7.45} $$

This adds another requirement: $b^*d = -1$. Finally, let's look at the problematic amplitude

$$_x\langle\downarrow|\downarrow\rangle_y \Rightarrow \frac{1}{2}\begin{pmatrix}1 & -1\end{pmatrix}\begin{pmatrix}1\\ d\end{pmatrix} = \frac{1-d}{2}. \tag{7.46}$$

Since we want the resulting probability (absolute-value-squared of the amplitude) to be equal to 0.5, we require that

$$\frac{1}{4}(1-d^*)(1-d) = \frac{1}{2}. \tag{7.47}$$

It is easy to check that $d = \pm i$ satisfies Eq. (7.47). For either choice, we obtain

$$\frac{1}{4}(1-d^*)(1-d) = \frac{1}{4}(1-d^*-d+d^*d) = \frac{1}{4}(1+i-i+1) = \frac{1}{2}. \tag{7.48}$$

However, when we deploy $b^*d = -1$ (requirement from Eq. (7.45)), we find that we need to use $d = -i$ to get the standard quantum mechanics form of Y. In this case, the two amplitudes b and d have only imaginary parts; they are called "pure imaginary."

Try It 7.17 Challenge: Separate d into real and imaginary parts, $d = f + ig$, and use that form in Eq. (7.48) along with $d^*d = 1$ to prove that $d = \pm i$ is the correct solution with $f = 0$ and $g = 1$.

Let's put all of this together in both the state right-vector form and in the column vector form using the conventional choice $d = -i$:

$$\begin{aligned}|\uparrow\rangle_y &= \frac{1}{\sqrt{2}}(|\uparrow\rangle_z + i|\downarrow\rangle_z) \doteq \frac{1}{\sqrt{2}}\begin{pmatrix}1\\ i\end{pmatrix}_z \\ |\downarrow\rangle_y &= \frac{1}{\sqrt{2}}(|\uparrow\rangle_z - i|\downarrow\rangle_z) \doteq \frac{1}{\sqrt{2}}\begin{pmatrix}1\\ -i\end{pmatrix}_z.\end{aligned} \tag{7.49}$$

Using these new coefficients, we should check that we can get a sensible answer when we calculate the expression in Eq. (7.42):

$$_x\langle\downarrow|\downarrow\rangle_y \doteq \frac{1}{2}\begin{pmatrix}1 & -1\end{pmatrix}\begin{pmatrix}1\\ -i\end{pmatrix} = \frac{1+i}{2}. \tag{7.50}$$

The probability is given by the absolute-value-squared $(1/4)(1-i)(1+i) = 1/2$, which is what we expect.

Try It 7.18 Show that the other combinations of probabilities work out for spin-up and spin-down in the x and y states and in the z and y states.

Although we used a two-state spin system to demonstrate the need for complex amplitudes in quantum mechanics, the same results show up whenever you have three or more "incompatible" (sometimes called "complementary") properties. For example, we might have three two-state operators U, V, and W. Incompatible, as mentioned previously, means that the operators do not commute: $UV \neq VU$, $UW \neq WU$, and $VW \neq WV$. The corresponding pairs of basis states are $|u_0\rangle$, $|u_1\rangle$, $|v_0\rangle$, $|v_1\rangle$, and $|w_0\rangle$, $|w_1\rangle$. (The three states don't need to be associated with directions in the physical, three-dimensional world.) If measurements using one basis set give no information about measurements with respect to the other two basis sets, then the properties (and the corresponding operators) are said to be maximally incompatible.

If the properties are maximally incompatible, as described in the previous paragraph, at least some of the state amplitudes need to be complex. Sometimes the term "noncommuting properties" is used.

For a spin-1/2 system, the commutation relations of the spin angular momentum operators are

$$[S_x, S_y] = i\hbar S_z \quad [S_z, S_x] = i\hbar S_y \quad [S_y, S_z] = i\hbar S_x. \tag{7.51}$$

Try It 7.19 Use the matrix form of the spin angular momentum operators to show that they satisfy Eq. (7.51).

7.7 The Quantum Uncertainty Relation

We know that many measurements of quantum system properties lead to probability distributions of results. In fact, if a quantum system has two (or more) measurable properties that are "incompatible" (as mentioned in the previous section), the system cannot be described by a state in which all the properties are precisely known. It means that the properties cannot all have distributions of results that have standard deviations equal to zero. The standard deviation is the most commonly used measure of the breadth of a probability distribution. In this section, we derive a famous and profound relationship between those standard deviations: the Heisenberg uncertainty relation.

Recall from elementary probability and statistics that the standard deviation of the distribution of values of property A is defined as the square root of the variance from the mean (average) of that property. We follow the physics convention and denote the standard deviation of property A by ΔA, so the definition of the standard deviation is

$$\Delta A = \sqrt{\operatorname{Var} A} = \sqrt{\left\langle (A - \langle A \rangle)^2 \right\rangle}, \tag{7.52}$$

where Var A means the variance of A. The last term is the square root of the average squared difference between a value and the mean. In short, the standard deviation is just a number

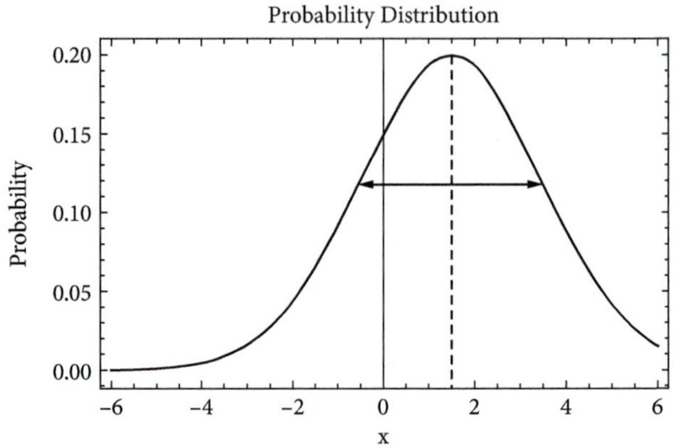

Fig. 7.9 A graph of a probability distribution. The mean of the distribution is 1.5 (vertical dashed line). The horizontal arrow has a length of about 2 standard deviations for the distribution. The standard deviation for this distribution is 2.

that characterizes the spread of the probability distribution (see Fig. 7.9). Most of the analysis that follows is parallel to what you learned in statistics, but we point out where things are different.

Try It 7.20 The standard deviation is sometimes written as $\Delta A = \sqrt{\langle A^2 \rangle - \langle A \rangle^2}$. Show that this form is equivalent to Eq. (7.52).

We will now show that if a quantum system has two operators that do not commute, that is, $AB\,|\psi\rangle \neq BA\,|\psi\rangle$, the product of the standard deviations of the two operators has a lower bound that depends on the average value of the commutator $[A, B]$ of the two operators where $[A, B] = AB - BA$ and on the quantum state of the system represented by $|\psi\rangle$. The result is known as the Heisenberg uncertainty relation, named after Werner Heisenberg, one of the pioneers of quantum mechanics in the 1920s and 1930s. The uncertainty relation connects the product of the standard deviations for the two operators to the mean value of the commutator of those operators:

$$(\Delta A)_\psi (\Delta B)_\psi \geq \frac{1}{2} \langle C \rangle_\psi = \frac{1}{2} \left| \langle [A, B] \rangle_\psi \right|, \qquad (7.53)$$

where we have added subscripts to remind us that the standard deviation and the average value of the commutator depend on the quantum state ψ. Some authors use the term uncertainty principle, but we argue that it is best thought of as a derived result and not a fundamental principle. In fact, we find that Eq. (7.53) is a straightforward derivation in linear algebra.

The important takehome message is that if two operators do not commute, it is not possible to have a quantum state in which both the standard deviations are zero. Having zero standard deviations is needed to have zero spread in the measured values of those properties.

Let's examine Eq. (7.53) in detail. First, because of the greater-than symbol, we see that the product of the standard deviations may be larger than the lower bound, depending on the state of the system. Second, if the average value of the commutator is not zero, neither of the standard deviations can be zero. However, one of them may be larger than the other. Third, if the lower bound on the product of standard deviations is zero, then it is possible (but not necessary) that one or both standard deviations are zero. In that case, we say that system is described by a precise value of the property represented by the associated operator.

Math Aside If we want to make a fair comparison between the standard deviations, we should make sure the two operators have the same units. For most of our analyses, we won't need to do that.

Derivation of the Uncertainty Relation

We shall now show that the uncertainty relationship is related to fundamental linear algebra features of the vector space representations of quantum states. First, let's look at what is called the *triangle inequality* of vectors in a vector space. Later, we bring in the operators A and B. We use a two-dimensional example, but the results apply to any number of dimensions. The *triangle inequality* is a relationship among the lengths ("norms") of three vectors, say, $\vec{X}$, $\vec{Y}$, and $\vec{Z}$ as illustrated in Fig. 7.10.

The norms are defined by $\left\|\vec{X}\right\| \equiv \left(X_1^2 + X_2^2 + \cdots X_n^2\right)^{1/2}$, the square root of the sum of the squares of the vector's components, where n is the dimensionality of the vector space. This is just a generalization of the standard way of calculating the magnitude of a vector in physics or engineering.

Math Aside Strictly speaking, the norm defined above is called the Euclidean norm or the ℓ^2 norm. It is possible (and sometimes useful) to use other definitions of the norm. For example, the ℓ^4 norm sums the fourth powers of the vector components and then takes the fourth root of that result. For our purposes, we want to use the Euclidean norm.

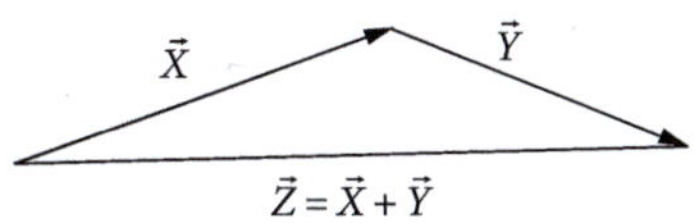

Fig. 7.10 Three vectors in a two-dimensional vector space.

In terms of the norms, the *triangle inequality* reads

$$\left\|\vec{X}\right\| + \left\|\vec{Y}\right\| \geq \left\|\vec{Z}\right\| = \left\|\vec{X} + \vec{Y}\right\|. \tag{7.54}$$

Try It 7.21 Explain qualitatively why the *triangle inequality* works. Hint: Consider some limiting (extreme) cases.

We now use the square of Eq. (7.54) to write

$$\left[\left\|\vec{X}\right\| + \left\|\vec{Y}\right\|\right]^2 \geq \left\|\vec{X} + \vec{Y}\right\|^2 = \left\|\vec{X}\right\|^2 + \left\|\vec{Y}\right\|^2 + \left|\langle\vec{X}|\vec{Y}\rangle + \langle\vec{Y}|\vec{X}\rangle\right|, \tag{7.55}$$

where, anticipating replacing the vectors $\vec{X}$ and $\vec{Y}$ with quantum state vectors, we used Dirac notation to express the scalar product of the two vectors on the right side of the equation. After expanding the left side of Eq. (7.55) and canceling identical terms on each side of the inequality, we obtain

$$2\left\|\vec{X}\right\| \left\|\vec{Y}\right\| \geq \left|\langle\vec{X}|\vec{Y}\rangle + \langle\vec{Y}|\vec{X}\rangle\right|. \tag{7.56}$$

Eq. (7.56) (or its equivalent) is known as the Cauchy–Schwarz inequality in linear algebra. We have used the form that is appropriate for a complex vector space in which the inner products may be complex numbers.

Note that the sum of the scalar products is surrounded by absolute value signs (single vertical lines) because the scalar products are just numbers. We could have used the double-vertical lines because a scalar can be viewed as a one-dimensional vector (only one component) for which the norm is the same as the absolute value of that component.

So far, our analysis has focused on vectors. You should have noted that the uncertainty relation involves average values of operators. We bridge that "divide" by using two operators A and B to define the vectors $\vec{X} \rightarrow |X\rangle \equiv A|\psi\rangle$ and $\vec{Y} \rightarrow |Y\rangle \equiv iB|\psi\rangle$, where $|\psi\rangle$ is the quantum state vector for the system. The imaginary i is needed to express the results in terms of the commutator of the two operators (see Eq. (7.61)). Since both A and B are supposed to represent observable properties, we assume that they are Hermitian operators. With these definitions and assumptions, we have

$$\begin{aligned} \|X\|^2 &= \langle X|X\rangle = |\langle\psi|A^{\dagger}A|\psi\rangle| = |\langle\psi|AA|\psi\rangle| = |\langle\psi|A^2|\psi\rangle| \\ \|Y\|^2 &= \langle Y|Y\rangle = |\langle\psi|B^2|\psi\rangle| \;. \end{aligned} \tag{7.57}$$

For the sake of simplicity, we assume that the average values for the two operators are zero:

$$\langle A\rangle_{\psi} = 0 \qquad \langle B\rangle_{\psi} = 0. \tag{7.58}$$

If the averages are not equal to zero, we may define new operators that lead to the same uncertainty relation (see Try It 7.22). When the averages are zero, Eq. (7.57) yields

$$\begin{aligned} \|X\| &= \left(\langle\psi|\, A^2 \,|\psi\rangle\right)^{1/2} = (\Delta A)_\psi \\ \|Y\| &= \left(\langle\psi|\, B^2 \,|\psi\rangle\right)^{1/2} = (\Delta B)_\psi. \end{aligned} \tag{7.59}$$

To complete the derivation, we now evaluate the scalar (inner products) on the right side of Eq. (7.56):

$$\langle X|Y\rangle = i\langle\psi|AB\,|\psi\rangle \qquad \langle Y|X\rangle = -i\langle\psi|BA\,|\psi\rangle, \tag{7.60}$$

which gives

$$\langle X|Y\rangle + \langle Y|X\rangle = i\langle\psi|AB - BA\,|\psi\rangle. \tag{7.61}$$

Using Eqs. (7.59) and (7.61) in Eq. (7.56) yields

$$(\Delta A)_\psi(\Delta B)_\psi \geq \frac{1}{2}\left|\langle[A,B]\rangle_\psi\right|, \tag{7.62}$$

which is just what we needed—the Heisenberg uncertainty relation in Eq. (7.53). End of proof.

Try It 7.22 If the average values of the operators in the state $|\psi\rangle$ are not zero, we may define new operators $\bar{A} = A - \langle A\rangle_\psi \mathrm{I}$ (as usual I is the identity operator) with an analogous definition of $\bar{B}$. Show that $(\Delta A)^2 = (\Delta\bar{A})^2$ and $(\Delta B)^2 = (\Delta\bar{B})^2$ and that $[\bar{A},\bar{B}] = [A,B]$. Argue that the uncertainty relations for the two sets of operators are the same.

Try It 7.23 Let's use a spin-1/2 system as an example. Suppose that the two operators whose uncertainty relation we need to explore are S_x and S_y and that the state is $|\psi\rangle = |\uparrow\rangle_z$, the spin-up state along the z-axis. Use the column and row vector forms of the state and the matrix representation of the operators to evaluate both sides of Eq. (7.62). Compare your results with Fig. 7.4. Hints: (a) From earlier in this chapter, we know that $\langle S_x\rangle = 0 = \langle S_y\rangle$ for this state. (b) The commutator of the two operators is $[S_x, S_y] = i\hbar S_z$.

Try It 7.24 Repeat the previous Try It but now with the two operators S_x and S_z. The commutator of the two operators is $[S_z, S_x] = i\hbar S_y$. (a) Show that for the spin-up state, we obtain $\Delta S_z = 0$. (b) Show that the right side of Eq. (7.62) for this choice of operators is equal to zero.

The result of Try It 7.24 should be surprising because it says that in this case the lower bound for the product of the two uncertainties is 0 even though the two operators do not commute. Note that the standard deviation for S_z is zero, but the standard deviation for S_x is not zero. So, we don't have both standard deviations equal to zero, and the uncertainty relation conclusion about precisely determined properties is still true.

The crucial conclusion is that the quantities in the uncertainty relation depend on the state of the system for which they are evaluated. You cannot automatically assume that the lower bound for the product of the standard deviations is not zero even when the operators do not commute. In this example, the system's quantum state is assumed to be an eigenstate of one of the operators and that led to the surprising result.

Try It 7.25 To convince yourself that the uncertainty relation results depend on the state, repeat the previous Try It with $|\psi\rangle = |\uparrow\rangle_y$, that is, the state is spin-up along the y-direction.

Classical Fourier Transform Analog of the Quantum Uncertainty Relation

While the quantum uncertainty relation may seem strange, it is important to know that classical wave physics has similar relationships between pairs of "complementary" wave properties. These relationships emerge from what is called Fourier analysis, the quantum version of which is presented in Chapter 15. The classical version is an implementation of Fourier's theorem: A time-dependent function $A(t)$ can be decomposed into a Fourier transform function $\tilde{A}(f)$, which is a function of the wave frequency f, which is the number of wave cycles per unit time. The decomposition is evaluated by the Fourier integral

$$\tilde{A}(f) = \frac{1}{\sqrt{2\pi}} \int_{-\infty}^{\infty} A(t) e^{-i2\pi f t} dt. \tag{7.63}$$

We may think of $\tilde{A}(f)$ as a measure of the overlap between the wave $A(t)$ and the basis function $e^{-i2\pi ft}$, which we recognize as a combination (Euler's formula) of a cosine and a sine function. $\tilde{A}(f)$ can also be thought of as a frequency distribution akin to the probability distribution in Fig. 7.5, and we may define a spread in frequencies Δf in analogy to the standard deviation of the probability distribution.

We may also run the calculation the other way around. Given the Fourier transform function, we may integrate it to find the time dependence of the wave function.

$$A(t) = \frac{1}{\sqrt{2\pi}} \int_{-\infty}^{\infty} \tilde{A}(f) e^{+i2\pi f t} df. \tag{7.64}$$

Let's now apply these ideas to a wave "pulse" or burst, which lasts for only a short period of time. A wave that is nonzero only for some range of times (Δt) is called a wave packet (see Fig. 7.11). The Fourier transform result is a relationship between Δf and Δt:

$$(\Delta f)\,(\Delta t) \geq n_{ft}, \tag{7.65}$$

where the numerical value of n_{ft} depends on the exact definitions of the two "uncertainties" Δf and Δt. The important point is that the lower limit is not zero. In other words, a longer duration wave packet allows for a smaller spread of frequencies.

Eq. (7.64) shows that we may think of the wave packet as a weighted sum of sines and cosines (embedded in the complex exponential function) with the weight given by the Fourier transform amplitudes $\tilde{A}(f)$.

We can make a similar argument about how the wave function oscillates as a function of position, which is important in the so-called spatial Fourier transforms, often used in image analysis as well as in quantum physics.

Figure 7.12 shows a freeze-frame image of a wave in which the wave disturbance $A(x)$ is plotted as a function of position, in this case, along an x-axis. We can perform a spatial Fourier transform in analogy with Eq. (7.63) resulting in a function $\tilde{y}(k)$ as a function of k, the number of wave cycles per unit distance. Consequently, k is called the "wave number"

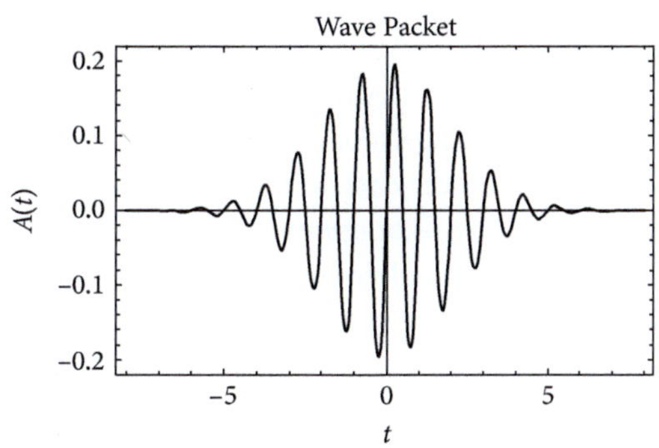

Fig. 7.11 A plot of a wave packet as a function of time.

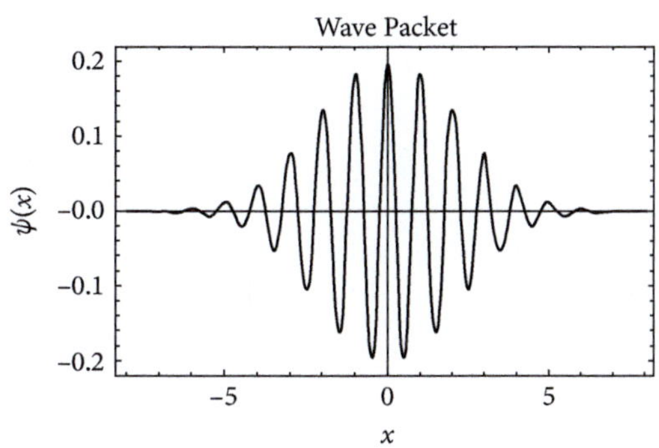

Fig. 7.12 A plot of a wave packet versus position (think of a flash photo or a freeze-frame from a video recording).

associated with the wave. Also, we may define a spread in k values Δk. Fourier analysis gives us a relationship between Δk and Δx, the spatial spread of the wave packet along the x-axis:

$$(\Delta k)(\Delta x) \geq n_{kx}, \tag{7.66}$$

where again the numerical value depends on the details of the definitions of the uncertainties. This spatial uncertainty analogy shows up when we talk about quantum state functions that depend on position.

Minimum Uncertainty States and Squeezed States

As shown above, the uncertainty principle (Eq. (7.62)) is expressed as an inequality. In general, different quantum states will have different products of the standard deviations for a given pair of operators. The quantum state (or states) that minimizes the product is called a *minimum uncertainty state* $|\psi_{\text{min}}\rangle$. For that state, Eq. (7.62) becomes

$$(\Delta A)_{\psi_{\text{min}}}(\Delta B)_{\psi_{\text{min}}} = \frac{1}{2}\left|\langle[A,B]\rangle_{\psi_{\text{min}}}\right|. \tag{7.67}$$

For the minimum uncertainty state, the product of the standard deviations is as small as possible.

In many cases, the two standard deviations (once adjustments in units have been implemented) are the same. But for some kinds of measurements, it is useful to have one of the standard deviations be smaller than the case when, in some sense, the standard deviations are equally distributed between the two operators. For example, we might want to measure one of the properties as precisely as possible (that means, the standard deviation should be relatively small), and we are willing to accommodate a much larger standard deviation of the other property. In such cases, we say that we have a *squeezed state*: We have squeezed down the standard deviation for one property while allowing the other standard deviation to increase. The product of the standard deviations is still described by the uncertainty relation. Such squeezed states are becoming widely used in quantum information systems and in precision measurement of gravitational waves, for example. Figure 7.13 shows a pictorial representation of a squeezed state's standard deviations.

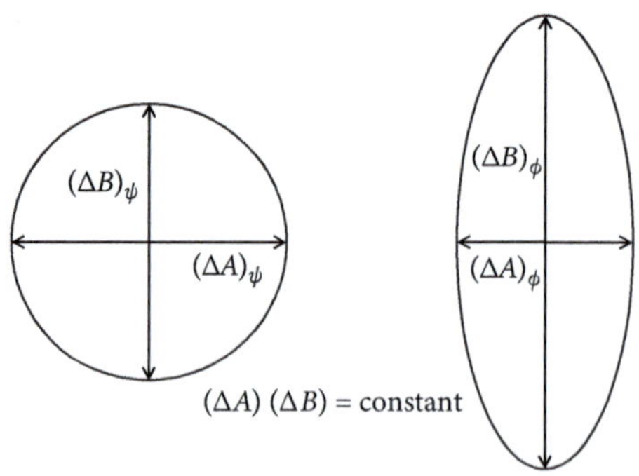

Fig. 7.13 On the left, the standard deviations of the two properties A and B are equal for the state $|\psi\rangle$. On the right, for the state $|\phi\rangle$ the standard deviation for A has been squeezed with a concomitant increase in the standard deviation for B.

The physical operations needed to produce a squeezed state depend on the physics of the qubits. Several methods have been developed for photon qubits and for spin qubits (see Further Reading).

7.8 Chapter Summary

- Spin states are a paradigmatic example of qubits with the two basis states spin-up and spin-down. They offer a test bed to practice and review the fundamentals we have covered so far.
- Spin-1/2 states provide an example of the difference between physical states and states in state space. A particle flipping from a spin-up to a spin-down has, in a sense, changed its physical spin orientation by 180° (but see the caveat about spin-up and spin-down at the end of Section 7.3). In the quantum state space, however, the change is only 90° because the abstract state-space basis vectors are orthogonal.
- Spin states can be operated on by quantum gates. For example, the NOT gate changes the spin state to its orthogonal partner. The Hadamard gate acting on a basis spin state creates a state of superposition of both spin states.
- There are many more candidates for qubits other than spin and polarization, but they all follow the same principles that we have laid out in Chapters 4–7.
- The Bloch sphere provides a way of visualizing the effects of the complex phase difference for amplitudes in a spin-1/2 qubit system.
- We showed that when a quantum system has three or more noncommuting operators, it is necessary to use complex amplitudes in building superposition states and the matrices associated with some operators will have complex number entries (matrix elements). In short, complex numbers are an essential part of quantum mechanics.
- The quantum uncertainty relation describes limitations on how precisely properties of a quantum system can be specified for a given quantum state. "Specified precisely" means that the standard deviation of the distribution of measurement results is zero.

7.9 Further Reading

For more information on spin in quantum mechanics, we recommend

Michael G. Raymer, *Quantum Physics: What Everyone Needs to Know* (Oxford University Press, Oxford, 2017). This book also has a fine discussion of polarization states of photons.

Daniel F. Styer, *The Strange World of Quantum Mechanics* (Cambridge University Press, Cambridge, 2000).

For other treatments of the need for complex coefficients in quantum mechanics, see

R. Karam, "Why are complex numbers needed in quantum mechanics," *American Journal of Physics* **88**(1), 39–45 (2020).

Jarrett L. Lancaster and Nicholas M. Palladino, "Testing the necessity of complex numbers in traditional quantum theory with quantum computers," *American Journal of Physics* **93**, 110–120 (2025).

More on the uncertainty relation:

Berthold-Georg Englert, Marlan O. Scully, and Herbert Walther, "The duality in matter and light," *Scientific American* **271**(6), 86–92 (1994). This paper provides a good explanation of the uncertainty relation and its manifestation in the quantum mechanics of particles and waves.

For more information on squeezed states:

Christopher C. Gerry and Peter L. Knight, *Introductory Quantum Optics* (Cambridge University Press, Cambridge, 2005). Chapter 7 provides a clear and illuminating discussion of squeezed states.

Your Basis, My Basis

8

Mathematics is the art of giving the same name to different things.

Jules Henri Poincaré, *Science et Méthode* (1908)

8.1 Taking Stock

Let's take a moment to review what we have done in the previous chapters. It is important to keep in mind that most of what we have been talking about with quantum computing and quantum information can all be formulated in terms of the computational (standard) basis states, which, in some sense, are independent of the physical nature of the qubits. They could be spin-1/2 particles, such as electrons, protons, or appropriate atoms; they could be photons, or they could be circulating superconducting currents or more. From now on, we will stick pretty much to the basis state language and only bring in the physical details like spin-1/2 particles or photons when we need explicit examples.

In Chapter 4, we showed how to represent an abstract qubit state vector in terms of a superposition of two basis vectors $|A_o\rangle$ and $|A_1\rangle$:

$$|S\rangle = a_0 |A_0\rangle + a_1 |A_1\rangle . \tag{8.1}$$

This state can also be expressed in the column vector form

$$\begin{pmatrix} a_0 \\ a_1 \end{pmatrix}_A = a_0 \begin{pmatrix} 1 \\ 0 \end{pmatrix}_A + a_1 \begin{pmatrix} 0 \\ 1 \end{pmatrix}_A . \tag{8.2}$$

Note that we have put subscripts on the column vectors to remind us of the basis we are using. A is the general basis name, while A_0 and A_1 label the two basis states. The state coefficients a_0 and a_1 are just numbers, and we noted that $|a_0|^2$ is the probability of getting the state $|A_0\rangle$ as a measurement result and $|a_1|^2$ is the probability of getting the state $|A_1\rangle$ when we make a measurement associated with the A basis.

We also noted that we can choose different basis vectors (states) for a given type of qubit. The different basis vectors correspond to different configurations of preparation and measurement devices—for example, the orientation of a polarizing beam splitter for

Quantum Computing and Quantum Physics. Alice Flarend and Robert Hilborn, Oxford University Press.
 DOI: 10.1093/oso/9780197904930.003.0008

linear polarized light. Using different basis states to analyze qubits turns out to be a critical component in quantum information processing and quantum computing.

In this chapter, we show you how to set up a relatively simple way of translating from the representation of the qubit's state in one set of basis states to the representation in another. We slow down the pace and provide more detail in this chapter than we have done previously because the results are so important to quantum computing (QC) and quantum information science (QIS).

We work through several examples that we use in later chapters. We also find that using different basis sets to observe qubits prepared in the same quantum state allows us to learn more about the quantum state. For QIS, we also learn that using two (or more) basis sets gives us a method to find out if someone is eavesdropping on Alice's communications with Bob. One of the promising uses of QIS is developing more secure communications processes.

8.2 Changing Basis States

Before we get into the QIS issues, let's tackle the fundamental issue of representing quantum states using different sets of basis vectors. In fact, a superposition state in one basis set will be a single basis state in some other appropriately chosen basis set. More importantly, as we have mentioned, we get more information from a quantum state if we make measurements on it relative to several basis sets. For example, we might start with the description of the polarization state for photons using horizontal and vertical polarization basis states but want to make measurements in terms of +45° and −45° polarization orientations. So, how do we express a qubit state in terms of these new basis states?

Let's warm up by considering ordinary two-dimensional vectors, like force vectors in physics. The actual physical interpretation is not important, but it may be helpful to have a concrete example in front of us. If you are familiar with dot products and changing unit vectors for describing ordinary vectors, you should feel free to jump ahead to Section 8.3.

Let's look at Fig. 8.1, where we can see the x and y components of the vector $\vec{F}$. We also used some of this in Chapter 4, but a review won't hurt. First, the vector $\vec{F}$ can be thought of as a weighted superposition of unit vectors pointing along the x- and y-axes:

$$\vec{F} = f_x\,\hat{x} + f_y\,\hat{y}. \tag{8.3}$$

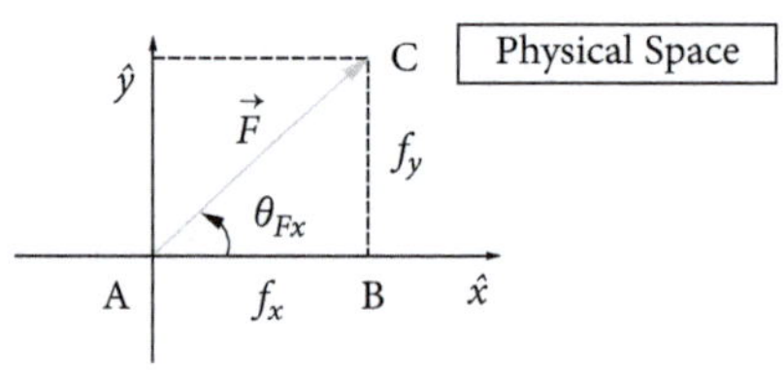

Fig. 8.1 A vector $\vec{F}$ expressed with components f_x and f_y along the $\hat{x}$ and $\hat{y}$ directions, respectively, where we used the common "^" (caret) notation to indicate a vector of unit length. ABC is a right triangle.

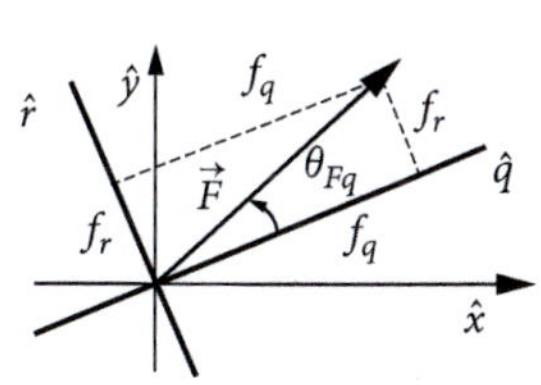

Fig. 8.2 The vector $\vec{F}$ and its components relative to the qr set of axes.

Those coefficients are the weights associated with how much of the vector $\vec{F}$ is along each unit vector direction. How do you find those coefficients? You could project the vector $\vec{F}$ along each unit vector direction by drawing a line perpendicular to the unit vector from the tip of $\vec{F}$ to the appropriate axis. You then use a ruler to measure f_x and f_y. Of course, the accuracy of your results depends on how carefully you construct and measure those lines. In many cases, we would like to use a formula to find f_x and f_y. Not only does that allow us to postpone putting numbers into the equations, but it also allows us to use math to derive interesting relationships among the results.

Thinking back to the definitions of cosines and sines and looking at the right triangle ABC in Fig. 8.1, you may recall that the coefficients f_x and f_y can be written as

$$f_x = \left\|\vec{F}\right\| \cos\theta_{Fx} \qquad f_y = \left\|\vec{F}\right\| \sin\theta_{Fx}.$$

Note if the length of the vector $\left\|\vec{F}\right\| = 1$, which is the choice used in QC and QIS as discussed in Chapter 4, then the x component of the vector is just $\cos\theta_{Fx}$ and the y component is $\sin\theta_{Fx}$. By the Pythagorean theorem, we have

$$f_x^2 + f_y^2 = \left\|\vec{F}\right\|^2. \tag{8.4}$$

Now let's figure out how this works out if we use a different set of unit vectors $\hat{q}$ and $\hat{r}$ as shown in Fig. 8.2. We need to focus on the new set of axes, $\hat{q}$ and $\hat{r}$, and ignore the original $\hat{x}$ and $\hat{y}$ directions. We find the representation of our vector $\vec{F}$ and the coefficients in the qr basis in the same way we found x and y components of $\vec{F}$, but now $\vec{F}$ lies at an angle θ_{Fq} from the q-axis:

$$\vec{F} = f_q\hat{q} + f_r\hat{r}, \tag{8.5}$$

so we can immediately write

$$f_q = \left\|\vec{F}\right\| \cos\theta_{Fq} \qquad f_r = \left\|\vec{F}\right\| \sin\theta_{Fq}. \tag{8.6}$$

This is where math notation becomes a bit slippery; the right sides of Eqs. (8.3) and (8.5) look different, but they are just two different ways of representing the same vector $\vec{F}$.

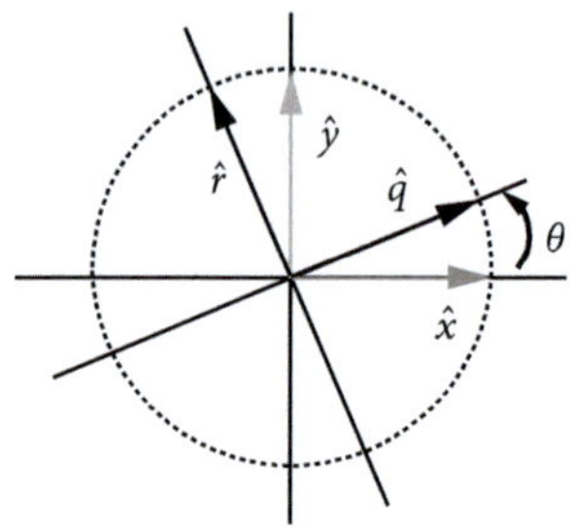

Fig. 8.3 This plot shows the two sets of unit vectors: (1) $\hat{x}$ and $\hat{y}$ and (2) $\hat{q}$ and $\hat{r}$. The dotted circle has a radius = 1. θ is the angle between $\hat{x}$ and $\hat{q}$.

In going from one set of basis axes to the other, $\vec{F}$ doesn't change—only the components along the axes change. To drive home that point, we can write

$$\vec{F} = f_q\hat{q} + f_r\hat{r} = f_x\hat{x} + f_y\hat{y}. \quad (8.7)$$

The problem we want to tackle is the following: given the coefficients f_x and f_y in the (x, y) basis, what are the coefficients f_q and f_r in the (q, r) basis.

We begin by looking at how the unit vectors in the two bases are related to each other, ignoring the vector $\vec{F}$ for the moment. Once we have that information, it is relatively easy to find the relationships among $\vec{F}$'s components.

In Fig 8.3, we show the two sets of unit basis vectors and the angle θ between them. The dashed circle has a radius = 1 and reminds us that each of the basis vectors has a length 1. Note that the qr set of axes is rotated relative to the xy-axes by the angle θ. We use the convention preferred by physicists: The angle θ is taken to be positive if the rotation is counterclockwise. Figure 8.3 shows the geometry that allows us to write $\hat{q}$ and $\hat{r}$ in terms of $\hat{x}$ and $\hat{y}$:

$$\begin{aligned} \hat{q} &= \cos\theta\,\hat{x} + \sin\theta\,\hat{y} \\ \hat{r} &= -\sin\theta\,\hat{x} + \cos\theta\,\hat{y} \end{aligned} \quad (8.8)$$

or the other way around

$$\begin{aligned} \hat{x} &= \cos\theta\,\hat{q} - \sin\theta\,\hat{r}. \\ \hat{y} &= \sin\theta\,\hat{q} + \cos\theta\,\hat{r} \end{aligned} \quad (8.9)$$

Try It 8.1 Check that Eqs. (8.8) and (8.9) make sense if $\theta = 0°$ and $\theta = 90°$ and check that the minus signs are in the correct positions. Remember that a $0°$ rotation is no rotation at all and that a $90°$ rotation means that the horizontal and vertical axes are interchanged.

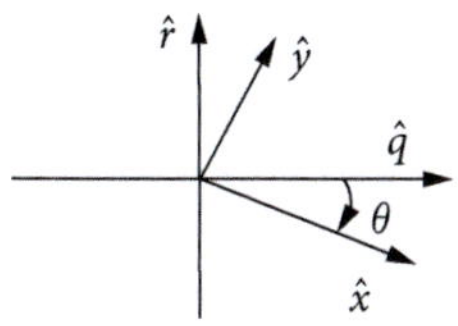

Fig. 8.4 The *xy*-axes have been rotated clockwise by the angle θ relative to the *qr*-axes.

Eq. (8.9) can be derived from Eq. (8.8) with a bit of algebra and trigonometry. Alternatively, we can be clever and think of the *xy*-axes as rotated *clockwise* by the angle θ relative to the *qr*-axes as shown in Fig. 8.4. Given our convention, the clockwise angle is counted as negative. Remembering that $\sin(-\theta) = -\sin\theta$ and that $\cos(-\theta) = \cos\theta$ is all we need to go from Eq. (8.8) to Eq. (8.9) without doing any algebra. We need to be aware that some authors use the opposite convention (clockwise angles are positive), which means that they have minus signs in different places than we do.

We now replace the cosines and sines by the dot (scalar) products among the basis unit vectors. From Fig. 8.3 and remembering that the unit vectors all have length 1, we find

$$\hat{q}\cdot\hat{x} = \cos\theta \quad \hat{q}\cdot\hat{y} = \sin\theta \quad \hat{r}\cdot\hat{x} = -\sin\theta \quad \hat{r}\cdot\hat{y} = \cos\theta. \tag{8.10}$$

Using the results in Eq. (8.10), we may rewrite Eqs. (8.8) and (8.9) as

$$\begin{aligned}\hat{q} &= \hat{x}(\hat{x}\cdot\hat{q}) + \hat{y}(\hat{y}\cdot\hat{q})\\ \hat{r} &= \hat{x}(\hat{x}\cdot\hat{r}) + \hat{y}(\hat{y}\cdot\hat{r})\end{aligned} \tag{8.11}$$

and

$$\begin{aligned}\hat{x} &= \hat{q}(\hat{q}\cdot\hat{x}) + \hat{r}(\hat{r}\cdot\hat{x})\\ \hat{y} &= \hat{q}(\hat{q}\cdot\hat{y}) + \hat{r}(\hat{r}\cdot\hat{y})\end{aligned} \tag{8.12}$$

Eqs. (8.11) and (8.12) are easy to remember once you recognize the pattern of unit vectors.

From Eqs. (8.11) and (8.12), we can easily find the relationship among the vector components in Eq. (8.7) by recalling that $f_x = \vec{F}\cdot\hat{x}$ with similar relationships for the other components. So, we take the dot product of $\vec{F}$ with each of the equations in (8.11) and (8.12) to obtain

$$\begin{aligned}f_q &= \vec{F}\cdot\hat{q} = (\vec{F}\cdot\hat{x})(\hat{x}\cdot\hat{q}) + (\vec{F}\cdot\hat{y})(\hat{y}\cdot\hat{q})\\ f_r &= \vec{F}\cdot\hat{r} = (\vec{F}\cdot\hat{x})(\hat{x}\cdot\hat{r}) + (\vec{F}\cdot\hat{y})(\hat{y}\cdot\hat{r})\end{aligned} \tag{8.13}$$

$$\begin{aligned}f_x &= \vec{F}\cdot\hat{x} = (\vec{F}\cdot\hat{q})(\hat{q}\cdot\hat{x}) + (\vec{F}\cdot\hat{r})(\hat{r}\cdot\hat{x})\\ f_y &= \vec{F}\cdot\hat{y} = (\vec{F}\cdot\hat{q})(\hat{q}\cdot\hat{y}) + (\vec{F}\cdot\hat{r})(\hat{r}\cdot\hat{y})\end{aligned} \tag{8.14}$$

We note that Eqs. (8.13) and (8.14) translate into their quantum cousins simply by replacing the dot products with Dirac brackets. We provide the details in Section 8.3.

There are a lot of variations in notation throughout science and math. You just need to get used to that and feel comfortable adjusting the notation to match what you like the best. It's a little like spelling the same word different ways, such as grey versus gray. Those two words look different but mean the same thing.

8.3 Changing Quantum Basis States

Now that we have ordinary vectors under control, we're ready to apply the same ideas to quantum basis state vectors in state space. You might be wondering why we need to change basis states if any set of basis states can be used to represent the quantum state. Indeed, all basis sets are equally fine. However, as discussed later, making measurements based on projections along different state-space basis sets gives us important information about a quantum state that we can't get from measurements with just one basis set. Almost all interesting QIS and QC methods make use of multiple basis sets. An example of this is changing the orientation of the polarizer for the light experiments as shown in Chapter 5.

To make the connection with classical vectors and their unit vectors obvious, let's use similar symbols to get started. Suppose that $|F\rangle$ is a quantum state vector and the unit vectors $|x\rangle$ and $|y\rangle$ are basis state vectors in a state space (Fig. 8.5).

For now, we restrict ourselves to basis state changes that can be expressed in terms of a single angle θ. Later, we show you what to do in more general cases.

Suppose the qr-axes are rotated relative to the xy-axes as shown in Fig. 8.6. What is the state-space equivalent of Eq. (8.8) which relates the two sets of basis vectors? It's just what you might expect by comparing Figs. 8.3 and 8.6:

$$\begin{aligned} |q\rangle &= \cos\theta\,|x\rangle + \sin\theta\,|y\rangle \\ |r\rangle &= -\sin\theta\,|x\rangle + \cos\theta\,|y\rangle, \end{aligned} \tag{8.15}$$

where the angle θ is the state-space angle between the corresponding state-space basis vectors. Looking at this from a quantum or linear algebra perspective, we are writing $|q\rangle$ and $|r\rangle$ as superpositions (linear combinations) of the states $|x\rangle$ and $|y\rangle$. We can easily find the "reverse" relationships:

$$\begin{aligned} |x\rangle &= \cos\theta\,|q\rangle - \sin\theta\,|r\rangle \\ |y\rangle &= \sin\theta\,|q\rangle + \cos\theta\,|r\rangle. \end{aligned} \tag{8.16}$$

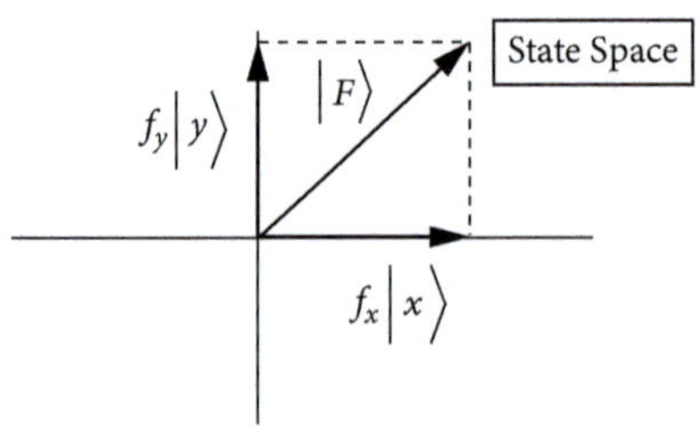

Fig. 8.5 A state-space diagram with basis states $|x\rangle$ and $|y\rangle$ and a state vector $|F\rangle$.

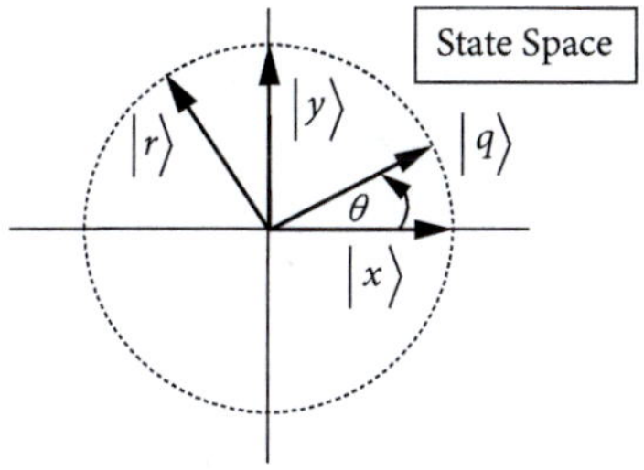

Fig. 8.6 A state-space diagram with two sets of basis vectors. The dashed circle has a radius = 1.

In writing Eq. (8.15), we looked at Eq. (8.8) and then used what we whimsically call the *fundamental theorem of mathematics*: If a result holds for symbols x, y, and z, then changing the symbols to a, b, and c still gives a valid result. Of course, the symbols must represent the same type of mathematical object: If x is a vector, then so must be a. But as long as you respect that rule, you have a very powerful theorem.

To complete the analogy with classical vectors, let's replace the cosines and sines in Eqs. (8.15) and (8.16) with the appropriate Dirac brackets. For example, $\cos\theta = \langle x|q\rangle$. With these replacements, we obtain

$$\begin{aligned} |q\rangle &= |x\rangle \underbrace{\langle x|q\rangle}_{\cos\theta} + |y\rangle \underbrace{\langle y|q\rangle}_{\sin\theta} \\ |r\rangle &= |x\rangle\langle x|r\rangle + |y\rangle\langle y|r\rangle \end{aligned} \tag{8.17}$$

$$\begin{aligned} |x\rangle &= |q\rangle\,\langle q|x\rangle + |r\rangle\,\langle r|x\rangle \\ |y\rangle &= |q\rangle\,\langle q|y\rangle + |r\rangle\,\langle r|y\rangle. \end{aligned} \tag{8.18}$$

Note that we have written Eqs. (8.17) and (8.18) in such a way that it is easy to remember various terms. You should explain the pattern to yourself or a friend.

Our final step is to write the equivalent of Eqs. (8.13) and (8.14) for a quantum state $|\psi\rangle$. The "obvious" strategy is to left multiply each of the equations in (8.13) and (8.14) with the left vector $\langle\psi|$. It gives us relationships among the projections (the Dirac brackets) along various basis state vectors:

$$\begin{aligned} \langle\psi|q\rangle &= \langle\psi|x\rangle\,\langle x|\psi\rangle + \langle\psi|y\rangle\,\langle y|q\rangle \\ \langle\psi|r\rangle &= \langle\psi|x\rangle\,\langle x|r\rangle + \langle\psi|y\rangle\,\langle y|r\rangle, \end{aligned} \tag{8.19}$$

or the other way round

$$\begin{aligned} \langle\psi|x\rangle &= \langle\psi|q\rangle\,\langle q|x\rangle + \langle\psi|r\rangle\,\langle r|x\rangle \\ \langle\psi|y\rangle &= \langle\psi|q\rangle\,\langle q|y\rangle + \langle\psi|r\rangle\,\langle r|y\rangle. \end{aligned} \tag{8.20}$$

Note the order in which various symbols appear. Once you see the pattern, you can write down these equations immediately. They, and Eqs. (8.13) and (8.14), are all you will ever need to relate sets of basis states and projections of quantum states on various basis states.

Math Aside We can rearrange Eq. (8.19) in a way that makes remembering it even easier. Here is the "trick": Pull all of the left vector symbols $\langle\psi|$ to the far left and all of the (q, r) basis state right vectors to the far right. We then have

$$\langle\psi|q\rangle = \langle\psi|\underbrace{\left[|x\rangle\langle x| + |y\rangle\langle y|\right]}_{I}|q\rangle$$
$$\langle\psi|r\rangle = \langle\psi|\underbrace{\left[|x\rangle\langle x| + |y\rangle\langle y|\right]}_{I}|r\rangle \tag{8.21}$$

The terms in the square brackets inside the quantum sandwich are projection operators $|x\rangle\langle x|$ and $|y\rangle\langle y|$ for the (x, y) basis states. Their sum is equivalent to the identity operator. Eq. (8.20) can be rewritten in a similar way using the projection operators for the (q, r) basis states.

Try It 8.2 Use the projection operator method in the previous Math Aside to derive the results in Eq. (8.20).

Let's now tie all this to qubit quantum states and explicitly show what happens when we have a change of basis states in calculating measurement probabilities. The basic procedure we describe is preparing a qubit in a state expressed in one set of basis states and then carrying out a measurement on that state with a measurement device whose set of basis states is different from the basis set used in preparing the state. The overall process is the following:

1. Express the qubit state as a superposition of the basis states used to prepare the state.
2. Transform the state description from the preparation basis states to the measurement basis states using the equivalent of Eq. (8.19). (We assume that we know the brackets $\langle x|q\rangle$, $\langle y|q\rangle$, $\langle x|r\rangle$, and $\langle y|r\rangle$ that relate the two sets of basis states from the equivalents of Eqs. (8.17) and (8.18). Note that those brackets are independent of the state $|\psi\rangle$.)
3. Use the absolute-values-squared of the measurement state coefficients to calculate the probabilities for the measurement outcomes of each of the measurement basis states.

These processes will be used in almost every QIS and QC application.

Here is a specific example. Suppose that the qubit state $|\psi\rangle$ is prepared as a superposition state in the xy basis set:

$$\begin{aligned}|\psi\rangle &= a_x|x\rangle + a_y|y\rangle \\ &= |x\rangle\langle x|\psi\rangle + |y\rangle\langle y|\psi\rangle.\end{aligned} \tag{8.22}$$

We decide we want to measure the state using the qr basis, so we use Eq. (8.18) in Eq. (8.22) to express the xy basis vectors in terms of the qr basis vectors:

$$\begin{aligned}|\psi\rangle &= |x\rangle\langle x|\psi\rangle + |y\rangle\langle y|\psi\rangle \\ |\psi\rangle &= \langle x|\psi\rangle\underbrace{\left(|q\rangle\langle q|x\rangle + |r\rangle\langle r|x\rangle\right)}_{|x\rangle} + \langle y|\psi\rangle\underbrace{\left(|q\rangle\langle q|y\rangle + |r\rangle\langle r|y\rangle\right)}_{|y\rangle} \\ &= |q\rangle\underbrace{\left(\langle q|x\rangle\langle x|\psi\rangle + \langle q|y\rangle\langle y|\psi\rangle\right)}_{\langle q|\psi\rangle} + |r\rangle\underbrace{\left(\langle r|x\rangle\langle x|\psi\rangle + \langle q|y\rangle\langle y|\psi\rangle\right)}_{\langle r|\psi\rangle}.\end{aligned} \tag{8.23}$$

Note that in the third line of Eq. (8.23), we grouped the terms for $|q\rangle$ and $|r\rangle$ so that we can easily find the state coefficients—the terms inside the parentheses.

Try It 8.3 Write out the details of Eq. (8.23) starting from Eq. (8.22) and using the basis conversions in Eq. (8.17) or (8.18) (you decide which to use) in Eq. (8.22).

Our final step is to use the Born rule to calculate the probabilities that the measurement outcome states will be $|q\rangle$ or $|r\rangle$. We use $P(|q\rangle)$ for the probability that the measurement outcome state is $|q\rangle$. In our examples so far, the coefficients are real numbers; therefore, we could just square them, but in general they might be complex numbers, so we use the absolute-value-squared. The two probabilities are

$$\begin{aligned} P(|q\rangle) &= |\langle q|x\rangle \langle x|\psi\rangle + \langle q|y\rangle \langle y|\psi\rangle|^2 \\ P(|r\rangle) &= |\langle r|x\rangle \langle x|\psi\rangle + \langle r|y\rangle \langle y|\psi\rangle|^2. \end{aligned} \tag{8.24}$$

8.4 Three Basis Sets: 0°, 60°, and −60°

After all of that abstract math, let's apply what we have learned to the change of basis for linearly polarized photons, which we employ in Chapter 17. And, in fact, the change of basis states for photons is an operation commonly used in many quantum information processes.

We consider three sets of state-space basis vectors, one with the $|\text{hlp}\rangle$ along the direction labeled 0° and the other two with the corresponding basis vectors rotated by 60° and −60° from the initial direction 0°. We label the basis vector sets as follows:

A. Basis vectors: $|\text{hlp}\rangle_{0°}$ and $|\text{vlp}\rangle_{0°}$

B. Basis vectors: $|\text{hlp}\rangle_{60°}$ and $|\text{vlp}\rangle_{60°}$

C. Basis vectors: $|\text{hlp}\rangle_{-60°}$ and $|\text{vlp}\rangle_{-60°}$.

We can use Eq. (8.15) to relate the various basis vectors. For example, we obtain

$$\begin{aligned} |\text{hlp}\rangle_{60°} &= \cos 60°|\text{hlp}\rangle_{0°} + \sin 60°|\text{vlp}\rangle_{0°} = \frac{1}{2}|\text{hlp}\rangle_{0°} + \frac{\sqrt{3}}{2}|\text{vlp}\rangle_{0°} \\ |\text{vlp}\rangle_{60°} &= -\sin 60°|\text{hlp}\rangle_{0°} + \cos 60°|\text{vlp}\rangle_{0°} = -\frac{\sqrt{3}}{2}|\text{hlp}\rangle_{0°} + \frac{1}{2}|\text{vlp}\rangle_{0°}. \end{aligned} \tag{8.25}$$

Try It 8.4 Check Eq. (8.25) by using the state-space diagram and the column vector representations. Then, use Eq. (8.15) to write the basis vectors in terms of each other. Verify that the transformed basis states have unit length.

We will also need the equations linking the −60° basis states to the 0° basis states. As mentioned before, you should be able to write those down immediately by replacing 60° in Eq. (8.25) with −60°.

Try It 8.5 Carry out the replacement as suggested. Keep that result handy for later work.

Let's see what happens to the measurement probabilities with these basis vector transformations. If we prepare a photon in the $|\text{vlp}\rangle_{60°}$ state, for example, and then have it interact with a linear polarization measurement device operating with 0° basis states, then the measurement device will give hlp in the 0° basis with probability $\left(-\sqrt{3}/2\right)^2 = 3/4$ and vlp for the 0° basis with probability $(1/2)^2 = 1/4$.

Try It 8.6 Check those numbers and verify that they are correct.

In Chapter 17, we demonstrate that the results developed in this section can be used to show that quantum systems have probability correlations that cannot be correctly described by classical physics even though the classical predictions are based on what seem to be obvious and intuitive ideas about the nature of objects and their properties. In other words, these quantum correlations show that our naive assumptions about the nature of reality are wrong.

8.5 **Spin-1/2 States**

Now let's apply these ideas to a spin-1/2 system with two different measurement bases rotated by 90° in real space (45° in state space) (see Fig. 8.7). This is a combination we will see frequently in QIS and QC. We noted the differences between the physical-space angles and the state-space angles in Chapter 7. Recall from Chapter 7 that this angle relation is one of the peculiar features of spin-1/2 systems in quantum mechanics. In quantum state language, for a spin-1/2 system a common measurement basis set will be the spin-up and spin-down states associated with the physical-space z-direction (0° basis states). With

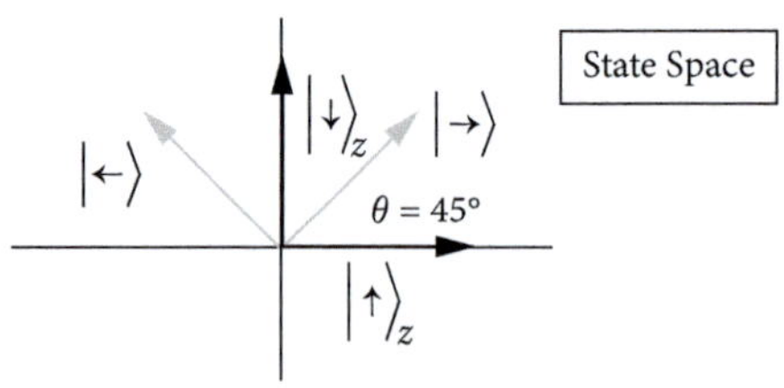

Fig. 8.7 A state-space representation with spin-up and spin-down 0° basis states $|\uparrow\rangle_z$ and $|\downarrow\rangle_z$ and with 90° (physical space) basis states $|\rightarrow\rangle$ and $|\leftarrow\rangle$.

$2\theta = 90°$, the measurement direction is now the physical-space horizontal direction, and we indicate those basis states as $|\rightarrow\rangle$ and $|\leftarrow\rangle$ instead of the up and down arrows. Note that the arrows indicate the physical-space direction, not the state-space direction of the basis states (see Fig. 8.7).

The basis states $|\rightarrow\rangle$ and $\langle\leftarrow|$ can of course be written as a superposition of the spin-up and spin-down states using the state-space angle θ:

$$|\rightarrow\rangle = \cos\theta|\uparrow\rangle_z + \sin\theta|\downarrow\rangle_z$$
$$|\leftarrow\rangle = -\sin\theta|\uparrow\rangle_z + \cos\theta|\downarrow\rangle_z. \tag{8.26}$$

Let's look at a specific example: suppose Bob decides to use a measurement basis for the spin direction that is at angle 90° (counterclockwise) in physical space from the standard z-direction (0°). In state space, the rotation angle is then just 45°, half of 90° as shown in Fig. 8.7.

Try It 8.7 Use Eq. (8.26) to show that for $2\theta = 90°$, we obtain

$$|\rightarrow\rangle = \frac{1}{\sqrt{2}}|\uparrow\rangle_z + \frac{1}{\sqrt{2}}|\downarrow\rangle_z$$
$$|\leftarrow\rangle = -\frac{1}{\sqrt{2}}|\uparrow\rangle_z + \frac{1}{\sqrt{2}}|\downarrow\rangle_z. \tag{8.27}$$

Try It 8.8 Show using Eq. (8.18) that the following give the "reverse" connections:

$$|\uparrow\rangle_z = \frac{1}{\sqrt{2}}|\rightarrow\rangle - \frac{1}{\sqrt{2}}|\leftarrow\rangle$$
$$|\downarrow\rangle_z = \frac{1}{\sqrt{2}}|\rightarrow\rangle + \frac{1}{\sqrt{2}}|\leftarrow\rangle. \tag{8.28}$$

Try It 8.9 Are the vectors $|\rightarrow\rangle$ and $|\leftarrow\rangle$ eigenvectors of the Pauli X operator? What are the corresponding eigenvalues. Compare the vectors $|\rightarrow\rangle$ and $|\leftarrow\rangle$ with the eigenvectors of X given in Eq. 7.9. These results show that the eigenvectors of an operator may not be unique.

To get some practice with changing basis states, let's rewrite some of the previous results using the column vector form of the basis state vectors. To keep things straight, we add subscripts to the column vectors to indicate the basis system being used. Eq. (8.27) is then written as

$$\begin{pmatrix}1\\0\end{pmatrix}_{90^\circ} = \frac{1}{\sqrt{2}}\begin{pmatrix}1\\0\end{pmatrix}_z + \frac{1}{\sqrt{2}}\begin{pmatrix}0\\1\end{pmatrix}_z = \begin{pmatrix}\frac{1}{\sqrt{2}}\\ \frac{1}{\sqrt{2}}\end{pmatrix}_z$$
$$\begin{pmatrix}0\\1\end{pmatrix}_{90^\circ} = -\frac{1}{\sqrt{2}}\begin{pmatrix}1\\0\end{pmatrix}_z + \frac{1}{\sqrt{2}}\begin{pmatrix}0\\1\end{pmatrix}_z = \begin{pmatrix}-\frac{1}{\sqrt{2}}\\ \frac{1}{\sqrt{2}}\end{pmatrix}_z . \tag{8.29}$$

The final equality in each line of Eq. (8.29) reminds us of the simple way of picturing the basis state: The entries in the column vectors are the "coordinates" of the tip of the basis state vectors in the standard basis state space.

Try It 8.10 Repeat the basis transformation used in Section 8.4 but now for three spin-1/2 basis sets, where the angles in state space are 0°, 120°, and -120°.

In Chapter 7, we observe that any spin-1/2 state can be written in terms of a colatitude angle and a phase angle associated with the Bloch sphere:

$$|\psi\rangle = \cos(\theta/2)\,|0\rangle + e^{i\phi}\sin(\theta/2)|1\rangle . \tag{8.30}$$

From Eq. (8.23), we can read off the Dirac brackets $\langle 0\,|\psi\rangle = \cos(\theta/2)$ and $\langle 1|\psi\rangle = e^{i\phi/2}$ $\sin(\theta/2)$.

Suppose we have another set of basis states $|q\rangle$ and $|r\rangle$ associated with the Bloch sphere. These basis states $|q\rangle$ and $|r\rangle$ can be written in terms of the computation basis states using the equivalent of Eq. (8.17) with $x \to 0$ and $y \to 1$:

$$\begin{aligned}|q\rangle &= \cos(\beta/2)|0\rangle + e^{i\phi_q}\sin(\beta/2)|1\rangle \\ &= |0\rangle\langle 0|q\rangle + |1\rangle\langle 1|q\rangle ,\end{aligned} \tag{8.31}$$

where β and ϕ_q are colatitude and phase angles for that basis state. It is easy to see that the other basis state $|r\rangle$ is

$$\begin{aligned}|r\rangle &= \sin(\beta/2)|0\rangle - e^{i\phi_q}\cos(\beta/2)|1\rangle \\ &= |0\rangle\langle 0|r\rangle + |1\rangle\langle 1|r\rangle .\end{aligned} \tag{8.32}$$

Try It 8.11 Show that the two basis states are orthogonal, that is, $\langle q\,|r\rangle = 0$ and that they are both normalized. Consequently, we have an orthonormal basis.

Our goal is to calculate the overlap brackets between the quantum state and the rq basis states. Those brackets then tell us what the state vector looks like in that basis. Here is where

Dirac notation is so helpful. The desired brackets can be written down quickly simply by left multiplying Eq. (8.17) by $\langle\psi|$:

$$\begin{aligned}\langle\psi|q\rangle &= \langle\psi|0\rangle\langle 0|q\rangle + \langle\psi|1\rangle\langle 1|q\rangle \\ \langle\psi|r\rangle &= \langle\psi|0\rangle\langle 0|r\rangle + \langle\psi|1\rangle\langle 1|r\rangle .\end{aligned} \tag{8.33}$$

All the ingredients can be read from Eqs. (8.30)–(8.32) and then using $\langle\Phi|\Psi\rangle = \langle\Psi|\Phi\rangle^*$ as needed.

Try It 8.12 Work out the details to express Eq. (8.33) in terms of θ and ϕ.

8.6 The Three-Polarizing-Sheet Experiment

We want to show you a simple experiment you can do at home. We believe this experiment should be performed by everyone who is interested in QC and QIS. We will use changes in basis states to illustrate the fundamental feature of quantum mechanics: superposition states. All you need are three sheets of linear polarizing material and a source of light. Your eyes will serve as the measurement detector. As mentioned before, for the polarizing sheets you could use polarizing sunglasses. Relatively inexpensive linear polarizing sheets are available from many suppliers online. Some suggestions are provided in the Further Reading section of Chapter 4.

First, you need to find the transmission axis directions for the polarizing sheets. This is easy to do using the setup shown in Fig. 4.5, where you view light reflected from a horizontal surface such as a shiny floor or the surface of water in a lake or pond. That light will have predominantly horizontal linear polarization. View the light reflected from that surface with the polarizing sheet placed between the horizontal surface and your eyes, with the sheet perpendicular to the line from the surface to your eyes. Then rotate the polarizing sheet around that line until you see the maximum amount of light coming through the polaroid sheet. The transmission axis of the polarizing sheet is now parallel to that horizontal surface. Usually (but not always), the edges of the polarizing sheet are cut parallel or perpendicular to that axis. Use a marker or a small piece of tape to mark that axis. Do that for all three polarizing sheets. In what follows, we call that direction the hlp direction for each sheet.

The next step is to place sheet A, with its transmission axis horizontal, in front of an illuminated surface. (You could put sheet A in front of a lamp or flashlight, but an illuminated flat surface provides a more uniformly distributed light intensity.) Then arrange sheet B, placed between A and your eyes, with its transmission axis perpendicular to that of sheet A (Fig. 8.8, center image). You should find that almost no light from the illuminated surface comes through that combination.

We can describe that situation using our linear polarization basis states. Because we will be dealing with three different polarizing sheets, we will change our state labels from hlp

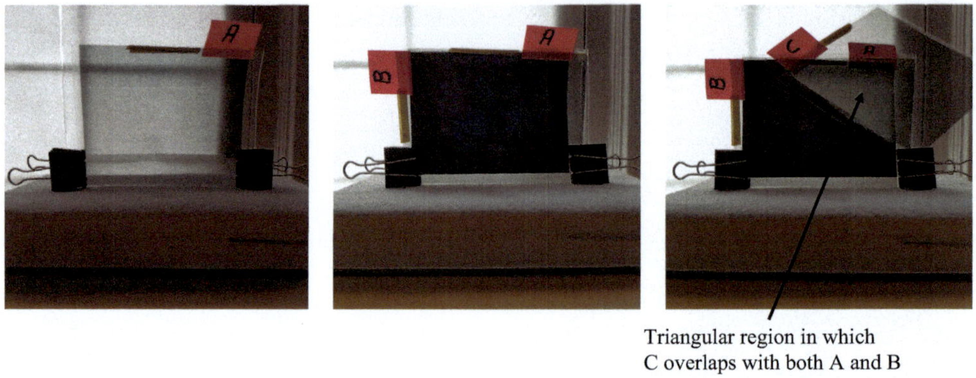

Fig. 8.8 The three-polarizing-sheet experiment. On the left is polarizing sheet A with its transmission axis horizontal, indicated by the piece of tape at the top of the sheet. In the center, polarizing sheet B, with its transmission axis vertical, is placed between A and the camera. On the right, sheet C, with its transmission axis at about 45°, is placed *between* sheets A and B. The arrow points to the region in which sheet C overlaps sheets A and B. The gray vertical and horizontal bars in the background are mullions in the window behind the sheets.
Source: Photo by R. Hilborn.

and vlp to trans and block. "Trans" is associated with the transmission axis of the sheet, which we previously labeled as hlp. Block is the old vlp. We will use subscripts on the ket symbols to indicate which of the three sheets A, B, and C we are concerned with.

The light from the illuminated surface first hits sheet A, whose transmission axis is horizontal. It means that the light exiting sheet A is described by the state vector $|\text{trans}\rangle_A$. Some people would say that sheet A has collapsed the light's state vector from whatever it was before entering A to $|\text{trans}\rangle_A$ for the light that gets through the sheet. We prefer to say that we use $|\text{trans}\rangle_A$ to describe the polarization state of the light after it has passed through sheet A, so we don't have to worry about the mechanism by which sheet A changed the light's polarization state.

Sheet B has its transmission axis vertical. It means that it will not transmit any light with the state vector $|\text{trans}\rangle_A$ since the basis states $|\text{trans}\rangle_A$ and $|\text{trans}\rangle_B$ (equivalent to $|\text{block}\rangle_A$) are orthogonal, which is so far, so good. Just what we expect for "crossed" linear polarizers.

Now comes the exciting part. Place the third polarizing sheet C *between* sheets A and B. (Fig. 8.8, right side image). You should see something truly amazing. (It is amazing if you think about it in the right way, of course.) You should see (Fig. 8.8) that light now comes through the region where sheet C is inserted between A and B. We find that amazing because adding sheet C between sheets A and B has somehow allowed light to get through even though sheets A and B have their transmission axes perpendicular and block all the other light. How can that happen? Before answering that question, you should check that putting sheet C between the light source and sheet A or putting sheet C between sheet B and your eye does not allow light to get through the combination.

To answer the "how" question, let's think about it this way: For light coming from sheet A to get through sheet B, sheet C must change the linear polarization of the light so that at least some of the light has linear polarization along the transmission axis of sheet B. To keep the explanation relatively simple, let's assume that you have oriented the transmission axis of sheet C so that it is at 45° with respect to transmission axes of A and B. Of course, light that comes through C has its linear polarization direction along the transmission axis of C, and the appropriate polarization state is represented by $|\text{trans}\rangle_C$.

Now let's use polarization state vectors to describe the situation more formally. After the light leaves sheet A, its linear polarization state is $|\text{trans}\rangle_A$. From Eq. (8.15), we can write that state as a superposition of the linear polarization states of sheet C:

$$|\text{trans}\rangle_A = \frac{1}{\sqrt{2}}\left(|\text{trans}\rangle_C + |\text{block}\rangle_C\right). \tag{8.34}$$

We see that there is a linear polarization component of that light along the transmission axis $|\text{trans}\rangle_C$ of C. The amplitude squared = ½, so we predict that 50% of the light from A gets through C. (In reality, it will be a bit less than that because the inexpensive polarizing sheets absorb a bit of the light. But that does not affect the essence of the experiment.)

Try It 8.13 Translate Eq. (8.15) into Eq. (8.34). Hint: Evaluate the appropriate cosines and sines and change the names of the states.

Next the light will enter sheet B. We write the state $|\text{trans}\rangle_C$ as a superposition of basis states of sheet B:

$$|\text{trans}\rangle_C = \frac{1}{\sqrt{2}}\left[|\text{trans}\rangle_B + |\text{block}\rangle_B\right]. \tag{8.35}$$

So, we see that when the light encounters sheet B, it has a component along the transmission direction $|\text{trans}\rangle_B$ of sheet B. We predict that 50% of the light from C will emerge from sheet B. So, overall with the transmission axis of sheet C at 45° to those of A and B, we predict that $0.5 \times 0.5 = 0.25$ (25%) of the light that gets through A will emerge from C. That is what you should see on the far-right part of Fig. 8.8.

Let's summarize what we have shown: The light entering sheet C from A can be described by a superposition state of the C basis states, so there is a probability of light getting through sheet C. The analogous statement is then true for the light entering sheet B: It is in a state that can be written as a superposition of linear polarization basis states of B. If there is no sheet C, the light transmitted through A is described by the state $|\text{trans}\rangle_A = \cos\theta_{AB}|\text{trans}\rangle_B + \sin\theta_{AB}|\text{block}\rangle_B$. But since $\theta_{AB} = 90°$, there is no projection of $|\text{trans}\rangle_A$ along the transmission state $|\text{trans}\rangle_B$ of polarizer B. Hence, none of the light gets through the "crossed" polarizer combination of A and B.

Now let's generalize the result to any angle between the A and C transmission axes. We rotate sheet C about an axis perpendicular to the plane of the sheet to see what happens as the angle between the transmission axis of C and those of A and B changes. You should see a smooth change in the amount of light that gets through. For the situation in which the A and B transmission axes are perpendicular, the more general case can be expressed in terms of the angle θ_{AC} between the A sheet and C sheet transmission axes:

$$I_B = \frac{1}{4} I_A \sin^2 2\theta_{AC}. \tag{8.36}$$

We have used I_A and I_B to indicate the intensity of light that emerges from sheet A and from sheet B, respectively, in the three-sheet setup.

Try It 8.14 Rotate sheet C relative to A and qualitatively check that Eq. (8.36) describes what is going on. Hints: $\sin 0° = 0$, $\sin 90° = 1$, and $\sin 180° = 0$.

It's not too hard to derive Eq. (8.36). Recall that in Chapter 5 we show that when we rotated just two polarizing sheets, the state amplitudes were multiplied by the cosine of the angle of rotation in state space, and the transmitted light intensities change as the cosine-squared. For linearly polarized light, that angle is the same angle by which we rotate the polarizing sheets relative to each other in physical space. In fact, we observed that the result for the new amplitude is just the old amplitude multiplied by the cosine of the angle by which the state vectors were rotated. Using a_A for the amplitude of the light coming through sheet A, with similar terms for sheets B and C, we have

$$\begin{aligned} a_C &= a_A \cos\theta_{AC} \text{ and } a_B = a_C \cos\theta_{BC} \\ a_B &= a_A \cos\theta_{BC} \cos\theta_{AC}. \end{aligned} \tag{8.37}$$

From the geometry of the setup, we know that $\theta_{AB} + \theta_{CB} = 90°$. From trigonometry, we remember that $\cos\theta = \sin(90° - \theta)$ and $\sin 2\theta = 2\sin\theta\cos\theta$ (the "double angle" formula). Using these results, we find that the second line of Eq. (8.37) becomes

$$a_B = a_A \sin\theta_{AB} \cos\theta_{AB} = \frac{a_A}{2} \sin 2\theta_{AB}\ . \tag{8.38}$$

The final step is to recall that the probability of a photon being detected is given by the square of the amplitude. For light sources with a lot of photons, the intensity of the light is proportional to that probability. Assembling all those details gives us the result stated in Eq. (8.36).

What you have seen is that linearly polarized light seems to be well described by superposition states. These effects were known long before the quantum nature of light (photons) was discovered. The prequantum explanations are based on the connection between the linear polarization direction and the direction of the electric field (a vector) according to the wave model of light.

The demonstration by itself does not show that the same "superposition effects" would work for individual photons: the particle model of light. However, experiments have been performed in which photons are detected individually, and the results show that the superposition quantum state model is correct. If you had the equipment to observe the light photon by photon for this experiment, you would have direct evidence that the superposition state vector and probability interpretation of the squares of the amplitudes are correct descriptions of how nature behaves. The classical electric field model is fine when you are dealing with large numbers of photons in your light beam, but to account for the behavior of individual photons and the associated probabilities, we need quantum mechanics.

We should also point out that this experiment demonstrates that polarizing sheets can be used both as part of measurement devices and state preparation devices. In particular, we can say that sheet A prepares the photons it transmits in the state associated with its transmission axis. Similar statements apply for sheets B and C. We could also say that sheet B can be used to measure the probability that photons coming from sheet C have a component of their linear polarization along the transmission axis of B.

8.7 Relative Phases Revisited

In Chapter 5, we showed that an overall sign or overall phase factor for quantum state does not change the predictions of measurement results based on that state vector. But we restricted ourselves to cases in which the basis state conversion had equal weights to the two original basis states and the relative phase appeared in only one of the amplitudes. In this section, we use the more general basis changing tools developed in this chapter to show how signs and relative phase factors affect the predictions of measurement outcomes when we change basis states.

We label the two sets of basis states (x, y) and (q, r) as we did in previous sections. For the convenience of readers, we repeat the essential results here.

We express the Dirac brackets as

$$\begin{aligned}\langle\psi|q\rangle &= \langle\psi|x\rangle\langle x|q\rangle + \langle\psi|y\rangle\langle y|q\rangle \\ \langle\psi|r\rangle &= \langle\psi|x\rangle\langle x|r\rangle + \langle\psi|y\rangle\langle y|r\rangle .\end{aligned} \tag{8.39}$$

By the Born rule, the relevant probabilities for getting the two new basis states as measurement outcomes are

$$P(|q\rangle) = |\langle\psi|q\rangle|^2 \qquad P(|r\rangle) = |\langle\psi|r\rangle|^2 . \tag{8.40}$$

The most general connection between the two sets of qubit basis states may be written as

$$\begin{aligned}|q\rangle &= \cos\theta|x\rangle + e^{i\phi}\sin\theta|y\rangle \\ |r\rangle &= -\sin\theta|x\rangle + e^{i\phi}\cos\theta|y\rangle .\end{aligned} \tag{8.41}$$

Note that the complex exponentials express the relative phase between the $|x\rangle$ and $|y\rangle$ contributions to the new basis states. We have chosen to write them in the terms involving $|y\rangle$.

The parameter θ sets the magnitudes of the amplitudes of the basis set. It is not necessarily the state-space angle between $|q\rangle$ and $|x\rangle$ unless both sets of amplitudes are real numbers. We want to examine the more general case where the amplitudes might be complex numbers.

Try It 8.15 Verify that the q, r basis states are orthonormal if the $|x\rangle$ and $|y\rangle$ basis states are orthonormal.

We now use the basis state transformation equations from Eq. (8.39) to obtain

$$\langle\psi|q\rangle = \langle\psi|x\rangle\cos\theta + \langle\psi|y\rangle\, e^{i\phi_{qy}}\sin\theta, \tag{8.42}$$

which then gives us the probability of getting $|q\rangle$ as the measurement outcome:

$$\begin{aligned} |\langle\psi|q\rangle|^2 &= \left[\langle\psi|x\rangle\cos\theta + \langle\psi|y\rangle\, e^{i\phi}\sin\theta\right]\left[\langle x|\psi\rangle\cos\theta + \langle y|\psi\rangle\, e^{-i\phi}\sin\theta\right] \\ &= |\langle\psi|x\rangle|^2\cos^2\theta + |\langle\psi|y\rangle|^2\sin^2\theta \\ &\quad + \langle\psi|x\rangle\langle y|\psi\rangle\, e^{-i\phi}\cos\theta\sin\theta + \langle\psi|y\rangle\langle x|\psi\rangle\, e^{+i\phi}\cos\theta\sin\theta. \end{aligned} \tag{8.43}$$

Note that the last two terms in Eq. (8.43) are each other's complex conjugate. Using the standard result for complex numbers $z + z^* = 2\,\mathrm{Re}(z)$, we may rewrite Eq. (8.43) as

$$\begin{aligned} |\langle\psi|q\rangle|^2 &= |\langle\psi|x\rangle|^2\cos^2\theta + |\langle\psi|y\rangle|^2\sin^2\theta \\ &\quad + 2\sin\theta\cos\theta\,\mathrm{Re}\left(\langle\psi|x\rangle\langle y|\psi\rangle\, e^{-i\phi}\right). \end{aligned} \tag{8.44}$$

Try It 8.16 Derive the equivalent expression for $|\langle\psi|r\rangle|^2$. Hint: $|\langle\psi|q\rangle|^2 + |\langle\psi|r\rangle|^2 = 1$.

These expressions are moderately complicated, but we are interested in knowing if they contain enough information to allow us to find the real and imaginary parts of the original state amplitudes $\langle\psi|x\rangle$ and $\langle\psi|y\rangle$. Those amplitudes allow us to calculate all the information embodied in the quantum state vector $|\psi\rangle$. We might think there are four unknowns. However, as mentioned before, we may always choose one of the amplitudes to be real. As a result, we have only three unknowns. It means that if we use at least two different measurement sets of basis states (different values of θ and ϕ), we can get all the details of the original state amplitude, except for a possible overall phase factor.

That process of taking "pictures" of a quantum state with different basis states is called "quantum tomography," in analogy with medical tomography, where medical images taken from different directions are combined to give a three-dimensional picture of the region being imaged. For a quantum system, we need to prepare the system in the same original state for each picture because the quantum measurement process, as discussed in Chapter 5, destroys the original state. Fortunately, that does not occur for medical imaging.

In real-life situations, the more appropriate description of the quantum state of a system is constructed using the density matrix approach, which is discussed in Chapter 22. That method allows for quantum states that are mixtures of pure quantum states—the types that we have discussed so far.

The takehome message is that states with different magnitudes of the basis states and with different relative signs can lead to different probabilities for measurement outcomes, or conversely, we can measure relative signs and the magnitudes of the amplitudes by making measurements with different basis states. As shown in later chapters, many quantum algorithms make use of relative phases.

8.8 Bennett–Brassard-84 Encryption Protocol

Now that we understand changes of basis states, we want to reward you by looking at an actual application of these ideas in QIS. The method is called the Bennett–Brassard-84 (BB84) Encryption Protocol after its inventors Charles Bennett and Gilles Brassard, who developed it—surprise—in 1984. The basic idea is that Alice and Bob, located some distance apart, want to establish a secret key (a string of 0s and 1s) to encrypt messages. Alice needs to send the key to Bob, but we are worried that someone is eavesdropping on our communications. The process that we use is an example of quantum key distribution.

Encryption is a big business in the digital age, and there are many, many different encryption schemes. Let us give you an example of one that we made up. Suppose the message we want to send is a plain English text. We might encrypt the message by replacing each letter in the message by the letter located a certain number of places to the right in the alphabet. For an example, we might replace an "a" in the message with a "d" three places to the right of "a" in the alphabet. This process is called a substitution encryption. Unfortunately, it is relatively easy to crack. But you can make the encryption more secure by varying how many places to the right you go to find the replacement letter. That is where the key comes in. Again, there are many ways you might do this, but here is a simple one. As mentioned before, a key is often just a string of 1s and 0s. Let's suppose we arrange the key's 1s and 0s into groups, each with 3 bits. So, our key might look like

$$110 \quad 001 \quad 101 \quad 100 \quad 111 \quad 010\ldots \tag{8.45}$$

We then carry out the replacement of the first letter in the original message using the number represented by the three bits. For a three-bit binary number, these go from 0 to 7. The key in Eq. (8.45) tells us to replace the first letter of the message with the letter six (=110 in binary) places to the right, the second letter uses the letter one place to the right, the third letter uses the letter five places to the right, and so on. Then when the encrypted message is received, the key can be used to replace the letters with the letter located the appropriate number of places to the left. That process allows you to reconstruct the original message.

If you choose the key bits randomly, you will have an encryption method that is difficult to crack.

Of course, the key must be kept from everyone except those who are worthy or deemed worthy to have access to the secret message, namely Bob and Alice. The BB84 Protocol is a way to use the properties of quantum states to allow Bob and Alice to generate a shared key while checking that no one else has gained access to it.

Try It 8.17 Suppose that Alice and Bob decide to use a classical communication channel with ordinary 0s and 1s, that is, the only states are 0 and 1. Describe what happens if someone is eavesdropping on that channel by intercepting a bit, recording its value, and then sending the same bit value to Bob. Can Alice and Bob detect that eavesdropping?

The BB84 Quantum Protocol is based on the idea of sending qubits whose states are described by superpositions in one basis set but not in another. If both Bob and Alice measure the qubit's state using the set of basis states in which there is no superposition, the measurement results are guaranteed to be the same. If, however, Bob measures the qubit state using a basis set different from the one Alice used, the mathematics that we just learned will come into play and Bob's measurement result will not always match Alice's result. Let's see how this works.

We start with a long list of randomly chosen classical 0s and 1s. Alice sends Bob a linearly polarized photon qubit for each bit in the list. Alice will prepare the qubit state in either the 0° or the 45° linear polarization basis, chosen at random and with equal probability. If the bit Alice wants to send is a 0, Alice prepares the qubit in the $|0\rangle$ (hlp) state in that chosen basis and launches it to Bob. If the bit Alice wants to send is a 1, Alice prepares the qubit in the $|1\rangle$ state (vlp) in Alice's chosen basis and sends that qubit off. We continue until Alice has sent the entire sequence of bits, choosing the preparation basis at random for each bit. Alice records both the basis used and the bit that is sent. An example of this is presented in Table 8.1.

When Bob receives one of Alice's photons, Bob chooses at random one of the two basis sets to observe the linear polarization state of the qubit. That is, Bob is choosing a measurement basis with no communication with Alice. Bob won't always match her choice of basis states because Bob doesn't know in advance what basis states she used to prepare the qubit state.

Bob observes the polarization state when the photon arrives and writes down his choice of measurement basis: a 0 if the observation is hlp and a 1 if it is vlp relative to the chosen basis. To be concrete, think of using a polarizing beam splitter as the measurement device. Table 8.2 shows a sample of the information Bob records.

If Bob is using a measurement basis that is the same one Alice used to prepare the qubit, Bob's result will agree with what Alice sent. However, for the 0° and 45° basis sets, if Bob chose the other basis set, 50% of the time Bob will get 0 and 50% of the time he will get 1, no matter what Alice sent. Look at the first four entries in Tables 8.1 and 8.2. The measurement basis states are the same only once and, of course, Bob observed the same bit that Alice sent.

Table 8.1 Alice's record. The top row indicates the basis set that Alice chose, and the second row is the classical bit she wants to send in her message. Bases are 0° and 45°.

Basis	0	45	0	45	45	0	0	0	45	45
Bit	1	0	1	1	0	1	1	0	0	0

Table 8.2 Bob's record. The top row indicates the basis set Bob chose, and the second row is the result of his observation. Bases are 0° and 45°.

Basis	45	0	0	0	45	45	0	45	0	45
Bit	1	1	1	0	0	0	1	1	0	0

They did not choose the same measurement basis for six of the entries, but for two of the six, they still had matching results. The results in Tables 8.1 and 8.2 are just a 10 run sample. The averages of those results do not exactly match those from hundreds of trials. For a long series of trials, Bob's results will agree with Alice's only 75% of the time. However, they don't yet know which runs have matching basis states and which don't.

Where did 75% come from? There are two parts to the answer. First, given the way they choose basis sets, Alice and Bob choose the same basis set only 50% of the time. When they do, they get the same bit results. They agree 100% of the time.

To see how we get to 75%, we need to look at the other 50% of the runs where they don't choose the same basis set. In that situation, the measurement results agree 50% of the time and disagree 50% of the time. Now let's put this together to find the overall probability of the runs that lead to their getting the same results. Let's write this as the product of probabilities. For example, the probability that Alice and Bob get the same results given that they use the same basis states is 1 and the probability that they use the same basis states is 0.5.

$$
\begin{aligned}
&P(\text{A \& B same results})\\
&\quad = P(\text{A \& B same results given the same basis states})P(\text{same basis states})\\
&\qquad + P(\text{A \& B same results given different basis states})P(\text{different basis states})\\
&\quad = 1 \times 0.5 + 0.5 \times 0.5 = 0.75.
\end{aligned}
$$

Overall, Alice and Bob should agree 75% of the time and disagree 25% of the time.

To show where the different basis state probabilities come from, we may use the math we learned about superpositions with different basis states. For example, suppose Bob is going to use the 45° basis for his measurements but Alice used the 0° basis to prepare the qubit state. Alice decides to send a vlp photon in that basis. Then, we can use Eq. (8.22) to relate the state for a vlp photon in the 0° basis to hlp and vlp states in the 45° basis:

$$
|\text{vlp}\rangle_{0^\circ} = \cos 45^\circ |\text{hlp}\rangle_{45^\circ} + \sin 45^\circ |\text{vlp}\rangle_{45^\circ} = \frac{1}{\sqrt{2}}|\text{hlp}\rangle_{45^\circ} + \frac{1}{\sqrt{2}}|\text{vlp}\rangle_{45^\circ}. \tag{8.46}
$$

Using the Born rule, we find that the probability that Bob gets hlp is ½, that is, 50% of the time for those conditions, Bob will observe a horizontally polarized photon in the 45° basis.

Try It 8.18 Examine the results in Tables 8.1 and 8.2 and see if they agree with the claims in the previous paragraphs. Remember that the tables give the results only for a limited set of trials. Good statistics needs many trials.

Now, let's suppose that Alice and Bob want to have a secret key with N bits. The following procedure will do two things: (1) determine if an eavesdropper has been listening into the communications and (2) determine the key to be used for subsequent communications. Alice transmits $4N$ qubits using the procedures described above. Why $4N$? We explain this in just a moment. The quick answer is that we are going to throw away ¾ of the results and keep only ¼. You will soon see why after we address eavesdropping.

Alice and Bob then send each other the sequence of basis sets chosen, keeping the bits sent and received secret (the lower rows in Tables 8.1 and 8.2). They go through the list of basis sets and keep only the results for which they both selected the same basis set. This will leave us with about $2N$ results.

Since we sent the list of basis state choices over an insecure communication channel, it is possible that an eavesdropper might have listened in or tapped the channel. However, the eavesdropper only gets information about the bases used to prepare and read the bits and nothing about the actual bits in the key. Knowing that we used the 45° basis to generate a particular bit in the key is as useful as knowing two spies are using Dostoevsky's *Crime and Punishment* to generate the key. You still do not know how they are using the hundreds of pages in the book to generate and read the cypher, so the information is useless.

But there is an advantage to this quantum protocol: We can use the results to check if someone is eavesdropping. It works like this: From the $2N$ results where we used the same basis, we select N entries and share that list of results (key bits) using an unencrypted, insecure communication channel (e.g., our smartphones). We expect that if no one is eavesdropping or otherwise messing with our qubit communication channel (the one we used to send the photons), if Alice sent a 1, Bob should have received a 1, and if she sent 0, Bob should have received a 0.

What happens if someone is indeed eavesdropping on the qubit channel? The eavesdropper wants to determine the sequence of 0s and 1s that Alice is sending to find the secret key. So, the eavesdropper lets the qubit that Alice sent interact with a measurement device using either the 0° or 45° basis set. On average, half the time the eavesdropper will choose the basis states Alice used and will successfully read the 0 or 1 which Alice sent. The eavesdropper can then launch the qubit onto Bob without being detected. Bob observes the bit he would have seen if the eavesdropper had not been eavesdropping.

However, half the time the eavesdropper chooses the "wrong" basis set (the one Alice didn't use) and has only a 50% chance of getting the right 0 or 1 because the qubit will be in a state of superposition with respect to the eavesdropper's basis. In this scenario, when the eavesdropper sends a qubit onto Bob, the measurement result will be what Alice sent only 50% of the time.

In summary, we would expect to have 100% agreement between Alice's bits and Bob's bits in the N selected runs if there is no eavesdropping. If they find that result, they will then keep the remaining N bits secret and use them as the encryption key. However, if the eavesdropper has been eavesdropping and using 0° and 45° basis states, they will get only about 75% agreement instead of having 100% agreement, assuming that the eavesdropper has been intercepting every photon. In other words, about ¼ of the time, we will get different results. It is ¼ because the probability of the eavesdropper's choosing the wrong basis states is ½, and given the wrong basis, the probability of sending the wrong bit to Bob is ½.

Why couldn't the eavesdropper measure the quantum state describing the photon Alice sent and then resend it to Bob? Here is where quantum mechanics enters. You need to recall the *no-cloning theorem* mentioned in Chapter 5. There we showed that there is no way to copy a quantum state with one observation. So, your scheme won't work. If you are eavesdropping on the qubit channel, Alice and Bob can detect you and then choose not to use that sequence of measurements to define our key. They will just repeat the process until they get results that indicate no significant eavesdropping. The bottom line is that if people are eavesdropping on the qubit channel, they can use the measurement results to detect them.

The eavesdropper might try to be subtle and not measure every qubit that goes by. Perhaps that person measures only every other qubit. Do you think the eavesdropping will go undetected? Do you get enough information about the bits being sent to figure out the key?

That's a moderately complicated question in probability theory. Alice and Bob will likely be able to detect the "part-time" eavesdropping because the agreement between what Bob observes and what Alice sent will drop below 100%. If they detect eavesdropping, they can always drop that sequence of bits and try another to generate a key. But deciding how much eavesdropping they can tolerate depends on exactly how they are using the key. That is a much more complex problem.

Let's look at the encryption procedure in various ways to check the conclusions we have drawn.

For example, suppose Alice chose the 0° basis and sent a 1, but the eavesdropper chose the 45° basis, as shown in Fig. 8.9. Alice's 1 is described by a superposition of the eavesdropper's states giving that person a 50% chance of reading a 1 and a 50% chance of reading a 0. Suppose the eavesdropper reads a 1 in the 45° basis and sends that onto Bob. If Bob has chosen the 45° basis, he will get a 1, which matches what Alice sent but with the "wrong" basis (it doesn't match Alice's). They will throw away that result when they compare our basis lists. If Bob chooses the 0° basis (the one Alice chose), he has a 50% chance of getting

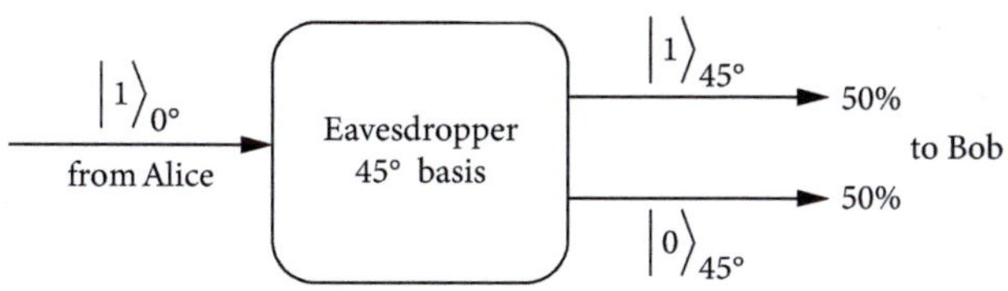

Fig. 8.9 A schematic diagram of what happens when Alice chooses to send 1 in the 0° basis, but the eavesdropper uses a 45° basis for measurements. In QIS, the eavesdropper is often called Eve.

Table 8.3 The eavesdropping results when Alice, the eavesdropper, and Bob have chosen the same basis set (either 0° or 45°). Alice's bit is the one she wants to send to Bob. Bob's bit is the result of his measurement.

Alice's basis	Alice's bit	Eavesdropper basis	Eavesdropper bit	Bob's basis	Bob's bit
0°	0	0°	0	0°	0
0°	1	0°	1	0°	1
45°	0	45°	0	45°	0
45°	1	45°	1	45°	1

Table 8.4 The eavesdropping results when Alice and the eavesdropper have chosen different basis sets.

Alice's basis	Alice's bit	Eavesdropper basis	Eavesdropper bit	Bob's basis	Bob's bit
0°	0	45°	0, 1 (50:50)	0°	0, 1 (50:50)
0°	1	45°	0, 1 (50:50)	0°	0, 1 (50:50)
45°	0	0°	0, 1 (50:50)	45°	0, 1 (50:50)
45°	1	0°	0, 1 (50:50)	45°	0, 1 (50:50)

a 1 and a 50% chance of getting a 0. If he gets a 1, they won't know that eavesdropping has occurred, but if he reads a 0, then we know there was eavesdropping.

Can the eavesdroppers know that they are measuring a superposition state or a single basis state? No, they will just read their measurement device, and by using the *measurement principle*, the qubit will end up being described by the measurement basis state corresponding to the observation. You cannot know about the state of the qubit before you measure it. The eavesdropper then sends that qubit to Bob where Bob measures it with his chosen basis. If Bob chooses the same basis as the eavesdropper, his measurement will match the eavesdropper's measurement 100% of the time. If Bob's basis is the other one, then the qubit is in a superposition state of Bob's basis states and he has a 50% chance of matching what was sent.

Table 8.3 summarizes the eavesdropping situation when the eavesdroppers, Bob, and Alice have used the same basis states: either 0° or 45°. Note that Bob's measurements always yield the bit Alice sent. In Table 8.4, we have listed the four cases in which the eavesdropper was unlucky and chose a basis different from Alice's and Alice sent either a 0 or 1. Bob has a 50% chance of getting a result that matches what Alice sent. Note that we have selected just those results in which Bob used the same basis Alice used.

To summarize, when Bob and Alice look at just the cases in which they have chosen the same basis states, they should always agree on the bit sent and received if there is no eavesdropping. The cases listed in Table 8.4 show that when the eavesdropper and Alice have chosen different basis sets, Bob will have only a 50% chance of getting the bit Alice sent because his basis set will not match the eavesdropper's.

Try It 8.19 Bob's basis was given along with Alice's and the eavesdropper's in Table 8.4. Suppose instead that Bob chose the other basis. Would the result be any different? If so, how?

Try It 8.20 There are, of course, other basis sets beyond those given in Tables 8.3 and 8.4. For example, there is the basis in which the polarizers are turned by $60°$. If you replace the $45°$ basis with a $60°$ basis, will the results of Table 8.3 change? If so, how? What about Table 8.4? Hint: Recall that $\cos(45°) = \sin(45°) = 1/\sqrt{2}$ while $\cos(60°) = 1/2$ and $\sin(60°) = \sqrt{3}/2$.

If we determine that someone is eavesdropping on the communication channel, we can try to send the message again. Of course, if the eavesdropper is persistent, we will realize that the communication channel is corrupted, and we will have to send the message some other way.

There is always some probability that Bob will get the same result Bob would have gotten without eavesdropping, but the probability that all N cases work out that way gets smaller as N gets larger. Let's suppose the person is eavesdropping on every qubit sent from Alice to Bob and that Alice and Bob compare N cases in which they had chosen the same basis set. For $N = 7$, there is less than a 1% chance that all of Alice's and Bob's results agree. But the probability is not zero, and so they could get complete agreement even with eavesdropping about 1% of the time if the whole process were repeated many times. It means that 99% of the time they would have at least some results that disagree. If $N = 20$, the chance for getting all matches with eavesdropping going on is about 10^{-4}%, and they will almost certainly detect that eavesdropping is happening.

We got those results quickly by thinking about the question in a different way and that leads to relatively easy calculations. Here is how we figured it out. The situation is like a coin toss for the eavesdropper: What is the chance of tossing a coin and getting a long run of heads? If we think of eavesdropper's choosing the same basis set that Alice had chosen as a "head," we need the probability of getting N heads in a row. The calculation is pretty easy because we know that the probability of getting N heads is just the probability of getting one head (0.5) multiplied by the probability of getting another head (0.5), and so on. So, the probability of getting N heads is just $1/2^N$. After dealing with binary numbers for a long time, we remember $1/2^7 = 1/128$; this probability is just a bit less than 1%. We also remember that $1/2^{10} = 1/1024 \approx 10^{-3} = 0.1\%$ and finally $1/2^{20} = \left(1/2^{10}\right)^2 \approx 10^{-6} = 0.0001\%$. So, using the coin analogy and simple numerical results, we got the answer.

To get a more precise estimate of the probability that the eavesdropper is not detected, we should also include the probability of choosing the "wrong" basis states, but nonetheless having Bob's measurement results match what Alice sent. We know that the eavesdropper chooses the basis set that does not match Alice's about half the time, and in these cases, Bob's measurements would give the right bit result about half the time. So, the probability of each of those "correct" bits being measured is 0.25. This leads to an interesting probability calculation: the two possibilities (the eavesdropper chooses the matching basis and the eavesdropper chooses the other basis, but Bob nevertheless gets the right bit from his measurement) could occur in any order. Working out the details would take us rather far afield. In practical terms, if the number of bits sent is reasonably large (typically in the 100s), the possibility of *not* detecting the eavesdropper is so small that we need not worry about it.

8.9 Changing Bases for Matrices

We have, thus far, focused on changing basis states for state vectors. But, there are other ingredients in our quantum stew that depend on basis states: the matrix representations of operators. Those operators act on vectors to produce other vectors. We know that the matrix representations of operators depend on the basis states used. For example, if we use the computational basis states $|0\rangle$ and $|1\rangle$, the matrix elements of an operator A are the quantum sandwiches formed with those basis states:

$$A \doteq \begin{pmatrix} A_{00} & A_{01} \\ A_{10} & A_{11} \end{pmatrix} = \begin{pmatrix} \langle 0|A|0\rangle & \langle 0|A|1\rangle \\ \langle 1|A|0\rangle & \langle 1|A|1\rangle \end{pmatrix}. \tag{8.47}$$

If we change to, say, the (q, r) basis sets, the matrix representation becomes

$$A \doteq \begin{pmatrix} A_{qq} & A_{qr} \\ A_{rq} & A_{qq} \end{pmatrix} = \begin{pmatrix} \langle q|A|q\rangle & \langle q|A|r\rangle \\ \langle r|A|q\rangle & \langle r|A|r\rangle \end{pmatrix}. \tag{8.48}$$

How are the matrix elements $\langle q|A|q\rangle$, etc., related to the $\langle 0|\, A\, |0\rangle$ matrix elements, etc.? Based on our experiences with changing basis sets for vectors, we might expect that the matrix elements in one basis can be written as a linear combination of the matrix elements in another basis. In fact, that is the case. In principle, you could replace $|0\rangle$ and $\langle 0|$ in the $\langle 0|\, A\, |0\rangle$ matrix element with $|0\rangle = |q\rangle\langle q\,|0\rangle + |r\rangle\langle r\,|0\rangle$ and an analogous expression with the left vector. This procedure, while straightforward, is algebraically messy.

It is better to borrow a method from linear algebra: Given a matrix A in some representation (basis set), you can convert it to another equivalent matrix (mathematicians say "similar matrix") in another representation by using a matrix operation

$$B = M^{-1}AM, \tag{8.49}$$

where M is any matrix that has an inverse. B is said to be similar to A in the sense that it has the same determinant, the same trace, the same eigenvalues, and the same rank (the dimension of the row space = the dimension of the column space = the number of linearly independent columns in the matrix). By choosing different Ms, you can generate a "family" of similar matrices. M is said to induce a *similarity transformation* on the matrix A.

For our purposes, we build the matrix M from the Dirac brackets that connect the basis vectors in the two representations. For example, for a 2×2 matrix A, the matrix M that converts from the computational basis to the (q, r) basis is

$$M = \begin{pmatrix} \langle 0|q\rangle & \langle 0|r\rangle \\ \langle 1|q\rangle & \langle 1|r\rangle \end{pmatrix} \qquad M^{-1} = \begin{pmatrix} \langle q|0\rangle & \langle q|1\rangle \\ \langle r|0\rangle & \langle r|1\rangle \end{pmatrix} \tag{8.50}$$

Try It 8.21 Challenge: Verify that the inverse matrix in Eq. (8.50) is correct by carrying out the matrix multiplication $M^{-1}M$ and seeing if you get the identity matrix I. Hint: Use the bracket "disassembly" illustrated in Eq. (8.21) and recall that $|0\rangle\langle 0| + |1\rangle\langle 1| = \mathrm{I}$.

One particular similarity transformation plays an important role in linear algebra and in quantum mechanics: the matrix that makes the matrix B a diagonal matrix. In this case, B is usually written as Λ (Greek uppercase lambda to remind us of eigenvalues), and the diagonal elements of Λ are the eigenvalues of the matrix A. The columns of the matrix M then consists of eigenvectors of A, and M is usually labeled as Q:

$$\Lambda \doteq \begin{pmatrix} \lambda_1 & 0 \\ 0 & \lambda_2 \end{pmatrix} \doteq Q^{-1}AQ. \tag{8.51}$$

We say that Q diagonalizes A. Diagonalization is an important operation in many areas in physics and engineering.

8.10 Chapter Summary

- This chapter developed the mathematics needed to express one set of basis vectors in terms of another. This is important in QIS because measuring a quantum state in terms of a basis that is different from the one in which it was prepared can yield additional information about the state.
- The following equations express a quantum state originally in basis A in terms of basis B. The two basis vector sets differ, in state space, by an angle θ_{AB}:

$$|A_0\rangle = \cos\theta_{AB}\,|B_0\rangle + \sin\theta_{AB}\,|B_1\rangle$$
$$|A_1\rangle = -\sin\theta_{AB}\,|B_0\rangle + \cos\theta_{AB}\,|B_1\rangle$$

In Dirac notation, the basis transformation reads

$$\langle A_0|\psi\rangle = \langle A_0|B_0\rangle\langle B_0|\psi\rangle + \langle A_0|B_1\rangle\langle B_1|\psi\rangle$$
$$\langle A_1|\psi\rangle = \langle A_1|B_0\rangle\langle B_0|\psi\rangle + \langle A_1|B_1\rangle\langle B_1|\psi\rangle.$$

- Changing the basis from A to B expresses the quantum basis states $|A_0\rangle$ and $|A_1\rangle$ as a superposition state in the other basis B. If a qubit is prepared in one of the A basis states, the absolute-value-squared of the coefficients of the B basis states give, according to the Born rule, the probabilities of obtaining measurement outcomes of that state with a measurement device designed to observe B basis set. These changes in the probabilities can be leveraged for applications.
- The power of measuring using a basis different from the preparation basis was illustrated using three linear polarizers. The light passes through a horizontal polarizer first, after which the light is described by a completely horizontal polarization state. If that light were to encounter a vertical polarizer, none of the light would get through because there is no projection of the completely horizontal state vector onto the vertical state basis vector. If, however, a 45° polarizer is inserted between the horizontal and vertical filters, light does indeed pass through the vertical filter. As can be seen from the 45° state space, a vlp state is an equal superposition of 45° basis states. This tells us that light will pass through with some probability given by the coefficients of that superposition. Next, according to the third polarizing sheet's state space, the 45° basis states are superpositions of the horizontal /vertical basis states. Therefore, there is a coefficient in the vertical component of that state vector, and some light will pass through the third filter. This probability explanation is necessary for single-photon measurements because the classical wave explanation involving an electric field vector is not applicable to a single photon.
- We showed that changing the algebraic sign of one of the amplitudes in a superposition state can lead to different measurement results if the state is measured using a different set of basis states. (There is no change in the measurements if both amplitudes are multiplied by -1.)
- If we allow for complex amplitudes, then multiplying one of the amplitudes in a superposition state by a complex exponential $e^{i\phi}$ will lead to different measurement outcomes—a generalization of the statement in the previous point. ϕ is called the relative phase of the two amplitudes.
- Quantum cryptography also utilizes the superposition that results from changing basis states. Alice and Bob both randomly choose from two basis sets to prepare (Alice) and read (Bob) the photons (or other qubits) they are using to communicate. If they both choose the same set, their measurement outcomes will agree 100% on the quantum state that describes the photon, and they can use their measurements to prepare a key with which to encrypt their message. They can openly communicate which basis they used because knowing the measurement basis alone does not tell us the actual measurement result.
- The danger in sending information is that it might be intercepted by an eavesdropper. To eavesdrop undetected, the eavesdropper needs to intercept Alice's photon, read its state, and then send a photon described by the same state onto Bob. If the eavesdropper reads and prepares the photon using a basis different from Alice, she will not have a

100% probability of sending Bob a photon described by the exact state as she received it because she cannot know the exact state. Since the eavesdropper is receiving a superposition state in terms of their basis states, their measurements cannot tell them the amplitudes of the components of the superposition. Hence, what they send onto Bob will lead in general to a change in outcome probabilities for Bob's measurements. To check for eavesdropping, Alice and Bob compare not only their basis sets but also the preparation and measurement results. If eavesdropping has occurred, they will not get the same results 100% of the time even when they choose the same basis states.

- Matrices can be expressed using different basis states, and the resulting matrices are connected by a similarity transformation $B = M^{-1}AM$. Similar matrices have the same trace, determinant, eigenvectors, and rank.

8.11 Further Reading

For more information on quantum tomography:

A.I. Lvovsky and M.G. Raymer "Continuous-variable optical quantum state tomography," *Reviews of Modern Physics* **81**(1), 299–332 (2009).

Quantum key distribution experiment:

Aayam Bista, Baibhav Sharma, and Enrique J. Galvez," A demonstration of quantum key distribution with entangled photons for the undergraduate laboratory," *American Journal of Physics* **89**, 111 (2021).

PART II

Multiqubit States

Multiqubit Systems, Entanglement, and Quantum Weirdness

9

They [atoms] move in the void and catching each other up jostled together and some recoil in any direction that may chance, and others become entangled with one another in various degrees according to the symmetry of their shapes and sizes and positions and order, and they remain together and thus the coming into being of composite things is effected.

Simplicius, *De Caelo*

9.1 Two-Qubit Quantum States

In this chapter, we talk about systems of two qubits. For useful quantum computing applications, we need systems with at least two qubits. How do we describe the states of those systems? What happens when we make measurements on the individual qubits? As shown later, systems with two (or more) qubits have properties that are both weird and useful as computational resources.

Before we talk about quantum states, let's think about general systems with two (or more) parts. We know that quantum states carry information about probabilities of results of various measurements, so let's look at some probabilities that involve two parts. From probability theory, we know that if two events are independent, then the probability of event 1 *and* event 2 happening is given by the product of the probabilities of each of the events $P(1 \text{ AND } 2) = P(1)P(2)$. For example, if we have a system that consists of a standard six-sided die and a coin, the probability of getting a 5 on the die AND getting heads on a fair coin is

$$P(5 \text{ AND Heads}) = P(5)P(\text{Heads}) = (1/6)) \times (1/2) = 1/12. \tag{9.1}$$

We assume that the die and the coin tosses are independent. If the coin were glued to one side of the die, they would not be independent and Eq. (9.1) would not apply.

As we know, in quantum mechanics, probabilities are calculated from quantum state vectors. It suggests that if we have two "independent" qubits (we'll need to figure out what

Quantum Computing and Quantum Physics. Alice Flarend and Robert Hilborn, Oxford University Press.
 DOI: 10.1093/oso/9780197904930.003.0009

independent means for qubits), then the quantum state vector $|S\rangle$ for the system of two qubits ought to be the product of the quantum state vectors for each of the qubits:

$$|S\rangle = \underbrace{|A\rangle}_{\text{Alice's state}} \otimes \underbrace{|B\rangle}_{\text{Bob's state}} = |A\rangle|B\rangle, \tag{9.2}$$

where $|A\rangle$ is Alice's state vector and $|B\rangle$ is Bob's state vector. The symbol $\otimes$ is used to indicate "multiplication" of two right vectors. This kind of product is called a "tensor product." We often omit that symbol, as we have done on the right side of Eq. (9.2) when the product form is obvious. In other situations, we use the symbol to help guide the eye in interpreting equations.

Note that the qubits could be entirely different kinds of physical systems (just like the die and the coin); Alice might use photon polarization and Bob might use neutron spin. Or Alice may use a photon that produces the sensation "blue light" in our eyes while Bob's photon produces the sensation "red light."

We also know that the quantum state vector of each qubit can be represented as a superposition of basis state vectors. So, if Alice uses a linearly polarized photon qubit and Bob uses an electron's spin-1/2 qubit of the type we looked at in Chapter 7, we can rewrite Eq. (9.2) as

$$|S\rangle = |A\rangle \otimes |B\rangle = \underbrace{\left[a_0|\text{hlp}\rangle + a_1|\text{vlp}\rangle\right]}_{\text{Alice's state}} \otimes \underbrace{\left[b_0|\uparrow\rangle + b_1|\downarrow\rangle\right]}_{\text{Bob's state}}. \tag{9.3}$$

You should recognize a superposition of linear polarization basis states for Alice's state and a superposition of spin-up and spin-down states for Bob's. If the state vector for the system can be written in the simple product form shown in Eq. (9.3), the two qubits are independent in the sense that the probability of Alice's observing a photon with vertical polarization is $|a_1|^2$ while the probability of Bob's observing an electron with spin-up is $|b_0|^2$. Moreover, the probabilities for Bob's observations are independent of Alice's observations and vice versa.

In most quantum computing applications, the two qubits in a two-qubit system are of the same physical type: for example, two photons, two electrons, two ions, or two superconducting circuits. From now on, we will assume, unless otherwise explicitly stated, that the qubits are of the same physical type.

What kind of state space does a two-qubit state occupy? Alice expresses her quantum state in terms of two basis state vectors while Bob may use a different pair of basis state vectors. This gives us a total of four basis state vectors, which tells us that two-qubit quantum states "live" in a four-dimensional state space. (You will pardon us if we don't draw a picture of that four-dimensional state space!) This space is called a product space because in a sense it contains the product of what Alice's state space contains and what Bob's state space contains. Its dimensionality is the product of Alice's state space dimensionality and Bob's state space dimensionality. In this case, the dimensionality becomes $2 \times 2 = 4$. If we combined three qubits, we would have a $2 \times 2 \times 2 = 8$ dimensional state space.

Table 9.1 Measurement results and system state vectors for a 2×2 product space.

Alice's result	Bob's result	System state vector
0	0	$\lvert 0\rangle_A \otimes \lvert 0\rangle_B = \lvert 0\rangle_A \lvert 0\rangle_B = \lvert 0_A 0_B\rangle$
0	1	$\lvert 0\rangle_A \otimes \lvert 1\rangle_B = \lvert 0\rangle_A \lvert 1\rangle_B = \lvert 0_A 1_B\rangle$
1	0	$\lvert 1\rangle_A \otimes \lvert 0\rangle_B = \lvert 1\rangle_A \lvert 0\rangle_B = \lvert 1_A 0_B\rangle$
1	1	$\lvert 1\rangle_A \otimes \lvert 1\rangle_B = \lvert 1\rangle_A \lvert 1\rangle_B = \lvert 1_A 1_B\rangle$

There is a way (not very elegant, but it is straightforward) to see that we need a four-dimensional state space: we simply write down all the possible results of measurements on Alice's and Bob's qubits. First, let's express the overall two-qubit state using the computational basis states rather than the polarization and spin basis states in Eq. (9.3). There are four individual basis states:

$$\lvert 0\rangle_A,\ \lvert 1\rangle_A,\ \lvert 0\rangle_B,\ \text{and}\ \lvert 1\rangle_B. \tag{9.4}$$

The subscripts tell us which of the qubits is associated with each state vector. Note that we have four different possible states, two for each qubit. When Alice makes measurements on her states, she gets 0 or 1 (using our usual computational basis form) and Bob can get 0 or 1, so together there are four possibilities:

In the rightmost column of Table 9.1, we have shown three different formats for the system state vectors that correspond to the measurement results listed in the first two columns. In the first format, the system state vectors are written as products of the individual state vectors (with labels A and B to distinguish Alice's from Bob's). The third, more succinct format uses a single state vector with two labels. We will always put Alice's label on the left and Bob's on the right. We should note that many authors do not use subscripts in this way and leave it to you to infer the appropriate basis states from the context. This is fine as long as we always agree to use the same order for the labels.

We can also see the need for the product basis states by examining the product of two states: one from the A state space and one from B:

$$\begin{aligned} \lvert\psi\rangle &= \left(a_0\lvert 0\rangle_A + a_1\lvert 1\rangle_A\right) \otimes \left(b_0\lvert 0\rangle_B + b_1\lvert 1\rangle_B\right) \\ &= a_o b_o\lvert 0\rangle_A\lvert 0\rangle_B + a_o b_1\lvert 0\rangle_A\lvert 1\rangle_B + a_1 b_o\lvert 1\rangle_A\lvert 0\rangle_B + a_1 b_1\lvert 1\rangle_A\lvert 1\rangle_B. \end{aligned} \tag{9.5}$$

We see that a state in the composite (product) state space is not the sum of vectors in the individual state spaces. It is the sum of *products* of the individual state vectors.

The crucial (and in some sense weird) point of this chapter is that there are system quantum states that describe qubits that are not independent. In these situations, what Bob observes is correlated with what Alice observes and vice versa. Such states are called "entangled" states and play a fundamental role in quantum information science (QIS) and quantum computing (QC). So, we take some time to understand how to determine if a state of a two-qubit system is entangled (or not) and how we describe such an entangled state. Then, in the following chapters, we see how entangled states can be prepared and manipulated for QIS and QC tasks.

9.2 General Two-Qubit States

To explore the properties of multiqubit states, we will start with a system of just two qubits. The generalization to more than two qubits is conceptually straightforward but is messy to write out in detail.

To be specific, we will use two spin-1/2 qubits (though they might be different kinds of spin-1/2 particles, e.g., one electron and one neutron) and write the system basis states in a form like that in Eq. (9.2), namely $|S\rangle = |A\rangle|B\rangle$. We will label Alice's qubits A and B for Bob's. We will also translate the formalism into the computational basis states $|0\rangle$ and $|1\rangle$.

For two spin-1/2 qubits, there are four possible combinations of spin-up and spin-down for observations along a chosen direction. It means, as we mentioned in the previous section, we have a four-dimensional state space and four basis states. There are several different ways of writing the basis states; we shall use all of them as we proceed. Table 9.2 shows three of the four basis states in four different formats. (You should fill in the last row of the table (see Try It 9.1).)

In Table 9.2, we have used four formal representations: Going from left to right, the first three formats are the same as those in Table 9.1: a product of individual state vectors, a single state vector for the system of two qubits, and a computational basis right vector for the combined system. The rightmost column introduces a set of four-entry column vectors whose entries are the projections of the state vectors along the four state-space directions.

Table 9.2 Three of the four basis states in four different formats.

Individual quantum state vectors in the spin basis	System vectors in the spin basis	System vectors in the computational basis	System column vectors
$|\uparrow\rangle_A|\uparrow\rangle_B$	$|\uparrow_A\uparrow_B\rangle$	$|00\rangle$	$\begin{pmatrix}1\\0\\0\\0\end{pmatrix}$
$|\uparrow\rangle_A|\downarrow\rangle_B$	$|\uparrow_A\downarrow_B\rangle$	$|01\rangle$	$\begin{pmatrix}0\\1\\0\\0\end{pmatrix}$
$|\downarrow\rangle_A|\uparrow\rangle_B$	$|\downarrow_A\uparrow_B\rangle$	$|10\rangle$	$\begin{pmatrix}0\\0\\1\\0\end{pmatrix}$
TBC	TBC	TBC	TBC

TBC means to be completed by the reader (see Try It 9.1).

The subscripts, when used, remind us that one of the qubits belongs to Alice and the other to Bob. Note that the computational basis vectors in the third column are in numerical order (in the binary system) going from the top row down: $|00\rangle$, $|01\rangle$, $|10\rangle$. This method of labeling states will be useful in describing multiqubit systems with a large number of qubits. In the four-entry column vector representation of the states in Table 9.2, the top column vector has a unit component along the first of the four state-space directions and 0 along the others, the next vector has unit length along the second state-space dimension and 0 along the others, and so on. As you become more proficient at quantum thinking, you should be able to switch back and forth among these representations.

Try It 9.1 Complete the fourth row of Table 9.2. Explain your reasoning.

The four-entry column vectors on the far-right column of Table 9.2 are examples of the tensor product of the individual qubit column vectors. The general rule for the tensor product of two column vectors is

$$\begin{pmatrix} a_1 \\ a_2 \end{pmatrix} \otimes \begin{pmatrix} b_1 \\ b_2 \end{pmatrix} = \begin{pmatrix} a_1 \begin{pmatrix} b_1 \\ b_2 \end{pmatrix} \\ a_2 \begin{pmatrix} b_1 \\ b_2 \end{pmatrix} \end{pmatrix} = \begin{pmatrix} a_1 b_1 \\ a_1 b_2 \\ a_2 b_1 \\ a_2 b_2 \end{pmatrix}. \tag{9.6}$$

Note that this is not the same as multiplying two square matrices although at first glance it may look like similar. You cannot multiply two 2×1 matrices using the matrix product rules. Tensor products of states are different animals. See Appendix: Quantum Toolkit for more on vector and matrix products.

9.3 Two-Qubit Basis Vectors Are Orthogonal

Remember that in a two-dimensional (single-qubit) state space, we wanted to have the basis state vectors perpendicular (orthogonal) to each other. This makes many of our calculations much easier to write out and to understand. It also reflects the fact that the basis states correspond to mutually exclusive measurement outcomes. We want to have that same orthogonality in our four-dimensional state space though it is admittedly more difficult to show that. We also need to remember that these are the basis state vectors that are perpendicular (in state space) and not the spins of Bob's and Alice's particles.

Try It 9.2 Use the inner product multiplication rule introduced in Chapter 4 (row vector times a column vector) to show that the four basis state column vectors in Column 4 of Table 9.2 are orthogonal to each other and also that they have unit length.

Try It 9.3 If we want to describe a spin-1/2 three-qubit state, how many basis states do we need? Hint: Write out all the possibilities of spin-up and spin-down.

Try It 9.4 Generalize these results to find how many basis states are needed for n spin-1/2 qubit states. Hint: For one qubit, we have a $2^1 = 2$-dimensional state space. For two qubits, we have $2^2 = 4$ dimensions.

While we are thinking about multiqubit states, we want to show you some notation that will be used when we are dealing with many qubits. To be general, let's suppose that we are working with n qubits. Then, we have 2^n basis states. It turns out to be useful to label those basis states and their state amplitudes with n binary bits, starting with $000\ldots0$ and ending with $111\ldots1$. Note that the numerically smallest label is $000\ldots0$ and the largest is $(111 \ldots 1)_2 = (2^n - 1)_{10}$, where we used subscripts to distinguish between binary and base 10 numbers.

In the QC and QIS literature, the symbol x is often used to denote the labels. Using that convention, a general superposition state of the n-qubit system can be written as

$$|\psi\rangle = \sum_{x=000\ldots0}^{111\ldots1} a_x|x\rangle. \tag{9.7}$$

We use uppercase Greek sigma (Σ) to indicate that we sum what follows with x ranging from 0 to $2^n - 1$. For example, for a two-qubit system, we have $2^n - 1 = 3$ and Eq. (9.7) becomes

$$|\psi\rangle = \sum_{x=0}^{11} a_x|x\rangle = a_{00}|00\rangle + a_{01}|01\rangle + a_{10}|10\rangle + a_{11}|11\rangle. \tag{9.8}$$

Try It 9.5 Write the equivalent of Eq. (9.8) for a three-qubit ($n = 3$) system and a four-qubit system. Compare these results to the ones in Table. 9.2.

9.4 Single Qubits and Product States

We should point out that product states are also used to describe the states of a single qubit (physical object) if that qubit has several distinct dynamical properties—properties that can be manipulated and change with time. In tech speak, such dynamical properties are called "degrees of freedom." A specific example will be helpful. Suppose the qubit is a photon. One of the dynamical properties is its linear polarization. We have seen how we can manipulate the linear polarization state by using polarizing filters, for example. But the photon has other dynamical properties: the direction or path in which it is traveling and its energy,

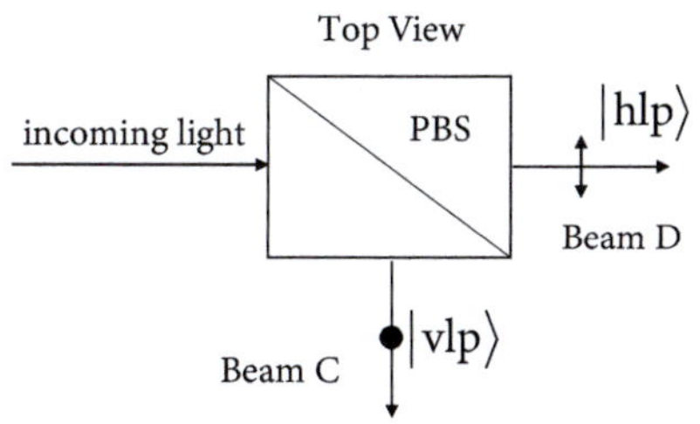

Fig. 9.1 A top view of a polarizing beam splitter (PBS).

which is proportional to the oscillation frequency of the corresponding classical wave. You may recall that a green photon has more energy than a red photon.

Taking those three dynamical properties into account, we write the state $|S\rangle$ of the photon as

$$|S\rangle = |\text{polarization}\rangle|\text{direction of propagation}\rangle|\text{energy}\rangle\,. \tag{9.9}$$

Each part of the product state might be a superposition state as we have seen for polarization. The overall state space for the qubit is then a tensor product state space.

If a particular dynamical property does not change in the situation we have set up, we can drop that part of the product state and ignore that part of the tensor product state space. This is often, but not always, the case with the photon's energy.

Let's return to our old friend the polarizing beam splitter (PBS), as shown in Fig. 9.1. As we discussed in Chapter 4, the PBS splits an incoming light beam into two beams traveling in different directions. This is one property. One of the outgoing beams is vlp light and the other is hlp, which is a second property. In this case, the state for a photon that has interacted with the PBS (assumed to be a 50:50 beam splitter for the sake of simplicity) is

$$|S\rangle = \frac{1}{\sqrt{2}}|\text{vlp}\rangle \otimes |\text{beam C}\rangle + \frac{1}{\sqrt{2}}|\text{hlp}\rangle \otimes |\text{beam D}\rangle\,. \tag{9.10}$$

Given this form, we know, even if we don't directly measure the polarization mode, that if we do measure the photon's path and detect it in beam C, its polarization state was vlp. Similarly, being in beam D is correlated with hlp polarization. We see that the PBS produces a correlation between the qubit's polarization state and its propagation direction state. In the next section, we explore similar correlations when our system has more than one qubit.

9.5 **Quantum Entanglement**

In this section, we show that if the state vector of the combined two-qubit system can be written as a product of the quantum state vectors of the initial subsystems (Alice's and Bob's) as illustrated in Eq. (9.2), then the probabilities of Alice observing spin-up or spin-down for her qubit are independent of what Bob observes for his qubit. You might ask if things could be otherwise. Hold onto your state-space vectors because we show that there are system states for which Alice's and Bob's observations are *not* independent. We use the spin-up and spin-down basis states to illustrate the issues, but you could use the computational basis states $|0\rangle$ and $|1\rangle$ as well.

Entanglement may appear weird to you as it did to Einstein but there is a way to think about it as just an extension of classical conservation laws. No one thinks it is weird that the law of conservation of momentum connects the final momenta of two cars after a collision based upon their total momentum before the collision. If we know the starting conditions of the system and then the momentum of one of the cars after the collision, we know the momentum of the other car even without measuring it. We do not normally use the word entangled in this context, but we could say that as a result of the collision, the momenta of the two cars become entangled (correlated). For a multiqubit quantum system, we can make similar predictions, but because it is a quantum system, there is not just a single measurement outcome. Instead, there are several, each with a probability of occurring. If we repeat the experiment making a measurement on just one of the qubits, there is no guarantee that we get the same results on the next run of the experiment. However, the entanglement of the qubits means that we get information about the qubit we did not measure. We explore this fact in more detail shortly.

To be quite general, we write the system state vector for the two-qubit system as a superposition of the four basis states listed in Table 9.2. (Recall that any state vector in the two-qubit vector space can be written as a linear combination of that space's basis vectors.) Three of the basis vectors are listed in the second column of Table 9.2 with the fourth from the results of your Try It:

$$|S\rangle = c_{\uparrow\uparrow}|\uparrow_A\uparrow_B\rangle + c_{\uparrow\downarrow}|\uparrow_A\downarrow_B\rangle + c_{\downarrow\uparrow}|\downarrow_A\uparrow_B\rangle + c_{\downarrow\downarrow}|\downarrow_A\downarrow_B\rangle. \tag{9.11}$$

We show (spoiler alert!) that if $c_{\uparrow\uparrow}c_{\downarrow\downarrow} = c_{\uparrow\downarrow}c_{\downarrow\uparrow}$ (the product of the amplitudes of the two outer terms is equal to the product of the amplitudes of the two inner terms), then the state vector can be written as shown in Eq. (9.3) as a product of the state vector for Bob's qubit and the state vector for Alice's qubit. In this case, Alice's and Bob's observations are independent. Remember when we use "outer" and "inner" coefficients in this way, we need to use our standard ordering of the state labels with "outer" corresponding to $|00\rangle$ and $|11\rangle$ in the third column of Table 9.2 and "inner" corresponding to $|01\rangle$ and $|10\rangle$.

The derivation of the criterion is not difficult. Essentially, we want to see if we can write the state vector in Eq. (9.11) as a product of two state vectors: one for Alice and one for Bob. In other words, we want to find the four amplitudes g, h, m, and n that satisfy

$$\begin{aligned}|S\rangle &= c_{\uparrow\uparrow}|\uparrow_A\uparrow_B\rangle + c_{\uparrow\downarrow}|\uparrow_A\downarrow_B\rangle + c_{\downarrow\uparrow}|\downarrow_A\uparrow_B\rangle + c_{\downarrow\downarrow}|\downarrow_A\downarrow_B\rangle \\ &= \underbrace{\{g|\uparrow_A\rangle + h|\downarrow_A\rangle\}}_{\text{Alice's qubit state}}\underbrace{\{m|\uparrow_B\rangle + n|\downarrow_B\rangle\}}_{\text{Bob's qubit state}}.\end{aligned} \tag{9.12}$$

Note that the second line came about by factoring out Alice's states as required for independent observations. Multiplying out the second line of Eq. (9.12) with the new variables and comparing that result to the first line, we find that we have four equations to satisfy:

$$gm = c_{\uparrow\uparrow} \quad gn = c_{\uparrow\downarrow} \quad hm = c_{\downarrow\uparrow} \quad hn = c_{\downarrow\downarrow}. \tag{9.13}$$

While this seems to get us nowhere, look carefully at the equations for $c_{\uparrow\uparrow}$ and $c_{\uparrow\downarrow}$. They have a factor of g in common, so dividing them will cancel out that variable. Similarly, $c_{\downarrow\uparrow}$ and $c_{\downarrow\downarrow}$ have a factor of h in common. This calculation yields $m/n = c_{\uparrow\uparrow}/c_{\uparrow\downarrow}$ and $m/n = c_{\downarrow\uparrow}/c_{\downarrow\downarrow}$. Further, both of those pairs have m/n in common, so we have a criterion for independent states:

$$c_{\uparrow\uparrow}c_{\downarrow\downarrow} = c_{\uparrow\downarrow}c_{\downarrow\uparrow}. \tag{9.14}$$

Thus, our criterion for the state in Eq. (9.11) to be a product state is Eq. (9.14): The product of the two "outer" state amplitudes equals the product of the two "inner" state amplitudes.

Conversely—and this is the critical point—if $c_{\uparrow\uparrow}c_{\downarrow\downarrow} \neq c_{\uparrow\downarrow}c_{\downarrow\uparrow}$, then the state vectors cannot be written as a simple product state and, as shown later, Alice's observations in general are correlated with what Bob observes and vice versa. The observations on the two qubits are not independent. In this case, the overall state vector of the system is called an *entangled state vector*. Or more succinctly, the system state is entangled.

Using the coefficient combinations in Eq. (9.14), we may define a quantitative measure of the "degree of entanglement" of a two-qubit state: $s = 2\left|c_{\uparrow\uparrow}c_{\downarrow\downarrow} - c_{\uparrow\downarrow}c_{\downarrow\uparrow}\right|$. We see that $s = 0$ for a separable (unentangled) state. It turns out that the maximum value of s is 1; in this case we say that the system state is maximally entangled.

This example shows that all entangled states are superposition states, but not all superposition states are entangled. The properties of entangled state vectors turn out to be rather weird (or at least very counterintuitive), but they are exceedingly important for QIS and QC.

N.B. Note that entanglement appears automatically when we consider superposition states for a multiqubit system. Entanglement, important as it may be, is not an additional principle of quantum mechanics. It is just a "feature" of superposition states in multiqubit systems.

Many people who work on QIS and QC say "Alice's qubit is entangled with Bob's qubit." In the language we have been using throughout this book, qubits are the physical systems (the photons, electrons, neutrons, circulating superconducting currents, . . .), and the quantum states are the formal descriptions used to predict the outcomes of measurements on those qubits. From that point of view, it is better to say that we use an entangled state to describe the system. This entangled state contains all details about the outcome of measurements on the system. We explain this issue in Chapter 17 where we discuss some of the fundamental, almost philosophical, issues surrounding the relationship between qubits and quantum states. As shown in this chapter, it is not always possible to say that a qubit "has" a quantum state in the way a car has a headlight. A lot of conceptual confusion can be avoided by distinguishing between the concept of qubit as a physical object and the concept of a quantum state. So, we will stick with saying "the system is described by an entangled state."

Try It 9.6 Below is a list of several two-qubit spin-1/2 states. Which ones are entangled states and which are unentangled states? Explain your reasoning.

$$
\begin{aligned}
|S_1\rangle &= \frac{1}{\sqrt{2}}|\uparrow_A\uparrow_B\rangle - \frac{1}{\sqrt{2}}|\downarrow_A\downarrow_B\rangle \\
|S_2\rangle &= \frac{\sqrt{3}}{2}|\uparrow_A\downarrow_B\rangle + \frac{1}{2}|\downarrow_A\uparrow_B\rangle \\
|S_3\rangle &= \frac{\sqrt{3}}{2}|\uparrow_A\downarrow_B\rangle + \frac{1}{2}|\uparrow_A\uparrow_B\rangle \\
|S_4\rangle &= \frac{\sqrt{3}}{2}|\uparrow_A\downarrow_B\rangle - \frac{1}{2}|\uparrow_A\uparrow_B\rangle. \\
|S_5\rangle &= \frac{1}{2\sqrt{2}}|\uparrow_A\uparrow_B\rangle + \frac{1}{2}|\uparrow_A\downarrow_B\rangle + \frac{1}{2}|\downarrow_A\uparrow_B\rangle + \frac{\sqrt{3}}{2\sqrt{2}}|\downarrow_A\downarrow_B\rangle \\
|S_6\rangle &= \frac{1}{2}|\uparrow_A\uparrow_B\rangle + \frac{1}{2}|\uparrow_A\downarrow_B\rangle + \frac{1}{2}|\downarrow_A\uparrow_B\rangle + \frac{1}{2}|\downarrow_A\downarrow_B\rangle \\
|S_7\rangle &= \frac{1}{2}|\uparrow_A\uparrow_B\rangle - \frac{1}{2}|\uparrow_A\downarrow_B\rangle + \frac{1}{2}|\downarrow_A\uparrow_B\rangle + \frac{1}{2}|\downarrow_A\downarrow_B\rangle
\end{aligned}
\tag{9.15}
$$

Partial Answer: $|S_3\rangle$ is not an entangled state because it can be factored into a simple product state $|S_3\rangle = |\uparrow_A\rangle\left\{\frac{\sqrt{3}}{2}|\downarrow_B\rangle + \frac{1}{2}|\uparrow_B\rangle\right\}$ with one factor for Alice and the other for Bob. We also see that the product of the two outer basis state amplitudes = 0 and the product of the two inner basis state amplitudes = 0. Since these products are equal, the state is not entangled.

We now investigate the probabilities for various measurement outcomes for a two-qubit system. As usual, the quantum state for a system contains all the information needed to calculate those probabilities, and those probabilities are both necessary and sufficient to predict all measurement outcomes for the system. However, with a two-qubit system, there are several different types of probabilities we can measure. A brief review of some basic probability theory will be helpful (see Appendix: Quantum Toolkit for more details).

First, we might ask for the probability, given the quantum state in Eq. (9.11), that Alice's measurement outcome is spin-up for her qubit and Bob's measurement outcome is also spin-up for his qubit. In probability language, we are asking for the *joint* probability $P(\uparrow_A \cap \uparrow_B)$ of these two events. The "upside-down" U means we are looking at the probability when both $\uparrow_A$ AND (in the Boolean logic sense) $\uparrow_B$ occur. To find that probability, we make use of the Born rule. Since the part of the state in which both qubits have spin-up

Table 9.3 The joint probabilities for Alice's and Bob's observations from Eq. (9.11).

$P(\uparrow_A \cap \uparrow_B)$	$P(\uparrow_A \cap \downarrow_B)$	$P(\downarrow_A \cap \uparrow_B)$	$P(\downarrow_A \cap \downarrow_B)$
$\lvert c_{\uparrow\uparrow}\rvert^2$	$\lvert c_{\uparrow\downarrow}\rvert^2$	$\lvert c_{\downarrow\uparrow}\rvert^2$	$\lvert c_{\downarrow\downarrow}\rvert^2$

is $|\uparrow_A\uparrow_B\rangle$, the probability is given by $|\langle\uparrow_A\uparrow_B|\psi\rangle|^2$. Recalling that the various two-qubit basis states are orthogonal, we see that the joint probability is given by

$$|\langle\uparrow_A\uparrow_B|\psi\rangle|^2 = |c_{\uparrow\uparrow}|^2. \tag{9.16}$$

The results for all the joint probabilities are given in Table 9.3.

Once we have the joint probabilities, we can use them to calculate the expectation (average) value of the property associated with some operator D. That expectation value is given by $\langle D\rangle = D_{\uparrow\uparrow}P(\uparrow \cap \uparrow) + D_{\uparrow\downarrow}P(\uparrow \cap \downarrow) + D_{\downarrow\uparrow}P(\downarrow \cap \uparrow) + D_{\downarrow\downarrow}P(\downarrow \cap \downarrow)$, where $D_{\uparrow\uparrow} = \langle\uparrow| D |\uparrow\rangle$, with analogous expressions for the other states.

Next, we can ask about "total probabilities." (These are sometimes called *marginal* probabilities because in the days before computerized spreadsheets, statisticians would add up the contributions to the probability and write the result in the page margins.) For example, what is the probability that Alice gets the measurement outcome $|\uparrow_A\rangle$ no matter what Bob gets? We can calculate that probability by summing the two joint probabilities:

$$P(\uparrow_A) = P(\uparrow_A \cap \uparrow_B) + P(\uparrow_A \cap \downarrow_B) = |c_{\uparrow\uparrow}|^2 + |c_{\uparrow\downarrow}|^2, \tag{9.17}$$

using the results in Table 9.3. Similarly, we find

$$P(\downarrow_A) = P(\downarrow_A \cap \uparrow_B) + P(\downarrow_A \cap \downarrow_B) = |c_{\downarrow\uparrow}|^2 + |c_{\downarrow\downarrow}|^2. \tag{9.18}$$

Note that $P(\uparrow_A) + P(\downarrow_A) = 1$ since the system state vector is normalized.

Try It 9.7 Show that $P(\uparrow_B) = |c_{\uparrow\uparrow}|^2 + |c_{\downarrow\uparrow}|^2$ and $P(\downarrow_B) = |c_{\uparrow\downarrow}|^2 + |c_{\downarrow\downarrow}|^2$.

For your convenience, we have gathered the total probability results in Table 9.4.

The other type of probability we need to consider is the *conditional* probability. For example, the conditional probability answers the question what is the probability that Bob gets a measurement outcome $|\uparrow_B\rangle$ given that Alice observed $|\uparrow_A\rangle$. The conditional probability in that situation is written as $P(\uparrow_B|\uparrow_A)$. In probability theory, Bayes's rule (also known as Bayes's theorem) connects the joint probability to the conditional probabilities and the total probability. For our example, Bayes's rule reads

$$P(\uparrow_A \cap \uparrow_B) = P(\uparrow_A|\uparrow_B)P(\uparrow_B) = P(\uparrow_B|\uparrow_A)P(\uparrow_A). \tag{9.19}$$

Table 9.4 The total probabilities for the state in Eq. (9.11).

$P(\uparrow_A)$	$P(\downarrow_A)$	$P(\uparrow_B)$	$P(\downarrow_B)$
$\lvert c_{\uparrow\uparrow}\rvert^2 + \lvert c_{\uparrow\downarrow}\rvert^2$	$\lvert c_{\downarrow\uparrow}\rvert^2 + \lvert c_{\downarrow\downarrow}\rvert^2$	$\lvert c_{\uparrow\uparrow}\rvert^2 + \lvert c_{\downarrow\uparrow}\rvert^2$	$\lvert c_{\uparrow\downarrow}\rvert^2 + \lvert c_{\downarrow\downarrow}\rvert^2$

Table 9.5 Conditional probabilities for a two-qubit system.

$P(\uparrow_A\vert\uparrow_B)$	$P(\uparrow_A\vert\downarrow_B)$	$P(\downarrow_A\vert\uparrow_B)$	$P(\downarrow_A\vert\downarrow_B)$
$\frac{\lvert c_{\uparrow\uparrow}\rvert^2}{\lvert c_{\uparrow\uparrow}\rvert^2+\lvert c_{\downarrow\uparrow}\rvert^2}$	$\frac{\lvert c_{\uparrow\downarrow}\rvert^2}{\lvert c_{\uparrow\downarrow}\rvert^2+\lvert c_{\downarrow\downarrow}\rvert^2}$	$\frac{\lvert c_{\downarrow\uparrow}\rvert^2}{\lvert c_{\uparrow\uparrow}\rvert^2+\lvert c_{\downarrow\uparrow}\rvert^2}$	$\frac{\lvert c_{\downarrow\downarrow}\rvert^2}{\lvert c_{\uparrow\downarrow}\rvert^2+\lvert c_{\downarrow\downarrow}\rvert^2}$
$P(\uparrow_B\vert\uparrow_A)$	$P(\uparrow_B\vert\downarrow_A)$	$P(\downarrow_B\vert\uparrow_A)$	$P(\downarrow_B\vert\downarrow_A)$
$\frac{\lvert c_{\uparrow\uparrow}\rvert^2}{\lvert c_{\uparrow\uparrow}\rvert^2+\lvert c_{\uparrow\downarrow}\rvert^2}$	$\frac{\lvert c_{\downarrow\uparrow}\rvert^2}{\lvert c_{\downarrow\downarrow}\rvert^2+\lvert c_{\downarrow\uparrow}\rvert^2}$	$\frac{\lvert c_{\uparrow\downarrow}\rvert^2}{\lvert c_{\uparrow\downarrow}\rvert^2+\lvert c_{\uparrow\uparrow}\rvert^2}$	$\frac{\lvert c_{\downarrow\downarrow}\rvert^2}{\lvert c_{\downarrow\uparrow}\rvert^2+\lvert c_{\downarrow\downarrow}\rvert^2}$

If we know the joint and total probabilities, we can calculate the conditional probabilities from Eq. (9.19). Using the previous results for the total and joint probabilities, we find

$$\begin{aligned} P(\uparrow_B|\uparrow_A) &= \frac{P(\uparrow_A \cap \uparrow_B)}{P(\uparrow_A)} = \frac{|c_{\uparrow\uparrow}|^2}{|c_{\uparrow\uparrow}|^2 + |c_{\uparrow\downarrow}|^2} \\ P(\uparrow_B|\downarrow_A) &= \frac{P(\downarrow_A \cap \uparrow_B)}{P(\downarrow_A)} = \frac{|c_{\downarrow\uparrow}|^2}{|c_{\downarrow\uparrow}|^2 + |c_{\downarrow\downarrow}|^2}. \end{aligned} \tag{9.20}$$

The conditional probabilities are listed in Table 9.5.

Try It 9.8 Read Eq. (9.19) paying particular attention to different types of probabilities that appear there. Apply Eq. (9.20) to a coin flip situation. Do the results make sense in terms of what you know about coin flip probabilities?

Try It. 9.9 Verify the results for the conditional probabilities in Table 9.5.

Independent Qubits and Separable States

If Alice's and Bob's qubits are independent, their measurement outcomes are not related. In informal language, for independent qubits, Bob's qubit doesn't care what happened to Alice's qubit. For that to work out, we need a certain relationship among the state coefficients in Eq. (9.11). From Eq. (9.20), we see that we need

$$P(\uparrow_B|\uparrow_A) = P(\uparrow_B|\downarrow_A) \Rightarrow \frac{|c_{\uparrow\uparrow}|^2}{|c_{\uparrow\uparrow}|^2 + |c_{\uparrow\downarrow}|^2} = \frac{|c_{\downarrow\uparrow}|^2}{|c_{\downarrow\uparrow}|^2 + |c_{\downarrow\downarrow}|^2}. \tag{9.21}$$

Multiplying out the far right of Eq. (9.21), we see that we need $|c_{\uparrow\uparrow}|^2|c_{\downarrow\downarrow}|^2 = |c_{\uparrow\downarrow}|^2|c_{\downarrow\uparrow}|^2$, essentially the square of the result we found previously (Eq. (9.14)) for separable states.

Try It 9.10 Verify that the following two-qubit state is a separable state:

$$|\psi_1\rangle = \frac{1}{2}\left[|\uparrow_A\uparrow_B\rangle + |\uparrow_A\downarrow_B\rangle + |\downarrow_A\uparrow_B\rangle + |\downarrow_A\downarrow_B\rangle\right]$$

For this state, calculate the following probabilities: $P(\uparrow_A \cap \uparrow_B)$, $P(\uparrow_A \cap \downarrow_B)$, $P(\downarrow_A \cap \uparrow_B)$, and $P(\downarrow_A \cap \downarrow_B)$. Then, calculate the four total probabilities $P(\uparrow_A)$, etc., as shown in Eq. (9.21) for independent qubits. Are these results expected or surprising? Explain your reasoning.

Try It 9.11 From probability theory we expect $P(\uparrow_A \cap \downarrow_B) = P(\uparrow_A)P(\downarrow_B)$ for independent qubits. Does this work for the state in the previous Try It.

Try It 9.12 Yet another criterion for separability is to require that the conditional probabilities be equal to the total probabilities for separable states. For the state in Try It 9.9, do we get $P(\uparrow_B|\uparrow_A) = P(\uparrow_B)$?

Entangled States

The unusual properties of *entangled* states are manifest if we calculate probabilities, particularly conditional probabilities, for an entangled state. Suppose Alice and Bob prepare a two-qubit system described by an entangled state. Alice keeps one of the qubits while Bob transports the other qubit (gently!) to a distant location to make sure there is no hidden connection between the two objects. We can calculate various conditional probabilities from the results given in Tables 9.3–9.5. Rather than looking at general cases, let's examine some specific entangled states.

The first example is the state

$$|\Psi_1\rangle = \frac{1}{\sqrt{3}}\left[|\uparrow_A\uparrow_B\rangle + |\uparrow_A\downarrow_B\rangle + |\downarrow_A\downarrow_B\rangle\right]. \tag{9.22}$$

For this state, the parameter $s = 2/3$; so, the state is an entangled state, but it is not a maximally entangled state. Using the results in Tables 9.3–9.5, we find that the total probabilities are given by $P(\uparrow_A) = 2/3$, $P(\downarrow_A) = 1/3$, $P(\uparrow_B) = 1/3$, and $P(\downarrow_B) = 2/3$. Let's look at just two of the conditional probabilities: $P(\uparrow_B|\uparrow_A) = 1/2$ and $P(\uparrow_B|\downarrow_A) = 0$. The two conditional probabilities are not equal and they are not equal to the total probabilities $P(\uparrow_B)$ and $P(\downarrow_B)$, typical results for entangled states. Both indicate that the measurement outcomes on the individual qubits, which may be far apart spatially, are correlated.

As we have shown before, these results were derived in a straightforward way from just basic probability and the Born rule. Nevertheless, the results have generated heated discussions over several decades. For example, some physicists say that $P(\uparrow_B|\uparrow_A) = 1/2$ and $P(\uparrow_B|\downarrow_A) = 0$ tells us that what Alice measures affects Bob's measurement outcomes even though the two qubits may be far apart. Others say that the results tell us that Alice's observation affects Bob's qubit or at least its quantum state (that's why the measurement outcome probabilities change). Almost all agree that whatever is going on can't be a physical process connecting one qubit to the other because the qubits may be so far apart that nothing traveling at or less than the speed of light can get from one measurement event to the other—a fundamental principle of Einstein's theory of relativity. We argue that the conditional probabilities express the statistical correlations that were built into the system when it was initially prepared in the entangled state. As long as no outside influences disturb that state, the correlations will be there no matter the spatial separation of the qubits. As shown in the next subsection, the only thing that changes after Alice's observation is the state we use to describe the system. This change occurs because Alice's observation gives us new information about the system. Nothing physical changes.

It seems that Einstein believed that the measurement of one of the qubits affected the other qubit in an entangled state. He found this so surprising that he later called it "spooky action at a distance." For a history of quantum entanglement and various interpretations it generated, see *The Age of Entanglement* in Further Reading.

Try It 9.13 For the state in Eq. (9.22) calculate the two conditional probabilities $P(\uparrow_A|\uparrow_B)$ and $P(\uparrow_A|\downarrow_B)$ and explain how those results support or do not support the conclusions given in the previous paragraph.

The second entangled state we examine is one of the famous Bell states

$$|\Phi^+\rangle = \frac{1}{\sqrt{2}}\left[|\uparrow_A\uparrow_B\rangle + |\downarrow_A\downarrow_B\rangle\right]. \tag{9.23}$$

(see Chapter 10 for more details about Bell states). For this state, the parameter $s = 1$ and we say that the state is maximally entangled. For this state, the total probabilities are $P(\uparrow_B) = 1/2$ and $P(\downarrow_B) = 1/2$. Two of the conditional probabilities are $P(\uparrow_B|\uparrow_A) = 1$ and $P(\uparrow_B|\downarrow_A) = 0$. In this case, the conditional probabilities tell us that Bob's measurement results are 100% correlated with Alice's. The 100% correlations are typical of maximally entangled states.

Try It 9.14 For the state in Eq. (9.23), calculate the total probabilities $P(\uparrow_A)$ and $P(\downarrow_A)$ and the conditional probabilities $P(\uparrow_A|\uparrow_B)$ and $P(\uparrow_A|\downarrow_B)$. Compare your results with those given in the previous paragraph. Explain any significant differences or similarities.

State-after-Measurement Approach to Entangled State Probabilities

There is another approach to entangled state probabilities that makes use of the state vector after a measurement has been carried out on one of the qubits. This is an application of Lüder's rule introduced in Chapter 5. To illustrate this approach, we use the general two-qubit quantum state in Eq. (9.11) with Alice's parts pulled to the left and Bob's expressed inside curly braces:

$$|S\rangle = \underbrace{|\uparrow_A\rangle}_{\text{Alice's}} \otimes \underbrace{\{c_{\uparrow\uparrow}|\uparrow_B\rangle + c_{\uparrow\downarrow}|\downarrow_B\rangle\}}_{\text{Bob's state, given Alice's}} + \underbrace{|\downarrow_A\rangle}_{\text{Alice's}} \otimes \underbrace{\{c_{\downarrow\uparrow}|\uparrow_B\rangle + c_{\downarrow\downarrow}|\downarrow_B\rangle\}}_{\text{Bob's state, given Alice's}}. \tag{9.24}$$

Try It 9.15 Write out Eq. (9.24) by hand starting from Eq. (9.11) to make sure you know how the terms are grouped.

We want to keep Bob normalized (or at least his quantum state vectors normalized), so we divide the terms inside curly braces by the norms of Bob's states and then multiply Alice's state vector by that length:

$$\begin{aligned}|S\rangle = &\underbrace{\sqrt{|c_{\uparrow\uparrow}|^2+|c_{\uparrow\downarrow}|^2}\,|\uparrow_A\rangle}_{\text{Alice}} \otimes \underbrace{\left\{\frac{c_{\uparrow\uparrow}|\uparrow_B\rangle + c_{\uparrow\downarrow}|\downarrow_B\rangle}{\sqrt{|c_{\uparrow\uparrow}|^2+|c_{\uparrow\downarrow}|^2}}\right\}}_{\text{Bob}} \\ &+ \underbrace{\sqrt{|c_{\downarrow\uparrow}|^2+|c_{\downarrow\downarrow}|^2}\,|\downarrow_A\rangle}_{\text{Alice}} \otimes \underbrace{\left\{\frac{c_{\downarrow\uparrow}|\uparrow_B\rangle + c_{\downarrow\downarrow}|\downarrow_B\rangle}{\sqrt{|c_{\downarrow\uparrow}|^2+|c_{\downarrow\downarrow}|^2}}\right\}}_{\text{Bob}}.\end{aligned} \tag{9.25}$$

Eq. (9.25) tells us that when Alice has an observation of spin-up (which occurs, as mentioned before with probability $|c_{\uparrow\uparrow}|^2 + |c_{\uparrow\downarrow}|^2$), Bob will get spin-up with probability $|c_{\uparrow\uparrow}|^2/\left(|c_{\uparrow\uparrow}|^2 + |c_{\uparrow\downarrow}|^2\right)$ and spin-down with probability $|c_{\uparrow\downarrow}|^2/\left(|c_{\uparrow\uparrow}|^2 + |c_{\uparrow\downarrow}|^2\right)$. These probabilities are just the conditional probabilities listed in Table 9.5. (As always, the probabilities are given by the *absolute-value-squared* of Bob's state vector coefficients.) Note that Bob's probabilities add to 1. Eq. (9.25) is a fairly complicated expression, but it will be a key element in quantum computation.

Try It 9.16 Write out Eq. (9.25) by hand and explain to yourself or a friend what various terms mean. If Alice has an observation of spin-down, what is the probability of Bob's observing spin-up? What is the probability of Bob's observing spin-down? Explain your reasoning.

The conclusion is that Bob's probabilities depend on whether Alice observed spin-up or whether she observed spin-down. Bob's results are correlated with Alice's observation. Einstein called this "spooky action at a distance" because the qubits may be spatially separated and there is no obvious physical connection between the qubits. A somewhat less flamboyant statement would be "there is a correlation between Alice's observation and

Bob's probabilities." Note that if we ask for Bob's probabilities (as we did in the paragraph after Table 9.3) when Alice does not make an observation, we see that they are different from Bob's probabilities when she does. The latter case is one of the so-called conditional probabilities (*if* Alice observes this, *then* Bob's probabilities are ...). The first case involves summing over Alice's two possibilities. It should not be surprising that the results are different. From this point of view, there is no action at a distance—spooky or otherwise.

The rule that we invoked in working out the probabilities for a two-qubit state is called the generalized Born rule because it is (surprise!) a generalization of the Born rule for single-qubit states. We also make use of the *generalized measurement principle* (Lüders rule), which we use in detail in the next section. Assuming that the measurement device states are the basis states in which the state vectors are prepared, we state the generalized Born rule as follows:

Generalized Born Rule: For a multiqubit state, the probabilities of the outcomes of measurements on the (normalized) state of only one of the qubits are the squares of the amplitudes for the appropriate basis states for that qubit.
Generalized Measurement Principle (Lüders Rule): After the measurement, the state of the system is the product of the basis state observed and the states associated with the other qubits.

9.6 Generalized Born Rule and Entanglement

Let's explore in more detail the generalized Born rule and the *generalized measurement principle* and how it gives us the probabilities for Bob's measurement results given that Alice has observed, for example, spin-up. After Alice's observation of spin-up, we have new information, and it means that the appropriate quantum state vector for the overall system is from Eq. (9.25):

$$|S \text{ after } \uparrow_A\rangle = \underbrace{|\uparrow_A\rangle}_{\text{Alice's state}} \otimes \underbrace{\frac{c_{\uparrow\uparrow}|\uparrow_B\rangle + c_{\uparrow\downarrow}|\downarrow_B\rangle}{\sqrt{|c_{\uparrow\uparrow}|^2 + |c_{\uparrow\downarrow}|^2}}}_{\text{Bob's state}}. \tag{9.26}$$

In other words, Alice's qubit is now described by the state $|\uparrow_A\rangle$ (the basis state associated with her measurement outcome), which we factored out in Eq. (9.26), and Bob's qubit is described by the remaining part of that equation. Bob's state is normalized. It seems reasonable to say that Alice's observation "caused" Bob's state to be the second part of the right side of Eq. (9.26). As shown later, the word "cause" is problematic for several reasons in this context.

An important observation: After a measurement on one of the qubits, the system state is no longer an entangled state. For example, the system state in Eq. (9.26) is a simple product state, not an entangled state. Measurement crushes entanglement. It sounds like a sports headline for the English Premier League.

Let's get back to our two-qubit analysis. Bob is getting anxious and decides to make a spin observation on his qubit. From Eq. (9.26), he should get spin-up with probability $|c_{\uparrow\uparrow}|^2/\left(|c_{\uparrow\uparrow}|^2 + |c_{\uparrow\downarrow}|^2\right)$. This is actually a conditional probability $P(\uparrow_B|\uparrow_A)$ and it agrees with the result in Table 9.5, as mentioned previously.

Try It 9.17 Explain why, under those same conditions, Bob will get spin-down with probability $|c_{\uparrow\downarrow}|^2/\left(|c_{\uparrow\uparrow}|^2 + |c_{\uparrow\downarrow}|^2\right)$. Compare this with the result in Table 9.5.

Try It 9.18 Show that Bob's probabilities add up to 1.

As mentioned before, probabilities say nothing about an individual observation. For each individual state preparation–observation cycle, Alice will get either spin-up or spin-down and Bob will get either spin-up or spin-down. It is only after many preparation–observation cycles that the numerical values of the probabilities emerge.

Now specifically look at the case in which Alice gets spin-down for her qubit. Then, the appropriate quantum state vector after the observation is

$$|S \text{ after } \downarrow_A\rangle = \underbrace{|\downarrow_A\rangle}_{\text{Alice's state}} \otimes \underbrace{\frac{c_{\downarrow\uparrow}|\uparrow_B\rangle + c_{\downarrow\downarrow}|\downarrow_B\rangle}{\sqrt{|c_{\downarrow\uparrow}|^2 + |c_{\downarrow\downarrow}|^2}}}_{\text{Bob's state}}. \tag{9.27}$$

Bob now has the probability $\frac{|c_{\downarrow\uparrow}|^2}{|c_{\downarrow\downarrow}|^2+|c_{\downarrow\uparrow}|^2}$ of getting spin-up and $\frac{|c_{\downarrow\downarrow}|^2}{|c_{\downarrow\uparrow}|^2+|c_{\downarrow\downarrow}|^2}$ of getting spin-down, as we claimed before.

Try It 9.19 Show that Bob's probabilities add up to 1 when Alice gets spin-down.

We see that by looking at the system state after one of the qubit's has been measured gives us conditional probabilities in complete agreement with the ones calculated via just the Born rule and the rules of probability theory without any consideration of changes in the system state as shown in Table 9.5.

Our entanglement analysis should be comforting in the sense that there are conditions under which Alice's and Bob's measurements are independent. Alas, this doesn't solve the problem of understanding the correlations between Bob's and Alice's observations when we do have an entangled state. From a mathematical point of view, if you prepare a superposition state for a two-qubit system by randomly choosing state amplitudes, you will almost certainly end up with an entangled state for which Eq. (9.14) is *not* satisfied. However, from a practical point of view, if Alice prepares her qubit in a quantum state here and Bob, in a distant location, prepares his qubit in a quantum state there, the overall system state will be a product state. To get an entangled system state, we need to act on both those qubits' states with some quantum gate or gates. That is both good news and bad news for QIS and QC. It is good because it tells us that it is relatively easy to produce specified entangled states by choosing the right quantum gates. It is bad because your qubit states might easily become entangled with the states of other qubits in the environment because of uncontrolled interactions, which consequently mess up the states you wanted to have for your QIS and QC algorithms.

Try It 9.20 Go through two scenarios in which Bob makes the first observations. What are the probabilities for Alice to find spin-up and spin-down in those cases? Partial answer: When Bob gets spin-up, Alice gets spin-up with probability $|c_{\uparrow\uparrow}|^2/(|c_{\uparrow\uparrow}|^2 + |c_{\downarrow\uparrow}|^2)$ and spin-down with probability $|c_{\downarrow\uparrow}|^2/(|c_{\uparrow\uparrow}|^2 + |c_{\downarrow\uparrow}|^2)$.

There is nothing special about either Alice or Bob when it comes to quantum measurements. Nevertheless, with an entangled state, the probabilities of getting spin-up or spin-down depend on what was observed previously for the other qubit. In essence, this is what is called a conditional probability statement. These conditional probabilities tell us about correlations among the results, given the entangled state. The key issue is that these are correlations among measurements on the qubits individually in a situation for which the quantum state applies only to properties of the complete system. For an entangled system state, we don't have quantum states for the qubits individually.

This result applies to any two-qubit system, not just spin-1/2 systems. You can easily check that by replacing the spin-up or spin-down basis states with linear polarization states or with computational basis states $|0\rangle$ and $|1\rangle$.

Try It 9.21 Rewrite the arguments in Try It 9.20 in terms of photon linear polarization states (horizontal and vertical polarization) and then with the computational basis states $|0\rangle$ and $|1\rangle$.

Try It 9.22 What happens if Alice and Bob make their observations simultaneously? Note: We implicitly treated this case in Table 9.3 and explained the results at the end of Section 9.5. Provide your own explanation of this situation.

Try It 9.23 Challenge: Show that the results of having Alice's measurements first, having Bob's measurements first, and having the two sets of measurements simultaneously are all consistent. Hint: If Alice measures first, from Eq. (9.25) the probability that she gets spin-up is $|c_{\uparrow\uparrow}|^2 + |c_{\uparrow\downarrow}|^2$. The probability that Bob then gets spin-down is $|c_{\uparrow\downarrow}|^2/\left(|c_{\uparrow\uparrow}|^2 + |c_{\uparrow\downarrow}|^2\right)$. Hence, the joint probability that Alice gets spin-up and Bob gets spin-down is the product of those two probabilities $\left(|c_{\uparrow\uparrow}|^2 + |c_{\uparrow\downarrow}|^2\right) \times |c_{\uparrow\downarrow}|^2/\left(|c_{\uparrow\uparrow}|^2 + |c_{\uparrow\downarrow}|^2\right) = |c_{\uparrow\downarrow}|^2$. Work through the other three possible combinations of outcomes.

By way of contrast, when most people first encounter entanglement, they think—quite reasonably—that there must be some hidden mechanism that causes the correlations we have described. But any such mechanism, if it exists, must be very strange and would violate well-established principles. For example, as far as we know, no signal from one system to another can travel faster than the speed of light $c = 3 \times 10^8$ m/s. In 2017, a team of Chinese scientists (Yin et al. 2017) used light from a laser on an Earth satellite and two ground-based light detectors sensitive to the polarization of the light. The two detectors were separated by hundreds of kilometers on Earth and were set up to demonstrate correlations among photons in an entangled state. The spatial separations were such that no signal could have traveled between the receiving stations in time for one set of measurements to influence the other. Nevertheless, the correlations of observations built into the entangled state were observed.

A bit of history: Erwin Schrödinger, whom we introduced in Chapter 4, invented the term entanglement. He noted that entanglement is *the* critical feature that distinguishes quantum mechanics from classical mechanics. Schrödinger's point is that if the two-qubit system has an entangled state, then it is not legitimate to say that Alice's qubit "has" a specific quantum state and Bob's qubit "has" its own specific quantum state. If the system state is entangled, there is only the system state and no individual states describes the individual qubits. In other words, the qubits are not independent and therefore cannot be said to have their own states. Therefore, it makes no sense to say that Alice's observations affect Bob's qubit's state. Bob's qubit did not have its own state to begin with.

Since entanglement is such a key concept in quantum mechanics, let's see how it plays out. The general two-qubit state in Eq. (9.24) is a good starting point. If we think back to the beginning of this chapter, we express a two-qubit state as a product state $|S\rangle = |A\rangle|B\rangle$. If the system state can be factored in this way—a state vector for Alice's qubit multiplied by a state vector for Bob's qubit—then, and only then, can we say that Alice's qubit is described by $|A\rangle$ and Bob's by $|B\rangle$. In the more general case shown in Eq. (9.24), we cannot factor the system state this way—a part associated with Alice's qubit and a part associated with Bob's qubit. In other words, for an entangled two-qubit state, we don't have a definite state vector for Alice's or Bob's qubit.

This might help us understand the entangled state situation: Another way to describe what is happening with an entangled state is to note that the results of measurements for one of the qubits are correlated with measurement results associated with the other qubit. Remember that quantum states embody the information needed to predict the probabilities of those various results. In particular, the entangled state carries information about the correlations among the probabilities of the individual qubits' measurement outcomes. When correlations are present, we cannot assign each qubit its own state. We take up the issue of correlated measurement outcomes and tests of quantum mechanics involving what are called Bell inequalities in more detail in Chapter 17. In the intervening chapters, we describe how entangled states show up in several QIS situations: quantum state teleportation, superdense coding, and entanglement swapping.

9.7 Three-Qubit Entangled States

There is a nice way of expressing the conditions for entangled states that allows us to extend the results to more than two qubits. These more general results are useful when we discuss quantum error correction in Chapter 13. This is also a nice example of where a careful choice of mathematical notation makes the results easier to remember. Building on Eq. (9.11), let's make two generalizations: We use computational basis states instead of spin states and we change the coefficients from those in Eq. (9.11) to symbols that are easier to associate with the basis states when we have more than two qubits. We now write the two-qubit system state as

$$|\Psi\rangle = a_{00}|00\rangle + a_{01}|01\rangle + a_{10}|10\rangle + a_{11}|11\rangle . \tag{9.28}$$

Recall the condition expressed in Eq. (9.14) for a state not to be entangled. Using the notation in Eq. (9.28), we write the "not entangled" condition as

$$a_{00}a_{11} = a_{01}a_{10}. \tag{9.29}$$

If we write out the expression for a two-qubit simple product state (unentangled state) with amplitudes for each of the single-qubit states labeled by a_0, a_1, b_0, and b_1 (note that there is only one subscript for each of these amplitudes), we obtain

$$\begin{aligned} |\psi\rangle &= \underbrace{(a_0|0\rangle + a_1|1\rangle)}_{\text{Alice's state}} \otimes \underbrace{(b_0|0\rangle + b_1|1\rangle)}_{\text{Bob's state}} \\ &= a_0b_0|0\rangle\otimes|0\rangle + a_0b_1|0\rangle\otimes|1\rangle + a_1b_0|1\rangle\otimes|0\rangle + a_1b_1|1\rangle\oplus|1\rangle \\ &= a_0b_0|00\rangle + a_0b_1|01\rangle + a_1b_0|10\rangle + a_1b_1|11\rangle. \end{aligned} \tag{9.30}$$

In Eq. (9.30), we have left out the A and B subscripts on the basis states. Comparing Eqs. (9.30) and (9.28), we see that the outermost product of state amplitudes and the middle product state amplitudes satisfy Eq. (9.29): $a_0b_0a_1b_1 = a_0b_1a_1b_0$. Each of the product state amplitudes appears once and only once on each side of the equation.

We can extend the results to handle three qubits, which means we have eight (2^3) basis states. The generalization of Eq. (9.28) is

$$\begin{aligned} |\Phi\rangle &= a_{000}\,|000\rangle + a_{001}\,|001\rangle + a_{010}\,|010\rangle + a_{011}\,|011\rangle \\ &\quad + a_{100}\,|100\rangle + a_{101}\,|101\rangle + a_{110}\,|110\rangle + a_{111}\,|111\rangle . \end{aligned} \tag{9.31}$$

If you write the three-qubit product state corresponding to Eq. (9.30) using Greek letters for the state coefficients to distinguish them from that state coefficients in Eq. (9.31), you obtain

$$|\Phi\rangle = (\alpha_0|0\rangle + \alpha_1|1\rangle) \otimes (\beta_0|0\rangle + \beta_1|1\rangle) \otimes (\gamma_0|0\rangle + \gamma_1|1\rangle) . \tag{9.32}$$

We won't multiply out all of the terms in Eq. (9.32), but you can see that the conditions to have the general three-qubit state expressible as a simple product state are

$$a_{000}a_{111} = a_{001}a_{110} = a_{011}a_{100} = a_{101}a_{010}. \tag{9.33}$$

If any of these conditions do *not* hold, then the state is an entangled state. Again, we see that if we choose the amplitudes randomly, we almost always get an entangled state. On the other hand, if we think about the practical implementation of quantum states on physically separated qubits, you will need to act on an initial product state with appropriate quantum gates to produce a specific entangled state that you want for some QIS and QC purpose.

Try It 9.24 Show that there is a pattern among the 0s and 1s in the subscripts of each pair of coefficients in Eq. (9.33).

Try It 9.25 Multiply out the terms in Eq. (9.32) and verify that they satisfy the conditions in Eq. (9.33).

As you might imagine, as the number of qubits in the system increases, it becomes more difficult to decide whether the system state is separable or entangled. Similarly, it is challenging to define an appropriate degree of entanglement. For a discussion of these issues, see Further Reading.

What if the conditions in Eq. (9.29) and (9.33) are almost satisfied but not quite. Is the system state only a little entangled? This is an intriguing question and one that is not addressed in most textbooks on quantum mechanics. Yes, qubits can have just a little bit of entanglement (no pun intended) and that will turn out to be an important concept when we deal with quantum error correction in Chapter 13. In more advanced books on quantum mechanics and quantum information science, there are fancier ways of describing the "degree of entanglement," but they would take us rather far afield. The following challenge will show you one way to introduce a "small amount" of entanglement. In Chapter 17, we again discuss the question of providing a measure of the amount of entanglement.

Try It 9.26 Challenge: Suppose the condition in Eq. (9.29) is not quite satisfied. Figure out a way to write the state in Eq. (9.30) as a product state added to a smaller amplitude entangled state. Hint: There are many ways to do this. For example, we might use

$$|\psi\rangle = a_0b_0|00\rangle + a_0b_1|01\rangle + a_1b_0|10\rangle + a_1b_1|11\rangle + \delta\,(|00\rangle + |11\rangle)$$

The first four terms on the right of the previous equation represent a simple product state, as we know from Eq. (9.30). The last part of the expression is by itself an entangled state.

(a) Explain how we know the last part of the state vector $\delta\,(|00\rangle + |11\rangle)$ is entangled.

(b) Write out the condition among the state amplitudes to ensure that the state is normalized.

Since the state vector has an entangled part, the overall state must be an entangled state, but we can change the amount of entanglement by changing δ.

(c) Go through the measurement arguments in Sections 9.5 and 9.6 to see how δ affects the results. We would expect that Bob's measurement results become less affected by Alice's measurement results as $\delta \to 0$.

9.8 Changing Basis States for Two-Qubit States

In Chapter 8, we demonstrate how to express a general quantum state using a variety of basis states. This change of basis states was important in calculating the measurement outcome probabilities when the state preparation basis states are different from the measurement basis states. Also, as shown in Chapter 17, using several basis states for an entangled state allows for important tests of the so-called Bell inequalities. You might also wonder if a change in basis states converts an entangled state into an unentangled state or vice versa.

In principle, the change-in-basis-state method is just what you might guess based on what is discussed in Chapter 8. Let's review the method.

As mentioned before, we write a single-qubit state in terms of two basis sets $|A_0\rangle$ and $|A_1\rangle$ and $|B_0\rangle$ and $|B_1\rangle$. We observed in Chapter 8 that we can express one set of basis states in terms of the other:

$$\begin{aligned}|A_0\rangle &= |B_0\rangle\langle B_0|A_0\rangle + |B_1\rangle\langle B_1|A_0\rangle \\ |A_1\rangle &= |B_0\rangle\langle B_0|A_1\rangle + |B_1\rangle\langle B_1|A_1\rangle\end{aligned} \tag{9.34}$$

and

$$\begin{aligned}|B_0\rangle &= |A_0\rangle\langle A_0|B_0\rangle + |A_1\rangle\langle A_1|B_0\rangle \\ |B_1\rangle &= |A_0\rangle\langle A_0|B_1\rangle + |A_1\rangle\langle A_1|B_1\rangle,\end{aligned} \tag{9.35}$$

In a two-qubit state, we have products of various individual qubit basis states. All we have to do is apply Eq. (9.34) or Eq. (9.35) to the appropriate right-vector basis states.

Rather than writing a general result, which turns out to be quite messy algebraically, let's explain it with an example. Using Dirac notation, as applied in Chapter 8, we make the calculation much less messy and most of the results can be written down as a matter of inspection (just follow the flow of the Dirac brackets). Suppose we have two qubits α and β and we prepare them in a state

$$|\psi\rangle = \frac{1}{\sqrt{2}}\left(|A_0\rangle_\alpha \otimes |A_0\rangle_\beta + |A_1\rangle_\alpha \otimes |A_1\rangle_\beta\right) \tag{9.36}$$

We want to change the basis states for qubit α from the A basis to the B basis. Let's break up the work into steps. First, we use the top line in Eq. (9.34) in the first part inside the parentheses in Eq. (9.36):

$$\begin{aligned}|A_0\rangle_\alpha \otimes |A_0\rangle_\beta &= \underbrace{\left[|B_0\rangle_\alpha \langle B_0|A_0\rangle + |B_1\rangle_\alpha \langle B_1|A_0\rangle\right]}_{|A_0\rangle_\beta} \otimes |A_0\rangle_\beta \\ &= |B_0\rangle_\alpha \otimes \left(|A_0\rangle_\beta \langle B_0|A_0\rangle\right) + |B_1\rangle_\alpha \otimes \left(|A_0\rangle_\beta \langle B_1|A_0\rangle\right),\end{aligned} \tag{9.37}$$

where we used $(a\,|F\rangle) \otimes |G\rangle = |F\rangle \otimes (a\,|G\rangle)$, valid for any number a, and grouped the terms according to the right vectors $|B_0\rangle_\alpha$ and $|B_1\rangle_\alpha$. We now do the same thing for the second term inside the parentheses in Eq. (9.36),

$$\begin{aligned}|A_1\rangle_\alpha \otimes |A_1\rangle_\beta &= \underbrace{\left[|B_0\rangle_\alpha \langle B_0|A_1\rangle + |B_1\rangle_\alpha \langle B_1|A_1\rangle\right]}_{|A_1\rangle_\alpha} \otimes |A_1\rangle_\beta \\ &= |B_0\rangle_\alpha \otimes \left(|A_1\rangle_\beta \langle B_0|A_1\rangle\right) + |B_1\rangle_\alpha \otimes \left(|A_1\rangle_\beta \langle B_1|A_1\rangle\right).\end{aligned} \tag{9.38}$$

Finally, we add Eqs. (9.37) and (9.38) and then group terms multiplying $|B_0\rangle_\alpha$ and $|B_1\rangle_\alpha$:

$$\begin{aligned}|A_0\rangle_\alpha \otimes |A_0\rangle_\beta + |A_1\rangle_\alpha \otimes |A_1\rangle_\beta &= |B_0\rangle_\alpha \otimes \underbrace{\left[|A_0\rangle_\beta \langle B_0|A_0\rangle + |A_1\rangle_\beta \langle B_0|A_1\rangle\right]}_{|B_0\rangle_\beta} \\ &\quad + |B_1\rangle_\alpha \otimes \underbrace{\left(|A_0\rangle_\beta \langle B_1|A_0\rangle + |A_1\rangle_\beta \langle B_1|A_1\rangle\right)}_{|B_1\rangle_\beta} \\ &= |B_0\rangle_\alpha \otimes |B_0\rangle_\beta + |B_1\rangle_\alpha \otimes |B_1\rangle_\beta.\end{aligned} \tag{9.39}$$

To get the final line in Eq. (9.39), we used Eq. (9.35) for the β qubit state. This result assumes that the brackets relating the two basis states are *real*. It may be surprising that the state with the basis states changed only in the α subspace has the same form as the original state. Yes, that is unusual and certainly not true of entangled states in general (see Try It 9.22). We use Eq. (9.39) in the following section.

Try It 9.27 Apply the change in basis states to the state

$$|\psi\rangle = \frac{1}{\sqrt{2}}\left(|A_0\rangle_\alpha \otimes |A_0\rangle_\beta - |A_1\rangle_\alpha \otimes |A_1\rangle_\beta\right).$$

Does it maintain its form in different basis sets the way the state in Eq. (9.36) does? Do the same for the state $|\psi\rangle = \frac{1}{\sqrt{2}}\left(|A_0\rangle_\alpha \otimes |A_1\rangle_\beta - |A_1\rangle_\alpha \otimes |A_0\rangle_\beta\right)$.

To wrap up this section, we need to remember that many authors do not include the tensor product symbol and the subscripts on the state vectors. That is, you may see a state

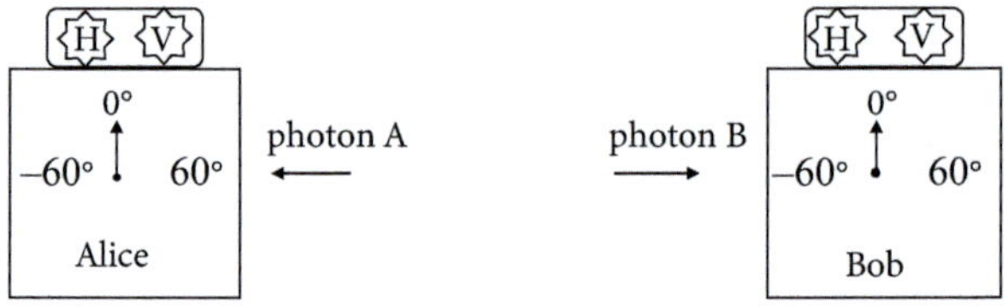

Fig. 9.2 Two photons interact with two different measurement devices, each of which has three choices for the orientation of the linear polarization detected by the devices. If "horizontal" polarization is detected for that measurement direction, the "H" light is lit on the device. If "vertical" polarization is observed, the "V" light is lit.

written as $|\psi\rangle = \left(1/\sqrt{2}\right)(|A_0\rangle\,|A_0\rangle + |A_1\rangle\,|A_1\rangle)$. It is up to you to keep the qubit states in the correct order.

9.9 Two Qubits and Three Measurement Bases

To get more experience with multiple bases and entanglement, let's work through an example of entangled states that might be observed with two measurement devices, each of which has three possible measurement bases. This may sound like a rather contrived situation, but it has been used as an experimental test of some of the fundamentals of quantum mechanics and variations on this model show up in a variety of QIS and QC applications. In Chapter 17, we use those results to illustrate one of the most profound results in quantum mechanics: Bell's theorem. A brief introduction of the setup is provided in Chapter 8 Section 4.

Figure 9.2 shows the setup. Two photons are launched toward two linear polarization measurement devices. Each device can be set to use a measurement basis along 0° (relative to some direction), 60°, or −60° relative to that direction. Let's assume that the two photons are prepared in the following two-qubit state:

$$|\psi\rangle = \frac{1}{\sqrt{2}}\left(|\text{hlp}\rangle_{0^\circ}|\text{hlp}\rangle_{0^\circ} + |\text{vlp}\rangle_{0^\circ}|\text{vlp}\rangle_{0^\circ}\right). \tag{9.40}$$

It is easy to see that this is an entangled state.

Try It 9.28 Explain how you know that the state in Eq. (9.40) is entangled.

Now we want to calculate the probabilities that Alice and Bob will record "H" or "V" for the three choices of measurement polarization directions. In particular, we want to find the fraction of the results that end up with Alice and Bob both recording H or both recording V. For now, we use the setup just to show how to change basis states and evaluate measurement probabilities for entangled states. Let's go through various possibilities.

(1) Both Alice and Bob use the 0° settings on their measurement devices. In this case, we simply read off the squares of the appropriate coefficients in the state since it is already written in the 0° basis. We need the amplitude of the $|\text{hlp}\rangle_{0^\circ}|\text{hlp}\rangle_{0^\circ}$ term to get the probability of HH and the amplitude of the $|\text{vlp}\rangle_{0^\circ}|\text{vlp}\rangle_{0^\circ}$ term for the probability of VV. Each amplitude is $1/\sqrt{2}$. So, the probability is ½. Thus, the probability of getting the same result (either HH or VV) is 1. This is another example of the correlations that come along with entangled states.

(2) Both Alice and Bob use 60° settings. There are two ways to figure this out. (I) We follow the procedures from Chapter 8 giving the state in terms of the amplitudes in the 60° basis:

$$\begin{aligned}|\text{hlp}\rangle_{0^\circ} &= \frac{1}{2}|\text{hlp}\rangle_{60^\circ} - \frac{\sqrt{3}}{2}|\text{vlp}\rangle_{60^\circ} \\ |\text{vlp}\rangle_{0^\circ} &= \frac{\sqrt{3}}{2}|\text{hlp}\rangle_{60^\circ} + \frac{1}{2}|\text{vlp}\rangle_{60^\circ}.\end{aligned} \tag{9.41}$$

We now substitute Eq. (9.41) into Eq. (9.40):

$$\begin{aligned}|\psi\rangle = \frac{1}{2}&\left[\left(\frac{1}{2}|\text{hlp}\rangle_{60^\circ} - \frac{\sqrt{3}}{2}|\text{vlp}\rangle_{60^\circ}\right)\otimes\left(\frac{1}{2}|\text{hlp}\rangle_{60^\circ} - \frac{\sqrt{3}}{2}|\text{vlp}\rangle_{60^\circ}\right)\right] \\ &+\left[\left(\frac{\sqrt{3}}{2}|\text{hlp}\rangle_{60^\circ} + \frac{1}{2}|\text{vlp}\rangle_{60^\circ}\right)\otimes\left(\frac{\sqrt{3}}{2}|\text{hlp}\rangle_{60^\circ} + \frac{1}{2}|\text{vlp}\rangle_{60^\circ}\right)\right] \\ =& \frac{1}{\sqrt{2}}\left[|\text{hlp}\rangle_{60^\circ}|\text{hlp}\rangle_{60^\circ} + |\text{vlp}\rangle_{60^\circ}|\text{vlp}\rangle_{60^\circ}\right].\end{aligned} \tag{9.42}$$

Note that the structure of the state is just like the 0° case we looked at before, so we end up with the same results: Alice and Bob get the same results with probability 1.

The second method (II) uses the general result in Eq. (9.39). We immediately recognize that when Alice and Bob choose the same basis set, no matter what its orientation is, we get identical results with probability 1. So, we know that this holds for the choice of the −60° basis as well.

(3) Alice uses the 0° basis and Bob uses the 60° settings.

This one is a bit trickier to analyze because the measurement device directions are not the same. But the calculation proceeds in a similar way. We use Eq. (9.34) to change the basis set for Bob's qubit's state:

$$\begin{aligned}|\psi\rangle &= \frac{1}{\sqrt{2}}\left[\underbrace{|\text{hlp}\rangle_{0^\circ}}_{\text{Alice}}\otimes\underbrace{\left(\frac{1}{2}|\text{hlp}\rangle_{60^\circ} - \frac{\sqrt{3}}{2}|\text{vlp}\rangle_{60^\circ}\right)}_{\text{Bob}} + \underbrace{|\text{vlp}\rangle_{0^\circ}}_{\text{Alice}}\otimes\underbrace{\left(+\frac{\sqrt{3}}{2}|\text{hlp}\rangle_{60^\circ} + \frac{1}{2}|\text{vlp}\rangle_{60^\circ}\right)}_{\text{Bob}}\right] \\ &= \frac{1}{\sqrt{2}}\left[\frac{1}{2}|\text{hlp}\rangle_{0^\circ}|\text{hlp}\rangle_{60^\circ} + \frac{1}{2}|\text{vlp}\rangle_{0^\circ}|\text{vlp}\rangle_{60^\circ}\right] - \frac{\sqrt{3}}{2}|\text{hlp}\rangle_{0^\circ}|\text{vlp}\rangle_{60^\circ} + \frac{\sqrt{3}}{2}|\text{vlp}\rangle_{0^\circ}|\text{hlp}\rangle_{60^\circ}.\end{aligned} \tag{9.43}$$

Table 9.6 Settings 0°, 60°, and −60° and the fraction of measurements with those settings for which Alice and Bob get the same results.

Settings	0° 0°	60° 60°	−60° −60°	0° 60°	60° 0°	0° −60°	−60° 0°	60° −60°	−60° 60°
Same	1	1	1	1/4	1/4	1/4	1/4	1/4	1/4

From Eq. (9.43), we read off that Alice and Bob will get HH with probability $\left(1/\left(2\sqrt{2}\right)\right)^2 = 1/8$ and VV with probability 1/8. So, for these settings, we will get matching results (VV or HH) with probability ¼. The results will be the same if we interchange the basis sets with Alice choosing 60° and Bob choosing 0°. We could also substitute −60° for 60°, and the results will be the same.

Try It 9.29 Convince yourself that the claims about various choices of basis sets are correct.

(4) Alice uses the −60° basis and Bob uses the 60° settings. Proceeding with the appropriate substitutions, we get

$$
\begin{aligned}
|\psi\rangle = \frac{1}{\sqrt{2}}\Bigg[&\bigg[\underbrace{\left(\tfrac{1}{2}|\text{hlp}\rangle_{-60^\circ} + \tfrac{\sqrt{3}}{2}|\text{vlp}\rangle_{-60^\circ}\right)}_{\text{Alice}} \otimes \underbrace{\left(\tfrac{1}{2}|\text{hlp}\rangle_{60^\circ} - \tfrac{\sqrt{3}}{2}|\text{vlp}\rangle_{60^\circ}\right)}_{\text{Bob}}\bigg] \\
&+\bigg[\underbrace{\left(-\tfrac{\sqrt{3}}{2}|\text{hlp}\rangle_{-60^\circ} + \tfrac{1}{2}|\text{vlp}\rangle_{-60^\circ}\right)}_{\text{Alice}} \otimes \underbrace{\left(+\tfrac{\sqrt{3}}{2}|\text{hlp}\rangle_{60^\circ} + \tfrac{1}{2}|\text{vlp}\rangle_{60^\circ}\right)}_{\text{Bob}}\bigg]\Bigg] \\
= -\frac{1}{2\sqrt{2}}&\left[|\text{hlp}\rangle_{-60^\circ}|\text{hlp}\rangle_{60^\circ} + |\text{vlp}\rangle_{-60^\circ}|\text{vlp}\rangle_{60^\circ}\right] \\
&+\frac{\sqrt{3}}{2\sqrt{2}}|\text{vlp}\rangle_{-60^\circ}|\text{hlp}\rangle_{60^\circ} - \frac{\sqrt{3}}{2\sqrt{2}}|\text{hlp}\rangle_{-60^\circ}|\text{vlp}\rangle_{60^\circ}.
\end{aligned}
\tag{9.44}
$$

Again, we are looking for the cases in which Alice and Bob both get H or both get V. The probability that they both get H is $\left(1/\left(2\sqrt{2}\right)\right)^2 = 1/8$ with the identical result for V. Hence, the probability for getting any identical result is ¼.

Since Bob and Alice are equivalent, at least as far as quantum mechanics is concerned, we get the same results if Alice chooses the 60° basis and Bob chooses −60°.

We can now pull together all these results to find out how often (on average) Alice and Bob get the same results, namely HH or VV. Table 9.6 lists the fractions of the runs that yield the same results.

Since there are nine settings, the overall average fraction of the runs that give the same measurement results is 4.5/9 = 0.5. The bottom line is that Alice and Bob will get the same measurement results 50% of the time if they choose equally among the three basis settings. As discussed in Chapter 17, you will see why it was important to do this math with different measurement basis sets for multiqubit states.

Try It 9.30 Work through at least one of the remaining setting cases to verify that Bob's claims about how they worked out are correct.

9.10 Chapter Summary

- Most quantum applications involve systems of more than one qubit. Multiqubit states can be built up as superpositions of products of single-qubit states. For n qubits, we need 2^n basis states.
- The product of two state vectors is called a tensor product and is often denoted with the symbol $\otimes$. For example, $|\psi\rangle = |A\rangle \otimes |B\rangle$, or without the symbol $|\psi\rangle = |A\rangle|B\rangle$.
- Entangled states cannot be factored into simple product states. The inseparability means there are correlations among the outcomes of the measurements on the two qubits.
- You can use quantum gates acting on multiqubit systems to produce entangled states. Qubit interactions with their environment can wipe out the entangled state you produced with quantum gates.
- All entangled states are superposition states, but not all superposition states are entangled states.
- We can distinguish entangled from nonentangled states mathematically by a relationship among the state amplitudes.
- Entangled states imply correlations among the measurements carried out on the individual qubits. It is important to note that the correlations in the measurement outcomes are not due to an interaction that occurs between the qubits during the measurement.
- In a two-qubit system, measurements kill off entanglement. After an observation of one of the qubits, the state of the composite system is no longer an entangled state.
- Entangled system states do not assign separate states to each qubit. Only the system can be said to have a state.
- We can change basis states for multiqubit states by using the methods developed in Chapter 8 on each of the qubit states individually.

9.11 Further Reading

A detailed discussion of the generalized Born rule and how to construct system states after one qubit has been measured is found in N. David Mermin, *Quantum Computer Science* (Cambridge University Press, Cambridge, 2007).

R. Horodecki, P. Horodecki, M. Horodecki, and K. Horodecki, "Quantum Entanglement," *Reviews of Modern Physics* **81**, 865–942 (2009). This paper provides a comprehensive view of the theory of quantum entanglement and how it provides resources for QC and QIS.

A survey of early experiments that demonstrated entanglement distribution and tests of Bell's theorem with special attention to the loopholes that have been closed over the years:

Alain Aspect, "Closing the Door on Einstein and Bohr's Quantum Debate." https://physics.aps.org/articles/v8/123.

The 2017 entangled polarization photon experiment (Yin et al. 2017):

J. Yin, et al., "Satellite-Based Entanglement Distribution over 1200 Kilometers," *Science* **356**, 1140–1144 (2017).

For an intriguing "fictionalized" history of quantum entanglement and the arguments it created, see Louise Gilder, *The Age of Entanglement: When Quantum Physics Was Reborn* (Vantage Books, New York, 2009).

Quantum Circuits and Multiqubit Applications

10

Therefore, send not to know
For whom the bell tolls,
It tolls for thee.

John Donne, *For Whom the Bell Tolls*

10.1 Two-Qubit Gates

We are now ready to learn how to carry out operations on entangled and other two-qubit quantum states. We consider only gates that have two inputs and two outputs. As we noted in Chapter 6, such gates can be reversible. It means we can use the gates to act on states to produce new states and then apply them again to revert back to where we started, if so needed. We then apply these gates to *entanglement swapping, the no-cloning theorem, superdense coding,* and *quantum state teleportation.*

10.2 CNOT Gates

Let's start with the CNOT gate, as introduced for classical logic gates in Chapters 2 and 3. The quantum circuit schematic is shown in Fig. 10.1.

Figure 10.1 shows what we might expect for a quantum CNOT gate based on what we know about a classical-computing CNOT gate. We will see that the correspondence is correct if the states $|A\rangle$ and $|B\rangle$ are the computational basis states $|0\rangle$ and $|1\rangle$, but the results are more interesting (and more useful) for more general quantum states. Recall that the upper-qubit line is called the control qubit state and the lower-qubit line is the target or controlled qubit state.

First, we need to translate from classical bits 0 and 1 to quantum state vectors $|0\rangle$ and $|1\rangle$. The logic table for the classical CNOT gate is presented in Table 10.1. Table 10.2 shows the corresponding table for a quantum CNOT gate acting on the computational basis states.

When a CNOT gate processes (operates on) the computational basis states, the table looks just like that for the classical-computing CNOT logic gate.

Quantum Computing and Quantum Physics. Alice Flarend and Robert Hilborn, Oxford University Press.
 DOI: 10.1093/oso/9780197904930.003.0010

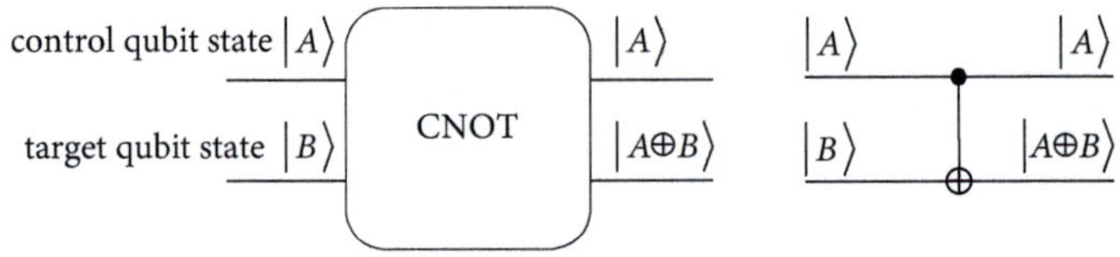

Fig. 10.1 A quantum controlled-NOT gate. On the left, a gate box representation is shown. On the right, the commonly used circuit representation is shown. ⊕ means XOR (see Chapter 2).

Table 10.1 CNOT truth table for classical bits *A* and *B*.

Input		Output	
A	*B*	*A*	$A \oplus B$
0	0	0	0
0	1	0	1
1	0	1	1
1	1	1	0

There is, of course, a matrix representation of the CNOT gate. Because we have four possible basis states, the matrix is a 4×4 matrix (four rows and four columns):

$$\text{CNOT} \doteq \begin{pmatrix} 1 & 0 & 0 & 0 \\ 0 & 1 & 0 & 0 \\ 0 & 0 & 0 & 1 \\ 0 & 0 & 1 & 0 \end{pmatrix}. \tag{10.1}$$

Try It 10.1 In Chapter 6, we stated four conditions the matrix associated with quantum gates should satisfy. Check that the CNOT matrix satisfies those conditions.

We should check that this matrix does what it is supposed to do. As an example, let's try out the fourth entry in Table 10.2 with the matrix operating on the column vector for the state $|1_A\rangle|1_B\rangle$

$$\begin{pmatrix} 1 & 0 & 0 & 0 \\ 0 & 1 & 0 & 0 \\ 0 & 0 & 0 & 1 \\ 0 & 0 & 1 & 0 \end{pmatrix}\begin{pmatrix} 0 \\ 0 \\ 0 \\ 1 \end{pmatrix} = \begin{pmatrix} 0 \\ 0 \\ 1 \\ 0 \end{pmatrix}, \tag{10.2}$$

which is just what we want: the control state is $|1_A\rangle$ and the target input state is $|1_B\rangle$. On the right side of Eq. (10.2), the upper output state is $|1_A\rangle$, unchanged from the input, and the target output state is $|0_B\rangle$.

Let's review the steps in converting from state vectors to column vectors. You may remember from the previous chapter that the system computational basis states for a

Table 10.2 The truth table for a quantum CNOT gate acting on computational basis states.

Input		Output	
A	B	A	$A \oplus B$
$\|0\rangle$	$\|0\rangle$	$\|0\rangle$	$\|0\rangle$
$\|0\rangle$	$\|1\rangle$	$\|0\rangle$	$\|1\rangle$
$\|1\rangle$	$\|0\rangle$	$\|1\rangle$	$\|1\rangle$
$\|1\rangle$	$\|1\rangle$	$\|1\rangle$	$\|0\rangle$

two-qubit system are $|00\rangle$, $|01\rangle$, $|10\rangle$, and $|11\rangle$. The column vector on the left side of Eq. (10.2) corresponds to $|11\rangle$ as shown in Table 9.2. When CNOT acts on that state, Table 10.2 indicates that the result is $|10\rangle$, which corresponds to the column vector on the right of Eq. (10.2). Note that the control "1" label on the left in $|01\rangle$ does not change because it represents the control bit.

Try It 10.2 Check the other entries in the table using the same method. This is a good exercise in matrix multiplication.

Life gets more interesting if we consider more complicated input quantum state vectors. Let's assume that the control state $|A\rangle$ is a superposition state:

$$\begin{aligned} |A\rangle &= \frac{1}{\sqrt{2}}|0_A\rangle + \frac{1}{\sqrt{2}}|1_A\rangle \\ |B\rangle &= |0_B\rangle, \end{aligned} \tag{10.3}$$

where $|A\rangle$ is Alice's control state and $|B\rangle$ is Bob's target state. Using the state product rule, we see that the overall initial ("input") state of the system is

$$|A\rangle|B\rangle = \frac{1}{\sqrt{2}}[|0_A\rangle + |1_A\rangle]|0_B\rangle. \tag{10.4}$$

You should be able to tell very quickly if that input state is an entangled state or not. We write this out in the general two-qubit state form:

$$\begin{aligned} &c|0_A\rangle|0_B\rangle + d|0_A\rangle|1_B\rangle + e|1_A\rangle|0_B\rangle + f|1_A\rangle|1_B\rangle \\ &= \frac{1}{\sqrt{2}}|0_A\rangle|0_B\rangle + 0|0_A\rangle|1_B\rangle + \frac{1}{\sqrt{2}}|1_A\rangle|0_B\rangle + 0|1_A\rangle|1_B\rangle. \end{aligned} \tag{10.5}$$

We observed in Chapter 9 that we need the product of the "outer" coefficients to be different from the product of the "inner" coefficients to have an entangled state. Eq. (10.5) has $cf = 0$ and $de = 0$ for these products. So, it is *not* an entangled state.

What is the output system state after the CNOT gate acts on the input state in Eq. (10.4)? We can see from Table 10.2 that the first term on the right side of Eq. (10.4) is not changed under the action of the CNOT gate because the control part is $|0_A\rangle$, but the second term becomes $\frac{1}{\sqrt{2}}|1_A\rangle|1_B\rangle$ because the control part is $|1_A\rangle$. Thus, the output state of the system is

$$|S_{\text{output}}\rangle = \frac{1}{\sqrt{2}}|0_A\rangle|0_B\rangle + \frac{1}{\sqrt{2}}|1_A\rangle|1_B\rangle. \tag{10.6}$$

Surprise! This state is an entangled state. Note, as usual for an entangled state, that the output state cannot be written as a product of an Alice state and a Bob state. The CNOT gate can act on an unentangled quantum state vector and produce an entangled state vector.

Try It 10.3 Check that Eq. (10.6) does indeed describe an entangled state.

This is such an important result that we should verify it by using the matrix form of the operation. The column vector form of the state in Eq. (10.4) is

$$|S_{\text{input}}\rangle \doteq \frac{1}{\sqrt{2}}\begin{pmatrix}1\\0\\1\\0\end{pmatrix}. \tag{10.7}$$

CNOT acting on this state is given in the matrix form by

$$\frac{1}{\sqrt{2}}\begin{pmatrix}1&0&0&0\\0&1&0&0\\0&0&0&1\\0&0&1&0\end{pmatrix}\begin{pmatrix}1\\0\\1\\0\end{pmatrix} = \frac{1}{\sqrt{2}}\begin{pmatrix}1\\0\\0\\1\end{pmatrix} \doteq |S_{\text{output}}\rangle. \tag{10.8}$$

You should convince yourself that the column vectors are correct and that the matrix multiplication in Eq. (10.8) gives the correct answer.

The first quantum CNOT gate was implemented by David Wineland and his team at the National Institute of Standards and Technology in Boulder, Colorado, in 1995. The group used trapped ions as qubits. Wineland went on to win the Nobel Prize in Physics in 2012.

To see how the CNOT turns an unentangled state into an entangled one, let's look at the action of CNOT on the basis states written in the combined right vector form. For example, we write $|0_A\rangle|0_B\rangle = |00\rangle$. Then, for a general two-qubit input state, we may write

$$\begin{aligned}&\text{CNOT}\{c|00\rangle + d|01\rangle + e|10\rangle + f|11\rangle\}\\&= c|00\rangle + d|01\rangle + e|11\rangle + f|10\rangle.\end{aligned} \tag{10.9}$$

The e and f terms are interchanged under the operation of the CNOT gate.

Try It 10.4 Use the matrix form of the CNOT gate in Eqs. (10.1) and (10.2) to check the result given in Eq. (10.9). Hint: Write the column vector in Eq. (10.2) with entries c, d, e, and f.

It might seem that you could just flip the e and f state entries, but you need to be sure that the basis states are expressed in the standard order with Alice's state on the left and Bob's on the right with the labels increasing in numerical order (00, 01, 10, 11) (in the binary form). As long as you do that, the shortcut method is just fine.

Before moving on to quantum circuits, let's look at other controlled gates. For example, we have a controlled-Z gate, labeled CZ. Its matrix representation is

$$\mathrm{CZ} \doteq \begin{pmatrix} \mathrm{I} & 0 \\ 0 & Z \end{pmatrix} = \begin{pmatrix} 1 & 0 & 0 & 0 \\ 0 & 1 & 0 & 0 \\ 0 & 0 & 1 & 0 \\ 0 & 0 & 0 & -1 \end{pmatrix}. \tag{10.10}$$

In the first matrix in Eq. (10.10), I, as usual, is the identity matrix and Z is the Pauli Z matrix. We have assumed that the control bit is associated with the upper input register wire and the target state with the lower input register wire (see Fig. 10.1).

Try It 10.5 Work out the action of the CZ gate on each of the four possible input states listed on the left in Table 10.2. Then, show that you get the same results if you make the lower register line the control line and the upper register the target. This symmetry leads some authors to use ZZ for the controlled-Z gate symbol or they use the symmetric circuit diagram on the right of Fig. 10.2.

Note the pattern of matrix entries in CNOT and CZ: the identity matrix in the upper left and the gate being controlled in the lower right. In fact, we can construct a controlled version of any 2×2 unitary matrix U by using

$$\mathrm{CU} \doteq \begin{pmatrix} \mathrm{I} & 0 \\ 0 & \mathrm{U} \end{pmatrix}, \tag{10.11}$$

which is a 4×4 matrix.

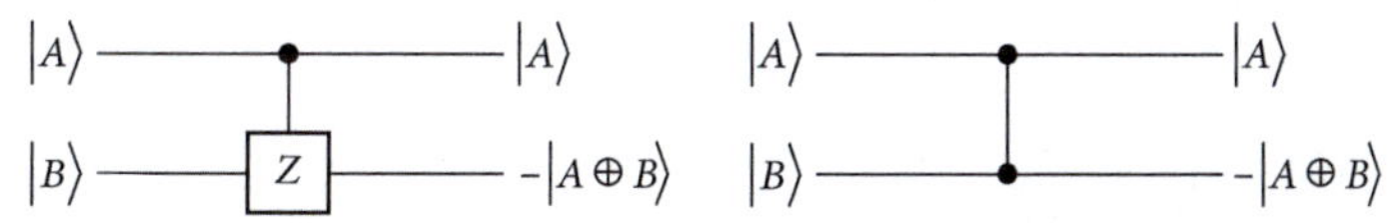

Fig. 10.2 Two circuit diagrams for the controlled-Z (CZ) gate. The diagram on the right is explored in Try It 10.5. Here, A and B may equal 0 or 1.

10.3 Quantum Circuits (Quantum Gate Arrays)

How do we combine the CNOT gate with other operations on quantum states to carry out some sort of quantum computing (QC) or quantum information science (QIS) algorithm?

Figure 10.3 shows a simple quantum computational circuit. These circuits are read from left to right as we did before. Usually, we start with all the qubits (four in this case) in the computational basis state $|0\rangle$. This is not necessary, but it is the most common choice. The states are then acted upon by various gates. Finally, on the far right, the resulting states meet up with measurement devices (usually in the computational basis) and the results are recorded on the output wires of the measurement gates.

Looking closer at the circuit, you can see that the top row qubit is a control for a CNOT gate with the second row qubit serving as the target. The second row qubit is also a target for a CNOT gate with the third row qubit as the control.

We should point out that the most current QC applications use this gate-array model, but there are other models for quantum computers. For example, there is "measurement-based quantum computation" and "topological quantum computation." For our purposes, we don't need to know about those other models because at least in principle they can be translated into the quantum circuit model (and vice versa). We describe some of these other models in Chapter 24.

10.4 The Bell Circuit

The circuit in Fig. 10.4, a combination of the Hadamard gate from Chapter 6, acting on a single qubit, and the CNOT gate, acting on two qubits, is named after John Stewart Bell (1928–1990). We call it a Bell circuit (BC). Bell's theories had (and still have) a major impact on our understanding of the conceptual foundations of quantum mechanics and on QC.

Figure 10.4 shows a BC composed of a Hadamard gate and a CNOT gate. We can also represent the BC as a gate box as shown on the right in Fig. 10.4. This gate plays a big role in various quantum algorithms. To understand what the BC does, recall that the Hadamard

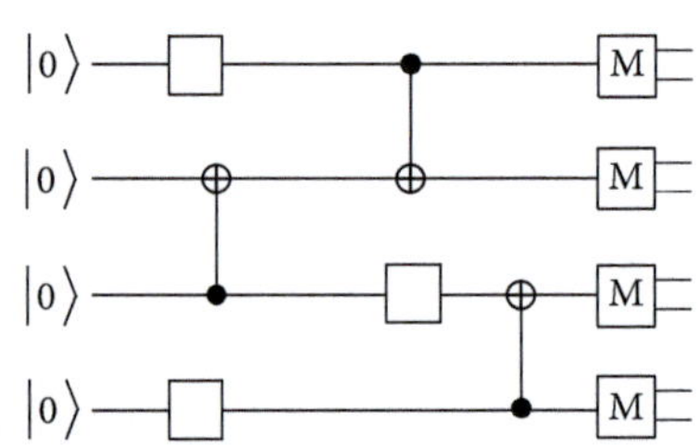

Fig. 10.3 A typical (but simple) four-qubit quantum computational circuit. The input states are on the left and are acted upon by various gates as you proceed to the right across the diagram. The open squares represent unspecified gates. The boxes on the far right are measurement devices.

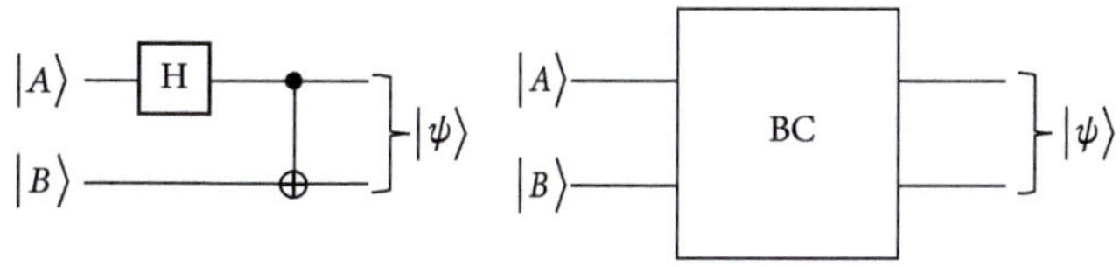

Fig. 10.4 The Bell circuit (BC). The large curly brace on the right side of each part of the diagrams reminds us that the output of the BC is an entangled state for the *system* of two output qubits, and it cannot be associated with either of the output qubit lines alone.

gate converts a computational basis state into a superposition state. For example, with $|A\rangle = |0\rangle$, we know that

$$\mathrm{H}\,|0\rangle = \frac{1}{\sqrt{2}}\,|0\rangle + \frac{1}{\sqrt{2}}\,|1\rangle\,. \tag{10.12}$$

In the BC, that superposition state is the control input for the CNOT gate. If $|B\rangle = |0\rangle$, the output of the BC is

$$|\psi\rangle = \mathrm{CNOT}\,[\mathrm{H}\,|0\rangle] = \frac{1}{\sqrt{2}}\,|00\rangle + \frac{1}{\sqrt{2}}\,|11\rangle\,, \tag{10.13}$$

which we recognize as an entangled state. We could also write this as

$$\mathrm{BC}\,|00\rangle = \frac{1}{\sqrt{2}}\,|00\rangle + \frac{1}{\sqrt{2}}\,|11\rangle = |\mathrm{Bell}_{00}\rangle = |\beta_{00}\rangle = |\Phi^{+}\rangle\,. \tag{10.14}$$

The resulting state is called a Bell state. (The last three terms of Eq. (10.14) give three different notations for the Bell state). We find the Bell state and its cousins (see Try It 10.6) quite frequently in what follows. The BC and its variants show up in many QIS and QC situations because they are gates that produce entangled states from nonentangled states. We cannot stress enough that entanglement plays a huge role in many QIS and QC applications, some examples of which we explore at the end of this chapter.

Try It 10.6 (a) Show that the following operations are implemented by the BC:

$$\begin{aligned}
\mathrm{BC}\,|01\rangle &= \frac{1}{\sqrt{2}}\,|01\rangle + \frac{1}{\sqrt{2}}\,|10\rangle = |\mathrm{Bell}_{01}\rangle = |\beta_{01}\rangle = |\Psi^{+}\rangle \\
\mathrm{BC}\,|10\rangle &= \frac{1}{\sqrt{2}}\,|00\rangle - \frac{1}{\sqrt{2}}\,|11\rangle = |\mathrm{Bell}_{10}\rangle = |\beta_{10}\rangle = |\Phi^{-}\rangle \\
\mathrm{BC}\,|11\rangle &= \frac{1}{\sqrt{2}}\,|01\rangle - \frac{1}{\sqrt{2}}\,|10\rangle = |\mathrm{Bell}_{11}\rangle = |\beta_{11}\rangle = |\Psi^{-}\rangle
\end{aligned} \tag{10.15}$$

(b) Show that the Bell states are normalized and orthogonal.

Try It 10.7 Is the BC reversible? Hint: Apply the BC to the state on the right side of Eq. (10.14).

Operators for Two-Qubit Systems

If we want to use matrices to represent the BC, we need 4×4 matrices and four-element column vectors because the circuit acts on two qubits. In particular, we need the matrix representation for the CNOT gate Eq. (10.1) and the matrix that represents the Hadamard gate acting on Alice's qubit while nothing happens to Bob's. The latter matrix is the *tensor product* of two matrices: the Hadamard gate acting on Alice's qubit and the identity operator acting on Bob's; so, we need to know how to form the tensor product of two matrices. Note that the tensor product of two matrices is different from the "matrix product" introduced in Chapter 5.

For any operators, F and G, each acting on a single-qubit state, each operator may be represented by a 2×2 matrix. How do we combine those matrices to represent the operator $F \otimes G$? The tensor product rule for 2×2 matrices is called the Kronecker product, the "natural" generalization of the tensor product rule for state vectors. The Kronecker product is also known as the "outer product" of two matrices. An operator F and its matrix elements are given by

$$F \Rightarrow \begin{pmatrix} F_{11} & F_{12} \\ F_{21} & F_{22} \end{pmatrix}, \tag{10.16}$$

where the subscripts refer to the row and column numbers $F_{\text{row column}}$. Using this notation, we write the tensor product of two operators as

$$F \otimes G \Rightarrow \begin{pmatrix} F_{11} \otimes G & F_{12} \otimes G \\ F_{21} \otimes G & F_{22} \otimes G) \end{pmatrix} = \begin{pmatrix} F_{11}G_{11} & F_{11}G_{12} & F_{12}G_{11} & F_{12}G_{12} \\ F_{11}G_{21} & F_{11}G_{22} & F_{12}G_{21} & F_{12}G_{22} \\ F_{21}G_{11} & F_{21}G_{12} & F_{22}G_{11} & F_{22}G_{12} \\ F_{21}G_{21} & F_{21}G_{22} & F_{22}G_{21} & F_{22}G_{22} \end{pmatrix}. \tag{10.17}$$

Let's work this out for the product $\mathrm{H} \otimes \mathrm{I}$. We know the matrices representing H and I can be expressed as

$$\mathrm{H} \doteq \frac{1}{\sqrt{2}} \begin{pmatrix} 1 & 1 \\ 1 & -1 \end{pmatrix} \text{and } \mathrm{I} \doteq \begin{pmatrix} 1 & 0 \\ 0 & 1 \end{pmatrix}. \tag{10.18}$$

Using those matrices in Eq. (10.17), we find

$$\mathrm{H} \otimes \mathrm{I} = \frac{1}{\sqrt{2}} \begin{pmatrix} 1 & 0 & 1 & 0 \\ 0 & 1 & 0 & 1 \\ 1 & 0 & -1 & 0 \\ 0 & 1 & 0 & -1 \end{pmatrix}. \tag{10.19}$$

Let's now apply these results to the BC with the input state $|00\rangle$. We want to evaluate the combination CNOT $[(\mathrm{H} \otimes \mathrm{I})\,|00\rangle]$. For the $\mathrm{H} \otimes \mathrm{I}$ part, we obtain

$$\frac{1}{\sqrt{2}}\begin{pmatrix} 1 & 0 & 1 & 0 \\ 0 & 1 & 0 & 1 \\ 1 & 0 & -1 & 0 \\ 0 & 1 & 0 & -1 \end{pmatrix}\begin{pmatrix} 1 \\ 0 \\ 0 \\ 0 \end{pmatrix} = \frac{1}{\sqrt{2}}\begin{pmatrix} 1 \\ 0 \\ 1 \\ 0 \end{pmatrix}. \tag{10.20}$$

Now we apply the CNOT matrix to the column vector on the right in Eq. (10.20):

$$\frac{1}{\sqrt{2}}\begin{pmatrix} 1 & 0 & 0 & 0 \\ 0 & 1 & 0 & 0 \\ 0 & 0 & 0 & 1 \\ 0 & 0 & 1 & 0 \end{pmatrix}\begin{pmatrix} 1 \\ 0 \\ 1 \\ 0 \end{pmatrix} = \frac{1}{\sqrt{2}}\begin{pmatrix} 1 \\ 0 \\ 0 \\ 1 \end{pmatrix} = \frac{1}{\sqrt{2}}\begin{pmatrix} 1 \\ 0 \\ 0 \\ 0 \end{pmatrix} + \frac{1}{\sqrt{2}}\begin{pmatrix} 0 \\ 0 \\ 0 \\ 1 \end{pmatrix}, \tag{10.21}$$

which is the matrix and column vector equivalent of Eq. (10.14). Note that the amplitudes for the lower two states in the column vector have been switched.

10.5 The Reverse Bell Circuit

The BC, like most of the gates we use, is reversible in the sense that if we use the output of the BC as input to first the CNOT gate and then send the upper qubit in Fig. 10.3 through the Hadamard gate, we end up with the original set of states. We call such a gate the reverse Bell circuit (RBC).

Figure 10.5 shows the circuit diagram for the RBC. As promised, it is just the reverse of the BC shown in Fig. 10.3.

The RBC acting on the output state in Eq. (10.14) begins with the CNOT:

$$\mathrm{RBC}\,[\mathrm{BC}\,|00\rangle] = \mathrm{H}\left[\mathrm{CNOT}\left\{\frac{1}{\sqrt{2}}|00\rangle + \frac{1}{\sqrt{2}}|11\rangle\right\}\right] = \mathrm{H}\left[\frac{1}{\sqrt{2}}|00\rangle + \frac{1}{\sqrt{2}}|10\rangle\right]. \tag{10.22}$$

Next, the Hadamard works only on the upper register, which is shown in Fig. 10.4 with the "H" box being only in the top line:

$$\mathrm{H}\left\{\frac{1}{\sqrt{2}}|00\rangle + \frac{1}{\sqrt{2}}|10\rangle\right\} = \frac{1}{2}(|00\rangle + |01\rangle + |00\rangle - |01\rangle) = |00\rangle = |0\rangle\,|0\rangle \tag{10.23}$$

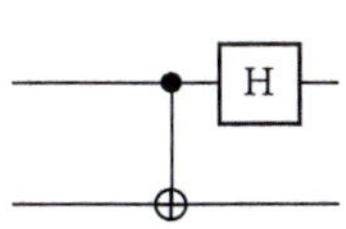

Fig. 10.5 The circuit diagram for the reverse Bell circuit.

In this case, the Hadamard gate replaces the initial input state with a superposition state. The second application of the Hadamard gate reduces the superposition back to a simple product state.

Try It 10.8 Apply the RBC to each of the superposition states in Try It 10.5. Does the RBC successfully return the original input states?

10.6 Entanglement Swapping

In Chapter 9, we found that entanglement is a rather strange quantum phenomenon with many counterintuitive notions. Here is an example that sets up an entangled state for two qubits that are always far apart and have no direct interaction with each other. No matter how strange it seems, this kind of entanglement is useful in several quantum cryptographic protocols.

We start with four qubits: *A*, *B*, *C*, and *D*. Qubits *A* and *B* are described by an entangled state while *C* and *D* are also described by an entangled state, but one that is completely independent of the *AB* entangled state. (That is, at this stage, there would be no correlations between measurements on the *AB* system and measurements on the *CD* system.) With the right choice of basis states for the measurements on *B* and *C*, we find that qubits *A* and *D* end up described by an entangled state even though they have never interacted directly. We say that the procedure has swapped entanglement from the *AB* and *CD* systems for entanglement in the *AD* system (Fig. 10.6).

How do we know that *A* and *D* are in fact described by an entangled state? As before, having qubits in an entangled state means that measurements on one of the qubits are correlated with measurements on the other qubit. So, individual measurements on *A* and *D* would reveal those correlations.

Let's work through an example of how entanglement swapping works. Suppose the qubits are photons and we prepare the pairs of qubits in the Bell state $|\text{Bell}_{00}\rangle$ for horizontal (hlp) and vertical (vlp) polarization (see Eq. (10.14)):

$$\begin{aligned} |AB\rangle &= \frac{1}{\sqrt{2}}\{|\text{hlp}\rangle_A|\text{hlp}\rangle_B + |\text{vlp}\rangle_A|\text{vlp}\rangle_B\} \\ |CD\rangle &= \frac{1}{\sqrt{2}}\{|\text{hlp}\rangle_C|\text{hlp}\rangle_D + |\text{vlp}\rangle_C|\text{vlp}\rangle_D\}. \end{aligned} \tag{10.24}$$

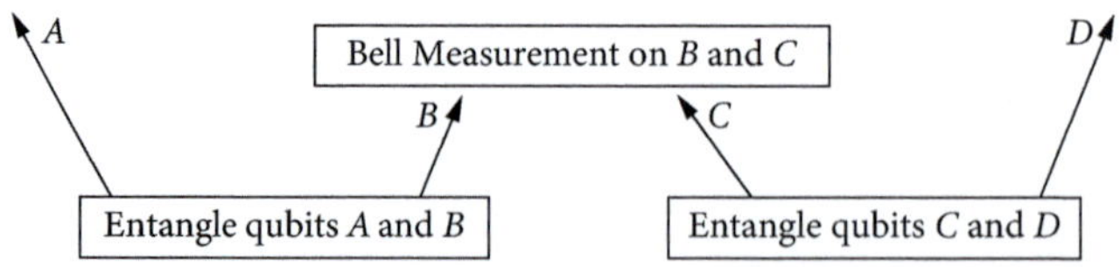

Fig. 10.6 A diagram of the entanglement swapping process. Qubits *A* and *B* are described by an entangled state. Qubits *C* and *D* are described by an independent entangled state. Qubits *B* and *C* are subject to a Bell measurement. The result is that system state for *A* and *D* is an entangled state.

Since *AB* and *CD* are prepared independently, the state for the overall *ABCD* system is a product of two two-qubit states:

$$|ABCD\rangle = |AB\rangle\,|CD\rangle\,. \tag{10.25}$$

Qubits *B* and *C* proceed to a measurement device whose measurement basis states are the four Bell states (Eq. (10.14) and Try It 10.5). Later briefly talk about how to build such a device for polarized photons.

To express what happens when the qubits *B* and *C* interact with the Bell state measurement device, we need to express the system state Eq. (10.26) as a superposition of Bell states for the *BC* qubit combination. To do that, we use Eq. (10.25) in Eq. (10.26) and expand the products of state vectors:

$$\begin{aligned}
|ABCD\rangle &= \frac{1}{2}\overbrace{\{|\text{hlp}\rangle_A\,|\text{hlp}\rangle_B + |\text{vlp}\rangle_A|\text{vlp}\rangle_B\}}^{AB\text{ entangled state}}\overbrace{\{|\text{hlp}\rangle_C\,|\text{hlp}\rangle_D + |\text{vlp}\rangle_C|\text{vlp}\rangle_D\}}^{CD\text{ entangled state}}\\
&= \frac{1}{2}\{|\text{hlp}\rangle_A\,\underline{|\text{hlp}\rangle_B\,|\text{hlp}\rangle_C}\,|\text{hlp}\rangle_D + \frac{1}{2}\,|\text{hlp}\rangle_A\,\underline{|\text{hlp}\rangle_B\,|\text{vlp}\rangle_C}\,|\text{vlp}\rangle_D\\
&+ \frac{1}{2}\{|\text{vlp}\rangle_A\,\underline{|\text{vlp}\rangle_B\,|\text{vlp}\rangle_C}\,|\text{hlp}\rangle_D + \frac{1}{2}\,|\text{vlp}\rangle_A\,\underline{|\text{vlp}\rangle_B|\text{vlp}\rangle_C}\,|\text{vlp}\rangle_D\,.
\end{aligned} \tag{10.26}$$

In the second and third lines of Eq. (10.26), we have grouped terms for the *B* and *C* qubits because they are the ones that will be subject to Bell measurements. Using Eqs. (10.14) and (10.15), the four underlined terms in Eq. (10.26) can be written in terms of the Bell states:

$$\begin{aligned}
|\text{hlp}\rangle_B|\text{hlp}\rangle_C &= \frac{1}{\sqrt{2}}|\text{Bell}_{00}\rangle_{BC} + \frac{1}{\sqrt{2}}|\text{Bell}_{10}\rangle_{BC}\\
|\text{hlp}\rangle_B|\text{vlp}\rangle_C &= \frac{1}{\sqrt{2}}|\text{Bell}_{01}\rangle_{BC} + \frac{1}{\sqrt{2}}|\text{Bell}_{11}\rangle_{BC}\\
|\text{vlp}\rangle_B|\text{hlp}\rangle_C &= \frac{1}{\sqrt{2}}|\text{Bell}_{01}\rangle_{BC} - \frac{1}{\sqrt{2}}|\text{Bell}_{11}\rangle_{BC}\\
|\text{vlp}\rangle_B|\text{vlp}\rangle_C &= \frac{1}{\sqrt{2}}|\text{Bell}_{00}\rangle_{BC} - \frac{1}{\sqrt{2}}|\text{Bell}_{10}\rangle_{BC}.
\end{aligned} \tag{10.27}$$

Using Eq. (10.27) in Eq. (10.26) and grouping the Bell states gives

$$\begin{aligned}
|ABCD\rangle &= \frac{1}{2}|\text{Bell}_{00}\rangle_{BC}\underbrace{\frac{1}{\sqrt{2}}\{|\text{hlp}\rangle_A\,|\text{hlp}\rangle_D + |\text{vlp}\rangle_A|\text{vlp}\rangle_D\}}_{|\text{Bell}_{00}\rangle_{AD}}\\
&+ \frac{1}{2}|\text{Bell}_{10}\rangle_{BC}\underbrace{\frac{1}{\sqrt{2}}\{|\text{hlp}\rangle_A\,|\text{hlp}\rangle_D - |\text{vlp}\rangle_A|\text{vlp}\rangle_D\}}_{|\text{Bell}_{10}\rangle_{AD}}\\
&+ \frac{1}{2}|\text{Bell}_{01}\rangle_{BC}\underbrace{\frac{1}{\sqrt{2}}\{|\text{hlp}\rangle_A\,|\text{vlp}\rangle_D + |\text{vlp}\rangle_A|\text{hlp}\rangle_D\}}_{|\text{Bell}_{01}\rangle_{AD}}\\
&+ \frac{1}{2}|\text{Bell}_{11}\rangle_{BC}\underbrace{\frac{1}{\sqrt{2}}\{|\text{hlp}\rangle_A\,|\text{vlp}\rangle_D - |\text{vlp}\rangle_A|\text{hlp}\rangle_D\}}_{|\text{Bell}_{11}\rangle_{AD}}\,.
\end{aligned} \tag{10.28}$$

Relabeling the *AD* Bell states indicated in Eq. (10.28) yields the final result:

$$\begin{aligned}|ABCD\rangle = &\frac{1}{2}|\mathrm{Bell}_{00}\rangle_{BC}\,|\mathrm{Bell}_{00}\rangle_{AD} + \frac{1}{2}|\mathrm{Bell}_{10}\rangle_{BC}\,|\mathrm{Bell}_{10}\rangle_{AD} \\ &+ \frac{1}{2}|\mathrm{Bell}_{01}\rangle_{BC}\,|\mathrm{Bell}_{01}\rangle_{AD} + \frac{1}{2}|\mathrm{Bell}_{11}\rangle_{BC}\,|\mathrm{Bell}_{11}\rangle_{AD}.\end{aligned} \tag{10.29}$$

The entire state can be written as a sum of products of Bell states. That result is an example showing that the Bell states can indeed be used as basis states for a two-qubit system.

Now let's see what happens if we subject the *BC* qubits to a measurement device whose basis states are the Bell states. This is an example of a measurement basis that is not tied to a simple physical property like spin-up or spin-down. Note that each of the possible Bell states appears in Eq. (10.29). The amplitudes for the various Bell states are all the same. So, we have a 25% probability of getting any one of them.

Suppose we carry out the Bell state measurement on the *B* and *C* systems and get $|\mathrm{Bell}_{01}\rangle_{BC}$. What is the state used to describe the system after that measurement? It should be the measurement basis state multiplied by the corresponding state for the *AD* qubits in Eq. (10.29). So, we should have $|\mathrm{Bell}_{01}\rangle_{BC}|\mathrm{Bell}_{01}\rangle_{AD}$. The *AD* state is an entangled state because it is a Bell state. This is surprising because *A* and *D* were never entangled before and never interacted directly.

Let's review what happened. We start with qubits *A* and *B* in an entangled state while *C* and *D* are in an independent entangled state. After the process, qubits *A* and *D* are described by an entangled state. We have swapped the entanglement from *AB* and *CD* to *AD* even though the *A* and *D* qubits have never directly interacted. This is truly strange. Nonetheless, if we carry out further measurements on qubits *A* and *D*, we find that those measurement outcomes are correlated, just as we would expect for an entangled state.

Some treatments of entanglement swapping take the process one step further. In these treatments, the *AD* entangled state is subject to additional operations to transform it into the Bell state that described the original *AB* and *CD* systems. In our example, this is the state $|\mathrm{Bell}_{00}\rangle$. It turns out that the appropriate operator (linear transformation) is one of the Pauli operators (or the identity operator if *AD* happened to end up being $|\mathrm{Bell}_{00}\rangle$). For details, see Lee, Ji, and Cheng (2024) in Further Reading.

How are Bell state measurements carried out? It turns out that you can set up a combination of beam splitters, polarizing materials, and photon detectors of the type we have described before. Photons described by each of the four Bell states will trigger a different combination of detectors (Further Reading; Pan et al., 1998; Jin et al., 2015; Kirby et al., 2016). From the pattern of the detector responses, we can deduce the Bell state that resulted from the measurement. These experiments were first carried out in 1998, and since then variations on those experiments have been used in many quantum communication protocols.

10.7 **No-Cloning Theorem**

As we dig deeper into quantum information, quantum cryptography, and quantum computing, we discover that if we could prepare a qubit in a general quantum state without knowing what that state is, a lot of things in QIS would be much easier. At the same time, we would lose much of the utility of qubits and quantum states. The notion of duplicating a general quantum state is called cloning. The formal statement of the impossibility of doing so is called the *no-cloning theorem*, as mentioned in Chapter 5.

First, let's specify what we would like to accomplish by cloning: We want to prepare a second qubit in (or more accurately, described by) a quantum state that is exactly the same state associated with a previously received qubit. We now prove the *no-cloning theorem*.

Cloning is an interesting task only if we do not know what the first qubit's quantum state is. Specifically, it means we do not know the state coefficients. If we knew the state coefficients and had sufficiently clever technical equipment, we could prepare the second qubit in a state identical to the state of the first qubit. To measure those coefficients, we would need to have multiple copies of the first qubit, all prepared in the same state. Then with a series of measurements, we find the coefficients—a process called quantum tomography—just like medical tomography in which a series of magnetic resonance imaging scans are assembled to provide a complete image of your liver, for example. As mentioned in Chapter 5, acting on a qubit with a measurement device gives us, at best, limited knowledge about the original state coefficients because the original state associated with the qubit may have been described by a superposition state. In general, we would have to use a series of observations, perhaps with two or more measurement basis choices, on a string of identically prepared qubits to get more information about the state.

As an example, let's think about a spin-1/2 qubit prepared in one of the two superposition states:

$$|S_{\pm}\rangle = \frac{1}{\sqrt{2}}\{|\uparrow\rangle \pm |\downarrow\rangle\}. \tag{10.30}$$

If we observe the spin orientation with a spin-up, spin-down measurement device, 50% of the time we get spin-up and 50% of the time we get spin-down with either of the superposition states. But with a single observation, we would only get one or the other result (spin-up or spin-down). And if we do not know the qubit was in a superposition state, we may think it had previously been in the basis state we measured. A single observation does not give us enough information to clone the state. Is it possible to devise a cloning procedure that requires just one operation on the original qubit? This process, according to standard quantum mechanics, is impossible. Let's see why.

Here's the general argument: Let's suppose that the original qubit's state is represented by

$$|\psi\rangle = a_0\,|A_0\rangle + a_1\,|A_1\rangle \tag{10.31}$$

in Alice's A basis. We would like to build a quantum gate that can operate on that state and produce a second qubit described by that same state while keeping the first qubit and its state intact. We also need to keep in mind that the only gates we want to use are reversible gates, so it means if we want two qubits out, we must put in two qubits. One of the input qubits will have the state described in Eq. (10.31)—the one we want to duplicate—while the other can be one of the basis states. It turns out that it doesn't matter what that other state is, so we choose something simple like $|0\rangle$. If you could create a cloning gate circuit C, it would look like the following:

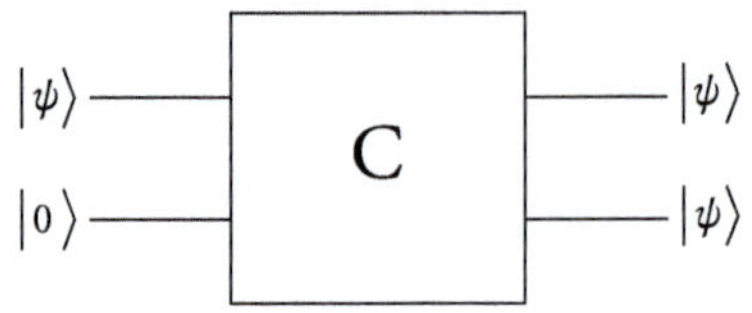

Fig. 10.7 A pictorial representation of what a cloning gate C circuit would look like and what it is supposed to do.

Figure 10.7 shows what we want a cloning gate to do if we could build one. We start with two input qubits. One is described by the state vector $|\psi\rangle$ that we would like to clone. Cloning means we want to produce another qubit in that same state, so we end up with two qubits described by the same quantum state vector. Let's work out what happens with three different input states for Alice's qubit.

1. Suppose $|\psi\rangle = |A_0\rangle$. We want the action of the cloning gate to be

$$\mathrm{C}\{|A_0\rangle\,|0\rangle\} = |A_0\rangle\,|A_0\rangle\,. \tag{10.32}$$

2. Suppose $|\psi\rangle = |A_1\rangle$. Then, we want

$$\mathrm{C}\{|A_1\rangle\,|0\rangle\} = |A_1\rangle\,|A_1\rangle\,. \tag{10.33}$$

3. Suppose the state is given by Eq. (10.31). Then, we want to obtain

$$\mathrm{C}\{a_0\,|A_0\rangle + a_1\,|A_1\rangle\}\,|0\rangle = \{a_0\,|A_0\rangle + a_1\,|A_1\rangle\} \quad \{a_0\,|A_0\rangle + a_1\,|A_1\rangle\}\,. \tag{10.34}$$

That is, in all three cases, we end up with two qubits each described by the same quantum state.

So far, there are no obvious problems. But let's look at Eq. (10.34) in two different ways. First, let's multiply out the right side of Eq. (10.34):

$$\mathrm{C}\{a_0\,|A_0\rangle + a_1\,|A_1\rangle\}\,|0\rangle = a_0^2\,|A_0\rangle\,|A_0\rangle + a_0a_1\,|A_0\rangle\,|A_1\rangle + a_1a_0\,|A_1\rangle\,|A_0\rangle + a_1^2\,|A_1\rangle\,|A_1\rangle\,. \tag{10.35}$$

Try It 10.9 Carry out the multiplication expansion to arrive at Eq. (10.35).

Now let's expand the left side of Eq. (10.34) knowing that the cloning gate, like all the other gates we are working with, is a linear operator (gate) that can be distributed to each part of a sum of state vectors. Applying that rule and referring to Eqs. (10.32) and (10.33) gives us

$$\begin{aligned}\mathrm{C}\{a_0\,|A_0\rangle + a_1\,|A_1\rangle\}\ |0\rangle &= \mathrm{C}\{a_0\,|A_0\rangle\ |0\rangle\} + \mathrm{C}\{a_1\,|A_1\rangle\ |0\rangle\} \\ &= a_0\,|A_0\rangle\,|A_0\rangle + a_1\,|A_1\rangle\,|A_1\rangle\,.\end{aligned} \tag{10.36}$$

Now we are in trouble.! We see immediately that Eqs. (10.35) and (10.36) cannot both be true in general because Eq. (10.35) has products of the state coefficients while Eq. (10.36) involves the state coefficients only to the first power. It means that we have a contradiction that follows from our assumption that we can "build" a cloning operation for quantum states. Therefore, the rules of logic tell us that we cannot build a device that meets our specifications.

There is one exception: If either $a_0 = 1$ and $a_1 = 0$ or vice versa, then the two equations are compatible. In this situation, the quantum state is actually one of Alice's basis states and not a state of superposition. If that is the case, the laws of quantum mechanics allow for cloning. This is an exception that most QC experts overlook or, at least, don't mention. But the *no-cloning theorem* says that for a general quantum superposition state—one in which the two sets of numerical values in Eqs. (10.34) and (10.35) are not compatible—cloning is not possible.

The *no-cloning theorem* prevents us from using the quantum equivalent of bit fanout, which is a key element in classical-computing architecture. On the positive side, it prevents eavesdroppers from reading the quantum state of a qubit Alice sends to Bob and replacing the intercepted qubit with another in the same quantum state. Remember how the lack of cloning foiled the eavesdropping attempt described at the end of Chapter 8. The eavesdropper could not clone a photon's linear polarization state and that led to being able to detect the presence of the eavesdropper.

Try It 10.10 Could you get around the theorem by using a different input state for the lower state on the input side of Fig. 10.6? To answer this question: use $|1\rangle$ as the lower state in Fig. 10.6. Also, try a superposition state in that input. Does the *no-cloning theorem* still hold?

10.8 **Superdense Coding**

We are now ready to describe a quantum information application that demonstrates an advantage over classical computing. Some QIS people call this "superdense coding," which sounds fantastic. It's great rhetoric, but the reality, though impressive, is more modest. Perhaps dense coding would be a better name, which is what we call it.

The basic idea behind dense coding is that you can send two classical bits with just one *single* qubit if you are clever about preparing the qubit so it is described by an entangled state and then having appropriate gates ready at the receiving end to process that qubit. It turns out that this method is a nice example of secure quantum communication. But it works only if the message sender has already sent you a properly prepared qubit.

Suppose Alice wants to send Bob two classical bits. The four possibilities are 00, 01, 10, or 11. We show that Alice can do this by sending Bob just one qubit if we do some preliminary work. First, Alice prepares two qubits C and D in the entangled Bell state

$$|\psi\rangle = |\text{Bell}_{00}\rangle = \frac{1}{\sqrt{2}}|0_C 0_D\rangle + \frac{1}{\sqrt{2}}|1_C 1_D\rangle. \tag{10.37}$$

Alice keeps one of the qubits and sends the other to Bob. You may wonder if that violates the *no-cloning theorem*. We seem to have two qubits described by the same state. This raises an important issue. The entangled state in Eq. (10.37) is the state of the two-qubit *system*. We can't assign a state to either of the qubits individually. All we have done is move one of the system parts (one of the qubits) from Alice's location to Bob's location. The system state stays the same. We don't violate the *no-cloning theorem* because we did not duplicate the entangled state.

Let's continue the argument: Alice now acts on her qubit with one of the four gates we introduced in Chapter 6—the identity gate I, the X gate, the Y gate, or the Z gate—depending on which of the pair of classical bits she wants Bob to receive. Here is how this works. If Alice wants to send 00, she applies the identity gate, which of course is equivalent to doing nothing to the two-qubit state. She then launches her qubit to Bob. When Bob receives the new qubit, he has an RBC act on the two-qubit system: the qubit he had already received and the qubit Alice just sent him. As shown before in Eq. (10.22), the output of the RBC when acting on the two states in Eq. (10.37) is the state $|0_C 0_D\rangle = |0_C\rangle\,|0_D\rangle$. Bob then measures each of the qubits using the computational basis and gets 0 and 0, just what Alice wanted him to see.

Now let's try another case. Alice wants to send 0 and 1 to Bob. She prepares the same two-qubit state described in Eq. (10.37) and, as before, sends on qubit (D) to Bob and keeps the other (C). She applies the X gate to her qubit C after which the two-qubit system state is

$$\mathrm{X}_C\left[\frac{1}{\sqrt{2}}|0_C 0_D\rangle + \frac{1}{\sqrt{2}}|1_C 1_D\rangle\right] = \frac{1}{\sqrt{2}}|1_C 0_D\rangle + \frac{1}{\sqrt{2}}|0_C 1_D\rangle. \tag{10.38}$$

Note that the X gate has flipped just the parts of the system state associated with Alice's qubit.

Alice launches her qubit to Bob. Then, he acts on it and the one he had received previously with the RBC:

$$\begin{aligned}\text{RBC}\frac{1}{\sqrt{2}}\{|1_C 0_D\rangle + |0_C 1_D\rangle\} &= \frac{\mathrm{H}_C}{\sqrt{2}}[\text{CNOT}\{|1_C 0_D\rangle + |0_C 1_D\rangle\}] \\ &= \frac{\mathrm{H}_C}{\sqrt{2}}\{|1_C 1_D\rangle + |0_C 1_D\rangle\}.\end{aligned} \tag{10.39}$$

Remember that C is the control bit and D is the target bit for the CNOT gate.

Next, the Hadamard gate acts on the C part of the entangled state. Let's recall the results from Chapter 6:

$$\begin{aligned} \mathrm{H}\,|0\rangle &= \frac{1}{\sqrt{2}}\,[|0\rangle + |1\rangle] \\ \mathrm{H}\,|1\rangle &= \frac{1}{\sqrt{2}}\,[|0\rangle - |1\rangle]\,. \end{aligned} \tag{10.40}$$

Using these results in Eq. (10.39) yields

$$\begin{aligned} \mathrm{RBC}\frac{1}{\sqrt{2}}\,\{|1_C 0_D\rangle + |0_C 1_D\rangle\} &= \frac{\mathrm{H}_C}{\sqrt{2}}\,\{|1_C 1_D\rangle + |0_C 1_D\rangle\} \\ &= \frac{1}{2}\,\{|0_C\rangle - |1_C\rangle\}\,|1_D\rangle + \frac{1}{2}\,\{|0_C\rangle - |1_C\rangle\}\,|1_D\rangle \\ &= |0_C\rangle\,|1_D\rangle\,. \end{aligned} \tag{10.41}$$

Bob then measures those qubits with the computational basis and gets 0 and 1. Success!

Try It 10.11 Work through the remaining two cases: sending 10 and 11. The answers are given below.

Answer for bits 10: To send 1 and 0, Alice applies the Z gate to her qubit and the two-qubit state becomes

$$\frac{1}{\sqrt{2}}\,|00\rangle - \frac{1}{\sqrt{2}}\,|11\rangle$$

The RBC produces $|1\rangle\,|0\rangle$, which Bob then measures and gets 1 and 0.

Answer for bits 11: To send 1 and 1, Alice applies the gates ZX to her qubit and the two-qubit state becomes

$$\frac{1}{\sqrt{2}}\,|01\rangle - \frac{1}{\sqrt{2}}\,|10\rangle$$

Application of the RBC to the two qubits gives $|1\rangle\,|1\rangle$ from which Bob measures 1 and 1.

This scheme is a nice example of secure quantum communication since Bob needs to have both qubits to decode the information Alice is sending to him. As shown before, if someone were eavesdropping and tried to intercept Alice's "modified" qubit, the scheme would not work because the eavesdropper would have to measure the qubit to get information about it and, in the process, modify the state. If the eavesdropper does not access Alice's qubit, there will be nothing to decode.

We do seem to get two bits by sending just one qubit but with a bunch of other operations required to get the result. There are preparatory steps you need to make this work. These steps involve sending a qubit, one that does not transmit any information. This example illustrates the general point that for some computational or information problems, an appropriately designed quantum algorithm has an advantage over the corresponding classical algorithm.

Try It 10.12 Challenge: Show that Alice can use any of the four Bell states, not just Bell_{00} state, for superdense coding. Does she need to adjust the operator (I, X, Y, or Z) that she applies to the state of the qubit she keeps to send the messages 00, 01, 10, or 11?

10.9 **Quantum State Teleportation**

Another good application is, in essence, the reverse of dense coding: Alice can send two classical bits that will allow Bob to prepare a qubit in a state that Alice has chosen. Note that she is not sending the qubit with the chosen state to Bob, just the two classical bits, and Bob will use these bits to prepare his qubit in the desired state. We call this *quantum state teleportation*. Note that this is not Star Trek teleportation. We are teleporting quantum state information, not an object—neither a qubit nor Commander Spock alas, there is no physical teleportation. That's why we emphatically call this *quantum state teleportation*. This strategy is used in some quantum algorithms for which you want a physically distant qubit to be described by the same quantum state as one of your local qubits.

Quantum state teleportation does not violate the *no-cloning theorem* because, as shown later, at any particular time there is only one qubit with the specified state description. We have teleported only the state information without teleporting the qubit. This is another example of why it is important to distinguish between the qubit—a physical system—and a state—an abstract description that contains information about the results of measurements.

Here is how Alice and Bob implement *quantum state teleportation*. Alice has a qubit described by a state $|\psi\rangle$. She doesn't need to know what that state is, but she will send Bob two classical bits that will enable him to prepare one of his qubits in the state $|\psi\rangle$. At no time does Alice send him the amplitudes for $|\psi\rangle$; in fact, she may not know the amplitudes. As before, we make use of entanglement to make the process work. There are several steps to the argument, but we have seen most of these operations before. See Fig. 10.8.

Let's go through each step in detail.

Step 1

Alice has a qubit whose state is given by

$$|\psi\rangle = a\,|0\rangle + b\,|1\rangle\,. \tag{10.42}$$

Alice does not need to know the numerical values of a and b.

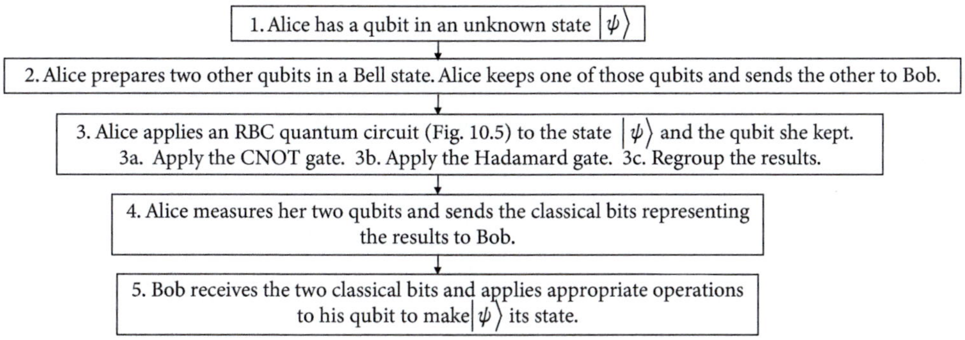

Fig. 10.8 A flowchart for the *quantum state teleportation* process.

Step 2

Alice prepares a separate two-qubit system in the by-now-familiar entangled Bell state

$$|AB\rangle = |\text{Bell}_{00}\rangle = \frac{1}{\sqrt{2}}|0_A0_B\rangle + \frac{1}{\sqrt{2}}|1_A1_B\rangle. \tag{10.43}$$

(Because we deal with three qubits, we use subscripts A, B, and ψ to keep track of which is which.) Alice keeps one of the AB qubits and sends the other to Bob. So, at this point, Alice has two qubits: one from Step 1 described by the state in Eq. (10.42) and the other from Step 2, which is her part of the two-qubit system described by Eq. (10.43).

Step 3

To find the appropriate classical bits to send to Bob so that he can prepare his qubit in the state $|\psi\rangle$, Alice subjects the state $|\psi\rangle$ and the entangled Bell state to a RBC with an additional circuit "wire" (see Fig. 10.9).

The state vector for the total system initial state on the left in Fig. 10.8 is

$$\begin{aligned}|\Phi_0\rangle = |\psi\rangle|AB\rangle &= \{a|0_\psi\rangle + b|1_\psi\rangle\}\left\{\frac{1}{\sqrt{2}}|0_A0_B\rangle + \frac{1}{\sqrt{2}}|1_A1_B\rangle\right\} \\ &= \frac{a}{\sqrt{2}}|0_A0_B\rangle|0_\psi\rangle + \frac{a}{\sqrt{2}}|1_A1_B\rangle|0_\psi\rangle + \frac{b}{\sqrt{2}}|0_A0_B\rangle|1_\psi\rangle + \frac{b}{\sqrt{2}}|1_A1_B\rangle|1_\psi\rangle.\end{aligned} \tag{10.44}$$

Let's unpack what we have written. The first line of Eq. (10.45) is just the usual product of state vectors. Since it is a simple product, not a superposition, the overall system state is *not* an entangled state. The second line of Eq. (10.45) repackages the state vectors. The double entry states, such as $|0_A1_B\rangle$, are the state vectors that will be affected by the CNOT gate. The single entry $|\,\rangle_\psi$ states are the control states.

Step 3a

The CNOT gate will now be used to switch the labels on the terms of the second line of Eq. (10.44) that multiply $|1\rangle_\psi$. The resulting system state in Fig. 10.9 (just before the Hadamard gate) will be

$$|\Phi_1\rangle = \frac{a}{\sqrt{2}}\{|0_A 0_B\rangle|0_\psi\rangle + |1_A 1_B\rangle|0_\psi\rangle\} + \frac{b}{\sqrt{2}}\{|1_A 0_B\rangle|1_\psi\rangle + |0_A 1_B\rangle|1_\psi\rangle\}. \quad (10.45)$$

Try It 10.13 Work out the details that lead to Eq. (10.45). Note that it is the A entries in the last terms (the ones multiplying $|1_\psi\rangle$) that are flipped by the CNOT gate.

Step 3b

Let's recall what the Hadamard gate does when acting on the computational basis states:

$$\begin{aligned} \mathrm{H}\,|0\rangle &= \frac{1}{\sqrt{2}}|0\rangle + \frac{1}{\sqrt{2}}|1\rangle \\ \mathrm{H}\,|1\rangle &= \frac{1}{\sqrt{2}}|0\rangle - \frac{1}{\sqrt{2}}|1\rangle. \end{aligned} \quad (10.46)$$

Noting that the Hadamard gate is acting only on the leftmost entry in the AB terms in Eq. (10.45), we arrive at the rather messy looking result

$$\begin{aligned} \mathrm{H}\,|\Phi_1\rangle = |\Phi_2\rangle &= \frac{a}{2}\{|0_A 0_B\rangle[|0_\psi\rangle + |1_\psi\rangle] + |1_A 1_B\rangle[|0_\psi\rangle + |1_\psi\rangle]\} \\ &+ \frac{b}{2}\{|0_A 1_B\rangle[|0_\psi\rangle - |1_\psi\rangle] + |1_A 0_B\rangle[|0\rangle - |1\rangle]\}. \end{aligned} \quad (10.47)$$

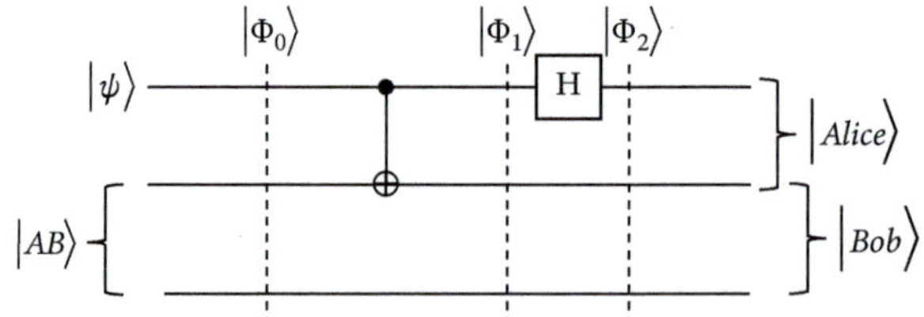

Fig. 10.9 The quantum circuit for quantum state teleportation. The upper wire is the control wire for the CNOT gate. The lower two wires represent the two qubits that are part of the entangled Bell state pair shared by Alice and Bob. The output on the lower right is the state of Bob's two-qubit system after the operation. The output on the upper right is the state of Alice's qubits after the operation. The vertical dashed lines indicate the locations of the states described in the text.

You can decipher what happened because we have grouped the terms in the following way: The first two sets of state vectors inside the curly braces multiplied by a are the ones that came from applying the Hadamard gate to the first term in the curly braces multiplied by a in Eq. (10.46). (Remember that the Hadamard gate transforms a basis state into a superposition state.) The second term comes from applying H to the second term in Eq. (10.46).

Try It 10.14 Work through the details in proceeding from Eq. (10.46) to Eq. (10.48).

Step 3c

Using $|0_A 0_B\rangle = |0_A\rangle |0_B\rangle$, we regroup the terms in Eq. (10.47) in the following way:

$$\begin{aligned} 2|\Phi_2\rangle = {} & |0_\psi\rangle |0_A\rangle [a|0_B\rangle + b|1_B\rangle] \\ & + |0_\psi\rangle |1_A\rangle [b|0_B\rangle + a|1_B\rangle] \\ & + |1_\psi\rangle |0_A\rangle [a|0_B\rangle - b|1_B\rangle] \\ & + |1_\psi\rangle |1_A\rangle [-b|0_B\rangle + a|1_B\rangle] . \end{aligned} \tag{10.48}$$

Note that the right side of Eq. (10.48) is a combination of the computational basis states for Alice's qubits and, in square brackets, variations on the state Eq. (10.42) Alice wants Bob to reproduce for Bob's qubit.

Step 4

Alice now measures the two qubits in her possession. Remember that one qubit has the unknown state $|\psi\rangle$ and the other is one of the qubits in the Bell state she prepared. The possible measurement results are (translated in the usual way to classical bits) 00, 01, 10, and 11. Table 10.3 indicates the list of possible results of Alice's measurement and the appropriate state vector for the qubit in Bob's possession. Remember that one of Alice's qubits is entangled with Bob's.

Table 10.3 The list of possible results of Alice's measurement and the appropriate state vector for the qubit in Bob's possession.

Alice's measurement results in classical bits	Alice's two-qubit state after the measurement	Bob's qubit's state after Alice's measurement
0,0	$\|0\rangle\|0\rangle$	$a\|0\rangle + b\|1\rangle$
0,1	$\|0\rangle\|1\rangle$	$b\|0\rangle + a\|1\rangle$
1,0	$\|1\rangle\|0\rangle$	$a\|0\rangle - b\|1\rangle$
1,1	$\|1\rangle\|1\rangle$	$-b\|0\rangle + a\|1\rangle$

Table 10.4 The gate Bob applies to his qubit depends on the two classical bits that Alice sends to him.

Alice sends	The gate that Bob applies to his qubit
0,0	I
0,1	X
1,0	Z
1,1	ZX

Step 5

After these manipulations, we see that Bob's qubit is described by a state similar to, but not the same as, the one Alice wants to teleport unless Alice's results are (0,0) as shown in the first line of Table 10.3. So, Alice sends Bob the two classical bits she obtained from her measurement. Bob then operates on the qubit he received from Alice to get its state to be the one Alice wants to transport. For example, if Alice's measurement results are (0,1), she sends Bob those two bits (in that order). After receiving the two bits (0,1), Bob knows he must operate on his qubit to get the desired state. Bob has figured out that if he gets (0,1), his qubit's state is the one in the third column, second row of Table 10.3. Bob operates on the qubit he received from Alice with the Pauli X gate (matrix), which interchanges the basis state vector labels. That is, it will change $a|1\rangle + b|0\rangle$ to $a|0\rangle + b|1\rangle$, just what we want. The required gate operations are listed in Table 10.4.

Try It 10.15 Use the properties of the Pauli gates (matrices) to verify that the gate operations listed in Table 10.4, when applied to Bob's qubit's state after Alice's measurements, will give the desired teleportation state.

Note that after Alice has made her observations, the state $|\psi\rangle$ is gone. Only after Bob applies the appropriate gate(s) to the qubit Alice sent him does the qubit's state change to $|\psi\rangle$. It means that the quantum state teleportation process does not violate the *no-cloning theorem*.

Why didn't Alice simply send the state $|\psi\rangle$ amplitudes to Bob? The problem is that in general Alice does not know how the qubit whose state is to be teleported was prepared. If Alice did, this method would work just fine, but it is limited to those cases where Alice knows all the details of the state $|\psi\rangle$. The quantum state teleportation method used here works whether Alice knows the numerical values of a and b or not. Once again, superposition and entanglement make the process work.

It is important to note that at the end of the teleportation process, Bob does not know the state amplitudes of $|\psi\rangle$ either. All he knows is that he followed the quantum rules and those guarantee that his qubit is now described by $|\psi\rangle$.

How might Bob check that the state teleported is the same as the state Alice started with? To check, Alice would need to use a series of prepare–measure cycles on her state $|\psi\rangle$ to find what it predicts for the probabilities of measurement outcomes. Bob would need to do the same after the teleportation process.

Try It 10.16 Suppose that in Alice's "unknown" state $|\psi\rangle = a\,|0\rangle + b\,|1\rangle$, we have $a = 1/2$ and $b = -\sqrt{3}/2$. Go through the steps of the teleportation process and show that Bob is able to produce the state $|\psi\rangle$.

Some authors claim that Bob's qubit "instantaneously jumps into one of the four states" listed on the right in Table 10.3 after Alice makes her measurement, and thusly, quantum state teleportation allows for instantaneous transmission of information over long distances. The flaw in that argument is that Bob's qubit state does not change until he receives the two classical bits and applies the appropriate gate and that takes time. There is no "instantaneous" transfer of information. But if Bob's qubit state changes immediately after Alice makes her measurement, Bob couldn't measure the state of his qubit before the classical bits arrive. If so, would we have instantaneous transmission of state information?

Try It 10.17 What is the answer to the previous question? Explain your reasoning.

We should emphasize one more time that quantum state teleportation does not involve the transport of any physical object (qubit) once the originally entangled qubits are set up, with Alice having one and Bob having the other. It is only the abstract quantum state information describing the qubit's state that is transported.

10.10 **SWAP Gates**

Before moving on to more quantum algorithms, we want to introduce one more two-qubit gate: the SWAP gate. As the name implies, the SWAP gate interchanges something between two local qubits without moving either of the qubits. That something is their quantum states. (This is to be distinguished from quantum state teleportation that we used to have a distant qubit be described by the same state as a local qubit. This algorithm required sending one of Alice's entangled state qubits to Bob.) Although the algorithms we discuss in the remainder of this book do not require SWAP gates, they are found in many other quantum algorithms so we thought to introduce them in this chapter. They also provide another nice example of using matrices to represent quantum gates (see Eq. (10.49) and Fig. 10.10).

Since we are dealing with two qubits, the basis states can be taken to be the four-entry column vectors as shown in Table 9.2. We need a 4×4 matrix to represent operations

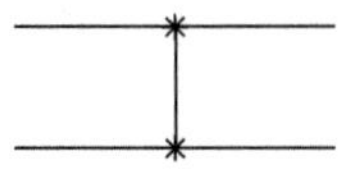

Fig. 10.10 The circuit symbol for the SWAP gate.

on these vectors. The SWAP operator that interchanges the states $|01\rangle$ and $|10\rangle$ has the following matrix representation in that column vector basis:

$$\text{SWAP}_{10} \doteq \begin{pmatrix} 1 & 0 & 0 & 0 \\ 0 & 0 & 1 & 0 \\ 0 & 1 & 0 & 0 \\ 0 & 0 & 0 & 1 \end{pmatrix}. \tag{10.49}$$

Try It 10.18 Use the column vector representation of the state $|01\rangle$ to show that $\text{SWAP}_{10}\,|01\rangle = |10\rangle$ and $\text{SWAP}_{10}\,|10\rangle = |01\rangle$. What happens if you apply the SWAP gate to the states $|00\rangle$ and $|11\rangle$? Explain why these results make sense.

Try It 10.19 Apply the SWAP_{10} gate to the two-qubit superposition state $|\psi\rangle = a\,|10\rangle + b\,|01\rangle$ with $|a|^2 + |b|^2 = 1$. Explain your results.

10.11 Chapter Summary

- The quantum CNOT gate is a two-qubit gate, which applies a NOT gate operation to one input state depending upon the state of the other input state. CNOT can be used to create an entangled state.
- The Bell Circuit (BC), consisting of a Hadamard gate acting on the control state followed by a CNOT, produces an entangled state. The Reverse Bell Circuit (RBC) has the gates in the opposite order, producing a product state from an entangled state.
- The *no-cloning theorem* states it is impossible to prepare a second qubit described by a quantum state that is exactly the same as the state associated with a previously prepared qubit whose state coefficients are not known. A general quantum state is a superposition state. One observation does not provide enough information to produce a second superposition state identical to the first.
- Superdense coding allows for two classical bits to be sent with a single qubit by leveraging an entangled Bell state. One qubit is kept by the sender and the other is sent to the receiver. The sender uses a gate on the first qubit, which is then sent to the receiver. The receiver applies an RBC to that qubit and measures the two qubits, yielding the desired classical bits.

- Quantum state teleportation allows a receiver to replicate a quantum state in a distant location using two classical bits and a qubit prepared as part of an entangled state. Only information is transported, not qubits. At no time does the process result in two qubits described by the same state, so the *no-cloning theorem* is not violated. The sender applies quantum gates to the qubit they kept and the qubit whose state is to be teleported. That combination is measured, resulting in two classical bits, which are sent to the receiver. The classical bits tell the receiver what gates to apply to their qubit to produce the desired state. Neither the sender nor the receiver needs to know the amplitudes and phases of the teleported state.

10.12 Further Reading

Quantum Gates

The Wikipedia article "Quantum logic gate" contains a thorough introduction to quantum gates, their names, and their circuit and matrix representations with a summary of their usage. https://en.wikipedia.org/wiki/Quantum_logic_gate

The Wikipedia article "List of quantum logic gates" contains an even more exhaustive list of quantum gates, but with less explanation of how they are used compared to the "Quantum logic gate" article. https://en.wikipedia.org/wiki/List_of_quantum_logic_gates

On Bell Measurements

Brian T. Kirby, Siddhartha Santra, Vladimir S. Malinovsky, and Michael Brodsky, "The utility of entanglement swapping in quantum communications," *Proc. SPIE 9980, Quantum Communications and Quantum Imaging XIV*, 99800L (13 September 2016); https://doi.org/10.1117/12.2237760

On Entanglement Swapping

Jian-Wei Pan, Dik Bouwmeester, Harald Weinfurter, and Anton Zeilinger, "Experimental Entanglement Swapping: Entangling Photons That Never Interacted," *Physical Review Letters* **80**, 3891–3894 (1998).

Rui-Bo Jin, Masahiro Takeoka, Ryosuke Shimizu, and Masahide Sasaki, "Highly Efficient Entanglement Swapping and Teleportation at Telecom Wavelength." *Nature Scientific Reports* **5**, 9333 (2015).

Peter Y. Lee, Huiwen Ji, and Ran Cheng, *Quantum Computing and Information: A Scaffolding Approach* (Polaris QCI Publications, 2024). Section 10.4 presents a thorough discussion of entanglement swapping.

On Entanglement Swapping Used in Quantum Communications

Brian P. Williams, Ronald J. Sadlier, and Travis S. Humble, "Superdense Coding over Optical Fiber Links with Complete Bell-State Measurements," *Physical Review Letters* **118**, 050501 (2017).

PART III

Quantum Algorithms

11 Quantum Computing Algorithms

Contrariwise, continued Tweedledee, if it was so, it might be; and if it were so, it would be; but as it isn't, it ain't. That's logic.

Lewis Carroll, *Through the Looking Glass*

11.1 Welcome to the Quantum Algorithm Zoo

We are finally ready to tackle some actual quantum computing algorithms. As shown later, quantum computing algorithms are not just classical computing algorithms modified to run on a quantum computer. Quantum algorithms are new ways of thinking about problems and are designed to exploit the properties of quantum states, quantum gates, and measurement devices.

We start with a few of the standard quantum algorithms, which were invented decades ago, primarily to show how we can use the features of quantum states, particularly entangled states, to implement algorithms more quickly or more efficiently than we can with classical computers. Many of these algorithms are, in a sense, designed to solve "made up" problems—problems selected because they are amenable to quantum algorithms and show some "advantage" over classical computational algorithms. There are many slightly different versions of these algorithms. We use versions that are fairly simple to understand.

The development of quantum algorithms is still in its infancy. Links to almost all the known quantum algorithms can be found on the website The Quantum Algorithm Zoo available at https://quantumalgorithmzoo.org and in the references cited in Further Reading. Many of the algorithms deal with esoteric topics in mathematics, but in a later chapter, we explore one that allows us to find the energy states of atoms and molecules. Many of the companies that have built working quantum computers are hoping that smart people like our readers will develop algorithms that allow these computers to do something useful that is not practical or possible with classical supercomputers.

As we noted, all the quantum computing algorithms we will look at have a common structure. As illustrated in Fig. 11.1, you start by preparing a set of qubits described by some initial state, usually one of the computational basis states. You then act on those qubits' states with various quantum gates. At the end, the resulting qubits interact with measurement devices, again usually employing the computational basis states. You then interpret the results of those measurements to answer the question the algorithm is designed to address.

Quantum Computing and Quantum Physics. Alice Flarend and Robert Hilborn, Oxford University Press.
 DOI: 10.1093/oso/9780197904930.003.0011

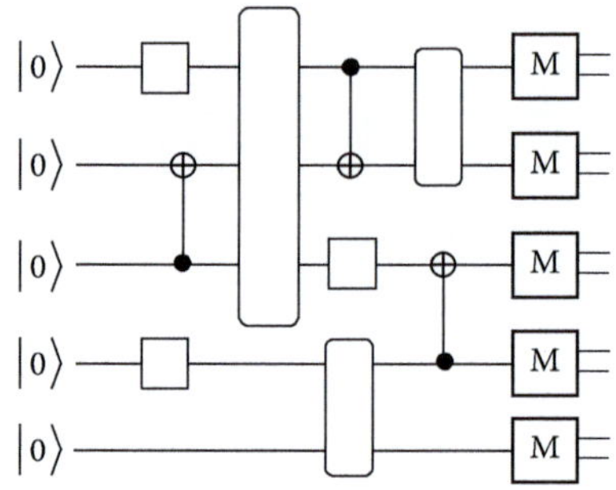

Fig. 11.1 A schematic of a typical quantum circuit (gate array) designed to implement a quantum computational algorithm. The blank square boxes are gates. The boxes labeled M are measurement devices. The other symbols are CNOT gates.

For now, we focus on the algorithms themselves. In Chapter 16, we describe how you translate those structures into commands that the quantum computer will execute—a quantum computer program.

11.2 The Deutsch Algorithm

In 1985, David Deutsch introduced one of the first quantum algorithms to show an advantage over classical computing algorithms. Many others, including Deutsch himself, subsequently modified the algorithm to make it more efficient and understandable.

Deutsch's algorithm asks a specific question about mathematical functions for which the independent variable (the input) is a binary digit, that is, 0 or 1, and the dependent variable (the output) is 0 or 1. Since there are only two possible input values and two possible output values, we have four possible functions. Customarily, these are labeled f_0, f_1, f_2, and f_3. Of course, you could use any four names you would like. We don't need to worry about the details of how the functions turn the inputs into the outputs. All we need to do is specify the output for each of the two possible inputs 0 and 1. Table 11.1 displays the results.

The functions are divided into two classes: (1) constant functions and (2) balanced functions. Constant means that the output is the same no matter what the input is. The other two give 0 for 50% of the inputs and 1 for the other 50% and are called balanced functions. For a one-qubit system, all nonconstant functions are balanced, as shown in Table 11.1. For a two-qubit system, we might have a function that gives $f(00) = 0$ and $f(01) = f(10) = f(11) = 1$. This function is nonconstant but not balanced. We restrict ourselves to constant and balanced functions.

Deutsch posed the following problem. Suppose someone gives you a box with an input and an output. Built into the box is the machinery needed to implement one of the four functions given in Table 11.1. How many times do you need to evaluate the function—by providing a number at the input and reading the output—to tell if the function is in the "constant" group or the "balanced" group? The answer seems easy: just try 0 at the input

Table 11.1 Constant and balanced single-qubit binary functions.

Input = 0	Input = 1	Type of function
$f_0(0)= 0$	$f_0(1)= 0$	Constant
$f_1(0)= 0$	$f_1(1)= 1$	Balanced
$f_2(0)= 1$	$f_2(1)= 0$	Balanced
$f_3(0)= 1$	$f_3(1)= 1$	Constant

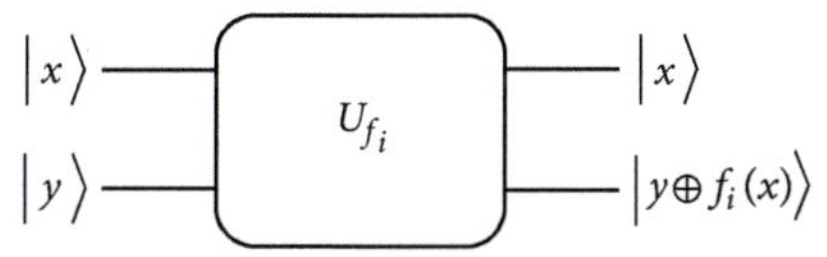

Fig. 11.2 The quantum gate that implements one of the four functions f_i, i = 0, 1, 2, or 3, listed in Table 11.1. See the text for the definitions of x and y in this context.

Table 11.2 The Exclusive OR (XOR) truth table.

Operation	Result
$0 \oplus 0$	0
$0 \oplus 1$	1
$1 \oplus 0$	1
$1 \oplus 1$	0

and record the output. Then try 1 at the input and record the output. If the two outputs are the same, we have a constant function, and if they are different, we have a balanced function. In fact, we could tell you which of the four functions is in the box once we have those results.

For classical computation, this reasoning is correct: you need two evaluations of the function box, with two different inputs, to answer the constant versus balanced question. What Deutsch did was to prove that, with appropriate quantum gates, you need to evaluate the function only once, not twice, to get the answer.

To make this work, we take advantage of the superposition form of qubit states. We need a quantum gate with two inputs and two outputs as shown in Fig. 11.2. The variables are x, y, and $y \oplus f_i(x)$, each of which are either 0 or 1. Remember that we want to use reversible gates and these require the same number of input rows on the left as we have output rows on the right. In Fig. 11.2, the upper row (sometimes called the *upper register*) state appears unchanged on the output side (on the right) while the lower row (the *lower register*) state is changed as indicated in the figure. Recall that $\oplus$ means "addition modulo 2" or "Exclusive OR," also known as XOR. Think of adding two classical binary digits and dropping any "carry" term if the result exceeds 1 as shown in Chapter 2. Table 11.2 repeats the XOR truth table, so we can have the results in front of us.

Let's suppose that we have the function f_1 inside the gate box in Fig. 11.2 and arbitrarily $x = 0$ and $y = 1$. From Table 11.1, we see that $f_1(0) = 0$. From Table 11.2, we read off $1 \oplus 0 = 1$, so $y \oplus f_1(0) = 1$ and the lower-register output state is $|1\rangle$.

This might be a good time to remind ourselves of a feature of the way we have written the lower-register output state $|y \oplus f(x)\rangle$. We have previously emphasized that what appears inside the state vector symbol is just a name to label the state. However, for many quantum algorithms, it is useful to have this name be a numerical value that is generated from other numerical values, here x and $f(x)$. We observed something like that when we labeled the four two-qubit state vectors as $|00\rangle$, $|01\rangle$, $|10\rangle$, and $|11\rangle$. The state labels increase in binary numerical order as we read from left to right.

Let's get back to the Deutsch function gate. Later we show you how to build the function gate to do this evaluation automatically. For now, let's assume that we have the function gate working properly. It turns out that we can write down the four possibilities for the results of acting on the computational basis states that leave out the unnecessary details of the functions. Let's suppose f denotes any one of the four functions. We also use the tensor product symbol $\otimes$ to separate the state vectors and make the equations easier to read.

If the gate U_f acts on the standard initial system product state $|x\rangle \otimes |y\rangle = |0\rangle \otimes |0\rangle$, we obtain

$$\begin{aligned} U_f\left[|x\rangle \otimes |y\rangle\right] &= U_f\left[|0\rangle \otimes |0\rangle\right] = |0\rangle \otimes |y \oplus f(x)\rangle \\ &= |0\rangle \otimes |0 \oplus f(0)\rangle = |0\rangle \otimes |f(0)\rangle . \end{aligned} \tag{11.1}$$

This works because $0 \oplus f(0) = f(0)$ no matter what $f(0)$ is: If $f(0) = 0$, we have $0 \oplus 0 = 0 = f(0)$. If $f(0) = 1$, then we have $0 \oplus f(0) = 0 \oplus 1 = 1 = f(0)$. These are the only possibilities.

You should also keep track of where the 0s come from in this example. In the example, both x and y are 0, but that will not always be the case. Sometimes, for example, x will be 1 and you will be evaluating $f(1)$ instead of $f(0)$. And then there are 0s in the example from $0 \oplus 0 = 0$, which is obtained by evaluating $y \oplus f(0)$. The binary value is the same, but the calculation is different. Knowing how the calculation works leads to understanding the algorithm.

We just did a single case of $|0\rangle \otimes |0\rangle$. Let's make it look simpler for other cases by introducing the notation $g(y) = 1 \oplus f(y)$. All the results from Eq. (11.1) can be summarized by

$$\begin{aligned} U_f\left[|0\rangle \otimes |0\rangle\right] &= |0\rangle \otimes |0 \oplus f(0)\rangle = |0\rangle \otimes |f(0)\rangle \\ U_f\left[|0\rangle \otimes |1\rangle\right] &= |0\rangle \otimes |1 \oplus f(0)\rangle = |0\rangle \otimes |g(0)\rangle \\ U_f\left[|1\rangle \otimes |0\rangle\right] &= |1\rangle \otimes |0 \oplus f(1)\rangle = |1\rangle \otimes |f(1)\rangle \\ U_f\left[|1\rangle \otimes |1\rangle\right] &= |1\rangle \otimes |1 \oplus f(1)\rangle = |1\rangle \otimes |g(1)\rangle \end{aligned} \tag{11.2}$$

for any of the four possible functions in Table 11.1.

Try It 11.1 Check the results in Eq. (11.2) by using Table 11.2.

What Deutsch was able to show—and this is typical in almost all quantum algorithms—is that if we use superposition states in place of the basis states as inputs to the quantum gate, the "magic" of quantum mechanics allows us to answer the question with just one application of the function gate.

Figure 11.3 shows the overall quantum circuit (gate array) for the Deutsch algorithm. In region 1, two Hadamard gates convert the initial states into superposition states—a general and crucial strategy in most quantum algorithms. In region 2, the function gate acts on those states to produce the output states. The upper row output state is acted upon by another Hadamard gate in region 3. And finally, in region 4, the measurement device tells us whether the upper row output state is $|0\rangle$ or $|1\rangle$.

The measurement step is another general feature of quantum algorithms. We read out the answer (or something logically linked to the answer) using measurement devices. For the Deutsch algorithm, the states have been cleverly set up so that the measurement result $|0\rangle$ in the upper register indicates that the function f is constant while $|1\rangle$ in the upper register shows that the function is balanced. The cleverness of the algorithm is in making that association work properly. We now go through the details to show how it works.

Let's assume that we prepare the circuit input state as the product of the two input states $|\psi_0\rangle = |1\rangle|1\rangle$. See Fig. 11.3 to find where the state labels $|\psi_j\rangle$ apply. Note that if you prefer to start with $|0\rangle|0\rangle$, you simply apply the quantum X gate (the equivalent of a NOT gate) to each of the states to obtain $|1\rangle|1\rangle$. To keep the diagrams as simple as possible, let's assume we have done that.

Next, we use our old friend the Hadamard gate on each of the input states to generate superposition states in both the upper register labeled A and the lower register labeled B. Recall that the action of the Hadamard gate on the computational basis state $|1\rangle$ is given by

$$H|1\rangle = \frac{1}{\sqrt{2}}\big(|0\rangle - |1\rangle\big). \tag{11.3}$$

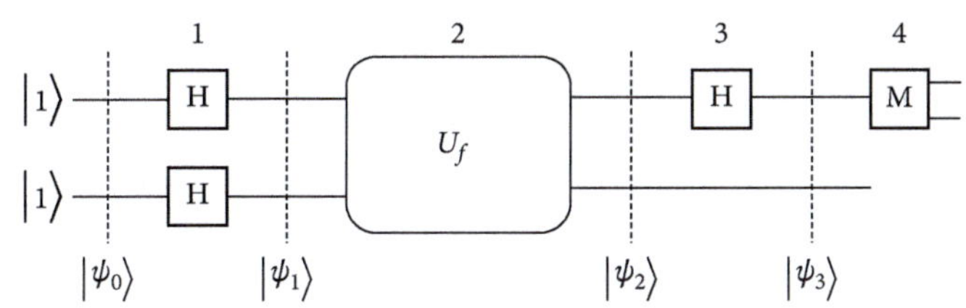

Fig. 11.3 The quantum circuit for the Deutsch algorithm. The vertical dashed lines indicate where various state labels apply. The four numbers above the circuit label different regions described in the main text.

The resulting system state is expressed as

$$|\psi_1\rangle = \mathrm{H}_A \otimes \mathrm{H}_B|1\rangle_A|1\rangle_B = \frac{(|0\rangle_A - |1\rangle_A)}{\sqrt{2}} \otimes \frac{(|0\rangle_B - |1\rangle_B)}{\sqrt{2}}$$
$$= \frac{1}{2}\left(|0\rangle_A|0\rangle_B - |0\rangle_A|1\rangle_B - |1\rangle_A|0\rangle_B + |1\rangle_A|1\rangle_B\right). \tag{11.4}$$

We have used the tensor product symbol ⊗ between the two Hadamard gate symbols to remind us that one of them acts on the *A* register state and the other acts on the *B* register state. Note that Eq. (11.4) contains all four possible states in our two-qubit system. This is sometimes called "quantum parallelism": We operate on all possible states for the system simultaneously, which is awesome. We come back to this issue later.

Try It 11.2 Is the state in Eq. (11.4) an entangled state? Hint: Review Eq. (9.10).

Now we are ready to have state $|\psi_1\rangle$ interact with the function gate in region 2 of Fig. 11.3. The function gate operates on the state given in Eq. (11.4). We have already listed the pieces in Eqs. (11.1) and (11.2).

Let's pull all those pieces together by writing out the $|\psi_2\rangle$ state after the function gate operation. The details of how to make the function gate are given in a later section. For now, let's focus on how we want the result to look. We drop the A and B subscripts and remember that the states on the left of a product are in the upper register, while those on the right are in the lower register:

$$|\psi_2\rangle = U_f(\mathrm{H} \otimes \mathrm{H}|1\rangle|1\rangle) = \left(\frac{1}{2}\underbrace{|0\rangle}_{\text{upper}} \otimes \underbrace{|f(0)\rangle}_{\text{lower}} - \underbrace{|0\rangle}_{\text{upper}} \otimes \underbrace{|g(0)\rangle}_{\text{lower}} - \underbrace{|1\rangle}_{\text{upper}} \otimes \underbrace{|f(1)\rangle}_{\text{lower}} + \underbrace{|1\rangle}_{\text{upper}} \otimes \underbrace{|g(1)\rangle}_{\text{lower}}\right). \tag{11.5}$$

The "trick" is to write the resulting state separately for constant and balanced functions. For constant functions, we have, by definition,

$$f(0) = f(1) \text{ and } g(0) = g(1). \tag{11.6}$$

For balanced functions, we have

$$f(0) \neq f(1) \text{ and } g(0) \neq g(1). \tag{11.7}$$

Furthermore, for the balanced functions, we obtain

$$f(1) = g(0) \text{ and } g(1) = f(0). \tag{11.8}$$

To see what is happening, let's check one possibility using Tables 11.1 and 11.2. For the balanced function, we have

$$f_1(0) = 0 \text{ and } g_1(0) = 1 \oplus f_1(0) = 1. \tag{11.9}$$

But we also have $g_1(1) = 1 \oplus f_1(1) = 0$. So, we see that $g_1(0) \neq g_1(1)$, $f_1(1) = g_1(0)$, and $g_1(1) = f_1(0)$. You can easily check the other results by writing out all the possible cases.

Try It 11.3 Verify the relationships stated in the paragraph that contains the "trick."

Let's go on and first assume that f is a constant function. Then, we replace $f(1)$ and $g(1)$ in Eq. (11.5) with $f(0)$ and $g(0)$, respectively. Carrying out those substitutions and doing some rearrangement yields

$$|\text{constant}\rangle = \frac{1}{2}\underbrace{(|0\rangle - |1\rangle)}_{\text{upper}} \otimes \underbrace{(|f(0)\rangle - |g(0)\rangle)}_{\text{lower}} = \frac{1}{2}(|0\rangle - |1\rangle) \otimes |F_f\rangle, \tag{11.10}$$

where we introduced $|f(0)\rangle - |g(0)\rangle = |F_f\rangle$ to simplify the notation further and to make the results more obvious. Following the same procedure for balanced fs gives us

$$|\text{balanced}\rangle = \frac{1}{2}\underbrace{(|0\rangle + |1\rangle)}_{\text{upper}} \otimes \underbrace{|F_f\rangle}_{\text{lower}}. \tag{11.11}$$

Try it 11.4 Check the details leading to Eqs. (11.10) and (11.11).

Note that the difference between Eqs. (11.10) and (11.11) is in the sign in the superposition state for the upper register.

There is a lot going on in this algorithm and for most of us it will be quite a cognitive overload to keep all of this in mind at once. The best strategy is to follow along with the steps. Once we are done with the algorithm, you should then go back through the steps to see how the pieces come together to answer the original question.

How can we use that difference to answer the Deutsch question? We now employ another "trick": Have the upper-register output state interact with another Hadamard gate. Remember that a Hadamard gate acting on a superposition state "un-does" the superposition and produces a basis state:

$$H\frac{1}{\sqrt{2}}\left(|0\rangle + |1\rangle\right) = |0\rangle \quad \text{and} \quad H\frac{1}{\sqrt{2}}\left(|0\rangle - |1\rangle\right) = |1\rangle. \tag{11.12}$$

So, in this case, we obtain two product states:

$$
\begin{aligned}
|\psi_3 \text{ constant}\rangle &= \mathrm{H}\,|\text{constant}\rangle = \frac{1}{\sqrt{2}}\,|1\rangle \otimes \left|F_f\right\rangle \\
|\psi_3 \text{ balanced}\rangle &= \mathrm{H}\,|\text{balanced}\rangle = \frac{1}{\sqrt{2}}\,|0\rangle \otimes \left|F_f\right\rangle .
\end{aligned}
\tag{11.13}
$$

The final "trick" is to have the upper-register state, after the Hadamard gate, interact with a measurement device set up to observe $|0\rangle$ or $|1\rangle$. Here is the grand conclusion: If the observation gives 1, then the upper row output state is $|1\rangle$. According to Eq. (11.13), we must have had a constant function f in the function box. If the observation gives 0, which indicates that the output state is $|0\rangle$, then the function f is in the balanced category. We have the answer. Sound the trumpets!

To emphasize some common features of quantum algorithms, let's now explain the Deutsch algorithm in more detail, taking Fig. 11.3 into consideration. You recall that the first two Hadamard gates converted the initial states into superposition states as described in Eq. (11.4). This is a general and crucial strategy in most quantum algorithms. The function gate then acts on those states to produce the output states shown in Eq. (11.5). Next, the upper-register state is acted upon by another Hadamard gate. This Hadamard gate produces a product state as described in Eq. (11.13). And finally, the measurement device allows us to determine whether the upper row output state is $|0\rangle$ or $|1\rangle$.

As mentioned before, this last step is another general feature of quantum algorithms. We read out the answer (or something logically linked to the answer) using measurement devices. For the Deutsch algorithm, the states have been cleverly set up so that the measurement result $|1\rangle$ in the upper register shows that the function f is constant while the measurement result $|0\rangle$ in the upper register indicates that the function is balanced. The cleverness of the algorithm is in making that association work properly.

Note in Eq. (11.13) the output state of the lower register is $\left|F_f\right\rangle$ for both measurement outcomes, so we don't learn anything else about the function. For example, if it were constant, we don't know whether it was f_0 or f_3.

Before we look at some further details of the Deutsch algorithm, we want to alert you to some terminology, which is often used for quantum algorithms that make use of one or more function gates. These function gates are often called "oracles" and having states interact with the oracle is called *querying the oracle*. For the Deutsch algorithm, we would say that the algorithm requires only one query to the oracle while the corresponding classical algorithm requires, as mentioned previously, two queries. So, if your definition of "advantage" is requiring fewer oracle queries, then the Deutsch algorithm has an advantage over the classical algorithm.

Figure 11.4 illustrates two additional pieces of jargon associated with quantum circuits. On the left, the figure shows circuit layers—gates aligned vertically constitute a circuit layer. On the right, the gates in the two registers are shifted horizontally relative to each other resulting in different layers. The convention is that the gates to the left are to be implemented

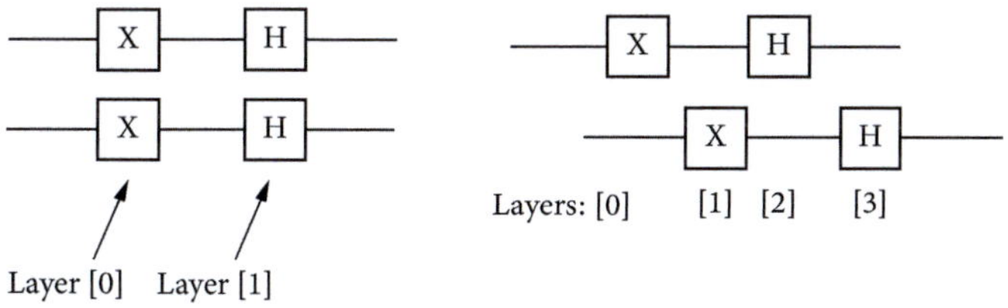

Fig. 11.4 Gates laid out vertically above one another constitute a quantum circuit *layer*. On the right, the gates in the lower register are shifted to the right, indicating that they should be implemented after the corresponding gates in the upper register. Such an arrangement is called circuit "scheduling."

before the corresponding gate in the other register. This is an example of circuit scheduling, which is important for complicated quantum circuits.

11.3 Implementing U_f with Simple Quantum Gates

This section is not directly needed for what follows, but we think it will help you to realize that the function box U_f could be constructed directly from just a few of the quantum gates we described earlier. Figure 11.5 (after Mermin, 2007, p. 42) shows you the circuit diagrams. Each circuit consists of two input wires and two output wires. A qubit state is associated with each wire. For example, in the top part of Fig. 11.4 each wire could represent either $|0\rangle$ or $|1\rangle$. Since these wires connect directly to the output, the states do not change.

In the second function circuit, the lower-register state is subject to a controlled-NOT gate, drawn as a controlled X gate, with the upper-register state acting as the control state.

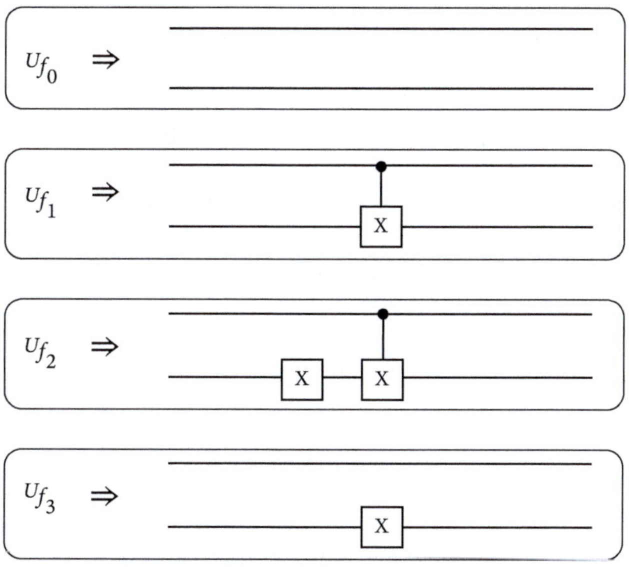

Fig. 11.5 Circuit diagrams for the four *f* function gates.

In this case, if the upper-register state is $|0\rangle$, then the lower-register state output is the same as the lower-register state input. If the upper-register state is $|1\rangle$, then the lower-register state is "flipped." For example, $|1\rangle$ becomes $|0\rangle$. This set of operations is just what is needed to implement function f_1 from Table 11.1.

Try It 11.5 Using the results from Chapter 6, verify that these circuits perform as promised.

11.4 The Simon Search Algorithm

As a second example of a classic quantum algorithm, we look at the Simon algorithm, first developed in the 1990s. While it has some general structural similarities to the Deutsch algorithm, it is also more complicated, so hang on to your gates. We follow the analysis given in Bernhardt's *Quantum Computing for Everyone* but with several additional explanations and examples to make the steps clearer. The strategy is to make use of superposition and entangled quantum states to create an algorithm that is more efficient than the corresponding classical algorithm.

The issue that the Simon algorithm addresses is constructing a method to find a "secret code" s that satisfies $f(x) = f(x \oplus s)$, where f is a function that takes a binary string x (a sequence of 0s and 1s) and generates a binary string for its output. Our condition for the secret code is that f generates the same output sequence when either x or $x \oplus s$ is the input. This means the function gives the same output for two different inputs. We exclude the string $s = 000\ldots$ since $x = x \oplus 000\ldots$ for any x and then obviously $f(x+0) = f(x)$. We want this to be an "if and only if" statement; so, if y is another binary sequence, we have $f(x) \neq f(y)$ if $y \neq x \oplus s$.

What does $x \oplus s$ actually mean? Let's illustrate the XOR addition of two sequences of 0s and 1s representing x and s. For example, we might have $x = 011100111$ and $y = 110001101$ as two binary strings of length 9. The symbol $\oplus$ means that we add the corresponding bits but drop any carry bit. Starting on the left with the first bits of both strings, we have $0 \oplus 1 = 1$, followed by $1 \oplus 1 = 0$, and so on. We then find that $x \oplus y = 101101010$. In short, if the two corresponding bits in the sequences are the same, the output is a 0, and if they are different, the output is 1. Since we are carrying out this operation bit by bit, we could also start on the right end of the sequence.

To be sure of what we are trying to accomplish with the Simon search algorithm, let's try the method with a two-bit sequence and $s = 01$. There are only four possible two-bit sequences, so we can easily write down all the possibilities.

Based on the first and last columns of Table 11.3, we see that for the secret code $s = 01$ to meet the $f(x) = f(x \oplus s)$ requirement, the function must satisfy

$$\begin{aligned} f(00) &= f(01) \\ f(10) &= f(11). \end{aligned} \tag{11.14}$$

Note that these are the only two possibilities for two-bit binary numbers.

Table 11.3 Secret code example with $s = 01$.

x	s	$x \oplus s$
00	01	01
01	01	00
10	01	11
11	01	10

For the Simon algorithm, we want to be general and use binary strings of length n for both the input and output of the function f. We don't need to know anything about the function except that you get the same binary string output (e.g., 11011100011...) if you add (bitwise) the string associated with s to the original sequence x.

The Simon algorithm works even if we don't know explicitly what f produces for each of the inputs. This is an important point. We observed a similar situation with the Deutsch algorithm: We did not need to specify the function to have an algorithm that works. The goal is to be as general as possible, and here it means we do not require a specific function for the algorithm.

The construction strategy is much like that used in the Deutsch algorithm. We let the states interact with quantum gates in such a way that a measurement of one of the output states gives us information that allows us to find the secret code. It turns out that the quantum gates don't give us the code s directly, but they produce equations that we can solve using classical computational techniques to find the code.

Let's first think about this from a classical computing point of view. We could try to find s by brute force: just choose an input sequence x at random and evaluate $f(x)$. Then, try another input sequence and keep repeating the process until we find a match. How many trials do we have to run? For example, for a string of length $n = 3$, the number of sequences is $2^3 = 8$, and for $n = 20$, we have $2^{20} = 1048576$ sequences.

Fortunately, we don't need to go through every possible sequence. First, we might get lucky and happen upon a match early on. At worst, we need to go through one more than half the total, that is $2^{n-1} + 1$ to be guaranteed to find a match. Note that the number increases exponentially with n, so it gets big quickly as n increases. We will see that the Simon algorithm is likely to give an answer with many fewer inputs, a number that grows only linearly with n. The Simon algorithm was the first to show that using a quantum computing algorithm can in principle give a tremendous advantage over classical computation.

It may seem strange that we said the Simon algorithm is *likely* to give an advantage. We shall see that probability plays an important role in the Simon algorithm. It also shows up in many other quantum algorithms because quantum mechanics is fundamentally probabilistic in nature. We will talk about this in detail after showing how the Simon algorithm works.

Try It 11.6 Use $n = 2$ and convince yourself that with the classical algorithm described in the previous paragraphs that you need to try two input numbers to find s.

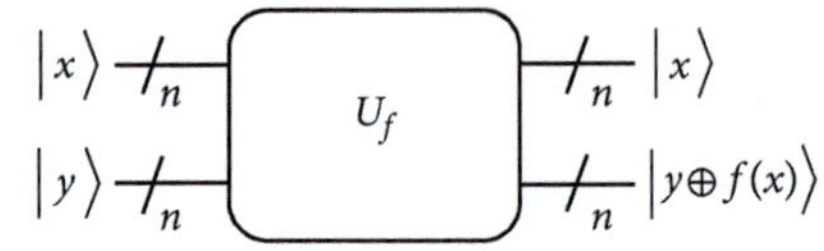

Fig. 11.6 The circuit diagram for the function gate for the Simon algorithm. The horizontal line with the diagonal slash $/_n$ and subscript *n* means we have *n* lines in that register. This figure is just the *n*-qubit version of Fig. 11.2.

Now we talk about the quantum method. First, we need to build a function box U_f of the type we used for the Deutsch algorithm, except in this case we have only one function to worry about. This seems to make it easier, but there is a catch. The complication is that we want to be able to handle bit sequences with n bits. This requires n input states for the 0s and 1s that specify x and n input states to link to the n bit sequences that are generated by f. It means we need a function box with $2n$ states on the input side and $2n$ on the output side.

In Fig. 11.6, we have drawn the Simon function gate. On the left are two registers of n input states, each with horizontal lines connecting to the box. On the right are two registers, each with n output states. The state names $(x, y, y \oplus f(x))$ each represent an n binary bit sequence.

If we used just the bare function gate, we would not have made any progress over a classical algorithm because to make sure we find the functions that match we would have to evaluate the function with up to $2^{n-1} + 1$ different inputs. However, we can again invoke the magic of quantum states by using a superposition state as the input to the top register connection to the function gate. In a sense, we have the function gate operate on all the possible basis states at once. Then by making measurements on the upper-register output states, we will get information that will help us find s.

You know what is needed to generate superposition states that include all the basis states: the Hadamard gate. For a single qubit, as shown before, the Hadamard gate generates the equal-amplitude superposition states, as illustrated in Eq. (11.3).

How do we generalize this to multiqubit states? Symbolically, we can write down a Hadamard gate that acts on n-qubit states as $\mathrm{H}^{\otimes n}$, where $\otimes$ reminds us that the basis states are tensor product states. For $n = 2$, the result is

$$
\begin{aligned}
\mathrm{H}^{\otimes 2}\,|00\rangle &= \frac{1}{2}\left(+\,|00\rangle + |01\rangle + |10\rangle + |11\rangle\right)\\
\mathrm{H}^{\otimes 2}\,|01\rangle &= \frac{1}{2}\left(+\,|00\rangle - |01\rangle + |10\rangle - |11\rangle\right)\\
\mathrm{H}^{\otimes 2}\,|10\rangle &= \frac{1}{2}\left(+\,|00\rangle + |01\rangle - |10\rangle - |11\rangle\right)\\
\mathrm{H}^{\otimes 2}\,|11\rangle &= \frac{1}{2}\left(+\,|00\rangle - |01\rangle - |10\rangle + |11\rangle\right).
\end{aligned}
\tag{11.15}
$$

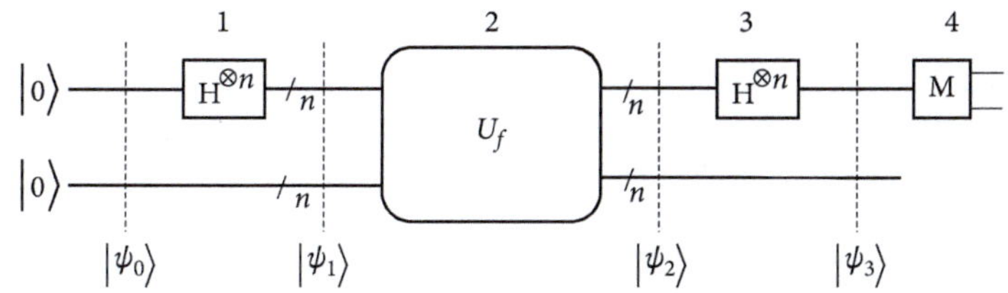

Fig. 11.7 The quantum circuit (gate array) for the n-qubit Simon algorithm. The system state $|\psi\rangle$ is indicated at four different positions indicated by the vertical dashed lines in the diagram. Although only one measurement device is shown at the top right of the diagram, there are actually n measurement devices. Each measurement device interacts with one qubit.

We show you how to derive these specific results once we have gone through the complete algorithm. But, if you look carefully at Eq. (11.15), you will see there is a pattern to the plus and minus signs: There are all plusses across the top row and down the first column, where the amplitudes for all the $|00\rangle$ states are $+1$. The minus signs are distributed the same way for row two and column two, and so on. Many of these features generalize to the n-qubit case.

Now let's get back to the Simon algorithm. The quantum circuit (gate array) is shown in Fig. 11.7. We start with the system state $|\psi_0\rangle$ on the far left. This state is a product of two sets of computational basis states:

$$|\psi_0\rangle = |0\rangle_n |0\rangle_n. \tag{11.16}$$

Here, we used the subscripts n on the right vectors to remind us that we are dealing with n-bit states. As mentioned before, we put the upper-register states on the left of each product, while the lower-register states go on the right.

As we read across the quantum circuit from left to right, the input states in the upper register in region 1 are acted upon by the Hadamard gate $H^{\otimes n}$ producing superposition states that connect to the function box. On the output side in region 3, the states in the upper register are acted upon by another Hadamard gate $H^{\otimes n}$, and then at the end in region 4, they interact with an n-qubit measurement device. Take a moment to look at Fig. 11.3 and see the similarities to the Deutsch circuit.

It will be helpful to work out the algorithm step by step for $n = 2$, which is simple enough that we can write down all the steps, but complex enough to show how the algorithm works. For two-qubit states, the input states on the far left of Fig. 11.7 in both the upper and lower registers are $|00\rangle$. From Eq. (11.15), we see that after the upper-register Hadamard gate acts on the input state, the state $|\psi_1\rangle$ for the system (upper register and lower register) will be

$$|\psi_1\rangle = (H^{\otimes 2}|00\rangle)|00\rangle = \frac{1}{2}\left(\underbrace{|00\rangle}_{\text{upper}} \otimes \underbrace{|00\rangle}_{\text{lower}} + |01\rangle \otimes |00\rangle + |10\rangle \otimes |00\rangle + |11\rangle \otimes |00\rangle\right). \tag{11.17}$$

Note that, as promised, we have the upper-register states on the left side of each of the products and the lower-register states, all $|00\rangle$, on the right side. If you prefer, you may add subscripts to remind you which group the states belong to.

In region 2, the system state then interacts with the function gate. Since the function gate (like all of our quantum gates) is linear in the states, we can act separately on each term in Eq. (11.17) and then add up the results. For example, when we act on the second term, we have $x = 01$ and $y = 00$, so the lower-register output state is $|00 \oplus f(01)\rangle = |f(01)\rangle$ because the upper register $x = 01$. Again ignoring the details of the function gate and assembling the results, we find that the system state after the function box is

$$\begin{aligned}|\psi_2\rangle &= U_f\left(H^{\otimes 2}|00\rangle\right)|00\rangle \\ &= \frac{1}{2}\left(\underbrace{|00\rangle}_{\text{upper}}\otimes\underbrace{|f(00)\rangle}_{\text{lower}}+|01\rangle\otimes|f(01)\rangle+|10\rangle\otimes|f(10)\rangle+|11\rangle\otimes|f(11)\rangle\right).\end{aligned} \tag{11.18}$$

Note that Eq. (11.18) contains the function f evaluated with each of the binary sequences associated with the two-qubit system, so in that sense we handle all of the possibilities at once. This is another example of the power of using superposition states. Note that Eq. (11.18) shows that the function gate produces an entangled state between the upper and lower registers. As usual, entangled states play critical roles in the quantum algorithm.

Try It 11.7 Check that the state in Eq. (11.18) is an entangled state.

We are almost done with the algorithm. In region 3, the top register qubits interact with the Hadamard gate to give us (using the results stated in Eq. (11.15)) the rather complex looking state

$$|\psi_3\rangle = \frac{1}{4}\begin{pmatrix} \underbrace{(+|00\rangle+|01\rangle+|10\rangle+|11\rangle)}_{\text{upper}}\otimes\underbrace{|f(00)\rangle}_{\text{lower}} \\ +(+|00\rangle-|01\rangle+|10\rangle-|11\rangle)\otimes|f(01)\rangle \\ +(+|00\rangle+|01\rangle-|10\rangle-|11\rangle)\otimes|f(10)\rangle \\ +(+|00\rangle-|01\rangle-|10\rangle+|11\rangle)\otimes|f(11)\rangle \end{pmatrix}. \tag{11.19}$$

Try It 11.8 Use Eq. (11.15) in Eq. (11.18) to obtain Eq. (11.19).

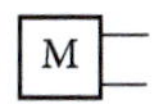

Fig. 11.8 The measurement device gives us output numbers (represented by the wires on the right side of the device) corresponding to the labels of the basis states presented to the device.

We can now rearrange Eq. (11.19), collecting terms associated with each of the upper-register states. The result is

$$|\psi_3\rangle = \frac{1}{4}\begin{pmatrix} \underbrace{|00\rangle}_{\text{upper}}\otimes\underbrace{\left(|f(00)\rangle+|f(01)\rangle+|f(10)\rangle+|f(11)\rangle\right)}_{\text{lower}} \\ +|01\rangle\otimes\left(|f(00)\rangle-|f(01)\rangle+|f(10)\rangle-|f(11)\rangle\right) \\ +|10\rangle\otimes\left(|f(00)\rangle+|f(01)\rangle-|f(10)\rangle-|f(11)\rangle\right) \\ +|11\rangle\otimes\left(|f(00)\rangle-|f(01)\rangle-|f(10)\rangle+|f(11)\rangle\right) \end{pmatrix}. \tag{11.20}$$

Note that Eq. (11.20) has all four possible function arguments with various combinations of plus and minus signs. Since we are looking for situations in which some of the function inputs give equal results $f(x) = f(x \oplus s)$, we expect that the terms in several rows might add up to 0, thus removing these upper-register states from the system state. Then when we make measurements on the upper-register qubits in region 4, the associated basis states never appear as results of the measurements. Though we haven't yet got to the details, we can guess that the measurement information results can then be used to find the secret code s. Each of the n measurement devices will provide two output numbers, which are the labels of the basis states.

In Fig. 11.8, we show one of the measurement devices. There will be one of these for each of the upper-register qubits. For the two-qubit example, we arrange the measurement devices to give an output of 00 for the $|00\rangle$ basis state, 01 for the $|01\rangle$ basis state, and so on. These two labels form a two-bit sequence $x = x_0x_1$.

But we are still not yet at the final answer. We need to show that if the measurement results $(x_o x_1)$ satisfy a mathematical relationship with the code sequence $s = (s_0 s_1)$, namely,

$$(x_0 \times s_0) \oplus (x_1 \times s_1) \equiv x \cdot s = 0, \tag{11.21}$$

where $\times$ means standard multiplication, then the corresponding row of functions does *not* add up to zero. The expression in Eq. (11.21) is called the "dot product" between the two binary sequences because it is similar to the definition of the ordinary vector dot product as a sum of the products of the corresponding components:

$$\vec{A} \cdot \vec{B} = A_x B_x + A_y B_y + A_z B_z, \tag{11.22}$$

with the ordinary plus sign replaced by bitwise XOR $\oplus$. It is easy to convince yourself that the result of $x \cdot y$ is either 0 or 1 no matter what the length of the bit sequences because the bit values are either 0 or 1.

One word of warning: The dot product of two binary sequences is not the same as a vector dot product. In fact, you can have a situation in which $x \cdot x = 0$. This occurs if x has an even number of 1s. It does *not* mean that x is orthogonal to itself.

Try It 11.9 Write out $x \cdot y$ in detail for the seven-bit sequences $x = 1010111$ and $y = 1110000$.

We have already discussed the quantum part of the algorithm; we now use some classical algorithms to wrap up the solution. Eq. (11.21) is a linear equation for the two unknowns s_0 and s_1. You may remember from linear algebra that for two unknowns you generally need two equations to get a definite solution. For the Simon algorithm, since we are excluding $s = 0$ and since the possible values for the solutions are just 0 and 1, it turns out that we need only one equation for the two-qubit situation. For n bits, we need $n - 1$ equations. Details aside, the crucial point is that from the measurement outcomes we can generate enough equations to solve for the binary bits in the code. There are a lot of tools for solving that system of equations. So, you can carry out that part of the algorithm using classical computational methods.

Let's work through the solution for a two-qubit example to make sure we know how the Simon procedure works. Suppose when we measure the upper-register output states, we find that the results are either 00 or 11. We know that 00 doesn't give us any information about s_0 and s_1; so, we use 11 in Eq. (11.21) with $(1 \times s_0) \oplus (1 \times s_1) = s_0 \oplus s_1 = 0$ from which we have $s_0 s_1 = 00$ or 11. But, remember that 00 is not allowed, so we conclude that $s = 11$. We are done! In this case, the solution of the linear equations did not require much work.

Let's summarize what we have done. In both the Deutsch algorithm and the Simon algorithm, we cleverly set up the state of the system so that entangled states, appropriately manipulated, will allow measurements that give us information from which we could calculate what we wanted to find, for example, the Simon secret code s. Quantum state entanglement was the key feature because entangled states have correlations among the measurements associated with various qubits of the system. These correlations let the quantum magic produce the desired result.

But there was one important difference between the Deutsch algorithm and the Simon algorithm. In the Deutsch algorithm, once we have the function gate set up, it takes only one measurement observation to find out if the function is constant or balanced. On the other hand, in the Simon algorithm, because the measurement observations occur "at random," it is possible that we might have repeated observations of the 0 basis states, and they give us no useful information for the Simon question. Or we may get the same measurement results repeatedly—just like getting a string of heads when we repeatedly flip a coin. This set of repeating results doesn't give us new information, that is, new equations to help solve for the bits in the code s. You can estimate the probability that you will get sufficient information after a certain number of measurements and, indeed, as the number of measurements increases you are highly likely to get enough information to answer the Simon question. But there is no guarantee. In principle, you could repeat the preparation–measurement cycle many times and still not get a definite answer. That is why we may say only that it is highly likely that the Simon algorithm will give an answer with fewer measurements than the classical computation; we are not guaranteed that it will happen.

11.5 Multiqubit Hadamard Gates

Since multiqubit Hadamard gates appear frequently in quantum information science (QIS) and quantum computing (QC) applications, in this section we explain how to construct these gates. Then, we show how the $x \cdot s = 0$ condition arises in the Simon algorithm. We won't need these details in subsequent chapters, so you should feel free to skip this section. But going through the analysis gives you some good background in quantum gates and state manipulation that may prove to be useful in more complex QIS and QC applications.

The matrix representation of a multiqubit Hadamard gate is built by arranging the single-qubit Hadamard matrices into larger matrices (more rows and columns). These more complicated structures carry out the equivalent operations on the multiqubit states. The idea is analogous to using products of state vectors to get state vectors for the composite system.

Let's start by reminding ourselves of the one-qubit Hadamard matrix:

$$H^{\otimes 1} = H \doteq \begin{pmatrix} 1 & 1 \\ 1 & -1 \end{pmatrix}. \tag{11.23}$$

For two-qubit states, the Hadamard gate for the system is a 4×4 matrix built using four one-qubit Hadamard matrices:

$$H^{\otimes 2} = \begin{pmatrix} H^{\otimes 1} & H^{\otimes 1} \\ H^{\otimes 1} & -H^{\otimes 1} \end{pmatrix} \doteq \begin{pmatrix} \begin{pmatrix} 1 & 1 \\ 1 & -1 \end{pmatrix} & \begin{pmatrix} 1 & 1 \\ 1 & -1 \end{pmatrix} \\ \begin{pmatrix} 1 & 1 \\ 1 & -1 \end{pmatrix} & \begin{pmatrix} -1 & -1 \\ -1 & 1 \end{pmatrix} \end{pmatrix} = \begin{pmatrix} 1 & 1 & 1 & 1 \\ 1 & -1 & 1 & -1 \\ 1 & 1 & -1 & -1 \\ 1 & -1 & -1 & 1 \end{pmatrix}. \tag{11.24}$$

It is a good exercise to apply this Hadamard matrix to each of the two-qubit basis states discussed in Chapter 9, which are shown here as column vectors,

$$\begin{pmatrix} 1 \\ 0 \\ 0 \\ 0 \end{pmatrix}, \begin{pmatrix} 0 \\ 1 \\ 0 \\ 0 \end{pmatrix}, \begin{pmatrix} 0 \\ 0 \\ 1 \\ 0 \end{pmatrix}, \begin{pmatrix} 0 \\ 0 \\ 0 \\ 1 \end{pmatrix} \tag{11.25}$$

to make sure it produces the superposition states in Eq. (11.19).

Try It 11.10 Work through the suggested matrix multiplications.

For a three-qubit system, since $2^3 = 8$, the Hadamard matrix needs eight rows and eight columns, so we construct it from four two-qubit Hadamard matrices:

$$H^{\otimes 3} = \begin{pmatrix} H^{\otimes 2} & H^{\otimes 2} \\ H^{\otimes 2} & -H^{\otimes 2} \end{pmatrix}. \tag{11.26}$$

You can see that the process of building these is very straightforward.

Try It 11.11 Write out the complete three-qubit Hadamard matrix. Hint: Follow the same procedure that was used in proceeding from Eq. (11.23) to Eq. (11.24). The answer is in Eq. (11.27).

$$H^{\otimes 3} \Rightarrow \begin{pmatrix} 1 & 1 & 1 & 1 & 1 & 1 & 1 & 1 \\ 1 & -1 & 1 & -1 & 1 & -1 & 1 & -1 \\ 1 & 1 & -1 & -1 & 1 & 1 & -1 & -1 \\ 1 & -1 & -1 & 1 & 1 & -1 & -1 & 1 \\ 1 & 1 & 1 & 1 & -1 & -1 & -1 & -1 \\ 1 & -1 & 1 & -1 & -1 & 1 & -1 & 1 \\ 1 & 1 & -1 & -1 & -1 & -1 & 1 & 1 \\ 1 & -1 & -1 & 1 & -1 & 1 & 1 & -1 \end{pmatrix}. \tag{11.27}$$

Now let's tackle the issue of using the dot product $x \cdot s$ to put the minus signs in the correct places in the Hadamard matrices and how that leads us to the code s. This is an interesting example of how changing notation and rearranging mathematical expressions lead to interesting conclusions. The key notion is that $x \cdot y = 0$ or 1 for any n-bit binary sequences and that $(-1)^{x \cdot y} = 1$ if $x \cdot y = 0$ and $(-1)^{x \cdot y} = -1$ if $x \cdot y = 1$.

To illustrate the general idea, let's return to the single-qubit Hadamard gate and apply it to the basis states $|0\rangle$ and $|1\rangle$. We put the two results together in one equation by using the variable x, which gives either 0 or 1 (the only two possibilities):

$$H\,|x\rangle = \frac{1}{\sqrt{2}}\left(|0\rangle + (-1)^x\,|1\rangle\right) = \sum_{y=0}^{1} (-1)^{x \times y}\,|y\rangle. \tag{11.28}$$

Here, $x \times y$ is ordinary multiplication. Note that including $(-1)^x$ and $(-1)^{x \times y}$ takes care of getting the correct signs when we apply the Hadamard gate.

In Eq. (11.28), the sum over y contains only two terms, and you should check for yourself that the signs come out correctly. Of course, if we used only one-qubit Hadamard gates, the condensed notation would not be useful. However, when we want to generalize to n qubits, the fancier notation makes the general expression much easier to write down and manipulate.

For n-qubit system, the action of $H^{\otimes n}$ on one of the x basis states is

$$H^{\otimes n}|x\rangle_n = \sum_{y=0}^{2^n-1} (-1)^{x \cdot y}|y\rangle_n. \tag{11.29}$$

Now we use this notation to rewrite the Simon algorithm states. After the first Hadamard gate in Fig. 11.7, the system quantum state is

$$|\psi_1\rangle = \frac{1}{2^{n/2}}\left(\sum_{x=0}^{2^n-1} |x\rangle_n\right) \otimes |0\rangle_n. \tag{11.30}$$

Let's take a close look at Eq. (11.30). Note that we have labeled the superposition states by the n-bit x values, which run from 00...0 to 11...1—all the possible n-bit values. This is another example of using numerical values to label states—something that is not necessary, but it turns out to be useful.

The next step is to have the system state interact with the function gate U_f after which the system state is

$$|\psi_2\rangle = \frac{1}{2}\sum_{x=0}^{2^n-1} |x\rangle_n |f(x)\rangle_n. \tag{11.31}$$

Now the upper-register part of the system state interacts with the second Hadamard gate to give

$$|\psi_3\rangle = \frac{1}{2^n}\sum_x \left(\sum_y (-1)^{x\cdot y}|y\rangle_n\right) \otimes |f(x)\rangle_n. \tag{11.32}$$

To unclutter the notation, we have left off the details of the limits of the sums. Both run over the full range of the n bit numbers 00 ... 0 through 11 ... 1, which is from $x, y = 0$ to $x, y = 2^n - 1$.

Here is the argument: We replaced the $(-1)^{x \times y}$ term in Eq. (11.28), where x and y were single bit numbers (0 or 1), with $(-1)^{x\cdot y}$, where x and y now are each n-bit numbers. In other words, we replaced the standard multiplication indicated by $\times$ with the dot product multiplication. We have already shown that the procedure works for $n = 2$. But let's look at this from a different point of view that allows us to handle larger n cases easily. The key idea is that we can use x and y to label the rows and columns of a matrix. If we then use $(-1)^{x\cdot y}$ for the matrix elements labeled by row and column, we automatically get the right plus or minus sign in the Hadamard matrix. Here is what it looks like for $n = 2$:

$$\begin{matrix} & y=00 & y=01 & y=10 & y=11 \\ \begin{matrix} x=00 \\ x=01 \\ x=10 \\ x=11 \end{matrix} & \multicolumn{4}{c}{\begin{pmatrix} (-1)^{x\cdot y} & (-1)^{x\cdot y} & (-1)^{x\cdot y} & (-1)^{x\cdot y} \\ (-1)^{x\cdot y} & (-1)^{x\cdot y} & (-1)^{x\cdot y} & (-1)^{x\cdot y} \\ (-1)^{x\cdot y} & (-1)^{x\cdot y} & (-1)^{x\cdot y} & (-1)^{x\cdot y} \\ (-1)^{x\cdot y} & (-1)^{x\cdot y} & (-1)^{x\cdot y} & (-1)^{x\cdot y} \end{pmatrix}} \end{matrix} = \begin{pmatrix} 1 & 1 & 1 & 1 \\ 1 & -1 & 1 & -1 \\ 1 & 1 & -1 & -1 \\ 1 & -1 & -1 & 1 \end{pmatrix} \tag{11.33}$$

For the $n = 3$ Hadamard gate matrix in Eq. (11.27) in the previous Try It, you can use the same method to find where the "correct" minus signs go. The rows and columns of that matrix can be labeled by the eight basis state labels corresponding to the values of x (the three-bit sequence):

$$000 \quad 001 \quad 010 \quad 011 \quad 100 \quad 101 \quad 110 \quad 111. \tag{11.33}$$

You should also note that these numbers run from 0 (on the left) to $2^n - 1 = 7$ (on the right). For an 8×8 matrix, there are 64 combinations of row labels x and column labels y. You should check that $(-1)^{x\cdot y}$ gives the sign that matches the ones in the explicitly constructed three-qubit Hadamard matrix. The first row and first column should be easy. The others are easy enough to write down and work out the dot products to check against the results given in Eq. (11.27).

Now let's move on with the rest of the Simon algorithm. Just as we did in proceeding from Eq. (11.19) to Eq. (11.20), we rearrange Eq. (11.32) to obtain

$$|\psi_3\rangle = \frac{1}{2^n}\sum_y |y\rangle_n \left(\sum_x (-1)^{x\cdot y}|f(x)\rangle_n\right). \tag{11.34}$$

The final step in the Simon circuit is the action of the n measurement devices on the upper-register state in region 4 of Fig. 11.7. The measurement device is designed to observe the basis states $|y\rangle_n$. Each measurement of the n-qubit upper register will yield a single value y. The probability of getting a specific value of y depends, of course, on the square of the term in parentheses in Eq. (11.34), which is in effect the amplitude for that state. So, let's play with that term to see how it gives us information about the code sequence s.

For the Simon algorithm, remember that we want to find the secret code s that satisfies $f(x) = f(x \oplus s) = z$, where z is one of the n-bit binary numbers. For each x_1 in the set $\{x\}$, there is one and only one x_2 in the set such that $f(x_1) = f(x_2) = z$, as specified in the second paragraph of Section 11.4. Using this notation, we rewrite the parenthetical terms in Eq. (11.34) as

$$\sum_z \left[(-1)^{x_1\cdot y} + (-1)^{x_2\cdot y}\right]|z\rangle. \tag{11.35}$$

Since $x_1 \oplus s = x_2$, the square bracket term in Eq. (11.35) is

$$\left[(-1)^{x_1\cdot y} + (-1)^{(x_1\oplus s)\cdot y}\right] = (-1)^{x_1\cdot y}\left[1 + (-1)^{y\cdot s}\right]. \tag{11.36}$$

For the dot product as we have defined it, the only two possibilities for $y \cdot s$ are zero (even) and 1 (odd). So, for any y that appears as the result of a measurement, we have

$$y \cdot s = 0 = (y_0 \times s_0) \oplus (y_1 \times s_1). \tag{11.37}$$

As mentioned in the previous section, a series of measurement results will yield several such equations, which in principle can be solved by standard algebraic methods to find s.

Don't worry if keeping track of these details has been difficult for you. It is always challenging to make sense of new notation and new ways of manipulating symbols. Of course, quantum mechanics itself has its conceptual challenges. But if we follow the quantum rules carefully, we get results that agree with experiments. This is quite amazing.

In some ways, the current state of quantum algorithms is what classical computer programming was like before high-level languages such as Fortran, C, Python, and so on were developed. We are sure that within a decade or so, we will have similar high-level languages for quantum computing. In the next chapter, we turn to some quantum algorithms that address more obviously interesting applications.

11.6 Chapter Summary

- The two quantum algorithms discussed in this chapter illustrate how the use of superposition and entanglement may make quantum computing more effective than classical computing by providing algorithms that work on all possible states at once. This does not mean that the calculation is done repetitively on each possible state, but rather the superposition of states interacts with quantum gates in a way that results in a single measurement having a high probability of containing an answer. As usual, this probability is given by the square of the coefficient of the state. Because there is a small but nonzero probability of a wrong answer—due to the probabilistic nature of measurements on superposition states—you may need to run an algorithm several times. But with a clever quantum algorithm, the process can be less computationally intensive than the corresponding quantum algorithm. Many quantum algorithms make use of entanglement to allow measurements on one part of the system's state to give information about another part of that state.
- The Deutsch algorithm analyzes whether a function is constant (always returns the same output regardless of input) or balanced (output depends on input). Table 11.1 shows the four possible outcomes for single-qubit inputs. A classical algorithm would need to evaluate the function twice to determine if it is constant or balanced. The Deutsch algorithm can determine the type of function with only a single evaluation of the function.
- The Deutsch algorithm circuit, like those for many other quantum algorithms, has an upper register and a lower register. The first step is to put both registers into states of superposition using a Hadamard gate. Next the function gate acts on the two superposition states and gives a different output depending on whether the function is constant or balanced. Finally, the upper register is acted upon again by a Hadamard gate to reverse the state of superposition into a single basis state. This state then interacts with a measurement device, and the measurement result can indicate whether the function was constant if the measurement yielded a 1 or balanced if the measurement result was 0.
- The Simon search algorithm finds a string s (the "secret code") that satisfies $f(x) = f(x \oplus s)$. The function f need not be specified.
- The approach is similar to the Deutsch algorithm in that a Hadamard gate is used to put both inputs into states of superposition but is different because we don't know the function in the Simon algorithm. The results of the superposition after the function gate is then acted upon by a second Hadamard gate that yields a matrix with a pattern of signs that ensures some of the results will cancel out and not appear in the final result, which is then sent to a measurement device that yields a string that is mathematically related to the secret code but is itself not the actual secret code.
- To determine the individual bits of the secret code, a classical algorithm is needed to solve this system of $n - 1$ equations. The advantage of the Simon quantum algorithm is that, as the number of bits in the secret code grows, the number of times the quantum

algorithm needs to be run, in order to get an answer, grows linearly, whereas for a classical algorithm, it grows exponentially.

11.7 Further Reading

Yohan Vianna, Mariana R. Barros, and Malena Ho-Meyal, "Classical realization of the quantum Deutsch algorithm," *American Journal of Physics* **86**, 914 (2018). This paper provides a good introduction to the Deutsch algorithm and shows how to use lenses and diode lasers to build a classical analog that implements the quantum algorithm.

N. David Mermin, *Quantum Computer Science, An Introduction* (Cambridge University Press, Cambridge, 2007). The simple circuits that implement the Deutsch functions are shown on p. 42.

Chris Bernhardt, *Quantum Computing for Everyone* (MIT Press, Cambridge, MA, 2019). Chapter 8 presents a nice, but compact introduction to the Simon algorithm.

More Quantum Algorithms

12

Let us record the atoms as they fall upon the mind.

Virginia Woolf, *The Common Reader* (1925)[1]

12.1 Introduction

We are now ready to tackle two more quantum algorithms, both of which have advantages over the corresponding classical algorithms. We also show how quantum computing (QC) can be used to find energies and shapes of complex molecules—a crucial step in finding new drugs and enhanced materials.

12.2 Grover Search Algorithm

The first algorithm we tackle in this chapter is the Grover search algorithm, which is a quantum algorithm that speeds up searching through databases. Lov Grover developed the algorithm in 1996. Although there are now many variations on and extensions of the original algorithm, we take a close look at how the original works because it uses a strategy that many other quantum algorithms use. We should point out that these algorithms are not useful for ordinary database searching because current quantum computers do not have enough qubits to compete even with your laptop. But the quantum algorithms do provide a proof of concept.

First, let's think about a classical search question. Suppose we have N playing cards, only one of which has a special purple dot on it. If we spread the cards out face down on a table, how many cards do we need to turn over to find the one with the purple dot? Well, we might be lucky and find it on the first try. More likely, we will have to turn over many cards before finding the one with the purple dot. In the worst case, we might turn over $N - 1$ cards without seeing the purple dot. If that happens, we know that the dot must be on the remaining card, but we still turn it over to make sure that no one has messed with the cards. It turns out that, *on average*, we need to flip just more than half of the cards to find the purple dot. The average is 2 for 3 cards, 2.5 for 4 cards, and 50.5 for 100 cards.

[1] https://www.brainyquote.com/authors/alice-walker-quotes

Quantum Computing and Quantum Physics. Alice Flarend and Robert Hilborn, Oxford University Press.
 DOI: 10.1093/oso/9780197904930.003.0012

Try It 12.1 Work out the probability problem for finding a marked card. Hint: The card problem is found in almost every introductory probability textbook.

For the quantum version, we search for a particular quantum basis state for a multiqubit system. For a system with n qubits, we have, as we know, $N = 2^n$ possible basis states. Suppose we want to find just one basis state, equivalent to the card with the purple dot. To set up the search mechanism, we label the states $|x\rangle$ with the binary number x ranging from 0 through $2^n - 1$ (see Table 9.2, third column).

We also need to "tag" the "target" state; so, the search mechanism has something to look for. Let's assume that there is a function $f(x)$ defined such that $f(x)= 0$ unless $x = w$, where $|w\rangle$ is the target state, and someone has arranged for $f(w)= 1$. Another way of putting this is that we give each of the basis states the amplitude $(-1)^{f(x)}/\sqrt{2^n}$. So, all the amplitudes have the same magnitude. All will be positive except for the amplitude of the target state $|w\rangle$, which is negative. This negative amplitude is the quantum equivalent of the purple dot in our card example.

Try It 12.2 Show that $(-1)^{f(x)}/\sqrt{2^n}$ satisfies the positive and negative amplitude criteria stated in the previous paragraph.

We want to develop an algorithm that starts with a superposition state with equal magnitude amplitudes for all the basis states and then manipulates the amplitudes to increase the one associated with $|w\rangle$ while decreasing the other amplitudes. The increased amplitude entails growth in the probability of observing the target state. When we carry out a measurement on the superposition state, if the target amplitude is larger (in magnitude) than the other amplitudes, we are much more likely to observe $|w\rangle$ than any of the other basis states. For a system with N states, Grover found that the number of iterations of the algorithm needed to identify the target state increases as $\sqrt{N}$, while, as we observed, the number of card "flips" for the classical algorithm increases proportional to N.

The Grover algorithm, like the Simon algorithm, is probabilistic. All we can calculate is the probability of getting the target state when we make measurements on a superposition of states. But we will see that probability can be very close to 1 under many circumstances.

To see how the Grover algorithm works, we use three methods: (1) a geometric picture of a two-dimensional slice of the N-dimensional state space. This geometric picture makes it relatively easy to understand what the algorithm does, and it allows us to prove many interesting results about the algorithm, including the probability that the measurement will give the target state. The second method translates the geometric approach into a quantum operator method. The third approach implements what is called "amplitude amplification," a method by which the amplitude (coefficient) associated with the target state increases while the amplitudes of all the other basis states decrease.

Geometric Approach

For the geometric picture, we use a state-space diagram with one axis along the $|w\rangle$ direction, the target state we want to find, and the other perpendicular to $|w\rangle$. This perpendicular direction represents a superposition of all the *other* basis states, each of which individually is perpendicular to $|w\rangle$. Let's call it $|\psi_{\text{other}}\rangle$. In the diagram, we also include the equal-amplitude superposition state

$$\begin{aligned} |s_0\rangle &= \frac{1}{\sqrt{2^n}}(|00\dots0\rangle + |00\dots1\rangle + \dots + |11\dots1\rangle) \\ &= \frac{1}{\sqrt{2^n}}\sum_{x=0}^{2^n-1}|x\rangle. \end{aligned} \tag{12.1}$$

How do we know which state has been tagged? We just have to recognize that someone tagged the target state by making its amplitude negative. We don't need to know the specific state that was tagged. The beauty of the algorithm is that it "finds" that state automatically. We show how this works in the examples that follow.

The Grover algorithm starts with $|s_0\rangle$, the equal-amplitude superposition of all the basis states in Eq. (12.1). In the geometric picture shown in Fig. 12.1, we note that, by construction, the angle between $|s_0\rangle$ and $|\psi_{\text{other}}\rangle$ is just a bit larger than 0°. The algorithm simply magnifies that slight difference to get the subsequent superposition state more aligned with $|w\rangle$.

It looks like it should be easy to get the superposition state vector aligned with the $|w\rangle$ direction; all we need to do is rotate $|s_0\rangle$ toward the $|w\rangle$ direction. But the problem is that the states are actually in an N-dimensional state space, and it takes many parameters to specify rotations in such a space. For example, for two dimensions we need one angle, for three dimensions two angles (think of it as how far up and how far over), for four dimensions three angles, etc. The matrix that implements the rotation operation gets correspondingly more complicated. So, if we can't figure out the rotations, what can we do to move the system state vector toward the $|w\rangle$ direction?

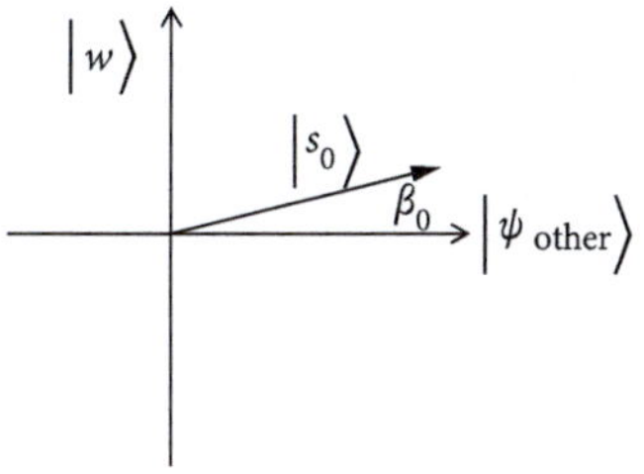

Fig. 12.1 Two-dimensional state-space representation of the "target" state $|w\rangle$, the superposition of all the other basis states $|\psi_{\text{other}}\rangle$, and the initial equal-amplitude superposition state $|s_0\rangle$. β_0 is the angle between $|\psi_{\text{other}}\rangle$ and $|s_0\rangle$. Note that the uniform (equal probability) superposition can be written as a linear combination of the "basis" states $|w\rangle$ and $|\psi_{\text{other}}\rangle$: $|s_0\rangle = |w\rangle\langle w|s_0\rangle + |\psi_{\text{other}}\rangle\langle\psi_{\text{other}}|s_0\rangle$.

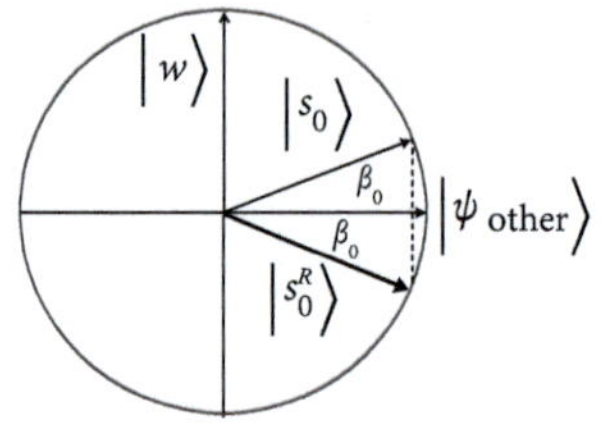

Fig. 12.2 The state-space projection from Fig. 12.1 with $|s_0^R\rangle$ (dark arrow), the reflection of the state vector $|s_0\rangle$ with respect to the $|\psi_{\text{other}}\rangle$ axis. The circle is the unit radius circle. All normalized state vectors have their tips on the circle (assuming that the tails are on the origin).

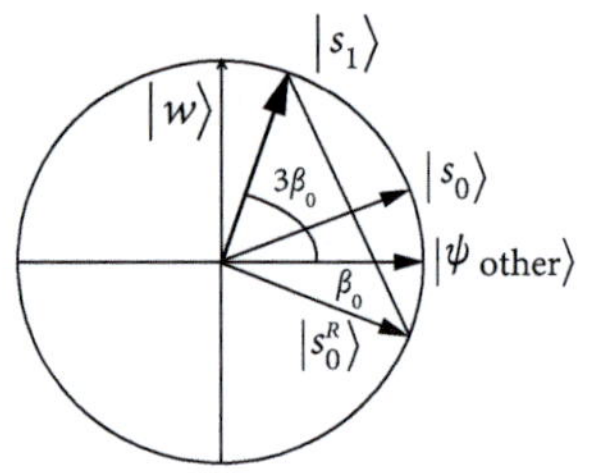

Fig. 12.3 $|s_1\rangle$ (dark arrow) is the result of reflecting $|s_0^R\rangle$ with respect to the $|s_0\rangle$ direction. $\beta_1 = 3\beta_0$ is the angle between $|\psi_{\text{other}}\rangle$ and $|s_1\rangle$.

The "trick" is to build the rotation as a series of reflections about directions in the two-dimensional state space. (By the way, the connection between reflections and rotations is frequently used in computer graphics.) Figure 12.2 illustrates how to construct the reflection. Draw a straight line (dashed in Fig. 12.2) from the tip of $|s_0\rangle$ perpendicular to the horizontal $|\psi_{\text{other}}\rangle$ axis. Continue that straight line until it intersects the unit circle. The reflected state vector $|s_0^R\rangle$ is represented by the dark arrow from the origin to the intersection point on the unit circle. The first reflection has the effect of multiplying the amplitude associated with $|w\rangle$ in the superposition $|s_0\rangle$ by -1. This step then sets up all the subsequent steps.

The second step, as illustrated in Fig. 12.3, reflects the state $|s_0^R\rangle$ with respect to the original $|s_0\rangle$ state direction to produce the state $|s_1\rangle$ represented by the dark arrow in Fig. 12.3. Examine Fig. 12.3 before reading further to see if you can figure out how $|s_1\rangle$ is situated. Hints: It is all about perpendicular lines and the procedure is similar to the first step's procedure: draw a line from the tip of the $|s_0^R\rangle$ vector perpendicular to the $|s_0\rangle$ vector. Continue that line until it intersects the unit circle. Then, draw an arrow from the origin to that intersection point. This arrow represents the state vector $|s_1\rangle$. Is it closer to being aligned with the desired target state vector $|w\rangle$? As shown in Fig. 12.3, the effect of these two reflections is a rotation of the superposition vector from the $|s_0\rangle$ to the $|s_1\rangle$ direction—closer to the desired target state vector $|w\rangle$.

Those two operations produce a new vector $|s_1\rangle$ that is rotated counterclockwise from $|s_0\rangle$. That new vector is at an angle β_1 relative to $|\psi_{\text{other}}\rangle$. In Fig. 12.3, we see that $\beta_1 = 3\beta_0$. We then repeat this process by performing two reflections of $|s_1\rangle$ to bring the superposition state even closer to the $|w\rangle$ direction. The process continues with a reflection about the $|\psi_{\text{other}}\rangle$ direction followed by a reflection about $|s_0\rangle$ until the angle $\beta_k \approx \pi/2$ (in radians). It will be helpful to introduce an integer k that specifies the relationship between β_k, the angle after k steps of the process, and β_0: $\beta_k = (2k+1)\beta_0$.

Try It 12.3 Convince yourself that β_k increases by $2\beta_0$ for each repetition of the two reflections.

Try It 12.4 The rotation by reflection also works for regular vectors. Draw a diagram, like the one below, with a regular three-dimensional vector $\vec{A} = A_x\hat{x} + A_y\hat{y} + A_z\hat{z}$ with $A_x = 3$, $A_y = 5$, and $A_z = 2$. If we want to reflect this vector with respect to the xy plane, we simply change the sign of the z component, which is perpendicular to the xy plane. Show on your diagram that this procedure gives the desired result. How would we reflect the original vector with respect to the xz plane?

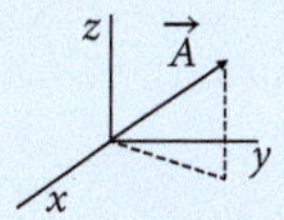

Quantum Operator Approach

We now show how to carry out these reflections/rotations formally using quantum operators (linear transformations). The geometry gives us a clue about how to do that. The first step is reflection with respect to $|\psi_{\text{other}}\rangle$. This corresponds to changing the sign of the amplitude for all states that are perpendicular to $|\psi_{\text{other}}\rangle$, which is just $|w\rangle$ because we have defined $|\psi_{\text{other}}\rangle$ that way. The operator that does this is labeled $R^{(\psi_{\text{other}})}$ because it involves reflections about the $|\psi_{\text{other}}\rangle$ direction

$$R^{(\psi_{\text{other}})}|s_0\rangle = R^{(\psi_{\text{other}})}\sum_{x=0}^{N}|x\rangle = \sum_{x=0}^{N}(-1)^{f(x)}|x\rangle = |s_0^R\rangle \tag{12.2}$$

So, the first step makes the amplitude of $|w\rangle$ negative and produces the superposition state $|s_0^R\rangle$.

The next step reflects the state vector $|s_0^R\rangle$ relative to $|s_0\rangle$ to produce $|s_1\rangle$. This reflection changes the sign of the part of $|s_0^R\rangle$ that is perpendicular to $|s_0\rangle$. How do we find a perpendicular part? To find the perpendicular component, we subtract the part that is parallel to $|s_0\rangle$ from $|s_0^R\rangle$:

$$|s_0^R\rangle_{\perp s_0} = |s_0^R\rangle - \underbrace{|s_0\rangle\langle s_0|s_0^R\rangle}_{\text{parallel to }|s_0\rangle}. \tag{12.3}$$

The reflection changes the sign of that perpendicular component. Adding the negative of the perpendicular component to the part parallel to $|s_0\rangle$ produces $|s_1\rangle$:

$$|s_1\rangle = -|s_0^R\rangle_{\perp s_0} + \underbrace{|s_0\rangle\langle s_0|s_0^R\rangle}_{\text{parallel to } |s_0\rangle} = -|s_0^R\rangle + 2\underbrace{|s_0\rangle\langle s_0|s_0^R\rangle}_{\text{parallel to } |s_0\rangle}. \tag{12.4}$$

The quantum operator that carries out this reflection is called $R^{(s_0)}$

$$\begin{aligned} R^{(s_0)} &= 2\,|s_0\rangle\langle s_0| - \mathrm{I} \\ &= 2\mathrm{P}_{s_0} - \mathrm{I}. \end{aligned} \tag{12.5}$$

We have defined a shorthand notation P_{s_0} for the projection operator $|s_0\rangle\langle s_0|$.

Try It 12.5 Work through the algebraic steps to proceed from Eq. (12.3) to Eq. (12.4).

We combine the two operators to form the Grover operator:

$$G = R^{(s_0)} R^{(\psi_{\text{other}})}. \tag{12.6}$$

Repeated application of G produces the sequence of superposition vectors that eventually are closely aligned to the target state $|w\rangle$:

$$\begin{aligned} &G\,|s_0\rangle = |s_1\rangle \\ &G\,|s_1\rangle = GG\,|s_0\rangle = |s_2\rangle \\ &\ldots \\ &G^k\,|s_0\rangle = |s_k\rangle. \end{aligned} \tag{12.7}$$

In terms of the angle β_0, we may write

$$G^k\,|s_0\rangle = \cos\left((2k+1)\beta_0\right)|\psi_{\text{other}}\rangle + \sin\left((2k+1)\beta_0\right)|w\rangle. \tag{12.8}$$

As the process continues, k increases and the coefficient multiplying $|\psi_{\text{other}}\rangle$ decreases while the coefficient of $|w\rangle$ increases as $(2k+1)\beta_0 \to \pi/2$. In terms of the probability of measurement outcomes, as k increases the probability for getting $|\psi_{\text{other}}\rangle$ decreases and the probability for getting $|w\rangle$ increases. In fact, the Born rule shows that the probability of getting $|w\rangle$ is $\sin^2((2k+1)\beta_0)$.

Let's see how this plays out for two qubits. We start with the equal-amplitude superposition of basis states $|s_0\rangle = (1/2)(|00\rangle + |01\rangle + |10\rangle + |11\rangle)$ and, for our example, choose the target state to be $|w\rangle = |01\rangle$. Our two-dimensional subspace is spanned by $|w\rangle$ and

$$|\psi_{\text{other}}\rangle = \left(1/\sqrt{3}\right)(|00\rangle + |10\rangle + |11\rangle). \tag{12.9}$$

The factor $1/\sqrt{3}$ ensures that $|\psi_{\text{other}}\rangle$ is normalized.

The first reflection $R^{(\psi_{\text{other}})}$ applied to $|s_0\rangle$ leaves the amplitudes of the basis states in $|\psi_{\text{other}}\rangle$ the same but changes the amplitude of $|01\rangle$ from ½ to −½. Hence, the state vector $|s_1\rangle$ after the first reflection is

$$|s_1\rangle = \frac{1}{2}\left(|00\rangle - |01\rangle + |10\rangle + |11\rangle\right). \tag{12.10}$$

For the next step, we need the Dirac bracket $\langle s_0|\, s_1\rangle$, which in this case gives

$$\langle s_0|\, s_1\rangle = (1/2)(1/2) + (1/2)(-1/2) + (1/2)(1/2) + (1/2)(1/2) = 1/2. \tag{12.11}$$

Try It 12.6 Put $|s_1\rangle$ in the column vector form and $\langle s_0|$ in the row vector form and calculate $\langle s_0|\, s_1\rangle$. Do you get the result in Eq. (12.11)?

We now use the equal amplitudes for $|s_0\rangle$, the amplitudes for $|s_1\rangle$ from Eq. (12.10), and $\langle s_0|\, s_1\rangle = 1/2$ from Eq. (12.11):

$$\begin{aligned}|s_1\rangle &= (1/2 - 1/2)\,|00\rangle + (1/2 + 1/2)\,|01\rangle + (1/2 - 1/2)\,|10\rangle + (1/2 - 1/2)\,|11\rangle \\ &= |01\rangle .\end{aligned} \tag{12.12}$$

So, if we measure the superposition state $|s_1\rangle$, we always get the "answer" 01, the label of our target state. This is a rather special case because we found the desired state after only once through the algorithm. This is unique for two-qubit systems, but it does show starkly the advantage of the quantum algorithm over the classical one. Remember that a two-qubit system has four basis states. If we had four playing cards, one with a purple dot, we might need to flip cards up to three times to find the purple dot card. In the quantum case, we did it in only one "flip" after the initial setup.

We show you a nice geometric way of explaining why we get the answer in one step for two qubits (see Fig. 12.3). After performing the two reflections, we see that the angle between $|s_2\rangle$ and $|\psi_{\text{other}}\rangle$ is $3\beta_0$, where β_0 is the angle between the initial uniform superposition state $|s_0\rangle$ and $|\psi_{\text{other}}\rangle$. For the two-qubit case, the angle β_0 is $\pi/6$ (Try It 12.6). After one application of the algorithm, the angle between $|s_1\rangle$ and $|\psi_{\text{other}}\rangle$ equals $3 \times (\pi/6) = \pi/2$. This shows that the superposition state points along the target state vector $|w\rangle$ and we are done.

Try It 12.6 Use the column and row vector forms for $|s_0\rangle = (1/2, 1/2, 1/2, 1/2)$ and $|\psi_{\text{other}}\rangle = \left(1/\sqrt{3}, 0, 1/\sqrt{3}, 1/\sqrt{3s}\right)$ to find $\langle s_0|\, \psi_{\text{other}}\rangle$, which is $\cos\beta_0$. Show that $\beta_0 = \pi/6$. Note: We need to have the two state vectors normalized to get the correct $\cos\beta_0$.

Estimates of the Grover Algorithm for N Basis States

Now let's look at the more general case with N basis states. Grover proved that the number of iterations needed for the quantum algorithm increases as $\sqrt{N}$, while, as shown before, the number of classical algorithm steps increases as N. We now take up that proof.

To see how the $\sqrt{N}$ number of steps comes in, note that the amplitude for the states in the superposition state $|\psi_{\text{other}}\rangle$ is $1/\sqrt{N}$. For each step, the amplitude for the target state $|w\rangle$ increases by about $1/\sqrt{N}$ (for large N). Thus after $\sqrt{N}$ steps, the amplitude is about equal to 1 for $|w\rangle$ and close to 0 for all the other states. The arguments below leading to Eq. (12.17) provide a more formal proof.

Going back to the general case, we see from Fig. 12.6 that each successive application of the reflection adds $2\beta_0$ to the angle between the state $|s_k\rangle$ and $|\psi_{\text{other}}\rangle$, effectively moving the superposition state further from $|\psi_{\text{other}}\rangle$ and closer to $|w\rangle$; so, after k iterations, the angle between $|s_k\rangle$ and $|\psi_{\text{other}}\rangle$ is

$$\beta_k = (2k + 1)\beta_0 \tag{12.13}$$

as shown in Fig. 12.4.

You can extend this method to get another estimate of how many iterations you need to use to get the state $|s_k\rangle$ to within some specified tolerance of the target state. As you might realize, the more qubits you have in the system, the closer the original equal superposition state $|s_0\rangle$ will lie along $|\psi_{\text{other}}\rangle$. So, more iterations of the algorithm are needed to obtain $|s_0\rangle$ pointing close to $|w\rangle$. Table 12.1 provides some numerical examples. The values of $\cos\beta_0$ are found from the Dirac bracket $\langle s_0|\, \psi_{\text{other}}\rangle$.

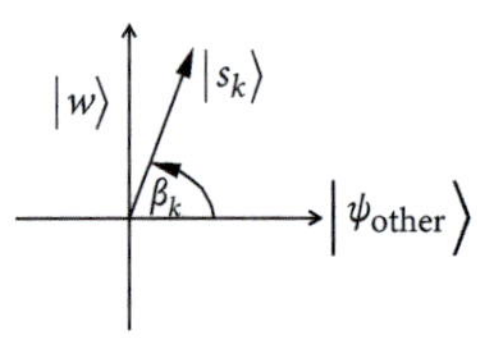

Fig. 12.4 A schematic diagram of the state vector $|s_k\rangle$ and the angle β_k after k iterations of the Grover algorithm.

Table 12.1 Angle results for n qubits. Results are provided with three significant figures.

n	$\cos\beta_0$	β_0 (degrees)
2	0.866	30.0
3	0.935	20.7
4	0.968	14.5
5	0.984	10.2
10	0.999	1.79
50	1.00	1.71×10^{-6}

The initial angle between the uniform superposition state $|s_0\rangle$ and $|\psi_{\text{other}}\rangle$ decreases rapidly as the number of qubits grows because with more qubits most of the basis states are in $|\psi_{\text{other}}\rangle$ with only one, $|w\rangle$, not included; so, subtracting $|w\rangle$ from $|s_0\rangle$ to produce $|\psi_{\text{other}}\rangle$ doesn't change the state vector very much. As the iterations proceed, the angle β_k increases as $\beta_k = (2k + 1)\beta_0$, where k is the number of iterations; so, it may take many iterations before the superposition state is aligned with $|w\rangle$.

Try It 12.7 Challenge: Show that for an n-qubit system (with $N = 2^n$), $\cos\beta_0 = \sqrt{1 - 1/N}$. Hints: For the uniform superposition state, the amplitude for each basis state is $1/\sqrt{N}$. The amplitude for the nonzero terms in $|\psi_{\text{other}}\rangle$ is $1/\sqrt{N-1}$ and there are $N - 1$ nonzero terms. Use this result to check the listings in Table 12.1.

We can use our geometric picture to estimate the probability of getting the target state if we carry out a measurement on the superposition state $|s_k\rangle$ after k iterations. The *measurement principle* shows that the probability of getting the target state is the square of the amplitude of the superposition state projected onto $|w\rangle$. So, we have

$$\text{Probability of observing } w = |\langle w\,|s_k\rangle|^2. \tag{12.14}$$

In Fig. 12.4, we see that $\langle w \mid s_k\rangle = \cos(\pi/2 - \beta_k) = \sin\beta_k$; so, we find

$$\begin{aligned}\text{Probability of observing } w &= \sin^2\beta_k \\ &= \sin^2\left[(2k+1)\beta_0\right].\end{aligned} \tag{12.15}$$

We found previously that $\cos\beta_0 = \sqrt{(N-1)/N}$ and hence $\sin\beta_0 = 1/\sqrt{N}$. If N is large, then the sine function will be small, and we can use the approximation $\sin\beta_0 \approx \beta_0 = 1/\sqrt{N}$.

Math Aside You can use a calculator or a spreadsheet to verify that $\sin\beta_0 \approx \beta_0$ for small β_0 if the angle is measured in radians. Or look at a graph of $\sin\beta_0$ versus β_0. For small β_0, the graph is almost a straight line with a 45° slope. There are more formal proofs, but these two methods should convince you that the result is correct.

Try It 12.8 How did we obtain $\sin\beta_0 = 1/\sqrt{N}$? Hint: From trigonometry we have $\cos^2\theta + \sin^2\theta = 1$ for any angle θ.

Combining these results, we see that the probability of observing $|w\rangle$ is approximately $\sin^2\left((2k+1)/\sqrt{N}\right)$ as mentioned previously. We get the largest probability when the argument of the sine function is equal to $\pi/2$. It means that we want to use k^* iterations, where k^* satisfies

$$(2k^* + 1)/\sqrt{N} \approx \pi/2 \tag{12.16}$$

or

$$\begin{aligned} k^* &\approx \pi\frac{\sqrt{N}}{4} - 1/2 \\ &\approx \pi\sqrt{N}/4. \end{aligned} \tag{12.17}$$

What happened to the ½ in the first equation of (12.17)? Because we are thinking about large N (for example, $N > 100$), ½ will be small compared to the $\sqrt{N}$ term so we can safely ignore it without significantly changing the result. This type of thinking happens a lot in math and science. The important conclusion is that Eq. (12.17) shows that the number of iterations grows as $\sqrt{N}$ for the quantum algorithm. We argued before that the classical algorithm requires about $N/2$ "flips" on average. This result illustrates the advantage of the quantum algorithm in reducing the number of iterations that are needed to find the target state.

Quantum Circuit for the Grover Algorithm

As shown with other quantum algorithms in the last chapter, it is customary to formulate the algorithm in terms of a function gate (an "oracle") because those function gates can be built from simpler gates, such as CNOTs, X, Y, Z, etc. For the Grover algorithm, the function gate is shown in Fig. 12.5. This is similar to the function gate for the Deutsch algorithm discussed in Chapter 11 (see Fig. 11.2) but with the upper register extended to n qubits.

The system state after the setup gate U_f has acted is

$$U_f\left(|x\rangle_n |y\rangle\right) = |x\rangle_n |f(x) \oplus y\rangle. \tag{12.18}$$

The setup gate is unique to the target state we are trying to find. That, in essence, is how we label the target state. In other words, the setup gate is constructed so that $f(w)= 1$, while $f(x)= 0$ for all the other xs. You may remember this is the function that was the subject of Try It 12.2.

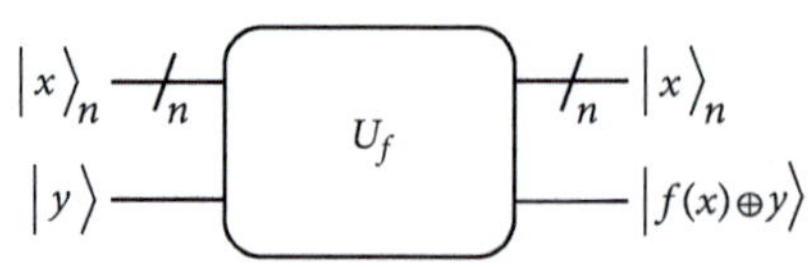

Fig. 12.5 The function gate for the Grover algorithm. The forward slash line with the subscript n indicates that there are n upper register lines.

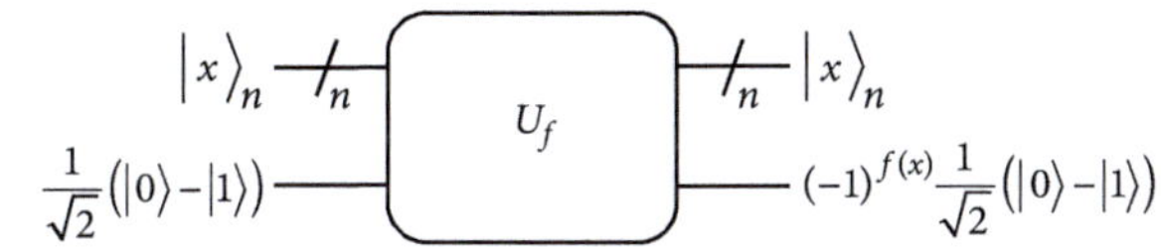

Fig. 12.6 An alternate representation of the Grover algorithm setup gate.

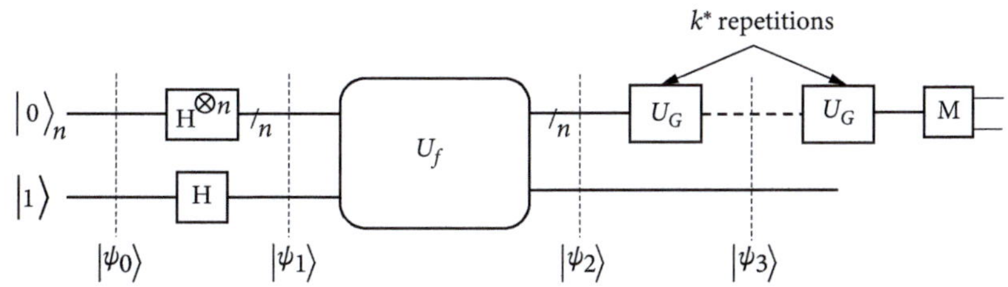

Fig. 12.7 The quantum circuit for the Grover algorithm. The vertical dashed lines indicate where each system state $|\psi_j\rangle$ occurs. There are k^* repetitions of the U_G gate.

The function gate is often defined in an alternate but equivalent version (see Fig. 12.6), with one that is more easily implemented in terms of simple quantum gates.

In the alternate version, we assume that the lower-register input state is the one-qubit superposition state $\frac{1}{\sqrt{2}}(|0\rangle - |1\rangle)$. You can then convince yourself that when $f(x)= 1$, the lower-register output results in interchanging the $|0\rangle$ and $|1\rangle$ terms, which is the same as multiplying the superposition state vector by -1.

Try It. 12.9 Challenge: Work out the details given in the previous paragraph.

Don't worry if those details are not yet clear to you. It took us a few minutes in a quiet place with a sharp pencil (with an eraser!) and several clean sheets of paper to convince ourselves of the details.

Figure 12.7 shows the quantum circuit for the Grover algorithm. Note again that this is just the n-qubit generalization of the Deutsch circuit shown in Fig. 11.3. The gate U_G implements the reflections of the state vectors that are described before. In practice, you would build in k^* repetitions of U_G to get close to the maximum probability of observing the target state when the measurement device works on the upper-register superposition state.

Let's go through the circuit for a two-qubit system. The upper-register state on the far left is then $|00\rangle$. This state interacts with an $H^{\otimes 2}$ Hadamard gate while the lower register interacts with a one-qubit Hadamard gate.

Step 1: For a two-qubit system, we have four basis states. The top Hadamard gate generates an equal-amplitude superposition of the four basis states. We also see that the lower-register Hadamard gate produces the superposition state $|y\rangle = \frac{1}{\sqrt{2}}(|0\rangle - |1\rangle)$ that is mentioned before.

At this stage, the system state is

$$|\psi_1\rangle = \underbrace{\{1|x\rangle\}}_{\text{upper register}} \otimes \underbrace{\left(\frac{|0\rangle - |1\rangle}{\sqrt{2}}\right)}_{\text{lower register}}. \tag{12.19}$$

Step 2: The function gate now acts on this state to produce the system state

$$|\psi_2\rangle = \underbrace{\{(-1)^{f(x)}|x\rangle\}}_{\text{upper register}} \otimes \underbrace{\left(\frac{|0\rangle - |1\rangle}{\sqrt{2}}\right)}_{\text{lower register}}. \tag{12.20}$$

In other words, the function gate has flipped the sign of the amplitude of the target state (the one with $f(x) = 1$), while leaving the other amplitudes unchanged. This is the equivalent of the first reflection of the superposition state in our geometric approach.

Step 3: We implement the equivalent of the second reflection whose main effect is to increase the amplitude of the target state while reducing the other amplitudes in the superposition state. This is what the gate U_G in Fig. 12.7 does.

As we observed with our geometric method and its translation into quantum operators (Eq.(12.6)), all we need is a gate to implement this operation. Let's assume we have such a gate. Chapter 4 of (Mermin, 2007) shows how to implement the Grover algorithm function gates using simple quantum gates. The U_G gate is applied k^* times after which the measurement of the upper-register states is highly likely to yield the target state. Because the result is probabilistic, you would most likely run the algorithm several times to check the results.

Flipping about the Mean and Amplitude Amplification

There is yet another algorithm that can be used to find the target basis state. Some authors call this set of operations "flipping about the mean." The mean m of the amplitudes in the superposition state is their sum divided by the number of basis states $N = 2^n$. So, for each amplitude a, we calculate $\delta = m - a$ and then subtract δ from a to obtain the new amplitude $a' = 2m - a$. Suppose we have four amplitudes $(1/2, -1/2, 1/2, 1/2)$. The mean is 1/4. The new amplitudes are shown in Table 12.2.

The result of flipping about the mean is to increase the amplitude of the target state (the one that started with the negative amplitude) while decreasing the amplitude of the other

Table 12.2 Flipping about the mean for a two-qubit state. The amplitude for the target state $|01\rangle$ has a shaded background.

Original a	1/2	−1/2	1/2	1/2
Mean m = 1/4 **New $a' = 2m - a$**	0	1	0	0

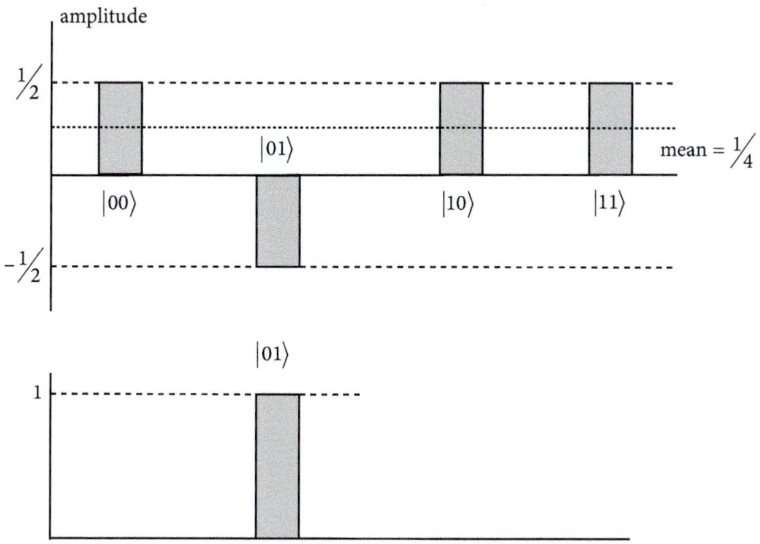

Fig. 12.8 A graphical representation of "flipping about the mean" for a two-qubit system. The amplitudes of the basis states are represented by the heights of the shaded rectangles. The target state $|01\rangle$ starts with a negative amplitude (top part of the figure). The lower part of the figure shows the amplitudes after flipping about the mean. $|01\rangle$ has an amplitude equal to 1. All the other amplitudes are 0.

states. We recognize, as mentioned before, a two-qubit system is special because the Grover algorithm finds the target state after only one iteration of the algorithm: the target state amplitude is 1 while the other amplitudes are 0.

We can also give a graphical representation of flipping about the mean. In Fig. 12.8, we represent the amplitudes of the basis states by the heights of rectangles. The target state $|01\rangle$ is given a negative amplitude, while all the other amplitudes are positive. The lower part of the figure shows the amplitude of the states after flipping about the mean.

Another graphical representation is often used for displaying state amplitudes. As shown in Fig. 12.9, we plot the amplitudes as columns in a rectangular grid. The amplitude is represented by the height of the column. Positive amplitudes go up, while negative amplitudes go down. Figure 12.9 shows the amplitudes for the uniform superposition state of a two-qubit system.

Figures 12.9–12.11 show the sequence of flipping about the mean for the two-qubit system with target state $|0\rangle\,|1\rangle$. Similar diagrams are widely used in QC to show state amplitudes after gates operate on the states. Comparisons with the output of an actual QC tells you about the gate "fidelity": how faithfully the QC has been able to implement the gate operation. In practice, the commonly used column graphs show what are called the probability density matrix elements for the system. These elements involve products of various state amplitudes rather than the amplitudes themselves. See Chapter 22 for more on the density matrix.

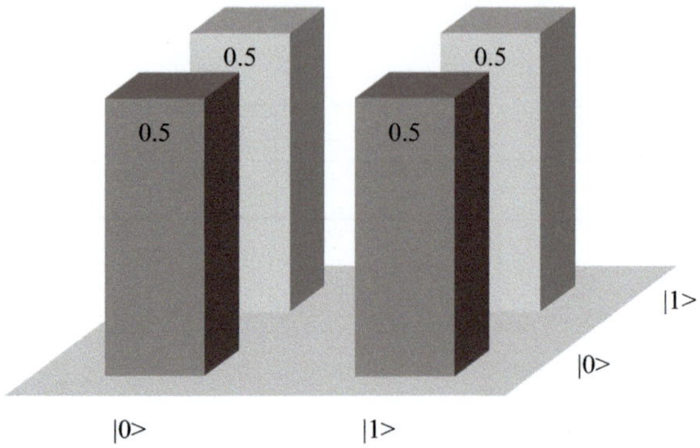

Fig. 12.9 The amplitudes for a uniform superposition state for a two-qubit system. The height of the column represents the amplitude. The actual numbers are indicated on the columns. The state labels are given by the product of the labels along the two "horizontal" directions. $|0\rangle|1\rangle$ is in the front row, far right.

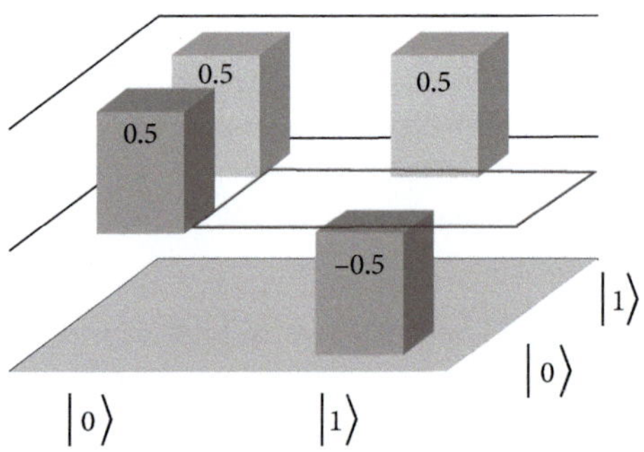

Fig. 12.10 The amplitudes for a two-qubit superposition state in which the target state $|0\rangle|1\rangle$ has been given a negative amplitude. The semitransparent rectangle near the middle indicates the amplitude = 0 plane.

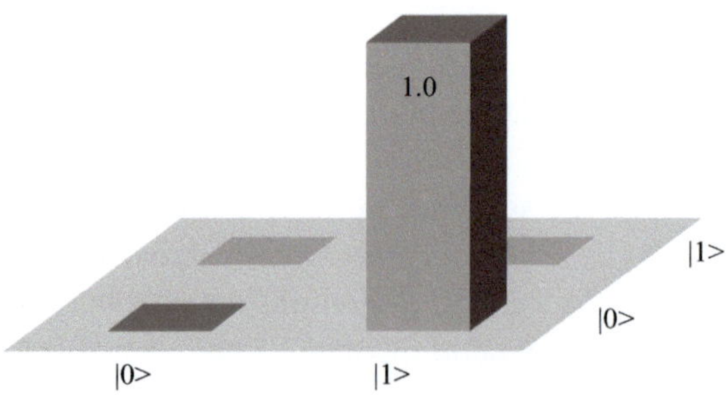

Fig. 12.11 The amplitudes for a two-qubit superposition state after one "flip about the mean." The target state $|0\rangle|1\rangle$ amplitude now equals 1 and all other amplitudes are 0.

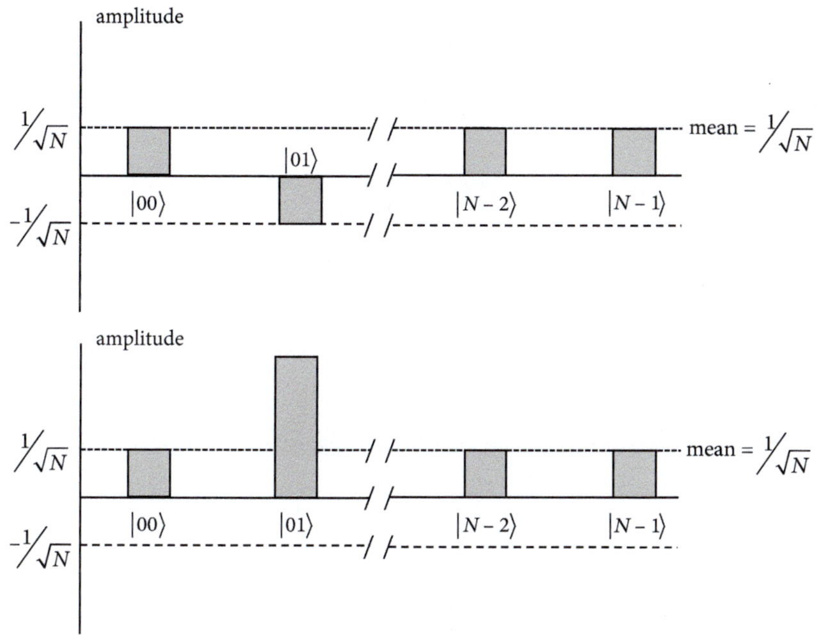

Fig. 12.12 A graphical representation of flipping about the mean for an n-qubit system.

Figure 12.12 shows a graphical representation of flipping about the mean when the number of states N is large. In this case, the initial mean amplitude is close to $1/\sqrt{N}$. After flipping about the mean, only the state with the negative amplitude has any substantial change in amplitude. If we make this increased amplitude negative and do another flip about the mean, this amplitude becomes even larger, and the other amplitudes decrease by a small amount. Continuing this process, we find that the target state amplitude eventually approaches 1 and the other amplitudes approach 0.

Let's express the flipping about the mean in terms of the operator in Eq. (12.7) acting on the state $|s_k\rangle$ to produce $|s_{k+1}\rangle$:

$$|s_{k+1}\rangle = R^{(s_0)} R^{(\psi_{\text{other}})} |s_k\rangle = 2|s_0\rangle \langle s_0|s_k\rangle - |s_k\rangle \,. \tag{12.21}$$

For a general set of qubit states, the new coefficients a'_j are given by

$$a'_j = 2m - a_j = 2 \times \left(\frac{1}{2^n} \sum_{i=0}^{2^n-1} a_i \right) - a_j = -\left(1 - \frac{1}{2^n}\right) a_j + \frac{1}{2^{n-1}} \sum_{i=0, i \neq j}^{2^n-1} a_i. \tag{12.22}$$

Looking at the first expression in Eq. (12.22), we see that if $a_j \approx m$, it will be almost the same after the operation. But the amplitude of the target state is initially $-m$ and after the operation we obtain $a'_j \approx 3m$. But remember that these approximations are valid only for large N.

For $n = 2$, we have

$$a'_j = 2m - a_j = \frac{1}{2}\sum_{i=0}^{3} a_i - a_j = -\frac{1}{2}a_j + \frac{1}{2}\sum_{i=0,i\neq j}^{3} a_i. \tag{12.23}$$

The matrix that carries out this calculation for a two-qubit system is

$$\begin{pmatrix} a'_0 \\ a'_1 \\ a'_2 \\ a'_3 \end{pmatrix} = \frac{1}{2}\begin{pmatrix} -1 & 1 & 1 & 1 \\ 1 & -1 & 1 & 1 \\ 1 & 1 & -1 & 1 \\ 1 & 1 & 1 & -1 \end{pmatrix}\begin{pmatrix} a_0 \\ a_1 \\ a_2 \\ a_3 \end{pmatrix}. \tag{12.24}$$

Try It 12.10 Use this matrix with the two-qubit state $\frac{1}{2}(|00\rangle - |01\rangle + |10\rangle + |11\rangle)$ and show that the results match with those provided in Table 12.2.

Now let's see how this works for a three-qubit system. In Table 12.3, we show the results when we choose the target state $|010\rangle$ in the third column.

Once again we see that the flipping about the mean leads to an increase in the amplitude of the target state and a decrease in the amplitudes of the other states. Figure 12.13 displays a column chart of the results presented in Table 12.3. Note that the target state amplitude is much larger than the amplitudes of the other states.

Table 12.3 Flipping about the mean for a three-qubit superposition state. The amplitudes for the target state have shaded backgrounds.

Original a	$1/\sqrt{8}$	$1/\sqrt{8}$	$-1/\sqrt{8}$	$1/\sqrt{8}$	$1/\sqrt{8}$	$1/\sqrt{8}$	$1/\sqrt{8}$	$1/\sqrt{8}$
Mean m	0.265							
$a' = 2m - a$	0.177	0.177	0.884	0.177	0.177	0.177	0.177	0.177

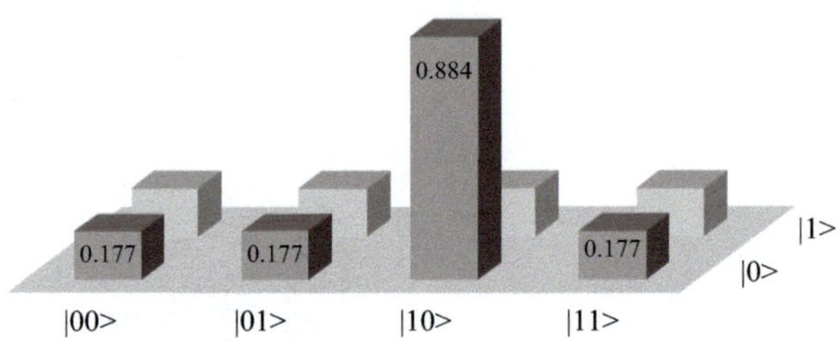

Fig. 12.13 A column chart representation of the results presented in Table 12.3. The height of a column is the amplitude for that state. The back row amplitudes are all 0.177. The labels were omitted for the sake of typographical simplicity.

Try It 12.11 Verify the numbers in Table 12.3. Carry out the next step by flipping about the new mean.

Let's pause for a moment and summarize what we have discussed about the Grover algorithm.

- The Grover algorithm allows us to find a basis state that has been chosen by making its amplitude negative while the amplitudes of all the other basis states are positive.
- We then follow a set of operations on the superposition state. The result of this set of operations is to increase the amplitude associated with the target state while reducing the amplitudes associated with the basis states that are not the target.
- The operations can be viewed as reflecting state vector amplitudes relative to the previous state in the sequence—in effect rotating the state vector closer to the target vector direction.
- From these results, we can estimate how many operations are needed to identify the chosen state. We found that the number increases as $\sqrt{N}$, where N is the number of qubit basis states in the system. The corresponding classical algorithm takes about N steps. Thus, the Grover algorithm demonstrates the advantage of the quantum algorithm over the classical one.
- Also, we observed how to implement these operators using a Grover algorithm quantum circuit.

Why do we need faster searches? Google can find anything in less than a second. There are several answers. First, Google gives you only a list of possible search outcomes. In most cases, you need to go through the list and find the ones closest to what you need. The Grover algorithm finds the best match automatically. The second answer may seem a bit tricky, but it illustrates an important principle: The Grover algorithm was not really meant to be a practical search algorithm. It simply shows that there is a quantum algorithm that is much faster than the corresponding classical one. This expands our horizons about what is possible with QC and why we might want to develop other, more practical quantum algorithms.

12.3 Quantum Computational Chemistry

One of the most promising applications of QC for near-term applications is the calculation of the structures and energies of molecules. Why is this important? One of the fundamental principles of chemistry is the link between form and function. For example, how well a drug molecule interacts with a target virus depends on the form (shape) of the molecule and how well it connects to the part of the virus the drug is designed to attack. Catalysts are another good example, where the shape of the catalyst materials determines the effectiveness of their ability to promote other chemical reactions. More generally, the properties of most complex materials depend on the shape of the molecules that make up the material. Of course, the chemical composition—which atoms are in the molecules—is also important.

The other key principle is that different shapes are associated with different energies, and it takes energy input or output for the molecule to change from one shape to another. Here is a specific example: When your eye "detects" light, a photon is absorbed by a rhodopsin molecule in the retina of your eye. The absorption of the photon's energy changes the rhodopsin from a more-or-less straight-line shape (called "cis") to a bent form (called "trans"). This change in shape leads to an electrical signal sent over your optic nerve from the retina to the brain, informing the brain that light has been detected. Knowing the energies associated with these two rhodopsin shapes is important in predicting the efficiency of light detection.

Scientists can calculate these energies through various methods based on quantum mechanics. The key idea is to think about how the atoms that make up the molecule come together to form the molecule. You may remember from your high school chemistry course that atoms have a central nucleus, made up of protons and neutrons, and that electrons are found in "clouds" centered on the nucleus. There is an attraction between the protons in the nucleus and electrons and a repulsion between various electrons and between the positively charged nuclei of different atoms. These interactions form part of the basis for molecular bonds and the energy associated with them.

In this process, some of the electrons that were associated with individual atoms rearrange themselves into the chemical bonds that hold the molecule together. In effect, some of the electrons stay around their original atomic nuclei; others are shared among several of the atoms forming molecular bonds. The task is to calculate the energy associated with these arrangements of atoms within the molecule. The configuration with the lowest energy is usually the most stable shape of the molecule.

The difficulty is that a typical molecule, particularly those that are of interest in biological settings, has hundreds of atoms and thousands of electrons. Keeping track of all these particles and how their energies change is quite a challenge. Nevertheless, chemists have developed various techniques that are reasonably successful in finding these energies.

But, to return to our drug example, if you want to explore dozens of molecules as potential therapeutic candidates, the computational time required strains even our best classical supercomputers. QC offers alternative ways of doing these calculations much more quickly so that less time is needed to explore various candidate molecules.

In recent years, quantum chemistry algorithms have been developed that combine classical computations with quantum computations. These have been applied so far only to relatively small molecules, but the techniques appear to be quite promising. Along the way, quantum computer scientists realized that to make these calculations work on an actual quantum computer, quantum error corrections need to be applied to account for the uncontrolled interactions of the qubits with other qubits and with stray electric and magnetic fields. Fortunately, these quantum error correction methods seem to be working as well.

Because quantum chemistry algorithms make use of concepts in quantum mechanics that are more advanced than others that we have used for describing QC, we give you just an overview of how the algorithms work, focusing on those features that quantum computational chemistry algorithms have in common with the other algorithms we have studied.

The key chemical concept is the notion of an orbital, which is a quantum state that describes an electron in the "cloud" of electrons around a nucleus. In performing these calculations, chemists identify possible orbitals for the electrons. But in the actual molecules, only some of the orbitals will have an electron in the described state; we say these orbitals are *occupied*. If there is no electron there, the orbital is *unoccupied*. In this picture, there are two possible states for the orbital: occupied and unoccupied, so it is not surprising that we can use qubit states to keep track of the orbitals. For more details, see Chapter 20 titled Number Operators and Number States.

Unfortunately, it is hard to calculate the exact orbitals except for the simplest molecules. In practice, we calculate approximate orbitals using many theoretical tools that have been developed over the past several decades. We then assume that the quantum state for the molecule is given by a superposition of these approximate states. Denoting the approximate orbitals by $|\phi_p\rangle$, where p is an index to keep track of the orbitals, we write the state for the electrons in the molecule as

$$|\psi\rangle = a_1|\phi_1\rangle + a_2|\phi_2\rangle + \ldots \tag{12.25}$$

As usual, the as are the amplitudes associated with each of the orbitals.

The next step is to calculate the average value of the energy associated with the state $|\psi\rangle$. We mentioned the attraction and repulsion of the protons and electrons, but we also need to take the kinetic energy of the electrons into account. There are other energy contributions from the magnetism associated with the electron's spin and some similar effects for the nuclei. But for our purposes, we may neglect them because these energies are much smaller than the electrical ones. In the quantum description of molecules, the energy, expressed as an operator (matrix), is called the Hamiltonian.

The quantum energy operator is named after William Rowan Hamilton, an Irish mathematician and physicist, who developed, among many other important mathematical tools, theoretical methods focusing on energy. He lived from 1805 to 1865. You should watch the YouTube video listed in Further Reading at the end of the chapter.

The crucial point is that the calculated energy depends on the structure of the molecule through the distances between the nuclei and through the amplitudes in Eq. (12.25). In most approaches to computational chemistry, we vary the amplitudes until we find the lowest value of the energy for that structure and then vary the structure to get the overall minimum energy. That lowest energy, the "ground-state" energy, is the signature of the most stable state of the molecule.

The name "ground state" came from the notion you may have heard about in your high school physics course: When an object (like a ball) interacts with Earth via gravity, there is

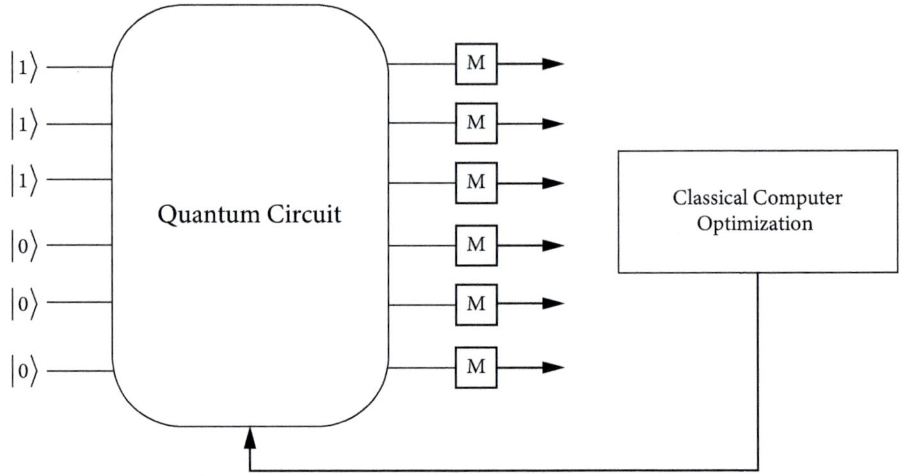

Fig. 12.14 Schematic diagram of the system used for quantum computational chemistry for a six-qubit system.

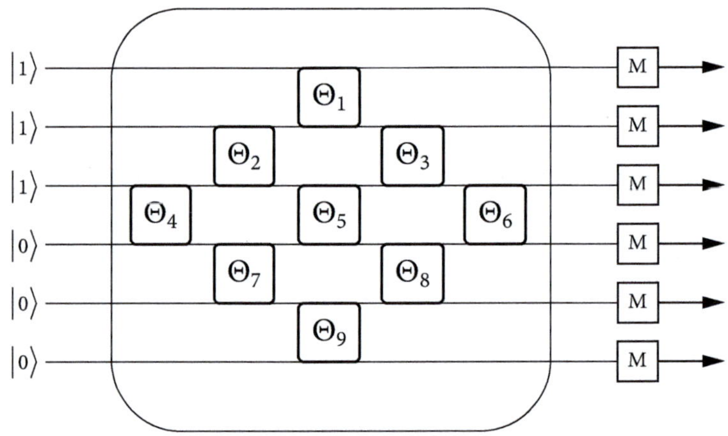

Fig. 12.15 A schematic diagram of the quantum circuit for quantum computational chemistry for a six-qubit system. The boxes with the symbols Θ_n are called rotation gates, and the symbols indicate parameters that are used to modify the qubit states using the two-qubit gates. In this diagram, six states are used to describe six molecular orbitals.

an energy associated with the position of the ball relative to the center of Earth. In everyday life, the closest the ball can get to the center of Earth is when the ball is on the ground, on the surface of Earth. That is the situation in which the gravitational energy for the Earth–ball system is smallest. Hence, the name "ground state" for the lowest energy state of any quantum system.

As mentioned before, we associate a qubit state with each of the orbitals and then use a quantum circuit, as illustrated in Figs. 12.14 and 12.15, to adjust the amplitudes. The gates labeled as Θ_n are sometimes called rotation gates because they contain, among other operations, rotation operators, as mentioned in Chapter 7. The rotation gates are different from the gates we have seen before because they are adjustable by changing the parameters Θ. The circuit diagram is simplified and does not show the connections that adjust the parameters.

We arrange the measurement devices so that when the qubits interact with the measurement devices, we get a 1 if the orbital is occupied and a 0 for an unoccupied orbital. From that information, we can use a classical computer to calculate the energy and to make decisions about how to vary the quantum circuit parameters shown in Fig. 12.15 to lower the energy. When we get to the minimum energy, small variations in the parameters will not lead to significant changes in energy; so, the method is called the variational quantum method. In quantum jargon, we say that we are searching for the energy eigenstates of the system. As we noted before, "eigen" is the German word for "characteristic." So, the implementation of quantum computers uses what is called the variational quantum eigensolver (VQE). To make all this work efficiently, we need to start with good guesses about the approximate orbitals. In practice, classical computers can provide these approximate orbitals to be used in the initial superposition state.

The quantum circuit shown in Fig. 12.15 has six qubits and was adequate to predict the energies of the two isomers of diazene H_2N_2, that is, two hydrogen atoms and two nitrogen atoms. If you check the periodic table of the elements online, you will find there are seven electrons for each nitrogen atom and one for each hydrogen atom, making for a total of 16 electrons in diazene. How can we make this work with just six qubits?

The "trick" is to recognize that most of the electrons associated with the nitrogen atoms stay close to the nitrogen nuclei. Only a few join the two electrons from the hydrogen atoms in rearranging themselves to make the chemical bonds; so, we only need to use these in our calculation. The other trick is to recognize that for a molecule like diazene the spin-up electrons are distributed almost exactly like the spin-down electrons; so, we decrease the number of required qubits in half.

The steps involved in the molecular orbital calculation are as follows:

1. Specify the molecule: structure, charge, etc.
2. Generate approximate orbitals to set an initial state.
3. Generate a quantum circuit that represents the initial molecular orbital as a superposition of the approximate orbitals with adjustable parameters.
4. Measure the output states from the quantum circuit and feed the results to a classical computer to calculate the energy and to find the changes to the parameters needed to go to a lower energy.
5. Continue the process until you reach a minimum in the energy for that structure.
6. Repeat with changes in structures (distances between the nuclei and angles between the chemical bonds).

In Step 4, a classical calculation uses the information about how this energy changed when the parameters of rotation matrices were varied. This allows us to estimate the parameter changes that will lower the energy. These new parameters are then used in the quantum algorithm again. The process continues until the minimum in the energy for that configuration of the molecule is found. Then, the structure is changed (e.g., in a diatomic molecule, the distance between the atoms is changed) to map out how that energy depends on the internuclear distance. The distance that gives the lowest of the energies is then the equilibrium internuclear distance for that molecule. For molecules with three or more atoms, you also have angles between the bond directions; so, you need to explore these as well to find the overall most stable structure.

Let's summarize what the quantum chemistry computational algorithm has in common with other algorithms:

1. Start with an initial state of the multiqubit system.
2. Act on that state with various "function boxes" and gates designed to implement the algorithm.
3. Subject the resulting state to a measurement device.
4. Based on the outcome from the measurement, make adjustments that will drive the multiqubit state closer to the desired target.

All of this sounds great—and it is—but as the number of two-qubit gates increases, the systematic noise in the QC increases. The number of "layers" of gates as shown in Fig. 12.14 dictates what is called the "depth" of the quantum circuit. As the depth increases, the level of "noise" in the quantum circuit increases. Here "noise" means that the amplitudes of the qubits change in unpredictable ways due to unwanted interactions with the environment, usually through electric and magnetic fields. Therefore, error mitigation methods of the types described in Chapter 13 must be deployed. The good news from recent work is that the error correction methods seem to work well for the quantum circuits used for quantum computational chemistry.

A recent paper (Rubin et al., 2020) shows how these methods can be applied to chains of up to 12 hydrogen atoms and to the diazene molecule H_2N_2, also written as $(NH)_2$. The diazene molecule is particularly interesting for quantum chemistry because, as mentioned previously, it has two shapes (isomers) with just slightly different energies. One isomer is called trans-diazene; it has the two hydrogen atoms on opposite sides of the N–N bond, while cis-diazene has the two hydrogen atoms on the same side. The quantum computational chemistry approach can account for this small energy difference. Although diazene is a relatively simple molecule, the reported results, including the effectiveness of the quantum error correction methods, are a cause for optimism that enhanced quantum computers, with more qubits, will be able to tackle much more complicated molecules.

If you would like to try quantum computational chemistry yourself, the IBM Qiskit website has a VQE that you can program to calculate the energy of simple diatomic molecules (molecules with two atoms) as a function of the distance between the nuclei of the two atoms. See, https://quantum.cloud.ibm.com/learning/en/courses/quantum-chem-with-vqe.

12.4 Chapter Summary

- Quantum algorithms are being developed in many fields, outperforming and working alongside classical computing methods.
- Searching for strings of numbers or text is commonplace. We do it all the time in our personal lives, searching for specific words on web pages or values on spreadsheets. The Grover algorithm implements a search among basis states by manipulating a superposition state so that the target state coefficient increases while the others decrease. Because the probability of measurement increases with the square of the coefficient, this process quickly increases the probability of finding the target state. Geometrically, this process reflects a pointer state vector across other vectors in state space. This state space is defined with the target state as one of the axes with the other axis being a vector that is the superposition of all the other state vectors. First, the uniform superposition vector is reflected across the nontarget axis and is then reflected across the initial superposition. This brings the state vector closer to the target state axis, which means that the target state coefficient has increased in the superposition. This process is then repeated until the target state coefficient is close to 1, implying that the probability of getting the target state upon observation is close to 1.
- Quantum algorithms have the power to model complex molecules to predict their shapes based on calculating changes in the ground-state energy of different configurations. The molecule will assume a shape that minimizes this energy. This is a computationally expensive process to do classically because the shape is affected by the interactions of each individual atomic particle with all the other particles in the molecule. The quantum algorithm uses, of course, superposition to more efficiently evaluate the changes in the orbital configuration based on changing input parameters of the model. A classical computer calculates the energy change for this model compared to previous models. This iterative method uses the results of the previous energy changes to determine the next configuration to test.

12.5 Further Reading

L. Grover, "Quantum Mechanics helps in searching for a needle in a haystack," www.arXiv.org: quant-ph/9706033. This paper is quite readable given what you have learned in this chapter.

Lov K. Grover, "From Schrödinger's equation to the quantum search algorithm," *American Journal of Physics* **69**, 769–777 (2001). This paper describes how Grover "discovered" his search algorithm.

B. Kain, "Searching a quantum database with Grover's search algorithm," *American Journal of Physics* **89**, 618–626 (2021). The paper includes an IBM Qiskit implementation of Grover's algorithm.

N. David Mermin, *Quantum Computer Science* (Cambridge University Press, Cambridge, 2007). Chapter 4 shows how to build the Grover algorithm function circuits from simple qubit gates.

A musical song about William Rowan Hamilton, done in the style of the recent musical *Hamilton*, is available at https://www.youtube.com/watch?v=SZXHoWwBcDc

Rubin et al., "Hartree–Fock on a superconducting qubit quantum computer" *Science* **369** 1084–1089 (2020). This paper describes quantum chemistry calculations for chains of hydrogen atoms and for diazene molecules.

The IBM Qiskit website url is https://qiskit.org/. The site has many tutorial materials and allows you to write a quantum computer program to run on either a quantum computer simulation or on one of the actual IBM quantum computers. The ground-state energy solver can be found at https://quantum.cloud.ibm.com/learning/en/courses/quantum-chem-with-vqe.

Cambridge Quantum Computing provides quantum software, a quantum development platform (t|ket)™), enterprise applications in the area of quantum chemistry (EUMEN), quantum machine learning (QML), and quantum augmented cybersecurity (IronBridge™). https://cambridgequantum.com/about/

Error Correction 13

> Delay is preferable to error.
>
> **Thomas Jefferson, Letter to George Washington, May 16, 1792**

13.1 Why Error Correction?

We now need to look at a general problem in computing and communications: error correction. Why error correction? Aren't computers supposed to follow precise rules and do exactly what we tell them to do? Ideally, yes, but we need to realize that for computers and other forms of digital information processing to work, we need to send information from one place to another. Sometimes the distances are quite short, say within your smartphone, but other times we use fiber optic links under the oceans, or we send the information via radio waves and satellites orbiting Earth. In traditional computing, this information is almost always in a sequence of 1s and 0s. But as the signals propagate, the radio waves, for example, might be contaminated with electrical noise from other sources such as power lines and lightning strikes. Fortunately, modern information technology has matured to the point that errors in those messages are infrequent. However, when they do occur, if they are not corrected, the received signal might be garbled to the extent that the output of the process is basically nonsense. Error detection and error correction, because of their importance, have evolved into rather sophisticated schemes.

For a classical computer, information is often sent in *clusters* of 1s and 0s, just 7 or 8 bits. The most common type of error in the transmission is to have one of the bits change from 0 to 1 or 1 to 0. This is called a "bit-flip" error. When we think about error correction for quantum information transmission, the situation becomes even more complex because the state of a qubit, the way we send quantum information, can change relatively easily due to interactions between the qubit and its environment. Not only can the basis states $|0\rangle$ and $|1\rangle$ be interchanged (the equivalent of a bit flip), but the amplitudes of a superposition state can change, thereby corrupting the information carried by the qubit's state because the predicted probabilities will change. We will now look at some simple examples that illustrate the basic features of error correction.

First, let's tackle classical error correction. The first step in error correction is to detect that an error has occurred. One way to do this would be to send the same message twice.

Quantum Computing and Quantum Physics. Alice Flarend and Robert Hilborn, Oxford University Press.
 DOI: 10.1093/oso/9780197904930.003.0013

We then compare the two messages. If they are the same, we are reasonably confident that no error occurred. This confidence is based on the relatively rare occurrences of errors in the transmission system. But since the errors are rare, we are wasting a lot of communication time by sending the message twice. A better scheme is to append another bit to the sequence that gives us some information about what is in the message. This additional "ancillary" bit is often called a "parity bit" because it is used to indicate that some number associated with the message is even or odd. In mathematics and computer science, "parity" almost always means even or odd.

Let's look at a specific example of the use of a parity bit in classical information transmission. Suppose the message we want to send is 7 bits long—0110101, for example. To aid in error detection, we can append a 1 to the right of the sequence if the number of 1s in the sequence is even and a 0 if the number of 1s is odd. For our example, we have four 1s in the message. Since four is an even number, we would append a 1 to the right and send the messages as 01101011.

When the message is received, the receiver counts the number of 1s and compares that number to the odd number expected (odd because we include the parity bit). If any of the original 1s flipped to a 0 or any of the original 0s flipped to a 1, there would be a disagreement between the expected number of 1s and the received number.

What if two bits flipped? Then we might not detect the error. This is true, but remember we assume that the errors occur only rarely, so the probability of having two errors in the same message is extremely small. It might happen, but it will be rare enough that we don't need to worry about that possibility.

To find which bit was affected, we must be clever. One option is simply to request that the message be resent if an error has been detected. But there are ways we can find the bit in which the error occurred and make the correction directly without having to communicate further with the sender.

One of the simplest schemes is to "pad" the bit that you want to send with other bits, in particular just add two more of the same bit. This is called repetition coding. For example, if you want to send 0 to Bob, you would actually send 000. If a one-bit error occurs during transmission, Bob will receive 100, 010, or 001. Here's the trick to spot the error: designate the bits as $b_0, b_1,$ and b_1. When Bob receives the three bits, he evaluates the parity tests $b_0 \oplus b_1$ and $b_0 \oplus b_2$. If the results are 0 and 0, then Bob knows there was no one-bit error. In fact, exactly the same process works if you want to send a 1. You launch 111 to Bob, he carries out the same binary addition without carry. If he gets 0 and 0, then he knows the message is ok in the sense that it does not contain a one-bit error.

Try It 13.1 Check this with the message bit = 1

What is more amazing is that if there is a one-bit error, the results of the binary additions let you know which bit was affected. For example, if $b_0 \oplus b_1 = 1$ and $b_0 \oplus b_2 = 0$, then we know that bit b_1 was the affected bit and we should flip it to get the correct bit sequence. This

Table 13.1 The corrective action required to correct a one-bit error.

$b_0 \oplus b_1$	$b_0 \oplus b_2$	Corrective action
0	0	No change
1	1	Flip b_0
1	0	Flip b_1
0	1	Flip b_2

is a variation on the parity test, but here the result of the binary addition tells us whether the two bits are the same or different.

Try It 13.2 (a) Verify that the example is correct. (b) Work out which bit needs to be flipped if Bob gets 01 and 11 from the parity tests. Hint: The answers are in Table 13.1.

Try It 13.3 Verify the results stated in Table 13.1 for the two possible cases: (a) the message sent was 000 but the message received was 010 and (b) the message sent was 111 and the message received was 011.

The take-home message is that the parity test allows you to find the bit where the error occurred, and the correction procedure does not depend on knowing the value of that bit.

13.2 Error Correction in Quantum Computing

We have already hinted at why error detection and correction are even more important for quantum information science (QIS) and quantum computing (QC) than they are for classical systems. The practical problem is that it is difficult to isolate a qubit completely from its environment, particularly if we want to act on the states describing the qubit to implement quantum algorithms. Moreover, the qubits we use (spinning electrons or neutrons, superconducting currents, trapped ions, and so on) are systems that have relatively small differences in the energies associated with the qubit quantum states. That means it doesn't take much jostling from interactions with nearby atoms and molecules to cause transitions between states (corresponding to classical bit flips) and even worse, even smaller amounts of jostling lead to changes in the amplitudes and relative phases of the components of a superposition state. We have seen how useful those superposition states are for quantum algorithms. For all these reasons, we must be prepared to detect and correct errors.

Couldn't we just measure the states describing the qubits and if they are not what we expected, we would conclude that an error had occurred? If the measurement shows that an error occurred, we simply rebuild the state with the correct amplitude. Right? Well, this assumes that we already know the state amplitudes. Also, we need to keep in mind that having a qubit interact with a measurement device simply projects the qubit's state onto one of the measurement device's basis states. In quantum systems described by a state of superposition, the measurement gives us only limited information about the state before the measurement. Consider photon polarization. Suppose we have a vertically polarized photon. If we are using a 45° measurement basis, the photon will be described by a state of superposition in terms of the 45° basis states. We have a 50/50 probability of getting either outcome in the 45° basis states. The same will occur if we use a horizontally polarized photon instead of a vertical one. In addition, after the measurement the appropriate state to describe the photon is the measurement basis state for the outcome of the observation. In most cases, the original photon state is gone. So, we need to develop a quantum error detection and correction procedure that does not involve measurements on those qubits that carry the message.

The quantum "trick" is to combine additional ancillary qubits similar to classic error correction with the message qubit or qubits in such a way that measurements on the additional qubits can indicate if an error has occurred and what to do to correct it. The only operations allowed on the message qubits are those of quantum gates; no interactions with measurement devices are permitted because the measurements in general change the state used to describe the qubit.

The quantum error detection and correction scheme we examine is analogous to the three-bit parity check procedure we looked at in classical computing. We embed the basis states, which of course form the core of any quantum message we want to transmit, into a three-qubit repetition system using the following rule:

$$|0\rangle \rightarrow |000\rangle \qquad |1\rangle \rightarrow |111\rangle . \tag{13.1}$$

A general state for a message is a linear superposition of those states: $|\psi\rangle = a|000\rangle + b|111\rangle$. You should note that this is an entangled state. As usual, employing entangled states is what makes the quantum algorithm work. For the sake of simplicity, let's first look at the quantum equivalent of one-bit-flip errors. For example, the states $|001\rangle$, $|010\rangle$, and $|100\rangle$ are single bit-flip errors associated with $|000\rangle$. Later, we show how to generalize the procedure for other types of errors.

The key concept behind quantum error correction is the notion that errors occur in a quantum state due to interactions with other qubits in our QIS system or qubits in the system's environment. These interactions, of course, can be mathematically described by operators acting on the states carrying the information we want to transmit. For example, the X operator acting on a state causes a bit flip. Let's see how that works with the state $|\psi\rangle = a\,|0\rangle + b\,|1\rangle$:

$$\mathrm{X}\,|\psi\rangle \doteq \begin{pmatrix} 0 & 1 \\ 1 & 0 \end{pmatrix}\begin{pmatrix} a \\ b \end{pmatrix} = \begin{pmatrix} b \\ a \end{pmatrix} \doteq |\psi'\rangle = b|0\rangle + a|1\rangle . \tag{13.2}$$

Once we have discovered that a bit flip has occurred, we may use an X gate to correct the state back to its original form.

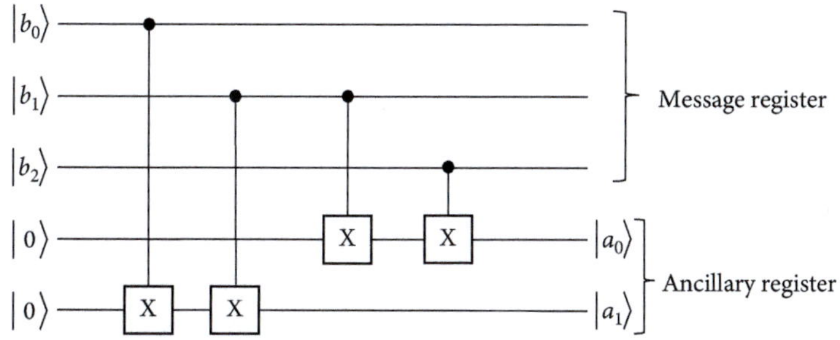

Fig. 13.1 A quantum circuit for detecting bit-flip errors. The message is carried by the top three qubit states called the message register. We assume that the register lines (qubits) have been subject to a communications channel or some other quantum processes that might induce errors.

Try It 13.4 Apply the operator X to the state $|\psi'\rangle$ in Eq. (13.2) and show that you get back to the original state $|\psi\rangle$.

Try It 13.5 Show that the operator Z creates a "phase" flip in the state $|\psi\rangle$. Here phase flip means that the amplitude b changes sign: $b \rightarrow -b$. Then, show that applying Z to the phase flipped state returns you to the original state.

These parallel results should not be surprising because of the properties of the Pauli operators. In particular $X_2 = XX = I$ and $ZZ = Z^2 = I$. Moreover, we know that any operator acting on a qubit can be written as a linear combination of the Pauli operators and the identity operator. So, the method can handle any kind of qubit error.

As mentioned before, we want to detect and correct these errors without having to carry out measurements on these message qubits because we know that measurements disturb the state of the system. To do this, we use the states of two ancillary qubits to detect the error and to correct it. First, let's think about the detection part.

The quantum circuit shown in Fig. 13.1 uses CNOT gates to connect the two ancillary qubit states to the three qubit states in our "message" register. It is not too hard to work out that if all three of the message register bits are the same, both the ancillary states will be $|0\rangle$. This is an indication that we don't have any one-bit-flip errors. Table 13.2 presents the possibilities for the circuit shown in Fig. 13.1.

The crucial point is that the ancillary qubit states allow us both to detect the error and to pinpoint its location among the three message qubit states because they flip to $|10\rangle$, $|01\rangle$, or $|11\rangle$ when there is an error that changes any of the message register states.

Let's get back to the question of what to do now that we have detected an error. How can we carry out error correction using gates? As shown before, for a quantum state, X is equivalent to a bit flip. So, to correct bit-flip errors, we just need to operate with X on

Table 13.2 One-bit-flip errors for the message $|000\rangle$ and the corresponding ancillary register states in Fig. 13.1.

Message register			Ancillary register	
$\|b_0\rangle$	$\|b_1\rangle$	$\|b_2\rangle$	$\|a_0\rangle$	$\|a_1\rangle$
$\|0\rangle$	$\|0\rangle$	$\|0\rangle$	$\|0\rangle$	$\|0\rangle$
$\|0\rangle$	$\|0\rangle$	$\|1\rangle$	$\|1\rangle$	$\|0\rangle$
$\|0\rangle$	$\|1\rangle$	$\|0\rangle$	$\|1\rangle$	$\|1\rangle$
$\|1\rangle$	$\|0\rangle$	$\|0\rangle$	$\|0\rangle$	$\|1\rangle$

Table 13.3 Toffoli gate truth table ($|\bar{a}\rangle$ means $|\text{NOT } a\rangle$).

$\|c_0\rangle$	$\|c_1\rangle$	$\|s\rangle$
$\|0\rangle$	$\|0\rangle$	$\|a\rangle$
$\|0\rangle$	$\|1\rangle$	$\|a\rangle$
$\|1\rangle$	$\|0\rangle$	$\|a\rangle$
$\|1\rangle$	$\|1\rangle$	$\|\bar{a}\rangle$

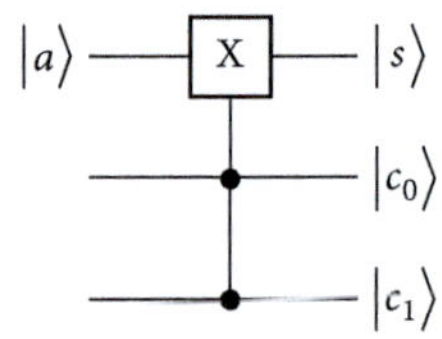

Fig. 13.2 The Toffoli gate circuit diagram. The lower two wires are the control states.

the appropriate state. Ideally, we want this process to be automated so that the errors are corrected without outside intervention.

One way to do this is to have the X operators be controlled by the ancillary states since the results in Table 13.2 link those states to the states that need to have their bits flipped if an error has occurred. With the CNOT gate, the control state determines whether states $|0\rangle$ and $|1\rangle$ are flipped. What we need now is a generalization of this gate to a gate with *two* control states. In other words, we need a doubly controlled-NOT gate, which is called a Toffoli gate.

Toffoli gates were invented in 1980 by Tommaso Toffoli, an Italian American electrical and computer engineer. Table 13.3 presents the truth table for the Toffoli gate, and Fig. 13.2 shows the quantum circuit diagram.

Try It 13.6 Convince yourself that Table 13.3 and the Toffoli gate circuit in Fig. 13.2 are consistent with one another.

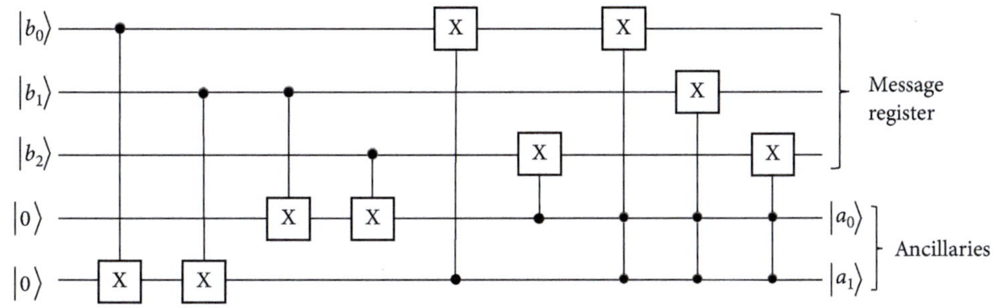

Fig. 13.3 Quantum error detection and error correction circuit.

To carry out the error correction for our one-bit-flip example, we put Toffoli gates on the three message qubit wires and make the control connections to the two ancillary state wires as shown on the right in Fig. 13.3. The X operations on the three message register qubits are all controlled by the two ancillary qubits.

Try It 13.7 Suppose Alice wants to send $|000\rangle$. Assume that the only errors are $|001\rangle$ and $|010\rangle$. Show that the correction circuit in Fig. 13.3 works as desired. Repeat the analysis for the message $|111\rangle$ and the one-bit error $|101\rangle$.

Try It 13.8 Challenge: Construct an argument showing that the correction circuit also works for errors in superposition states. Assume that the message is $a|000\rangle + b|111\rangle$ and the one-bit error is $a|001\rangle + b|110\rangle$. Hint: All the quantum circuit operators are linear operators.

It is also helpful to embed the message qubit into systems with more register qubit states (7 is a commonly used number). Furthermore, the error circuits may have more than two ancillary states. The references at the end of this chapter provide more details but be warned that the circuits and the corresponding formalism can get complicated.

The more general quantum error correction schemes are often just generalizations of what we have done so far. The other common errors are phase errors. For example, a state amplitude a_0 might turn into $-a_0$. This change is called a phase flip. For these kinds of errors, you can use Hadamard gates and measurements with respect to the Hadamard basis states, as described in the Wikipedia article on "Quantum Error Correction" in Further Reading at the end of the chapter.

Many other quantum error correction codes have been developed over the years. Some use 5, 7, or 9 qubits with various gates to detect and then correct bit flips and phase flips. These codes can be combined in various ways to produce so-called stabilizer codes. If the codes are extended to a two-dimensional array of qubits, they are called surface codes. Other types of codes, called bosonic codes, make use of number states and ladder operators, which are described in Chapter 20.

13.3 Logical Qubits

Since error correction is critical for the operation of quantum computers and quantum information, more generally, it would be nice to have quantum states with the error correction built in. Such states are called "logical qubits." (We guess the regular states would have to be called illogical states, but we are just being facetious.) Logical qubits are also called error-corrected qubits, which is a bit more informative. Figure 13.3 is in essence a logical qubit circuit. Logical qubits have been recently shown to work with real hardware (see Bluvstein et al. (2023) in Further Reading). In a logical qubit (not a physical single object but rather a type of state), the state information is spread over a number of redundant physical qubits (objects) with the goal of reducing the error rates. For example, if any one of the physical qubits fails, the design of the logical qubit ensures that the state information will not be affected. To make this work, it turns out that you must act on many physical qubits simultaneously. But it is a strategy that shows signs of being implementable on existing physical qubits. Of course, much work is needed to scale up the logical qubits so that a quantum computer becomes entirely fault-tolerant.

An analogy might be helpful. Think about a politician who is campaigning and trying to communicate a message to potential voters. If the politician is distracted by hecklers or is tired due to a busy travel schedule, these parts of an environment can affect the politician's articulation of the desired message. This environment can cause the voters to take away the wrong message. Sometimes the politician goes "off message" and creates errors in the message. Such a message system is not fault-tolerant. However, the politician is usually surrounded by staff and media experts whose job is to detect these errors and correct them. The staff and media experts are like the redundant physical qubits, who, in conjunction with the message qubit (the politician), provide redundancy and error checking and error correction. They use measurement devices such as polls and focus groups and the reports from news media to detect errors. They then need to have ways of correcting errors while not disturbing the politician directly.

13.4 Chapter Summary

- Correcting errors in transmitted information is another important computational task. Commonly, these errors are found by attaching additional bits to the message and then comparing all the bits. For example, instead of sending a single digit 1, three digits, 111, are sent. The three digits are then compared to check if they are different from one another. If they are, a bit-flip error has occurred.

- Qubit error detection presents a more complex problem. Merely comparing the bits does not work because measurement forces the qubit into the basis state of the measurement device, causing a loss of information about the original state of the qubit. This difficulty is overcome by using ancillary register states as the control states for a Toffoli gate, which acts on the message register states.
- The interactions that cause physical qubit errors can be modeled with the standard Pauli operators (X, Y, Z, and I). Likewise, these operators can be used in methods that correct the errors.
- Logical qubits are collections of physical qubits connected in such a way as to detect errors and correct them with little or no outside intervention.

13.5 Further Reading

The Wikipedia article on Quantum Error Correction provides a more detailed overview of error correction methods (https://en.wikipedia.org/wiki/Quantum_error_correction).

Error Correction Zoo (https://errorcorrectionzoo.org/) has detailed information on both classical and quantum error-correcting codes. This is a good place to look up jargon in the error correction business.

A physical implementation of logical (error-corrected) qubits is described in D. Bluvstein, et al. *Nature*, **626**, 58–65 (2023).

14 RSA Encryption and Number Theory

> Mathematics is the queen of the sciences and number theory is the queen of mathematics.
>
> **Attributed to Carl Friedrich Gauss**

14.1 RSA Encryption and Factoring

This chapter introduces the RSA encryption algorithm, a classical cryptographic procedure, variations of which are widely used in securing encrypted information. To understand the RSA encryption algorithm, we need a modest amount of number theory. Breaking the RSA encryption involves factoring large numbers. We show in Chapter 15 that the same number theory helps explain the famous Shor quantum algorithm, which involves factoring large numbers. That factoring, if implemented on a large scale, could break many currently used encryption codes.

We spend some time explaining the number theory results because many algorithms, both classical and quantum, make use of them. In fact, number theory shows how we reduce the factoring problem to the problem of finding the period of a periodic function. The Shor algorithm provides a slick way of using superposition states to find the period.

Why should we worry about factoring numbers? Isn't that something you do in middle school math and the SAT exam and never use again? It turns out that factoring is not just an intellectual or mathematical curiosity. The difficulty of finding the factors of large numbers lies behind some widely used classical encryption procedures, designed to make it difficult for anyone to decode digital messages. This is of obvious interest to financial, governmental, and medical institutions. If an efficient factoring scheme could be developed, these encryption procedures could be broken into.

The basic idea behind encryption is that you use an encryption "key" to code letters and numbers just the way the dots and dashes in the Morse code stand for letters of the alphabet or, at a more sophisticated level, what the Germans did during World War II with the famous Enigma machine, which changed the encryption key daily. If you can find the encryption key, you can decode the encrypted messages. Of course, for digital computers, the key will be a sequence of 0s and 1s. For example, you might have a key that contains 8 bits $\times$ 26 letters = 208 bits in which the successive 8-bit sequences each represent a letter in the English alphabet like the ASCII code that your keyboard uses to represent whatever

Quantum Computing and Quantum Physics. Alice Flarend and Robert Hilborn, Oxford University Press.
 DOI: 10.1093/oso/9780197904930.003.0014

symbol you push on the keyboard. We showed an example of an encryption key in the treatment of the BB84 encryption method discussed in Chapter 8.

The RSA encryption system, named after Riverst, Shamir, and Adleman, who developed the algorithm in 1977, takes a different tact. The RSA algorithm, an example of *public key distribution*, makes the encryption numbers public but keeps other information needed to decode the message secret. Unlike the Morse code, RSA is an asymmetric encryption algorithm because you use one number to encrypt a message and a different number to decrypt (decode) the encrypted message. One of the encryption numbers is a large integer that is a product of two large prime numbers. To decode the encrypted message, you need to know these prime numbers. In other words, you need to factor that large number. It turns out that finding these prime number factors is a difficult computational task and essentially impossible because of time constraints for a classical computer if the number has several hundred digits. The Shor algorithm (Chapter 15) uses quantum methods to do that factoring. So far, quantum computers have been able to factor only small numbers (for example, 15 and 21), but there is no fundamental barrier to scaling the process to large numbers. We just need quantum computers with more qubits.

Let's remind ourselves about the basics of prime numbers. A prime number is one that is evenly divisible (no remainder) only by 1 and by itself. The smallest prime numbers are 2, 3, 5, 7, 11, 13, and so on. Note that besides 2, all prime numbers are odd numbers. By the standard mathematical definition of prime numbers, 1 is not a prime number because it has only one factor and that factor is 1.

The Fundamental Theorem of Arithmetic states that all positive integers (also called natural numbers) except for the number 1 can be expressed as products of prime numbers, and furthermore, this product (except for the order of the prime numbers within the product) is unique. If the product contains only 1 and the number itself, then that number is a prime number. For example, the number 91 can be written as $7 \times 13 = 91$. Since 7 and 13 are both prime numbers, they are the only prime numbers whose product equals 91.

Before stating the RSA encryption algorithm, which itself is somewhat complicated, we explore a few aspects of number theory used in the algorithm. As mentioned before, number theory is the mathematics of the integers 0, 1, 2, 3, ... For most of our work, we need only positive integers. It sounds simple, but there are several results that are far from obvious. But, with some persistence and attention to detail, we can work our way through the steps.

14.2 Number Theory Warm-Up

An important number theory tool for RSA encryption and for many other classical algorithms is the mod function, which we like to call the remainder function. Most mathematicians call it the modulus function or the modulo operation. Let's explore it before looking at the RSA algorithm.

We start with some simple arithmetic. Remember we are dealing only with positive integers. For any two positive integers x and y (assume $x < y$), we can always find an integer Q that is the largest integer such that $Qx < y$. Then, we introduce the remainder $R = y - Qx$

when you divide y by Qx. In other words, we have $y = Qx + R$, a linear relationship between x and y. In the language of everyday arithmetic, Q is the quotient, x is the divisor, and y is the dividend—the number to be divided.

Try It 14.1 For $x = 5$ and $y = 22$, show that $Q = 4$ and $R = 2$. Check that $y = Qx + R$.

The mod (or modulus) function gives you the remainder part without needing to specify the quotient. This will be handy in dealing with large numbers. The mod function for x and y is written as

$$R = y \mod x, \tag{14.1}$$

where R is the remainder after dividing y by $Q\,x$. Mathematicians like to say "R equals y, modulo x."

Actually, we use the mod function every day when we talk about clock time: three hours after ten o'clock (13 o'clock) is 1 o'clock because we use "mod 12" for clock time. 1 is the remainder: $1 = 13 \mod 12$.

For a fixed value of x, the remainders go through a cycle from 0 (for numbers evenly divisible by x) to $x - 1$ as y increases. This cyclic nature plays a role in what follows.

Fortunately, most computer languages and spreadsheets have a mod function to do the calculation for you. Often, they take the form $R = \mathrm{MOD}[y, x]$, where y is the dividend, x is the divisor, and R the remainder. But if you are stuck on a desert island and need to calculate by hand, the arithmetic is straightforward: simply divide y by x. Then R is the integer remainder. Of course, for large numbers the division can be rather tedious.

Try It 14.2 For $x = 11$, calculate $R = y \mod x$ for $y = 0$ through 24. Is the result cyclic?

We need one more variation on the mod operation. If you think about it for a minute, you will realize that for a given divisor x, there are infinitely many dividends (y) that give the same remainder, each with its own value of Q. We note that the RSA encryption algorithm makes use of this fact.

Try It 14.3 Show that $3 = 15 \bmod 12$ and that $3 = 27 \bmod 12$; so, 15 and 27 have the same remainder $\bmod 12$. Note that $15 + 12 = 27$.

If $y \mod x = u \mod x$, that is, y and u have the same remainder when divided by x, then the difference between y and u is evenly divisible by x. (We normally use $|u - y|$ for the difference to restrict ourselves to positive integers.)

Let's do a numerical example. 17 and 32 have the same remainder for the divisor 5. In fact, $2 = 17 \bmod 5$ because $17/(3 \times 5)$ has the remainder 2 and so does $32/(6 \times 5)$. So, we write $32 \bmod 5 = 17 \bmod 5$ (the remainders are equal), and we note that the difference between the dividends $32 - 17 = 15$ is evenly divisible by 5. Numbers that have the same remainder for a given mod x are said to be congruent.

Try It 14.4 Find the following remainders: $23 \bmod 2$, $121 \bmod 2$, $15 \bmod 3$, $16 \bmod 3$, $19 \bmod 5$, and $24 \bmod 5$. Are any of the remainders the same for the given divisors? If so, check if the difference between the dividends is evenly divisible by the divisor.

Before we move on, it might be helpful to look at a graph that shows some of the features we have been talking about. Figure 14.1 displays a plot of the remainder R as a function of the dividend y with the divisor fixed at $x = 7$. Note that the plot is periodic in y. The values of y for which R is the same are said to be congruent (mod 7).

Try It 14.5 The difference between successive numbers that are congruent (mod 7) is called the "repetition period" (or just "period") of that congruence. What is the period of congruence for the results shown in Fig. 14.1?

What does periodic mean here? When we think of periodic, we usually imagine something that repeats itself in time: periodic appointments and the like. In number theory, periodic is similar, but we don't have to interpret the numbers as times or anything specific. In more formal terms, a function $f(x)$ is periodic with period r if $f(x) = f(x + r)$ for all x.

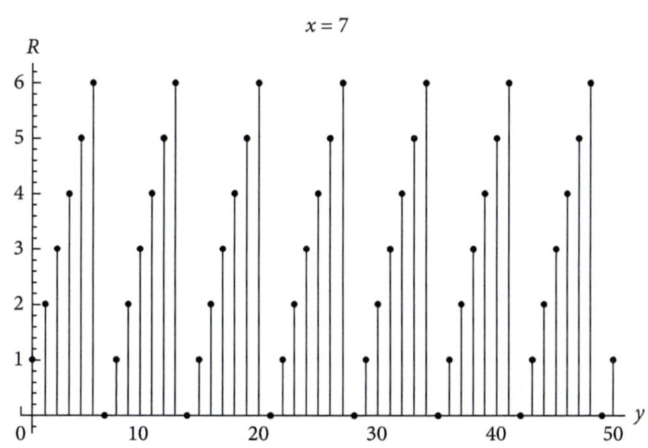

Fig. 14.1 A plot of $R = y \bmod x$ as a function of the dividend y, with $x = 7$.

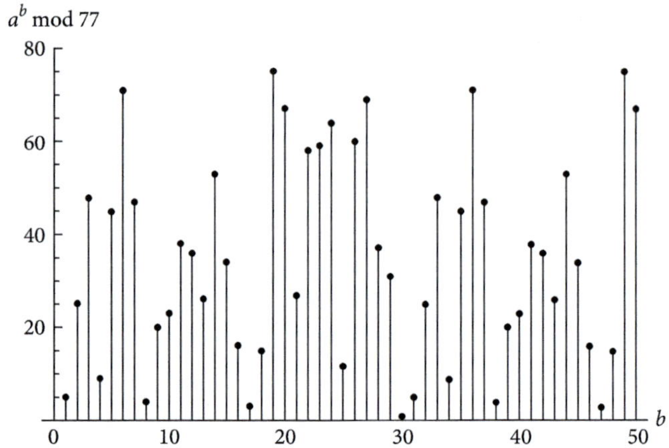

Fig. 14.2 A plot of $a^b \bmod 77$ with $a = 5$. The results are periodic as a function of the exponent b. See the main text for a hint about the periodicity.

Now let's look at functions with exponents. In particular, let's use $y = a^b$ and consider $a^b \bmod N$. We find that there are integers r such that

$$a^b \bmod N = a^{b+r} \bmod N. \tag{14.2}$$

The value of r that works in Eq. (14.2) depends on a and N. Of course, if r works, so does $2r$, $3r$, and so on. We say that $a^b \bmod N$ is a periodic function of the exponent b with period r. Eq. (14.2) is an example of "modular exponentiation," which plays a major role in RSA encryption and the Shor algorithm. In Section 14.4, we show how to solve this kind of equation.

To match the notation commonly used in quantum information science and quantum computing, we use lower case r to indicate the repetition period of a periodic function and keep uppercase R for remainders. So, here r means "repetition period."

Figure 14.2 displays a plot of $a^b \bmod 77$ as a function of b with $a = 5$, as a function of the exponent b. Note that the results are periodic with respect to b though this periodicity might not be obvious. To convince yourself that there is some periodicity there, look at the sequence of filled circles with b between 8 and 20 and then between 38 and 50.

> **Try It 14.6** Describe the results displayed in Fig. 14.2. Are the results periodic with respect to b? If so, what is the period r?

14.3 The RSA Algorithm

With that bit of number theory under our belts, we are ready to tackle the RSA algorithm. Note again that this is a purely classical algorithm. It makes no use of quantum ideas.

Here is how it goes: Bob wants to send Alice an encrypted message. The RSA algorithm has the following steps:

1. Alice creates a large number N by multiplying together two fairly large prime numbers p and q, so $N = pq$. They need to be large so the factors of N are not easily found. Alice keeps p and q a total secret.
2. Alice also creates an encryption key K with $K < N$ and checks that K has no common factors (other than 1) with $(p - 1)(q - 1)$. Alice sends the integers K and N to Bob, but she does not send p and q. Alice may use an unencrypted channel to send K and N. It does not matter if an eavesdropper gets them.
3. Let's assume that Bob expresses his message as a number m. Using the values of K and N Alice has sent him, Bob calculates another number c such that $c = m^K \bmod N$ and sends the number c to Alice. This is his encrypted message.
4. In the meantime, Alice has calculated another integer d such that

$$1 = Kd \bmod (p - 1)(q - 1), \tag{14.3}$$

which means that $(p - 1)(q - 1)$ divides $Kd - 1$ evenly, that is, with no remainder. Alice needs the two prime numbers p and q to calculate d. This is known only to Alice.
5. To decode the message, Alice solves $m = c^d \bmod N$.

Let's examine some of the details. We use just two prime factors because two are sufficient for our purposes. Remember the Fundamental Theorem of Arithmetic: Once we have chosen p and q, we know that they are the only two prime factors of N.

Why is the product $(p - 1)(q - 1)$ used in Steps 2 and 4? The answer is a bit complicated. We show you the details in Section 14.5. For now, let's read Eq. (14.3) to make sure we understand what it is indicating. It says that if we divide Kd by $(p - 1)(q - 1)$, we get the remainder 1, which, as shown before, is equivalent to $Kd = Q \times (p - 1)(q - 1)+1$ for some integer Q. In other words, $Kd - 1$ is evenly divisible by the product $(p - 1)(q - 1)$.

How do you check that there are no common factors when you choose K? Finding common factors is straightforwardly done with what is called Euclid's method or the Euclidean algorithm. Alternatively, many computer programming languages and spreadsheets have functions that allow you to find the "greatest common divisor (gcd)," the largest integer that divides both factors evenly. If that gcd is 1, then the two numbers have no common factors. We explain the Euclidean algorithm in Section 14.5.

Let's get back to the RSA algorithm. In Step 3, Bob sends Alice the number c. Since Alice is well versed in the RSA method, Alice knows that she can find the message string m from c by using

$$m = c^d \bmod N. \tag{14.4}$$

That is, the message (viewed as a binary number) is the remainder when c^d is divided by N. So, we are left with another modular exponentiation problem.

There is a lot of number theory behind these calculations, and it takes some patience and concentration to work through the arguments. The main point is that unless someone knows the secret number d in Step 4, which Alice calculated with secret prime factors p and q, there is no way to decipher Bob's message m from the number c he sends to Alice.

Let's try this with a numerical example. Suppose Alice chooses two prime numbers $p = 59$ and $q = 47$, for which $N = pq = 2773$. Next, Alice picks a number $K < N$ randomly and checks to see if K divides $(p - 1)(q - 1) = 2668$ evenly or not. If not, Alice is ready to go to the next step. If it does divide that product evenly, Alice picks another K and tries again. Note that if K divides the product evenly, then $K + 1$ will not. Let's suppose Alice picks $K = 17$, which does not divide 2668 evenly; so, Alice is ready to go on. Finally, Alice needs to compute an integer d such that $1 = Kd \bmod 2668$, from which Alice finds $d = 157$. How did Alice solve for d? There are several ways to do this. The easiest way is to use the Extended Euclid's Algorithm (Section 14.5). Also, many classical computer languages implement the process in a *modular inverse* (or similarly named) function.

Let's continue with the example. Next, Bob carries out the encryption calculation for the message m. Let's suppose that the message (in decimal digits) is 73. So, Bob calculates the integer c given by

$$c = m^K \bmod N = 73^{17} \bmod 2773 = 928. \tag{14.5}$$

Then, Bob sends $c = 928$ to Alice over an unencrypted channel. Note that Bob doesn't know (and doesn't need to know) the numbers p, q, and d, which Alice has and keeps secret.

Once Alice receives $c = 928$, she uses $d = 157$ to solve for m from $m = c^d \bmod N$ and finds $m = 928^{157} \bmod 2773 = 73$. Alice has decoded the message. Note that if an eavesdropper doesn't know d (which Alice has kept secret) or the factors p and q, there is no way to find m.

The process seems to work, but if you try to evaluate some of the exponentiations, your spreadsheet and your hand calculator will both give up because the results get too large. That's where modular exponentiation comes in. We discuss the details of how to handle modular exponentiation in Section 14.4 and modular inverses in Section 14.5.

14.4 Modular Exponentiation

We promised you that we would go through some of the mathematical details of the RSA algorithm. If you are not interested in the details, you may skip ahead to the next section.

Modular exponentiation means that we want to find c, given general integers a, b, and N, where

$$c = a^b \bmod N. \tag{14.6}$$

In other words, c is the remainder of dividing a^b by N. We could use a brute force method to find c: calculate a^b and then divide by N to find the quotient and the remainder c. The difficulty is that a, b, and N might be large numbers, which need to be stored as exact binary sequences, and a^b is extremely large and must be stored exactly. Also, carrying out the division by N is time-consuming. However, once we have chosen a and N, we can find c with a simple algorithm that avoids the large number issues:

1. Set $c = 1$ and $j = 0$.
2. Increase j by 1.
3. Find a new c by $c = a \cdot c \mod N$.
4. If $j < b$, go to Step 2. Otherwise, c is the solution to $c = a^b \mod N$.

Note that j serves as an index to keep track of the loops through the algorithm. A flowchart for the algorithm is shown in Fig. 14.3.

Try It 14.7 Use the algorithm to find $4^6 \mod 17$. This is easy to evaluate directly by hand, but it is a good exercise to see how the algorithm operates.

You will notice that as we go through the loop in the flowchart shown in Fig. 14.3, we build up c factors of a in Step 3, but the mod operation keeps the numbers relatively small

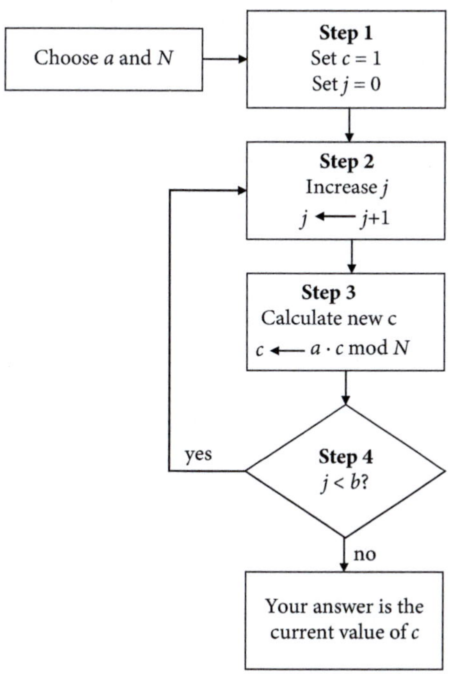

Fig. 14.3 Flowchart for modular exponentiation.

(less than N) at each step, although there may be many steps. In the straightforward version of the algorithm, there are b steps.

There is another "trick" that may drastically reduce the number of operations needed to perform modular exponentiation, while keeping the small number method described above. It is a combination of the previous method and a more general principle called *exponentiation by squaring*[1] (also known as *binary exponentiation*).

First, we convert the exponent b to binary form, that is, b can be written as a string of 1s and 0s:

$$b = \sum_{j=0}^{n-1} k_j 2^j, \tag{14.7}$$

where n is the number of bits required to represent b, and the coefficients k_j are either 0 or 1 depending on whether we need that particular exponent j.

An example might be helpful: suppose $b = 17$ (decimal). In binary form $b = 10001$ and $n = 5$. Hence, we may write

$$b = 10001 = 1 \times 2^4 + 0 \times 2^3 + 0 \times 2^2 + 0 \times 2^1 + 1 \times 2^0,$$

from which we read off

$$k_0 = 1,\ k_1 = 0,\ k_2 = 0,\ k_3 = 0,\ k_4 = 1.$$

Try It 14.8 Do the same analysis for $b = 15$.

Let's continue with our binary exponentiation. Using Eq. (14.7) and employing the sum rule for exponents $x^y x^z = x^{y+z}$, we can write a^b as

$$a^b = a^{\sum k_j 2^j} = a^{k_0 2^0} a^{k_1 2^1} \ldots a^{k_{n-1} 2^{n-1}}. \tag{14.8}$$

Why does this help? As we go through the algorithm loop, any time we have a $k_j = 0$, we can skip that multiplication step because $a^0 = 1$. This reduces the number of steps considerably in most cases.

Let's do an example of binary exponentiation. We choose the exponent $b = 11$ (decimal) = 1011 (binary). The algorithm uses a process similar to the previous modular exponentiation except that we have

$$a^b = a^{11_{10}} = a^{1011_2} = a^{1 \cdot 2^3} a^{0 \cdot 2^2} a^{1 \cdot 2^1} a^{1 \cdot 2^0}.$$

Thus, we need to do only four multiplications (instead of 11_{10}) because whenever there is a 0 in the exponent, we may skip that multiplication since it is equivalent to multiplying by 1.

[1] https://en.wikipedia.org/wiki/Exponentiation_by_squaring

This process is called binary exponentiation because we use powers of 2. There are efficient implementations of this method in common programming languages such as Python and Mathematica.

To wrap up, we point out that all the RSA encryption steps can be carried out efficiently using classical computers. Quantum computation comes in when we try to beat the encryption by finding the prime factors p and q to decipher the message as shown in Chapter 15.

14.5 Greatest Common Divisor and the Modular Inverse

In this section, we explain how to find the gcd of two integers a and b, a concept used in Section 14.3, and how to solve $1 = Kd \bmod M$ for d using a concept called *modular inverse*.

Greatest Common Divisor

Of course, you could always use a "brute force" method to find the gcd of two integers. First, find all the divisors of the first integer—looking for integers that divide a with no remainder. Then, do the same for b. Then, look at those divisors to find the divisors that are common to both and select from that subset the one that is the greatest. This works fine for relatively small numbers. But it takes a long time for larger numbers.

A more efficient way makes use of what is called Euclid's method (or the Euclidean algorithm), which was written up more than 2000 years ago in Euclid's famous book *Elements*. Many authors claim that this algorithm is the oldest known algorithm, but there is no historical evidence that Euclid himself invented it. What is important is that he made it readily available to subsequent generations of mathematicians. As mentioned before, gcd functions are available in most programming languages and spreadsheets, but it is nice to know that you could find the gcd "by hand" if you had to or to gain more experience in number theory.

Here is how the algorithm works:

1. Let a be the larger of the two integers, a and b. Then, find the remainder r_1 from dividing a by b. If the remainder is 0, then b is the gcd of a and b. If it is not, go to Step 2.
2. Divide b by r_1 to find the remainder r_2. If r_2 is not equal to 0, proceed to Step 3.
3. Divide r_1 by r_2 to find a new remainder r_3.
4. Continue these steps until the remainder is 0. The next-to-last remainder (the one that is nonzero) is the largest common divisor of a and b.

Here is an example: What is the gcd of 350 and 245? The steps are

$$\begin{aligned} a/b = 350/245 &= 1, \ r_1 = 105 \\ 245/105 &= 2, \ r_2 = 35 \\ 105/35 &= 3, \ r_3 = 0. \end{aligned} \tag{14.9}$$

This example took only three divisions. The last nonzero remainder is 35; so, the gcd of 350 and 245 is 35.

Try It 14.9 Draw a Euclidean algorithm flowchart diagram like the one for modular exponentiation. Use it to work out the examples in the following Try It.

Try It 14.10 Use the Euclidean algorithm to find the gcd of 77 and 33 and the gcd of 77 and 35.

You can also use the Euclidean algorithm as a series of algebraic expression steps:

$$\begin{aligned} &r_0 = a \quad r_1 = b \\ &\text{For } i \geq 1 \quad r_{i+1} = r_{i-1} - q_i r_i \text{ with } 0 \leq r_{i+1} < |r_i|\,. \end{aligned} \tag{14.10}$$

r_i decreases steadily until one of them, call it r_{k+1}, equals 0. Then, $\gcd(a, b) = r_k$.

Try It 14.11 Use the algebraic method for the numbers in Try It 14.10 and verify that you get the same result you got with the flowchart method.

If you are wondering why the Euclidean method works, you are going to turn into a mathematician because you always want to delve into the details of "why." This is a good habit to cultivate even if you are not going to be a professional mathematician. It turns out that the proof of the algorithm is not difficult. You can easily find several proofs with video explanations on the Internet. The basic idea is that you can work your way "backward" through the Euclidean algorithm steps. For example, in Eq. (14.9), 35 obviously divides itself evenly and it divides r_1 evenly, and consequently it must divide 350 and 245 evenly because $350 = 245 + 105$ and $340 - 245$ is a multiple of the gcd.

Modular Inverse

Next, we need to explain how you solve $1 = Kd \bmod M$ for d, where we use $M = (p - 1)(q - 1)$ to simplify the expressions that follow. First, some terminology: We assume that M and the product Kd are coprime (also called relatively prime or mutually prime). It means that the gcd of the two numbers M and Kd is 1. Examples of coprime pairs are (4, 7) and (8, 15).

Furthermore, if Kd satisfies

$$1 = Kd \bmod M, \tag{14.11}$$

then K is called the *modular inverse* (or *modular multiplicative inverse*) of d, mod M. Conversely, d is the modular inverse of K, mod M. The terminology is based on the analogy that if $x^{-1}x = 1$ for any number x, we call x^{-1} the inverse of x. Modular inverse is a bit trickier because the modular inverse of K is not equal to $1/d$ since we restrict ourselves to integers. The bottom line is that solving $1 = Kd \bmod M$ for d is a modular inverse problem.

To find the modular inverse, it is easiest to use *modular inverse* functions built into many programming packages. To understand what is going on, we show you how to find the modular inverse by hand. To find the modular inverse, we make use of Bézout's equation, which states that the gcd of a pair of integers A and B may be written as a linear combination of those integers:

$$\gcd(A, B) = mA + nB, \tag{14.12}$$

where the coefficients m and n are integers, with one of them possibly negative. As an example, we may write the gcd of 12 and 42 as

$$6 = 11 \times 12 - 3 \times 42. \tag{14.13}$$

We now tackle the general case. If we take mod B of both sides of Eq. (14.12), we obtain

$$\gcd(A, B) \bmod B = mA \bmod B + nB \bmod B. \tag{14.14}$$

We are interested in the case when $A = K$, $B = M$, and $m = d$ to connect with the variables used in the RSA algorithm. Since K and M are coprime (by assumption), we obtain

$$1 \bmod M = Kd \bmod M + nM \bmod M. \tag{14.15}$$

Since the last term in Eq. (14.15) is zero (M divides nM evenly) and $1 \bmod M = 1$, we obtain

$$1 = Kd \bmod M. \tag{14.16}$$

We have related Bézout's equation to the RSA modular inverse problem expressed by Eq. (14.11).

We now need to find a method to find the coefficient $m = d$ in the Bézout's equation. The most straightforward method is an extension of the algebraic form of the Euclidean algorithm Eq. (14.10). This is called, not surprisingly, the Extended Euclidean Algorithm.

It works by adding two new sequences to the steps

$$
\begin{aligned}
&r_0 = A \quad r_1 = B \\
&m_0 = 1 \quad m_1 = 0 \\
&n_0 = 0 \quad n_1 = 1 \\
&\text{For } i \geq 1 \quad r_{i+1} = r_{i-1} - q_i r_i \text{ with } 0 \leq r_{i+1} < |r_i| \\
&m_{i+1} = m_{i-1} - q_i m_i \\
&n_{i+1} = n_{i-1} - q_i n_i
\end{aligned} \tag{14.17}
$$

The calculation proceeds just as in the Euclidean algorithm until $r_{k+1} = 0$. Then, r_k (the last nonzero remainder) is the gcd, and m_k and n_k are the two Bézout coefficients. If you are interested in knowing why the Extended Euclidean Algorithm works, see the Further Reading section.

Try It 14.12 Challenge: Use the Extended Euclidean Algorithm to show how Alice solved $1 = Kd \mod 2668$ with $K = 17$ to find $d = 157$ in Section 14.3.

Although we had to go deeply into some details of number theory, the details are important if you want to understand the logic behind the Shor factoring algorithm (Chapter 15) and many other quantum algorithms.

14.6 Chapter Summary

- Encryption employs a key K that is typically used with other numbers to encode and decode the message. The RSA protocol for encryption involves factoring N, a product of two large prime numbers p and q. It also uses the value of d which satisfies

  $$1 = Kd \bmod (p-1)(q-1).$$

  The prime factors, p and q, and the number d are kept secret. K and N may be shared publicly, and the sender uses them to encrypt the message. For large N, classical computers are unable to do the calculations fast enough to determine p, q, and d in a practical amount of time.
- Modular exponential, the greatest common denominator, and the modular multiplicative inverse from number theory play important roles in many quantum algorithms.

14.7 Further Reading

"Fundamental Theorem of Arithmetic," Wikipedia provides nice explanations of the theorem, why it is important, and how it is proved.

There are many excellent YouTube videos on modular arithmetic, including congruence, greatest common divisors, and modular exponentiation.

The Wikipedia article on the Extended Euclidean Algorithm provides a detailed discussion of the algorithm and a proof of why it works.

M. Suhail Zubairy, *Quantum Mechanics for Beginners* (Oxford University Press, Oxford, 2020). Chapters 13 and 16 provide good introductions to Quantum Secure Communications.

15 Shor Factoring Algorithm and the Quantum Fourier Transform

> Mathematics compares the most diverse phenomena and discovers the secret analogies that unite them.
>
> Joseph Fourier, *The Analytical Theory of Heat* (1822)

15.1 The Shor Factoring Algorithm

The Shor factoring algorithm is perhaps the most famous of current quantum computing algorithms. The problem Shor wanted to solve is how to factor large numbers. This is an interesting problem because factoring large numbers allows you to break many encryption codes as mentioned in Chapter 14. We want to alert you, our dedicated readers, that this chapter is rather challenging. It uses results from number theory and some advanced math called the Fourier transform. We have worked hard to explain all the steps, but there are many steps. We provide examples and graphs to help build intuition about the results, but we know from our own experience it takes time to get your head wrapped around these ideas.

Once we get to the quantum part of what Shor did, we will see familiar tools from other quantum algorithms: superposition states, quantum function gates, and measurement devices. Nevertheless, expect to spend more time on this chapter than you did on the others. We hope that the results of your efforts will be satisfying in providing more insight into how encryption works and how quantum algorithms are engineered to take advantage of the nature of quantum states. But given the length of the arguments, we suggest that you feel free to try out the algorithms without going through all the details of the justifications for the steps. Once you see what the algorithm does, you can come back later to look at the detailed explanations with a better understanding of the context.

We are now ready to look at the Shor factoring algorithm and how it allows us to break the RSA encryption described in Chapter 14. The key feature is finding the two prime factors of N. This would seem to be a relatively straightforward thing to do. We just go through the list of prime numbers less than N and find the two whose product equals N. Then, we are done; we could use that information to break the RSA encryption. But, if you think about how many numbers you need to check for large values of N, you should be able to convince yourself that the task is hopeless even with today's supercomputers to help out. The prime counting function tells us that the number of prime numbers less than or equal

Quantum Computing and Quantum Physics. Alice Flarend and Robert Hilborn, Oxford University Press.
 DOI: 10.1093/oso/9780197904930.003.0015

to N is about $\pi(N) \sim N/\ln N$, where ln is the natural (base e) logarithm. For example, if $N = 10^{300}$, there are about $10^{300}/\ln(10^{300}) \approx 10^{297}$ prime numbers to try. Let's say that your supercomputer can check one number in 10^{-10} seconds. It would then take about 10^{287} seconds to complete the factoring. Of course, you might get lucky and find a prime factor early on, but, on average, it's going to take a long time. In round numbers, the age of the universe is 10 billion years. There are about 10^7 seconds in one year; so, the age of the universe in seconds is about 10^{10} years times 10^7 seconds per year = 10^{17} seconds, way smaller than the 10^{287} seconds needed to complete the factoring of 10^{300}.

As we have shown, several times before, the strategy is to reduce one problem to another one that we hope is simpler and less time-consuming. The basic idea behind the Shor algorithm is to use some number theory ideas that we discussed in the RSA encryption procedures in Chapter 14 to reduce the factoring problem to one for finding the period of a periodic function. The Shor algorithm also provides a quantum method to find this period for large numbers much more quickly than can be done on a classical computer. Once you have the period, it takes just a few relatively simple calculations to find the prime factors.

Let's go through the steps in the Shor algorithm and then do a few examples to see how they work. We then tackle the quantum method that speeds up the period-finding part. In Sections 15.4 and 15.5, we provide details about why the steps work. Our presentation was inspired by the one in Zubairy's *Quantum Mechanics for Beginners* (see Further Reading).

Just as a reminder, unless otherwise specified, all the numbers we deal with are positive integers. Here are the steps in the Shor factoring algorithm to find two prime factors of some number N. Note that they are close to the reverse of the RSA algorithm's steps.

1. Choose a random integer $a < N$. Check to see if a divides N evenly (no remainder). If it does, we have found a factor. That is unlikely; so, we proceed with the following steps.
2. Find the smallest integer r such that $1 = a^r \mod N$. In other words, we are looking for the smallest number r such that $a^r - 1$ is evenly divisible by N.
3. If r turns out to be an even number, proceed with Step 4. Otherwise, return to Step 1 and choose another random number a.
4. Calculate $s = a^{r/2} \mod N$. Note that we need r to be an even number to make this work.
5. Calculate the prime factors p and q from

$$\begin{aligned} p &= \gcd(s - 1, N) \\ q &= \gcd(s + 1, N), \end{aligned} \tag{15.1}$$

   where $\gcd(s - 1, N)$ means greatest common divisor—the largest number that divides both $s - 1$ and N exactly.
6. Check that the results satisfy $N = pq$.

We see that modular exponentiation plays a big role in the algorithm.

Try It 15.1 Construct a flowchart for the Shor factoring algorithm. Then, use the flowchart to work through the examples that follow.

Let's try out the algorithm with two examples. Here is one you could do in your head: $N = 15$. The two prime factors are 3 and 5; so, it will be easy to check that the algorithm gives the right answer. But let's follow the steps anyway.

1. Randomly choose $a = 2$. Obviously, a does not divide 15 evenly.
2. Find the smallest r such that $1 = 2^r \bmod 15$. We recognize that the smallest r that works is $r = 4$ because $2^4 = 16$ and $16/15 = 1$, remainder 1. We also check that $a^r - 1$ is evenly divisible by 15, which obviously works in this case.
3. r is even; so, we proceed to Step 4.
4. Calculate $s = a^{r/2} \bmod N = 2^2 \bmod 15 = 4$.
5. Find $p = \gcd(s - 1, 15) = \gcd(3, 15) = 3$ and $q = \gcd(s + 1, 15) = \gcd(5, 15) = 5$.
6. Check that p and q are factors of 15 by calculating $N = pq$.

So, we have successfully found the two prime factors of 15, namely, 3 and 5.

Try It 15.2 Use $N = 15$ and $a = 4$. Do the four steps work? If not, what is the problem? How can you get around that problem?

Now, let's try a more difficult example: $N = 77$. You could probably do this one in your head, but we want to try the Shor algorithm. You need a calculator or spreadsheet to calculate some of the modular exponentiation results.

1. Randomly choose $a = 4$, which does not divide 77 evenly. So, it is not a factor.
2. We find that $1 = 4^{15} \bmod 77$; so, the repetition period $r = 15$.
3. But r is not even; so, we need to try a different value of a. With $a = 5$, we find $1 = 5^{30} \bmod 77$. Since the repetition period $r = 30$ is even, we are ready to go on.
4. We then find $s = a^{r/2} \bmod 77 = 34$.
5. Find $p = \gcd(s - 1, 77) = \gcd(33, 77) = 11$ and $q = \gcd(s + 1, 77) = \gcd(35, 77) = 7$.
6. Since $7 \times 11 = 77$, the prime factors of 77 are 11 and 7, just the ones you might have obtained by trial and error. Note that we find the period of 30 in the data shown in Fig. 14.2.

You can see in Step 4 why we need r to be an even integer. We want to have $r/2$ be an integer and this is guaranteed if r is even.

15.2 Connecting Periodic Functions and Factoring

We now look at the general connection between finding the period of a function and factoring a number. At least for the authors, if someone had told us there is such a connection, we would have been highly skeptical. The concepts seem like they have no relationship. But number theory has many surprises of that sort. If you are willing to accept that connection, you may skip ahead to Section 15.3.

We now elaborate on and explain the Shor algorithm steps. To see how the connection comes about, we again assume $N = pq$ and follow the algorithm's steps to find p and q. The first step seems totally unmotivated: We choose an integer $a < N$ and check to see that a is not a factor of N, that is, N is not evenly divisible by a. If it is divisible by a, we have found a factor, and we are essentially finished. So, we assume that a is not a factor of N. Note that it is not necessary for a to be a prime number. It just can't divide N evenly. Mathematicians like to say that a and N must be *coprime*.

Once we have chosen a, the next step in connecting the periodic function to factoring is to find the smallest even integer r for which

$$a^b \bmod N = a^{b+r} \bmod N. \tag{15.2}$$

This is where the periodicity comes in. Note that in Eq. (15.2) you could replace r with $2r, 3r, \ldots$ They would all work. This is why we insist on finding the *smallest* integer r that works. We also require r to be positive; so, we can eliminate $r = 0$. (For $r = 0$, Eq. (15.2) is valid, but not very informative.) The function $f(b) = a^b \bmod N = f(b + r) = f(b + 2r)\ldots$ is periodic with repetition period r. Of course, the integer r depends on both a and N but not on b. We noted this periodicity in Fig. 14.2. Another example of the periodicity of a modular exponential function is shown in Fig. 15.1.

Try It 15.3 Find the period of the results displayed in Fig. 15.1. Hint: Look at the plot where b is in the range near [4,8] and search for the range of b where that pattern appears again.

Figure 15.1 shows why finding the period of such a function is not easy. If you try to find the period by picking a result for a particular value of b and searching for the next appearance of that result, you have many numbers to sort through. And this is for a relatively small value of N, 121 in this case. Imagine what the graph would look like for an N with 300 digits.

Now let's get back to the algorithm. If we choose $b = 0$ in Eq. (15.2), we obtain

$$1 = a^r \bmod N, \tag{15.3}$$

which means that there is some integer k such that

$$a^r - 1 = kN. \tag{15.4}$$

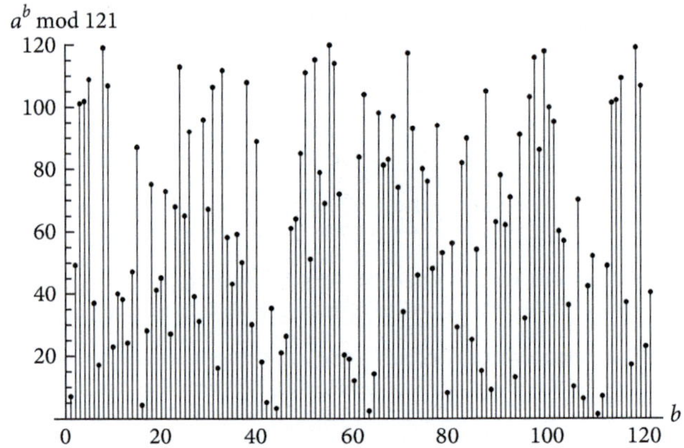

Fig. 15.1 A plot of a^b mod 121 with $a = 7$ plotted as a function of the exponent b.

Since we are requiring that r be even, we may write Eq. (15.4) as

$$0 = (a^r - 1) \mod kN = \left(a^{r/2} - 1\right)\left(a^{r/2} + 1\right) \mod kN. \tag{15.5}$$

Because Eq. (15.5) equals zero, $\left(a^{r/2} - 1\right)$ and $\left(a^{r/2} + 1\right)$ are factors of kN. Are they factors of N also? The first part is right. But the second part is not correct unless $k = 1$. Here is a simple example with $N = 5$. We know that 5 and 7 are factors of 35. Now let $kN = 35$ with $k = 7$ and $N = 5$. We see that 7 is not a factor of $N = 5$ even though it is a factor of $35 = kN = 7 \times 5$.

We now invoke some number theory arguments. Since r is the smallest even integer for which $a^r - 1$ is evenly divided by N, we conclude that the integers $\left(a^{r/2} - 1\right)$ and $\left(a^{r/2} + 1\right)$ are *not* evenly divided by N. It means that N must have a factor (excluding 1) in common with $\left(a^{r/2} - 1\right)$ and $\left(a^{r/2} + 1\right)$. Here is why: The second part of Eq. (15.5) shows that

$$(a^{r/2} - 1)(a^{r/2} + 1) = kN = kpq. \tag{15.6}$$

We see that we may write $(a^{r/2} - 1) = p \times k_p$ and $(a^{r/2} + 1) = q \times k_q$, where k_p and k_q are two integers whose product equals the integer k. We see that N and $\left(a^{r/2} - 1\right)$ have a common factor p and N and $\left(a^{r/2} + 1\right)$ have a common factor q.

We can find these common factors by finding the greatest common divisor of $\left(a^{r/2} - 1\right)$ and N and then the greatest common divisor of $\left(a^{r/2} + 1\right)$ and N. Recall that we discussed greatest common divisors in Chapter 14 and how you could use the Euclidean algorithm to find them. Also, many spreadsheets and computer programming packages include greatest common divisor as a function, often denoted by gcd[x, y] or GCD[x,y]. The prime factors of N are these greatest common divisors.

You might wonder why we used $s = a^{r/2} \mod N$ in Step 4 of the Shor algorithm. For the Shor algorithm, the number $a^{r/2}$ is often very large. Why does $s = a^{r/2} \mod N$ solve this problem? First, note that the prime factors p and q and gcd($a^{r/2} - 1, N$) will always be less than N. This suggests using a remainder (also less than N) to replace $a^{r/2}$; s is the appropriate remainder. Here is how this works.

Table 15.1 The exponent b and the remainder $R = 8^b \bmod 91$.

b	0	1	2	3	4	5	6	7	8	9	10
$R = 8^b \bmod 91$	1	8	65	57	1	8	65	57	1	8	64

From $s = a^{r/2} \bmod N$, we know that $a^{r/2} = c_s N + s$, where c_s is an integer (the quotient). It means we may write $a^{r/2} - 1 = c_s N + s - 1$ and $a^{r/2} + 1 = c_s N + s + 1$. Using these results, we write

$$\begin{aligned}\left(a^{r/2} - 1\right)\left(a^{r/2} + 1\right) &= (c_s N + (s-1))(c_s N + (s+1)) \\ &= c_s^2 N^2 + c_s N(s+1) + c_s N(s-1) + (s-1)(s+1).\end{aligned} \tag{15.7}$$

We know from Eq. (15.5) that the left side of Eq. (15.7) is divisible by N with no remainder. We see that the first three terms on the right of Eq. (15.7) are also divisible by N with no remainder. Following the argument in the sentence after Eq. (15.6), we conclude that $s - 1$ and N have a common factor p and that $s + 1$ and N have a common factor q. We find these common factors using the greatest common divisor function

$$p = \gcd(s-1, N) \text{ and } q = \gcd(s+1, N), \tag{15.8}$$

We can check to see if indeed the greatest common divisors are factors of N simply by multiplying them, and we can also check that they are prime numbers by using the greatest common divisor algorithm.

Let's work through a period-finding and factorization example. Suppose we want to find the prime factors of 91. First, we use $R = a^b \bmod N$ with $a = 8$ and to find the period r. Table 15.1 presents the numerical results with b in the range [0,10].

We see that the repetition period for $8^b \bmod 91$ is $r = 4$. Next, we calculate $\left(a^{4/2} - 1\right) = 63$ and $\left(a^{4/2} + 1\right) = 65$. (Note that a^r is a relatively small number, and Step 4 in the Shor algorithm gives $s - 1 = 63$ and $s + 1 = 65$.) For the final step, we calculate $\gcd[63, 91] = 7$ and $\gcd[65, 91] = 13$. The factoring result is that $7 \times 13 = 91$.

15.3 Period Finding and Fourier Analysis

Where do quantum algorithms come in? It is important to note that all the steps, except for finding the period r, can be carried out effectively with classical algorithms. So, how do we figure out the period using a quantum algorithm? For relatively small numbers N, we can use the equivalent of Fig. 15.1, but it should be obvious that the results soon become complicated enough that it is hard to see the period by eye. In Fig.15.1, the periodicity is hard to spot because the period is long compared to the range of b plotted.

A standard way to find periodicities in science, engineering, and technology is a mathematical tool called Fourier analysis. The basic idea is to compare the signal you want to analyze to sums of standard periodic functions: usually sines and cosines. You adjust the repetition period of the sines and cosines and find those periodicities that give you the best matches with your data. There are many efficient implementations of Fourier analysis in almost every modern scientific computing package. What Shor realized is that there are quantum algorithms that in principle will do the Fourier analysis in a much shorter time compared to the time required by a classical computer, particularly when it comes to large numbers with numerically large periodicities. Jean-Baptiste Joseph Fourier, who worked for Napoleon, did lots of wonderful mathematics, wrote books about Egypt, and recognized what we now call the greenhouse effect—how certain gases in the atmosphere can lead to climate warming. Another surprising tidbit: The mathematics behind Fourier analysis was mainly developed by others: Clairaut, Lagrange, Euler, Gauss, and others. What Fourier recognized is that the methods would work for any function, an assertion now known as Fourier's theorem.

If your only previous experience with sines and cosines was in trigonometry, you may wonder what they have to do with periodic functions. In trigonometry, we usually are worried about things like $\sin\theta$ and $\cos\theta$, where θ lies between 0° and 360° or, equivalently, between 0 and 2π radians. But it is perfectly fine to think of θ as a mathematical variable that can take on any numerical value. If you do that, you find that sine and cosine repeat whenever the variable goes through an entire circle from θ to $\theta + 2\pi$ if we use radian units. This periodic property is useful in both finding the period of other functions and then representing these functions as sums of sines and cosines. These are the basic ideas behind Fourier analysis, which is often used in classical signal processing and quantum algorithms.

Try It 15.4 Use a calculator or spreadsheet to show that

$$\cos(45°) = \cos(45° + 360°) = \cos(45° + 720°)$$

if your device uses degrees, or if your device uses radians

$$\cos(0.2) = \cos(0.2 + 2\pi) = \cos(0.2 + 4\pi).$$

Let's first look at traditional Fourier analysis (sometimes called the Fourier transform method) and how it allows us to find the period of a "signal." If you are familiar with Fourier analysis, you may skip ahead to the next section where we develop the quantum version.

In Fig. 15.2, we plot a pure cosine function and another periodic function with a period different from the pure cosine.

You can see that because the cosine's period does not match the period of the other function, their plots do not align very well. A way to show this mathematically is to multiply the two curves together and average the results. In this case, you would get a relatively small number because the positive parts of the product would tend to be canceled by the negative parts of the product. However, if you adjusted the period of the pure cosine, you could find a value for which the two curves match as shown in Fig. 15.3. Then, the average product would be reasonably large.

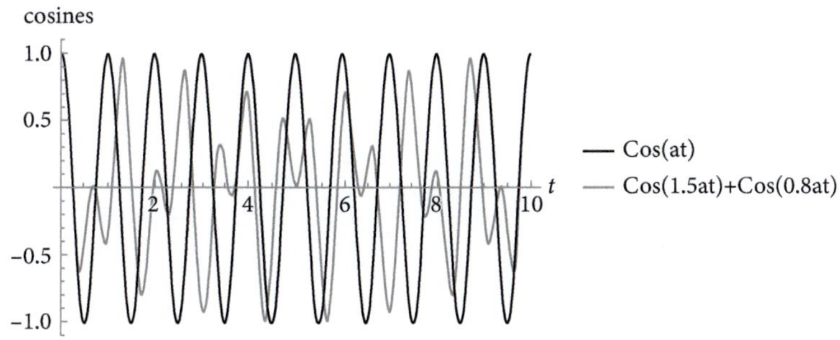

Fig. 15.2 A plot of a pure cosine function (dark curve) and another periodic function (gray curve).

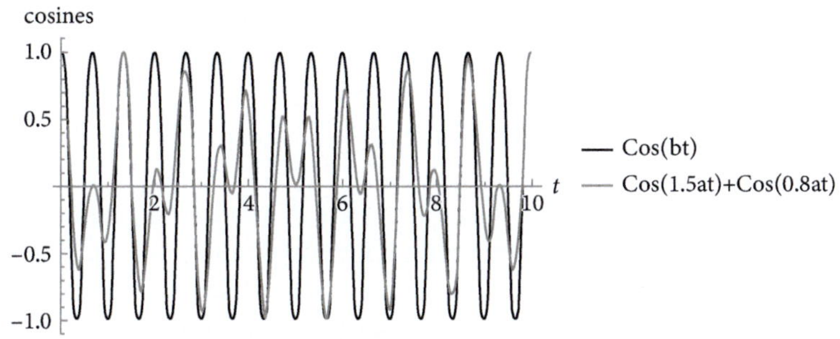

Fig. 15.3 The frequency of the pure cosine (dark curve) has been adjusted so that there is a good match between the pure cosine and the other periodic function.

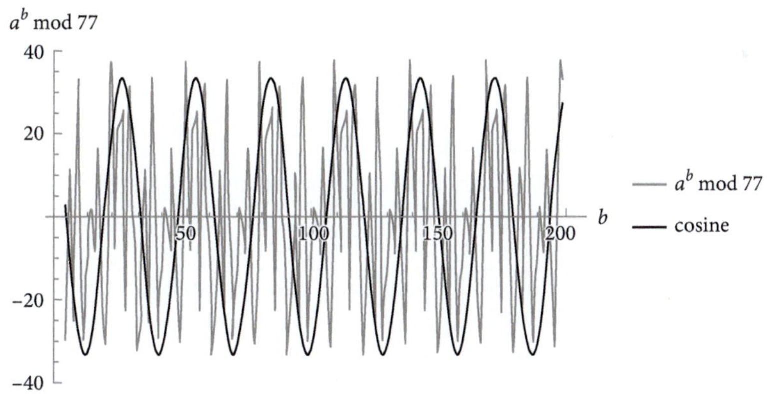

Fig. 15.4 A plot of a^b mod 77 with $a = 5$ and with b ranging from 1 to 200 along with a cosine function with a period of 30 (in units of b). The two periodicities match reasonably well.

Now let's apply this same idea to modular exponentiation. In Fig. 15.4, we plot a^b mod 77 with b running from 1 to 200. We subtracted the average value from each of the data points, so the function "oscillates" centered around 0 on the vertical axis, just the way cosines and sines do. We have also plotted a cosine function whose period is 30 (to match the period

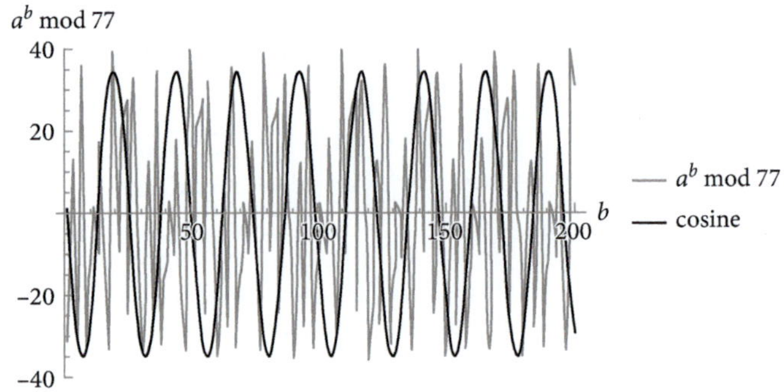

Fig. 15.5 A plot of a^b mod 77 with $a = 5$ and with b ranging from 1 to 200 and a cosine function with a period of 25 (in units of b). The two periodicities do not match very well.

for this case). We have moved the cosine function to the left a bit to match the "peaks and valleys" of the a^b mod 77 function.

If we calculate the product of the cosine times the a^b mod 77 "signal" and add up the results over the entire signal, we find that the sum of the products is large if the cosine is a good match to the signal and a smaller number if it is not. When the sum of the products is large, we know that the signal has a periodicity nearly the same as that of the cosine and that the peaks and valleys of the cosine are aligned with those of the signal. Why does this lead to a large number? If there is a good match, when the cosine is positive, we will multiply it by a positive signal (on average), and when the cosine is negative, we will multiply it by a negative signal (on average) giving, again, a positive result; so, overall the *product* contributions are positive. If the cosine period is not a good match for the signal, there will be more or less equal amounts of positive products and negative products giving an overall contribution near zero. Figure 15.5 shows a plot of this situation.

For the situations shown in Figs. 15.4 and 15.5, the sum of the products is about 840 for the first case with a period of 30 (in units of b). In the second case, the period of the cosine function is 25 and the sum of the products is about 3. So, we conclude that the signal has a period of about 30.

The full Fourier analysis method automatically runs through a range of periodicities to find the ones that best match the signals. Most often, both cosines and sines are used together. That method avoids having to do any left–right shifting to find the best match.

There are more formal methods for finding the cosine and sine matches that end up giving us a way to express a periodic function as a linear combination of the cosines and sines. We won't be using these results, but they are a nice illustration of an infinite-dimensional vector space in linear algebra. The role of the basis states is played by the sine and cosine functions. T is the period of the periodic function. The linear combination is given by

$$\begin{aligned} f(t) = f(t+T) &= a_0 + a_1 \cos(2\pi t/T) + a_2 \cos(4\pi t/T) \\ &+ \cdots + b_1 \sin(2\pi t/T) + b_2 \sin(4\pi t/T) + \cdots \end{aligned} \tag{15.9}$$

To complete the analogy with vector spaces, we say that the amplitudes (or coefficients)—the *as* and *bs*—are given by the "overlap" between the function $f(t)$ and the corresponding sine or cosine function, just as we used $\langle 0|\psi\rangle$ and $\langle 1|\psi\rangle$ to express the overlap between a quantum state $|\psi\rangle$ and the computational basis states. In the continuous-time Fourier case, because we have an infinite-dimensional state space, the overlap is calculated by evaluating an integral over the duration of the signal.

15.4 The Discrete Fourier Transform

In this section, we introduce what is called the *discrete* Fourier transform. This method was devised to work with discrete (distinct, unconnected) sampled signals that arise in all sorts of digital signal processing, including audio and image processing used on your electronic devices. They are called "discrete" because the data are discrete (separated) samples of an independent variable rather than continuous. In the following section, we describe how the discrete Fourier transform can be implemented by a quantum algorithm.

Let's focus on an audio signal that varies in time. The extension to spatial samples (pixels) for images is straightforward. In the discrete time case, a signal $f(t)$ is recorded at discrete times $t = \{t_0, t_1, \ldots, t_{N-1}\}$ for a total of N samples. If the samples are taken at equal time intervals Δt starting with $t_0 = 0$, we can write the sequence of times as $t_j = \{j\Delta t\}$ with $j = 0, 1, 2, \ldots, N-1$. DVD audio files use a sampling rate of 48,000 samples per second; so, the time Δt between samples j and $j+1$ is $1/(48{,}000 \text{ samples/sec}) = 0.000021$ seconds. Audio CDs and MP3 files use 44,100 samples per second though MP3 files may also use different sampling rates.

As mentioned previously, the basic idea of Fourier analysis is to multiply the signal by cosines and sines and add up the results over the duration of the signal with a large value indicating a good fit. In place of sines and cosines, we use the complex exponential function. You may recall that Euler's formula expresses the complex exponential as a linear combination of a cosine function and a sine function:

$$e^{i\theta} = \cos\theta + i\sin\theta. \tag{15.10}$$

For the discrete Fourier transform, the parameter θ is replaced by $2\pi jk/N$. The complex exponentials are then

$$\exp(i2\pi j\,k/N) = \cos(2\pi j\,k/N) + i\sin(2\pi j\,k/N)\,. \tag{15.11}$$

In Eq. (15.11), N is the number of discrete data points to be analyzed; j and k are integers that run from 0 to $N-1$. When $j\,k/N$ is an integer, $\exp(i2\pi j\,k/N) = 1$, its largest possible value.

Figure 15.6 shows a plot of the real part of $\exp(i2\pi j\,k/N)$ as a function of the index j. Note that the function has a maximum whenever $j\,k/N$ is an integer (including 0), as promised.

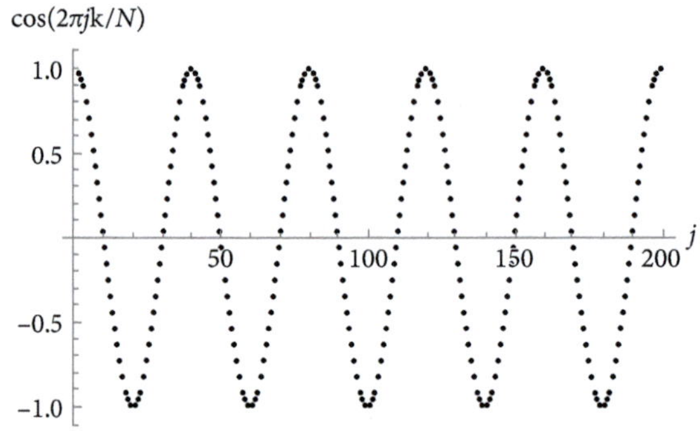

Fig. 15.6 A discrete-sample plot of the discrete function $\cos(2\pi j\, k/N)$ with $N = 200$ and $k = 5$ as a function of the index j.

Try It 15.5 Verify from Fig. 15.6 that the plot has a period of 40. Calculate k/N for this plot. Is that what you expect for a period of 40?

Using this mathematical machinery, we define the discrete Fourier transform of $f(t)$ as

$$\tilde{f}_k = \sum_{j=0}^{N-1} f(j\Delta t)\ \exp(i2\pi j\, k/N), \tag{15.12}$$

where $\tilde{f}_k$ is the so-called Fourier amplitude, which is labeled by the index k. Eq. (15.12) is just the sum of products of the signal at a particular time and the complex exponential at the same time. Here the time is indicated by the index j. In general, the Fourier amplitudes will be complex numbers. We can think of the complex exponential functions in Eq. (15.12) as a set of basis states with j as the independent variable.

Once we have found $\tilde{f}_k$ for all values of the index k, we may reverse the process and write the original function as a superposition (a linear combination) of the $\exp(-i2\pi j\, k/N)$ functions:

$$f(t_j) = \sum_{k=0}^{N-1} \tilde{f}_k\ \exp(-i2\pi j\, k/N). \tag{15.13}$$

In Eq. (15.13), $\exp(-i2\pi j\, k/N)$ plays the role of a member of a different set of basis states with k as the independent variable. The Fourier amplitude $\tilde{f}_k$ is the overlap between the function $f(t)$ and $\exp(i2\pi j\, k/N)$, calculated from Eq. (15.12).

In this formulation, we are using an N-dimensional state space with N basis states. The index k can be interpreted as the “frequency” associated with a particular basis state; that is, it represents the number of cycles per unit “time” as indicated by the j index.

An aside: In writing Eqs. (15.12) and (15.13), we have omitted "normalization" factors that are not important for our general argument. Some authors use $1/\sqrt{N}$ as a normalization factor in both Eqs. (15.12) and (15.13). Others use $1/N$ in only one of Eqs. (15.12) and (15.13). The important point is that $\tilde{f}_k$ for specific integers k will be large if $f(t)$ and $\exp(i2\pi j\,k/N)$ "line up" as j runs from 0 to $N-1$. These values of k then tell us about the periodicity of the signal $f(t)$.

In fact, though we won't go through the details, these basis states $\exp(i2\pi j\,k/N)$ are orthogonal and normalized in the sense that the product of two of them, summed over the range of the indices, gives 0 if the k indices are different (different basis states) and 1 if the k indices are the same (and if we have included the appropriate normalization factors).

All of this may seem rather abstract; so, let's show how this works in a case that might be familiar from introductory physics. Suppose we record an audio signal that is a mix of 440 Hz (middle A on the musical scale) and 572 Hz. Figure 15.7 shows the combined audio signal. Figure 15.8 shows a plot of $|\tilde{f}_k|^2$, which is called the discrete Fourier power spectrum of the signal and shows peaks at 440 and 572 Hz as expected. (Here, we won't worry about the details of converting from the integer k to the frequency in Hz. See end-of-chapter problems

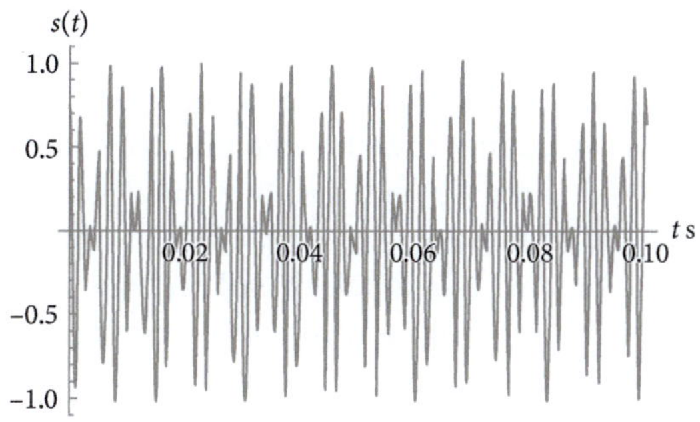

Fig. 15.7 A plot of a periodic function corresponding to a sum of two cosines.

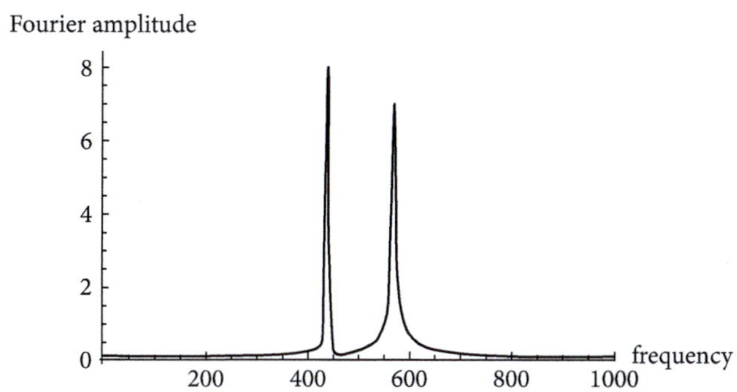

Fig. 15.8 A plot of the discrete Fourier "spectrum" of the signal in Fig. 15.7. The peaks occur at the two frequencies 440 Hz and 572 used to generate the signal in Fig. 15.7.

on the book's website.) We can think of the original audio signal as being a superposition of a 440 Hz signal and a 572 Hz signal. The period T of the 440 Hz signal is $1/(440 \text{ Hz}) = 0.0023$ s. The 572 Hz signal has a period of 0.0017 s.

15.5 Quantum Fourier Analysis

As mentioned before, the heart of the Shor algorithm is a quantum formulation of Fourier analysis that is much faster than the classical Fourier transform in finding the period of a periodic function. The quantum method is similar to the discrete Fourier transform for a classical signal.

We want to show how the quantum Fourier analysis allows us to find the smallest period r of the function

$$f(x) = a^x \bmod N. \tag{15.14}$$

with $f(x) = f(x + r)$, excluding $r = 0$ as mentioned previously. We have shown in Section 15.2 that if we know the period, we can find the factors of N. We have used x as the exponent to match what other authors have used and to link to the way we have previously labeled multiqubit basis states. As usual, a, x, r, and N are integers. As we proceed through the algorithm, you should pay attention to where we use superposition states and measurements, which we have seen to be key features of other quantum algorithms.

The first ingredient we need is the quantum implementation of the function $f(x) = a^x \bmod N$ as a quantum function gate U_f. We won't worry about the details of this implementation, but we note that it is usually based on binary modular exponentiation (see Section 14.5). As we have seen in other quantum algorithms, the function gate has two "registers" of input states: an upper register and a lower register as shown in Fig. 15.9.

As usual, we start with computational basis states at the left of the circuit. Both the upper and lower register states are n-qubit states with n chosen as the smallest integer so that $2^n \geq N$, the number we want to factor. As usual, the upper register states are then acted on by Hadamard gates to produce a superposition state consisting of equal contributions of all

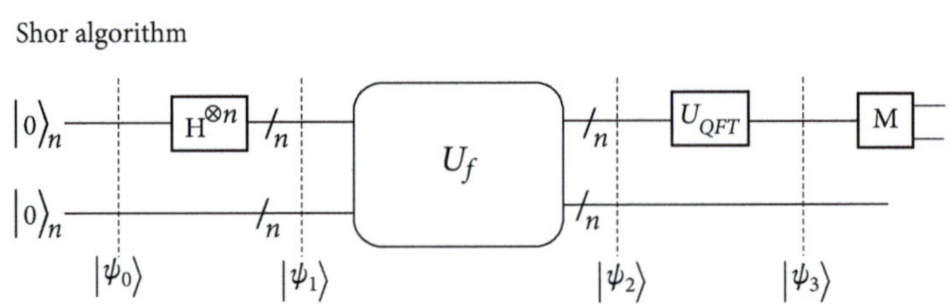

Fig. 15.9 A schematic of the Shor period-finding circuit with a function gate U_f, upper- and lower-register inputs and outputs, a gate U_{QFT} representing the quantum Fourier transform, and n measurement devices M for the upper register. Only one measurement device is shown.

the basis states labeled by the binary numbers x, where x is in the range $[0, N - 1]$. After the Hadamard gate acts on the upper-register states, the system state is

$$|\psi_1\rangle = H^{\otimes n}\,|\psi_0\rangle = \underbrace{\frac{1}{\sqrt{N}}\sum_{x=0}^{2^n-1}|x\rangle}_{\text{upper register}}\;\underbrace{|0\rangle_n}_{\text{lower register}}\;. \tag{15.15}$$

The modular exponentiation function gate U_f then acts on the state in Eq. (15.15) to produce the state

$$|\psi_2\rangle = U_f\,|\psi_1\rangle = \frac{1}{\sqrt{N}}\sum_{x=0}^{2^n-1}\underbrace{|x\rangle}_{\text{upper register}}\;\underbrace{|f(x)\rangle}_{\text{lower register}}\;, \tag{15.16}$$

where $f(x)$ is given by Eq. (15.14).

This state is a superposition of the products of all the basis states and the states labeled by the results of the periodic function in Eq. (15.14) evaluated at x. After we process this state with U_{QFT}, as described below, we let the upper-register states interact with measurement devices. The measurements' outcomes give us information on the desired period.

Next, we apply the quantum Fourier transform gate U_{QFT} to the upper-register states. The quantum Fourier transform gate's action is defined by its action on each of the basis states $|x\rangle$

$$U_{QFT}\,|x\rangle = \frac{1}{\sqrt{N}}\sum_{y=0}^{2^n-1}\exp\left(i2\pi xy/N\right)|y\rangle. \tag{15.17}$$

Note that we have replaced the j, k indices in Eq. (15.11) with the binary indices x and y. In Eq. (15.17), the U_{QFT} function acts on a particular basis state $|x\rangle$ to produce a superposition of all the basis states, labeled by y in the sum. Each basis state in the sum is multiplied by $\exp\left(i2\pi xy/N\right)$, itself a periodic function with the indices j and k replaced by the integers x and y.

The result of the quantum Fourier transform gate acting on $|\psi_2\rangle$ in Fig. 15.9 is

$$|\psi_3\rangle = U_{QFT}\,|\psi_2\rangle = \frac{1}{N}\sum_x\left[\sum_y\underbrace{\exp\left(i2\pi xy/N\right)|y\rangle}_{\text{upper register}}\right]\underbrace{|f(x)\rangle}_{\text{lower register}}\;. \tag{15.18}$$

Let's read off what is in the state. The function $\exp\left(i2\pi xy/N\right)|y\rangle$ is the representation of applying the quantum Fourier transform operator to the basis states $|x\rangle$ in the upper register. The $|f(x)\rangle$ parts are unchanged because nothing has operated on the lower-register states in progressing from $|\psi_2\rangle$ to $|\psi_3\rangle$.

Finally, we let the n upper-register states interact with the n measurement devices. The result will be some value of y, the state index integer. We shall see shortly that the most probable result of the upper-register y measurement will be $y = N/r$. Once we have that value of y, we can find the period r from $r = N/y$.

We now want to figure out the probability of getting a particular measurement value y. As with many other quantum algorithms, we don't always get precisely the answer we want by running the algorithm one time. But we want to arrange the algorithm, so the probability of getting the right answer is close to 1 every time. As promised, this result gives us information about the period r of the function a^x mod N, and we know from the analysis of the Shor factoring algorithm that it will allow us to factor the number N.

To find the probability, we follow the usual generalized Born rule for multiqubit states: We find the appropriate amplitudes, sum them, and then square the result to get the probability. The critical observation is that we have several amplitudes that contribute because the function $f(x)$ is periodic in x. That is, we have $f(x_0 + qr) = f(x_0)$, where q is any integer such that $x = x_0 + q\,r < N - 1$. Then, the probability of getting the result y from the upper-register measurement is

$$\frac{1}{N}\left|\sum_{x:\, f(x)=f(x_0)} \exp\left(i2\pi xy/N\right)\right|^2 = \frac{1}{q_{\max}N}\left|\sum_{q=0}^{q_{\max}} \exp\left(i2\pi[x_0 + qr]y/N\right)\right|^2, \tag{15.19}$$

where $q_{\max}$ is the largest integer less than $2^n/r$, approximately the number of periods of length r in the range $[0, 2^n - 1]$. The quantum efficiency in the algorithm occurs because we need to evaluate only a relatively small number of terms compared to the number that would be required with the classical discrete Fourier transform.

Try It 15.6 Find n for $N = 1000$.

This is pretty abstract; so, let's use the real part of $\exp\left(i2\pi[x_0 + qr]y/N\right)$:

$$\begin{aligned}\mathrm{Re}[\exp\left(i2\pi[x_0 + qr]y/N\right)] &= \cos[2\pi(x_0 + q\,r)y/N] \\ &= \cos[(2\pi x_0 y/N) + (2\pi\, q\, r\, y/N)].\end{aligned} \tag{15.20}$$

In the second line of Eq. (15.20), we see that the term $2\pi x_0 y/N$ is a constant that simply shifts the results when we sum over the integer q. It corresponds to shifting the cosine left or right but does not involve the periodicity. So, for our purposes, we just focus on $\cos[2\pi\, q\, r\, y/N]$. Recall that this is just like the function plotted in Fig. 15.6. Note that the general function $\cos(2\pi k)$ equals 1 if k is an integer. Analogously, the second term in Eq. (15.20) has its maximum value 1 when $r\,y/N$ is an integer, and it will be close to 1 when $r\,y/N$ is close to being an integer. This shows that the most probable result of the upper-register y measurement will be $y = N/r$ from which we find the desired period $r = N/y$ as mentioned earlier. So, we see that the quantum Fourier transform method is likely to give us the desired period, but there is some probability that it won't, much like the probabilistic Simon algorithm in

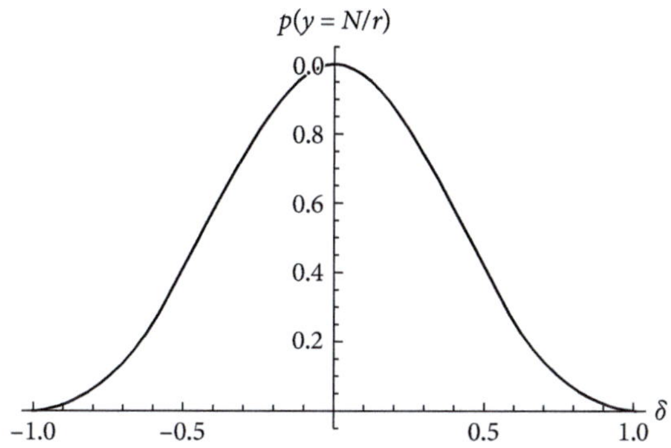

Fig. 15.10 A plot of the (unnormalized) probability given in Eq. (15.21) as a function of δ with $N = 10^7$ and $r = 30{,}000$.

Chapter 11. Of course, once you have a value for r, you should check that it is indeed a period of a^x mod N. This can be done by using the classical modular exponentiation algorithm (Section 14.4) to show that $1 = a^r \bmod N$, which is equivalent to showing that $a^r - 1$ is evenly divisible by N. You should also check that r is the smallest integer period by showing that $r/2$ does not work.

Let's get back to an important question. How likely are we to get the most probable result? It turns out that the sum in Eq. (15.19) in the general quantum Fourier transform method can be expressed as a geometric series. The result is that the probability can be given by

$$p(\delta) = \frac{1}{q_{\max} N} \frac{\sin^2(\pi\, \delta\, q_{\max}\, r/N)}{\sin^2(\pi\, \delta\, r/N)}, \tag{15.21}$$

where $\delta = y - N/r$ indicates how close the integer y is to N/r. A geometric series is a sum of terms, each of which is the previous term multiplied by a common factor R. In plain English, it means that R in the successive terms in the sum has exponents that range from 0 to $n - 1$. For example, in the series

$$a + aR + aR^2 + \cdots + aR^{n-1} = a\frac{1 - R^n}{1 - R}, \tag{15.22}$$

where R is the ratio of successive terms in the series and the expression on the right is the result of summing the series.

So, we see from Fig. 15.10 that the probability is sharply peaked when $\delta = 0$, that is when $y = N/r$, and the probability remains high even for $y = N/r \pm 0.5$; that is, the quantum Fourier transform is highly likely to give us the "correct" answer though there is some (smaller) probability that it will miss the target. In such a case, we ought to repeat the analysis several times to ensure we get the right answer.

Let's go through the details of the derivation of Eq. (15.21). We start by using complex exponentials to express the sum of amplitudes shown in the last term in Eq. (15.19) and repeated here as

$$\frac{1}{q_{max}N}\left|\sum_{q=0}^{q_{max}} \exp\left(i2\pi[x_0 + qr]y/N\right)\right|^2. \tag{15.23}$$

The sum in Eq. (15.23) can be written as

$$\sum_{q=0}^{q_{max}} \exp[i2\pi(x_0 + qr)y/N] = \exp[i2\pi x_0 y/N] \sum_{q=0}^{q_{max}} \exp[i2\pi qry/N], \tag{15.24}$$

where we temporarily dropped the normalization factors. In the last term of Eq. (15.24), we pulled the part of the exponential that does not depend on q out of the sum. When we calculate the absolute value squared to get the probability, that factor will drop out.

Next, we recognize that the sum in Eq. (15.24) is a geometric series with the common factor $e^{i2\pi ry/N}$. Using the result for the geometric series sum in Eq. (15.22), we find that the sum is

$$s = \sum_{q=0}^{q_{max}} e^{i2\pi qry/N} = \frac{1 - e^{i2\pi ryq_{max}/N}}{1 - e^{i2\pi ry/N}}. \tag{15.25}$$

To get the probability, we need the absolute value squared $s^*s = |s|^2$:

$$\begin{aligned} |s|^2 &= \frac{1 - e^{-i2\pi ryq_{max}/N}}{1 - e^{-i2\pi ry/N}} \frac{1 - e^{i2\pi ryq_{max}/N}}{1 - e^{i2\pi ry/N}} = \frac{2 - \left(e^{i2\pi ryq_{max}/N} + e^{-i2\pi ryq_{max}/N}\right)}{2 - \left(e^{i2\pi ry/N} + e^{-i2\pi ry/N}\right)} \\ &= \frac{1 - \cos\left(2\pi ryq_{max}/N\right)}{1 - \cos\left(2\pi ry/N\right)} \\ &= \frac{\sin^2\left(\pi ryq_{max}/N\right)}{\sin^2\left(\pi ry/N\right)}. \end{aligned} \tag{15.26}$$

In the first line of Eq. (15.26), we multiplied the geometric series sum by its complex conjugate and then multiplied binomials in the numerator and the binomials in the denominator, remembering that $e^{i\theta}e^{-i\theta} = 1$. In the second line, we used Euler's formula to replace the exponentials with cosines. Finally, in the last line, we used the trigonometry relationship $1 - \cos\theta = 2\sin^2\theta/2$ in both the numerator and denominator.

Remember that the Shor algorithm allows us to find the repetition period r of a function, where r is an integer. In Eq. (15.26), the sine functions will be largest (and the probability will be largest) when the arguments are $\pi/2$ times an odd integer. Since q is, by design, an integer, we want yr/N to be an integer. However, we can't guarantee that yr/N will be exactly an integer. But we do expect y to be close to N/r. So, we introduce a new variable $\delta = y - N/r$ and anticipate that the highest probability will occur when δ is close to 0. We make that substitution in the last line of Eq. (15.26) and use another trigonometry relationship $\sin(\theta + n\pi) = \pm\sin\theta$. Since we are squaring the sines in Eq. (15.26), it doesn't matter whether we have plus or minus.

Try It 15.7 Go through the algebra and trigonometry in Eq. (15.26) by multiplying the binomials in the numerators and the binomials in denominators.

Recalling the normalization factors from Eq. (15.23), we end up with our desired result:

$$P(\delta)= \frac{1}{q_{\max}N}\frac{\sin^2(\pi r\delta q_{\max}/N)}{\sin^2(\pi r\delta/N)}. \tag{15.27}$$

Try It 15.8 Make the suggested substitutions and check the results in Eq. (15.27).

Once we have found $y = N/r$ from Eq. (15.27), we can solve for the repetition period r. From r, we can find the prime factors of N—just what the Shor algorithm is designed to do.

15.6 **Quantum Fourier Transform Example**

We always understand a procedure better when we work out a simple example. Here is one from Zubairy (2020). Let's return to the Fourier transform and apply it to the state function in the Shor circuit in Fig. 15.9. In particular, let's look at the system state $|\psi_3\rangle$ after we apply the quantum Fourier transform to the upper-register states. The result, which is an entangled state, is expressed in Eq. (15.18) and repeated here as

$$|\psi_3\rangle = U_{QFT}\,|\psi_2\rangle = \frac{1}{N}\sum_x\left[\sum_y \underbrace{\exp(i2\pi xy/N)\,|y\rangle}_{\text{upper register}}\right]\underbrace{|f(x)\rangle}_{\text{lower register}}. \tag{15.28}$$

In Eq. (15.28), xy means ordinary multiplication. Note that when we sum over x for a fixed y, the argument of the exponential function changes in steps of $i2\pi y/N$.

To see what is going on, let's write this out for the case of $N = 8$ (a three-qubit system). For typographical simplicity, we label the states $|f(x)\rangle$ with the *decimal* values of the arguments. For example, we denote $|f(x = 010)\rangle$ as $|f(2)\rangle$. Here are the first few terms in the sums in Eq. (15.28) with $y = 0,\ 1,$ and 2:

$y = 0$

$$\frac{1}{8}|0\rangle\left[|f(0)\rangle + |f(1)\rangle \ldots + |f(7)\rangle\right] \tag{15.29}$$

$y = 1$

$$\frac{1}{8}|1\rangle\left[e^{i0}\,|f(0)\rangle + e^{i\pi/4}\,|f(1)\rangle + e^{i\pi/2}\,|f(2)\rangle \ldots + e^{i\pi 7/4}\,|f(7)\rangle\right] \tag{15.30}$$

Table 15.2 The arguments θ of the phase factors $e^{i\theta}$ (amplitudes) for the states in Eq. (15.28) along with the arrow representations of these phase factors (for example, $e^{i\pi/4}$ is represented by ↗ and $e^{i\pi/2}$ is represented by ↑).

$\vert y\rangle$	$\vert f(0)\rangle$	$\vert f(1)\rangle$	$\vert f(2)\rangle$	$\vert f(3)\rangle$	$\vert f(4)\rangle$	$f(5)$	$\vert f(6)\rangle$	$\vert f(7)\rangle$
$\vert 0\rangle$	0 →	0 →	0 →	0 →	0 →	0 →	0 →	0 →
$\vert 1\rangle$	0 →	$\pi/4$ ↗	$2\pi/4$ ↑	$3\pi/4$ ↖	$4\pi/4$ ←	$5\pi/4$ ↙	$6\pi/4$ ↓	$7\pi/4$ ↘
$\vert 2\rangle$	0 →	$\pi/2$ ↑	$2\pi/2$ ←	$3\pi/2$ ↓	$4\pi/2$ →	$5\pi/2$ ↑	$6\pi/2$ ←	$7\pi/2$ ↓
$\vert 3\rangle$	0 →	$3\pi/4$ ↖	$6\pi/4$ ↓	$9\pi/4$ ↗	$12\pi/4$ ←	$15\pi/4$ ↘	$18\pi/4$ ↑	$21\pi/4$ ↙
$\vert 4\rangle$	0 →	$4\pi/4$ ←	$8\pi/4$ →	$12\pi/4$ ←	$16\pi/4$ →	$20\pi/4$ ←	$24\pi/4$ →	$28\pi/4$ ←
$\vert 5\rangle$	0 →	$5\pi/4$ ↙	$10\pi/4$ ↑	$15\pi/4$ ↘	$20\pi/4$ ←	$25\pi/4$ ↗	$30\pi/4$ ↓	$35\pi/4$ ↖
$\vert 6\rangle$	0 →	$6\pi/4$ ↓	$12\pi/4$ ←	$18\pi/4$ ↑	$24\pi/4$ →	$30\pi/4$ ↓	$36\pi/4$ ←	$42\pi/4$ ↑
$\vert 7\rangle$	0 →	$7\pi/4$ ↘	$14\pi/4$ ↓	$21\pi/4$ ↙	$28\pi/4$ ←	$35\pi/4$ ↖	$42\pi/4$ ↑	$44\pi/4$ ↗

$y = 2$

$$\tfrac{1}{8}\,|2\rangle\left[e^{i0}\,|f(0)\rangle + e^{i\pi/2}\,|f(1)\rangle + e^{i\pi}\,|f(2)\rangle + \ldots + e^{i\pi 7/2}\,|f(7)\rangle\right]. \tag{15.31}$$

Let's put these amplitudes and the other amplitudes into Table 15.2.

Table 15.2 lists the arguments θ of the phase factors $e^{i\theta}$ (amplitudes) of the states in Eq. (15.28). We have not simplified fractions because the unsimplified forms make the pattern more obvious. For example, for the state $|3\rangle$ in the left column, the arguments of the phase factors for the $|f(j)\rangle$ state in the top row are integer multiples of $3\pi/4$ (starting with 0): $0\pi/4$, $3\pi/4$, $6\pi/4$, $9\pi/4$, etc. Note that for a particular row in the table, the argument increases by the entry in the $|f(1)\rangle$ column as you progress across the row. We have also included an arrow, which gives a graphical representation of the phase factor with the assumption that the arrow associated with $e^{i0} = 1$ points to the right along the horizontal direction. The arrows each rotate counterclockwise by the $|f(1)\rangle$ angle from the previous arrow's direction. You can add the amplitudes by adding up the arrows graphically as shown below.

We now invoke the periodicity of the function $f(x) = a^x \bmod N$. For our simple case, let's assume that $r = 2$ for some combination of a and N. In this case, we have

$$\begin{aligned} f(0) &= f(2) = f(4) = f(6) \\ f(1) &= f(3) = f(5) = f(7). \end{aligned} \tag{15.32}$$

As an example, we can use the results in Eq. (15.32) to rewrite Eq. (15.30) for $y = 1$ as

$$\tfrac{1}{8}\,|1\rangle\left[\left\{e^{i0} + e^{i\pi/2} + e^{i\pi} + e^{i3\pi/2}\right\}|f(0)\rangle + \left\{e^{i\pi/4} + e^{i\pi 3/4} + e^{i\pi 5/4} + e^{i\pi 7/4}\right\}|f(1)\rangle\right], \tag{15.33}$$

using the complex exponentials with phases. Alternatively, we may use arrows to indicate phases:

$$\tfrac{1}{8}\,|1\rangle\left[\{\rightarrow + \uparrow + \leftarrow + \downarrow\}\,|f(0)\rangle + \{\nearrow + \nwarrow + \swarrow + \searrow\}\,|f(1)\rangle\right]. \tag{15.34}$$

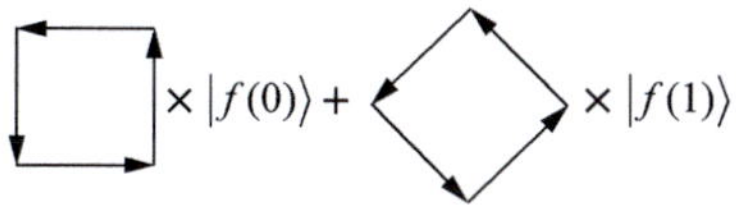

Fig. 15.11 The addition of the vectors inside the curly braces in Eq. (15.33).

If you evaluate the quantities inside the curly braces, you find that they give 0; the amplitude for measuring the $|1\rangle$ state in the upper register is 0. You can do that addition either by adding up the complex numbers associated with the complex exponentials or you can add up the arrows using the usual "tail-to-tip" rule. If the arrows form a closed figure when added tail-to-tip as shown in Fig. 15.11, the result is zero just like walking around the block brings you back to where you started. In particular, for the arrows in Eq. (15.33), we get the results shown in Fig. 15.11. This result is said to exhibit destructive interference among the state vectors. The net effect is that they add to 0.

You can use a spreadsheet to calculate the exponentials, but there is an easier way to see that the sum of the numerical value of the exponentials gives zero. You can use the sum of exponents rule $e^x e^y = e^{x+y}$ and $e^{i\pi} = -1$ to see that the terms inside the curly braces cancel pairwise. For example, in the second set of curly braces in Eq. (15.33), we find $e^{i\pi 7/4} = e^{i\pi} e^{i\pi 3/4} = -e^{i\pi 3/4}$.

Continuing with our example, if you work through the details, you find that the amplitudes for the upper-register states $|1\rangle$, $|2\rangle$, $|3\rangle$, $|5\rangle$, $|6\rangle$, and $|7\rangle$ are all 0 (destructive interference), meaning that the probability of getting these states as results of measurements on the upper-register states is 0. Only the states $|0\rangle$ and $|4\rangle$ have probabilities that are not zero, and in this case, both have probabilities equal to 50%. For these states, the arrows tend to line up in the same direction to give a large amplitude, often called constructive interference.

Let's see how the addition works for states $|0\rangle$ and $|4\rangle$. For the state $|0\rangle$, we have

$$\begin{aligned}
&\frac{1}{8}|0\rangle\left[\left\{e^{i0} + e^{i2\pi} + e^{i4\pi} + e^{i6\pi}\right\}|f(0)\rangle + \left\{e^{i\pi} + e^{i3\pi} + e^{i5\pi} + e^{i7\pi}\right\}|f(1)\rangle\right] \\
&= \frac{1}{8}|0\rangle\left[\{\rightarrow + \rightarrow + \rightarrow + \rightarrow\}|f(0)\rangle + \{\rightarrow + \rightarrow + \rightarrow + \rightarrow\}|f(1)\rangle\right] \qquad (15.35) \\
&= |0\rangle\left[\frac{1}{2}|f(0)\rangle + \frac{1}{2}|f(1)\rangle\right].
\end{aligned}$$

For the state $|4\rangle$, we find

$$\begin{aligned}
&\frac{1}{8}|4\rangle\left[\left\{e^{i0} + e^{i2\pi} + e^{i4\pi} + e^{i6\pi}\right\}|f(0)\rangle + \left\{e^{i\pi} + e^{i3\pi} + e^{i5\pi} + e^{i7\pi}\right\}|f(1)\rangle\right] \\
&= \frac{1}{8}|4\rangle\left[\{\rightarrow + \rightarrow + \rightarrow + \rightarrow\}|f(0)\rangle + \{\leftarrow + \leftarrow + \leftarrow + \leftarrow\}|f(1)\rangle\right] \qquad (15.36) \\
&= |4\rangle\left[\frac{1}{2}|f(0)\rangle - \frac{1}{2}|f(1)\rangle\right].
\end{aligned}$$

We should note that the amplitudes work out to be zero or nonzero only if N is evenly divided by r. In more general cases, some of the amplitudes will add to small numbers, but

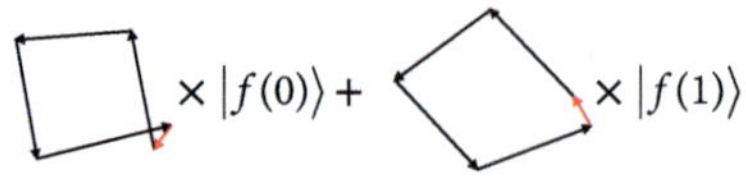

Fig. 15.12 The sum of the arrows may not form a closed circuit.

not quite equal to zero. In these cases, the arrows will almost form a closed figure but not quite. For these states, the probabilities for observing them upon measurement will be small but not zero. The resulting amplitude is given by the length of the vector from the tail of the first vector to the tip of the final vector, indicated by the short red lines in Fig. 15.12. These small amplitudes mean that the algorithm will not give the desired result exactly, but the probability for getting the correct result is close to 1.

How do we find the repetition period r, if we obtain $|0\rangle$ or $|4\rangle$ in the upper-register measurements? In general, if $f(a) = f(a + r)$ and if N is evenly divisible by r, the only y states with nonzero amplitudes are

$$|0\rangle,\ |N/r\rangle,\ |2N/r\rangle, \ldots, |(r-1)\,N/r\rangle\,. \tag{15.37}$$

Since the $|0\rangle$ state doesn't give us any useful information, we try $N/r = 4$ in our example and find $r = 2$, the desired period. If N is not evenly divided by r, then we need to use the methods embodied in Eq. (15.27).

Let's review the general ideas of the Shor algorithm. The Shor algorithm works by using the quantum Fourier transform to find the repetition period of the function $f(x) = a^x \bmod N$. We know from number theory that given a, N, and the period, we can find the prime factors of N, *which are the secret part of the encryption key*. The Shor algorithm cleverly makes use of superposition and entanglement to produce system states that give high probabilities to be detected in a measurement if the state label y satisfies $y \approx N/r$ and small otherwise. For the high probability states, the amplitudes add together constructively with the sum close to 1. For the low probability states, the amplitudes add together destructively with the sum close to 0. This is often called quantum interference.

15.7 Chapter Summary

- The Shor algorithm is a famous quantum algorithm that can find an encryption key faster by making use of the periodic nature of the remainder or mod function. In effect, the Shor algorithm allows us to break the RSA encryption and other encryption methods based on factoring large numbers, the basis for a large portion of our Internet security. The algorithm does this by making use of quantum Fourier transforms that approximate periodic functions as a series of cosine and sine functions. The periods of the cosine and sine functions give information about the period of the original function.

- In terms of quantum gates and a quantum circuit, the Shor algorithm has an upper register and a lower one with the upper register put into a state of superposition using Hadamard gates. The system state is acted on by a function gate U_f with $f(x) = a^x \bmod N$. The function is periodic $f(x) = f(x + r)$, where r is the period of the function. Next the upper register is operated on by a quantum Fourier transform gate. Finally, we send both registers to measurement devices from which we find $y = N/r$ and solve for $r = N/y$. Once we find the period r, we can find the prime factors p and q by straightforward classical computations.
- The IBM website https://quantum.cloud.ibm.com/learning/enhas is a good exposition of many topics in quantum computing and quantum information science. The URL for phase estimation is (as of March, 2026) https://quantum.cloud.ibm.com/learning/en/courses/fundamentals-of-quantum-algorithms/phase-estimation-and-factoring/phase-estimation-procedure

15.8 Further Reading

M. Suhail Zubairy, *Quantum Mechanics for Beginners* (Oxford University Press, Oxford, 2020). Chapters 13 and 16 provide good introductions to Quantum Secure Communications and the Shor Algorithm.

16 Programming Quantum Computers

> Programmer: A machine that turns coffee into code.
>
> **Anonymous**

16.1 Introduction

One way to think about programming quantum computing (QC) applications is to break the process into two steps. First, we develop an algorithm that makes use of superposition states, entanglement, quantum gates, and quantum measurements. As discussed in previous chapters, representing those algorithms as quantum circuits is something like writing flowcharts for classical computer programming. In these circuits and flowcharts, we describe the logic of the process we want to implement without worrying about the computer programming code needed to send the appropriate instructions to the computer. The second step is writing code that can be sent to the QC to implement the algorithm.

For the short-term future, it is highly likely that the commands you send to a quantum computer will be specific for each device because you need to communicate directly with that device's hardware. Most browser-accessible quantum computer systems provide you with a commonly available programming language such as Python with special instructions in another layer you do not see that handle the direct communications with the hardware. Eventually higher-level quantum programming languages will be developed akin to Python or C# running on classical computers; so, the typical user will not need to worry at all about how to communicate with the specific underlying hardware. That is certainly the situation in classical computers today, but within living memory, there was a time when a programmer needed to be aware of how memory was set up in a computer and which part of the central processing unit did what. One of the authors remembers working with some of the first minicomputers in the 1970s. You had to flip a set of mechanical toggle switches to get the machine up and running. Computer engineers quickly figured out how to build in those instructions using integrated circuits and more-or-less permanent computer memory units.

Figure 16.1 shows a simple quantum circuit that we use as an example of QC programming. The circuit contains the first two steps we just mentioned. You will recognize the familiar elements in this circuit: two qubits, their initial states, a Hadamard gate, a CNOT gate, and two measurement devices. We also included two classical bits that are used to

Quantum Computing and Quantum Physics. Alice Flarend and Robert Hilborn, Oxford University Press.
 DOI: 10.1093/oso/9780197904930.003.0016

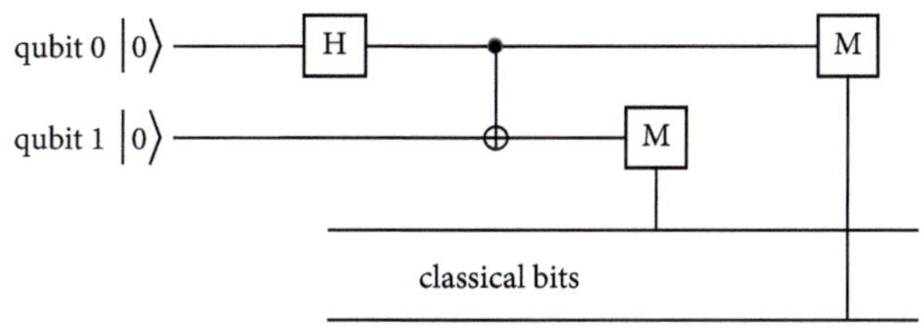

Fig. 16.1 The circuit diagram for the quantum programming example. H is a Hadamard gate; the boxes with Ms represent measurement devices. Two classical bits, used to record the outcome of the measurements, are shown at the bottom of the diagram.

record the outputs of the measurement devices. These outputs are passed along to a classical computer for further processing.

You should also note on the far right of Fig. 16.1 that the circuit makes computational basis state measurements on the circuit. The measurement outcomes are recorded by a classical computer. In the box below is an example of how we translate this circuit into programming code using the syntax from IBM's Qiskit website. The code sets up a quantum circuit with the two qubits. Then, the code adds a Hadamard gate (H gate) followed by a CNOT gate with qubit 0 in Fig. 16.1 acting as the control and qubit 1 acting as the target. The code also adds two classical bits to capture the measurement outcomes. The quantum circuit program is then run many times, in this case for a total of 1000 iterations. Statements following # are comments to tell you what is happening in various parts of the code, a common practice among programmers.

```
# Specify a quantum circuit consisting of two qubits.
circuit = QuantumCircuit (2,2)
# Include a Hadamard gate (h)acting on qubit 0
circuit.h(0)
# Include a CNOT (CX) gate with control qubit 0 and target qubit 1
circuit.cx (0,1)
# Include measurements based on the computational basis states
circuit.measure([0,1],[0,1])
#  assign  the  program  to  a  quantum  processor  and  exe-
cute the steps 1000 times to get a set of measurement results
myJob = execute(circuit, quantum processor, shots=1000)
# get the results
result = myJob.result()
# get the counts for each of the two possible measurement outcomes
counts = result.get_counts(circuit)
# print the results for further analysis
print("\nTotal count for 00 and 11 are:",counts)
# END
```

Of course, the details of the code syntax will vary from one programming setup to another, just like the differences among Python, C#, Mathematica, MATLAB, and other programming environments for classical computers. Eventually, many of the individual gate operations will be pulled together into higher-level programming structures. As a simple example, the combination of the Hadamard gate and a CNOT gate is so common, and it is likely to be put together into something like a BuildSuperposition function.

The primary differences compared to classical computer programming are in setting up the quantum circuit with qubits, gates, and measurement devices and in handling multi-qubit states, which as we know can be a bit tricky. Of course, the hard work is in developing the algorithm and its corresponding quantum circuit.

16.2 **Access to Quantum Computers**

While you are waiting to have your own QC and want to use more than a QC simulator, it is now possible to access remotely quantum computers operated by companies such as IBM, IonQ, and D-Wave. For example, Amazon AWS has set up an interface called Braket that allows access to QCs of other companies. These interfaces allow you to send instructions to the QC using special functions built into standard programming languages such as Python and C#. The companies also provide a lot of good background information on QC that should be accessible to you now that you have read this book. The Further Reading section contains links to these resources.

Since the browser-based quantum computer systems change their available commands and syntax from time to time, be sure to visit a site well in advance of the time you need to use it. Almost all these sites have tutorials to help new users get started.

16.3 Chapter Summary

- Quantum computer programming does not (yet) have universal programming languages akin to Python and C#. Currently, each elementary gate needs to be specified instead of choosing a complex operation such as creating a state of superposition. There are websites offering access to quantum computers with interfaces that use standard programming languages.

16.4 Further Reading

Eric R. Johnston, Nic Harrigan, and Mercedes Gimeno-Segovia, *Programming Quantum Computers* (O'Reilly, Sebastol, CA, 2019). This book aimed at experienced computer programmers.

"Cloud-based quantum computing," Wikipedia. An overview of what is available for online access to quantum computers.

Caveat: The links to websites in what follows may change frequently.

Companies and labs in the United Kingdom have developed an open-source universal quantum operating system that will enable programmers to write programs for quantum computers with different hardware structures: https://thequantumdaily.com/2021/09/23/uk-companies-tout-major-step-toward-universal-quantum-operating-system/

Google cirq https://quantumai.google/cirq provides a Python programming framework and a library of algorithms for designing, manipulating, and optimizing quantum circuits. It runs circuits on a QC simulator or actual Google QC. Tutorials are available, but they assume familiarity with quantum computing at about the level of the textbook by Michael Nielsen and Isaac Chuang, *Quantum Computation and Quantum Information* (Cambridge University Press, Cambridge, 2009).

IBM Qiskit https://qiskit.org provides access to IBM quantum computers and simulators as well as many tutorials on all aspects of QC.

AWS Braket https://aws.amazon.com/braket/ provides access to QCs at several different companies. Access must be purchased, but scientists at universities can get credits (grants). The system also provides Jupyter notebooks that come installed with the Amazon Braket Software Development Kit (SDK).

Microsoft Azure Quantum https://azure.microsoft.com/en-us/solutions/quantum-computing. In 2025, Microsoft launched the first information about its topological quantum computer. As of this writing, that quantum computer is not yet fully operational. However, the Microsoft Azure system provides a lot of information about quantum computers and their applications.

Sandia Labs scientists have developed an assembly language for their trapped-ion computer. "Coding for Qubits: How to Program in Quantum Computer Assembly Language," https://spectrum.ieee.org/tech-talk/computing/software/qscout-sandia-open-source-quantum-computer-and-jaqal-quantum-assembly-language.amp.html.

The company IonQ allows access to their quantum computers through a variety of cloud-based interfaces, including those of Google, IBM, Amazon, and Microsoft. See https://ionq.com/quantum-cloud.

Jed Brody and Gavin Guzman, "Calculating spin correlations with a quantum computer," *American Journal of Physics* **89**, 35–40 (2021). This article describes how to use the online-accessible IBM quantum computers to calculate the average value of "spin correlations" for two-qubit states. The authors assume that the reader is familiar with Pauli matrices, complex numbers, and rotation matrices. We have touched on all these topics but putting them together may be somewhat challenging for our readers. Nevertheless, our dedicated readers should be able to skim the article to get the basic idea of how to program and use a quantum computer to produce measurements of quantum properties of a system.

PART IV

Deeper Dive into Quantum Mechanics

Fundamental Quantum Issues

17

For every complex problem there is an answer that is clear, simple, and wrong.

H. L. Mencken

17.1 Introduction

In the previous chapters, we have introduced several different types of quantum algorithms and quantum information processing schemes. All these procedures heavily rely on quantum superposition states, particularly entangled states. In this chapter, we explore some of the conceptual issues raised by entanglement. Although understanding these issues doesn't directly impact our ability to use quantum algorithms, they do highlight the differences between classical computational resources and quantum computing (QC) resources.

We also consider how other fundamental aspects of quantum physics are connected to quantum information science (QIS). Although these topics are not directly related to QC, we hope you are intrigued enough about these fundamental issues that you will read on. Understanding the "why" gives you the tools to find new applications.

As we dig into those concepts, we ought to be a bit modest about our current state of knowledge of the quantum world and say, as far as we know, that is the way the world works. It would be highly presumptuous to claim that we have arrived at the final truth. And, of course, there are plenty of unanswered questions in the quantum physics world.

The history of science and technology is littered with now-defunct claims of achieving the final truth. On the other hand, no one has any idea of what an alternative to quantum mechanics would look like or how an alternative could reproduce the successes of quantum mechanics while offering a different picture of the nature of reality. But we must offer the caveat about that being a possibility.

17.2 Bell's Theorem and Quantum Weirdness

The first issue we look at is the nature of reality. This sounds really profound, but it is an issue that underpins the consequences of entangled states—the main ingredient in many QIS procedures. Thanks to the theoretical work of the late physicist John Bell and the experimental work of dozens of scientists around the world, we now have experimental proof

Quantum Computing and Quantum Physics. Alice Flarend and Robert Hilborn, Oxford University Press.
 DOI: 10.1093/oso/9780197904930.003.0017

Fig. 17.1 The participants in the Foundations of Quantum Mechanics Conference, Amherst College, Amherst, MA, June 10–15, 1990. John Bell is in the first row, far right. Author Bob is in the middle of the second row from the top.
Source: Photograph used by permission from Amherst College. Frank Ward, photographer.

that nature really does work in what seems to be the crazy way described by entangled states and that any efforts to sneak around the strictures of quantum mechanics are doomed to fail.

In 1990, author Bob was a participant in the Foundations of Quantum Mechanics Conference held at Amherst College (Fig. 17.1). Among the participants were John Bell, David Mermin, Anton Zeilinger, Danny Greenberger, and many other notables in the development of QIS and the interpretations of the foundations of quantum mechanics. In the 1960s, John Bell introduced what became known as Bell's theorem, which provided a way to test the quantum predictions of correlations among measurement outcomes against the predictions of classical physics. In 1990, only a few experiments testing Bell's theorem were available, and each of them had several loopholes that prevented them from being definitive tests of the theorem. Much of the discussion at the conference focused on what Bell's theorem and those experiments were actually telling us about the nature of reality. Sadly, John Bell passed away about four months later, but the conversation has continued.

David Mermin (see Further Reading) had developed a simple model that displayed, with a minimum of mathematics, the contradictions between what seems like a straightforward, commonsense prediction of the results of a set of measurements and the quantum predictions for these outcomes. In essence, Mermin's model is a specific instance of the general results embodied in Bell's theorem. We use a slight variation on Mermin's model to illustrate Bell's theorem.

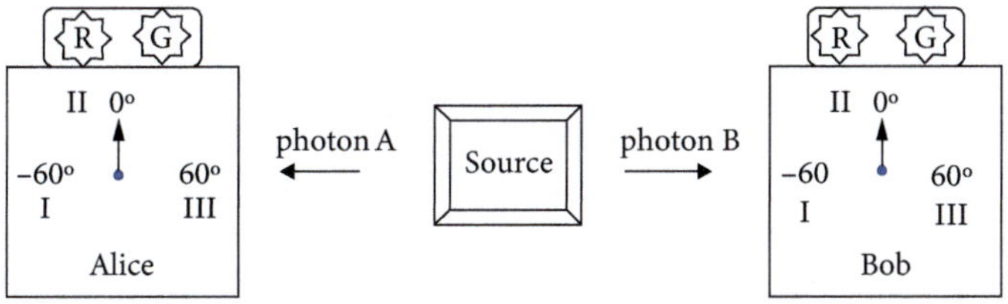

Fig. 17.2 A sketch of the apparatus used to illustrate Bell's theorem. The source emits two photons, one of which travels to the right and one travels to the left. They interact with the measurement devices and either a green light or a red light indicates the measurement outcome. Each measurement device has three different measurement bases available. The bases are labeled I, II, and III or −60°, 0°, and 60°.

Here is the scenario: Imagine that we have put together a device that emits two correlated qubits. Those two qubits later encounter two measurement devices as shown in Fig. 17.2. The devices are identical, and each has three settings for the measurement basis states. For now, we call the settings I, II, and III. Later we translate them to linear polarization direction settings for photons which is why they have degree values associated with them.

Here's how the measurement devices work. Each device has two lights: one red and one green that indicate the measurement results. When a single qubit interacts with one of the measurement devices, things are arranged such that only one light goes on—the red one or the green one—but never both, just as we have seen for polarizing beam splitters and photons. When the two qubits interact with the two measurement devices (one on the left and one on the right) set to the *same* settings—both I, both II, or both III—we find that both red lights go on or both green lights go on but in an apparently random pattern. On average, you get 50% red lights and 50% green.

The question is what determines whether the red lights or the green lights go on. A classical model does not allow for random results. There must be a definite cause. It seems obvious from that line of reasoning that the qubits must carry some instructions or have some property that determines how they interact with the measurement devices and determines which light goes on. For example, you might imagine that the qubits have an electrical charge: plus charge or minus charge. When the entity interacts with the measurement device, which we might assume is sensitive to the sign of the electrical charge, we could arrange that plus charge makes the green light go on while minus charge leads to the red light going on.

Since there are three settings for each device in the Bell experiment, it seems reasonable to think that the qubits have three properties (or instructions) given to them by the source to indicate which light should go on when they interact with the measurement devices. We represent these instructions as a set of boxes, for example |*R*|*R*|*G*|, with one box for each of the possible device settings. The *R* or *G* in the box means that there is some qubit property that leads the lit light to be red or green depending on the property and the measurement device setting. Reading from left to right, we have the I setting result (*R* in this example) and the

II setting result (*R* again), and on the right, the III setting result (*G*). These sets of instructions are sometimes called "hidden variables" because we have not been able to see them with available experiments. Thinking classically, it seems something like hidden variables must be there determining the measurement outcomes. We assume that the source, on any given run, gives the same instructions to each of the qubits; this assumption guarantees that there will be correlations among the outcomes of the measurements on the two qubits.

As part of the classical model, we assume that there are no other communication channels between the right and left devices. The devices respond solely to the interaction with their local qubit. This kind of model is called a "local realistic," or better, a "local deterministic" model: the results are determined (with no randomness) by those local instructions. The only reason the red and green lights seem to go on randomly is that with successive trials the source sends out qubits with different instruction sets and we don't know in advance, nor can we control, what the instruction set is. To account for the results when the analyzer settings are the same on the right and the left side, this model's entities would each have to have the same hidden variables [*R*][*R*][*R*] or [*G*][*G*][*G*] for each measurement setting, with each occurring 50% of the time.

Now let's look at what our local deterministic model predicts, for the eight possible instruction sets

[*R*][*R*][*R*] [*R*][*R*][*G*] [*R*][*G*][*R*] [*G*][*R*][*R*] [*R*][*G*][*G*] [*G*][*R*][*G*] [*G*][*G*][*R*] [*G*][*G*][*G*]

Also, the measurement devices can have different settings. There are nine possible setting combinations of those two devices. If we list Alice's device's setting on the left, Bob's device's setting on the right, the six possibilities for different settings are (I, II), (I, III), (II, I), (II, III), (III, I), or (III, II) and three more with the same settings (I, I), (II, II), or (III, III).

Since there are nine different combinations of measurement device settings and a total of eight sets of instructions, we fill in a 9×8 table with the lit light results predicted by our model. Remember that we assume that each pair of qubits is produced with the same set of three properties or instructions by the source to account for having the same lights lit when the settings are the same. The entries in the table labeled A mean the measurement results agree (both Alice and Bob see red or both see green). The ones labeled D mean that Alice sees red and Bob sees green or vice versa.

> **Try It 17.1** Fill in the final row of Table 17.1 and explain your reasoning. **Challenge:** Fill in all Table 17.1 boxes with R/R, G/G, R/G, or G/R to indicate which lights are lit in each case.

Now we are ready to analyze the frequency of those results. We ask a very simple question: What fraction of all the results listed in the table give rise to the same light being lit in both devices? The lights could be both red or both green. If you count the number of

Table 17.1 The results of running the Fig. 17.1 device. The top row indicates the possible qubit instruction sets. (Each qubit has the same instruction set as the other for each trial.) The first column on the left indicates the detectors' settings I, II, or III. The other boxes display A when both measurement results "agree" (both devices show a red light or both show a green light) and D when the results "disagree" (one shows red, and the other shows green). The last row is to be filled in by the reader (see Try It 17.1).

	R R R	G G G	R R G	R G R	G R R	R G G	G R G	G G R
I, I	A	A	A	A	A	A	A	A
II, II	A	A	A	A	A	A	A	A
III, III	A	A	A	A	A	A	A	A
I, II	A	A	A	D	D	D	D	A
I, III	A	A	D	A	D	D	A	D
II, I	A	A	A	D	D	D	D	A
II, III	A	A	D	D	A	A	D	D
III, I	A	A	D	A	D	D	A	D
III, II								

A results, you should get 48. Since we have 9 rows and 8 columns for a total of 72 possible results, we get the same lights lit 48/72 = 2/3 of the time.

We assume that we have constructed the source device to produce the instruction sets with equal probability and the measurement devices to use I, II, or III settings with equal probability. Then, if we look at the results of a long sequence of qubit launchings and measurement device detections, our model predicts that we should see two green lights or two red lights lit 2/3 of the time. Of course, it means we should see red/green or green/red 1/3 of the time. We could also check our prediction that each detector displays 50% red and 50% green lights on average.

Try It 17.2 Verify from the listings in Table 17.1 that our model predicts that each detector should, on average, display 50% red and 50% green lights lit.

It turns out we don't need to make any assumption about how often the source makes various instruction sets. Think about it in the following way. Let's look at the specific types of columns in Table 17.1: ones where the instruction is the same for all detector settings and ones where they are different. In Columns 1 and 2, we always get the same color lights lit in the two detectors regardless of their settings. So, if the source produces only $\boxed{R}\boxed{R}\boxed{R}$ or only $\boxed{G}\boxed{G}\boxed{G}$ or only some mixture of $\boxed{R}\boxed{R}\boxed{R}$ and $\boxed{G}\boxed{G}\boxed{G}$, we will have the same colors lit in every trial. That is, the probability P_{same} of getting lights of the same color is 1. This is the maximum value a probability can be. Now count the number of times we get the same colors for Columns 3–8 in Table 17.1 with different qubit instruction sets for each setting. Do you find that five out of the nine detector settings give lights of the same color? For those columns, the probability of getting two lights of the same color is 5/9, smaller than that for the first two columns. Look at each of the columns individually and note that they all indicate a probability of 5/9 for getting the same colors. If the source produces the instruction sets corresponding to any one of the third through eighth columns or even a mixture of those columns, we get the same color lights with a probability of 5/9.

Now let's put the result for the two types of columns together. This may take some thinking. For any mixture of all those instruction sets or all the columns, we always have a probability of the same color lights greater than or equal to 5/9. For our equal probability model, where all instruction sets are equally probable, we obtained a probability of 2/3, larger than 5/9. The overall results of our local deterministic model could be specified as an inequality: $5/9 \leq P_{\text{same}} \leq 1$. This type of inequality is called a Bell inequality.

What John Bell was able to show, which is now called Bell's theorem, was that any theory that purports that qubits (or any other quantum entity) have a "deterministic" and "local" set of properties independent of measurements gives predictions that satisfy a similar inequality. Bell's theorem gives a quantitative prediction we can test by doing an experiment.

It is now time to see what quantum mechanics predicts for the results of these measurements. We use linearly polarized photons as the qubits emitted by the source. We have had some experience calculating the results of quantum measurements on such photons, and almost all the experimental tests of Bell's theorem have been carried out with photons using setups that are similar to the one described in Fig. 17.2.

We set the source to prepare two linearly polarized photons in the entangled superposition state (a Bell state)

$$|\psi\rangle = \frac{1}{\sqrt{2}}|\text{hlp}\rangle|\text{hlp}\rangle + \frac{1}{\sqrt{2}}|\text{vlp}\rangle|\text{vlp}\rangle. \tag{17.1}$$

You recall that if this is the state of the two-qubit system, then it is not legitimate to say that one of the qubits has vlp and the other has hlp. To use the language we employed to build our local deterministic model, we see that quantum mechanics does not assign a definite property (or instruction set) to each photon. The state, as usual, tells us only the probabilities of observing hlp or vlp when the two qubits interact with linear polarization measurement devices.

In the photon version, the three measurement device settings correspond to the horizontal linear polarization direction set at −60°, 0°, and 60°, which are labeled on our measurement devices under I, II, and III. This amounts to having a choice of three different measurement basis states. If the green light is lit, the measurement device has detected hlp relative to the measurement setting axes, and if red is lit, it has detected vlp relative to these axes. In fact, we worked out the details of these basis states in Chapter 9 (Section 9.9). But let's review the results of choosing various measurement basis states because the results are fundamental to understanding the quantum results.

If both Bob and Alice choose the same polarization measurement basis states, that is, they choose the same linear polarization directions for measurements, then they will always get the same measurement results: 50% of the time they will find both observe "vertical" polarization (red light lit) and 50% of the time they will find both observe "horizontal" polarization (green light lit) relative to this measurement direction. They don't get red or green every time, but Bob and Alice will always have the *same* result: red or green.

What happens if Bob and Alice each choose different measurement directions? Then, they will not always get the same lights lit (either both red or both green). That is, they will sometimes get the same lights lit and sometimes different lights lit. In fact, we calculated the probabilities for those results in Chapter 9. We found that for different measurement basis settings, Alice and Bob both get red or both get green only 25% of the time. And averaging over all possible measurement device settings, we get the same color only 50% of the time; that is, according to quantum mechanics, the probability of our getting the same lights lit is 1/2.

Let's compare the predictions of the two models. We previously showed that our local, deterministic model predicts that we should get the same color with a probability $5/9 \leq P_{\text{same}} \leq 1$ with the exact value depending on the selection of the various possible instruction sets. The quantum model, on the other hand, predicts that the probability of getting the same color exceeds 1/4, with the lower limit occurring when the measurement device settings are different. If we average over all the measurement device settings, the probability for getting the same lights lit is 1/2, which is smaller than 5/9. In other words, we have distinctly different predictions from the local, deterministic model and from quantum mechanics.

When we carry out actual experiments to test Bell's theorem, we find that the probability that Alice and Bob observe the same color is (drumroll) 1/2. Which model do the experimental results support—the quantum model or the local, deterministic model? On the one hand, the experiments seem to support the quantum model. On the other hand, the local deterministic model seems so reasonable and matches our commonsense ideas about measurements. For many people, the obvious conclusion is that the experiments are wrong or flawed in some fundamental way—a quite understandable reaction if you rely on classical common sense. There are in fact some subtle aspects of the experiments that we need to think about carefully. But after several decades of work, all these questionable aspects have been dealt with and the precision of the experiments has been dramatically improved. There is no longer any reasonable doubt about the validity of the experimental results: The predictions of quantum mechanics for this setup are right (they agree with the experimental results) and those of a local, deterministic model are wrong.

Many scientists, including Einstein, thought there must be something wrong with quantum mechanics particularly because of its probabilistic nature. As Einstein put it, "God does not play dice." He thought that quantum mechanics was a useful theory, but that eventually someone would figure out a better theory—better in the sense of getting the right predictions but doing away with the probability part.

Let's now unpack the terms "deterministic" and "local." By "deterministic" we mean the commonsense notion that an object like a photon or an electron always has a definite value of its properties such as linear polarization direction or spin orientation and that these properties determine (no probability involved) how the object interacts with measurement devices (or with anything else for that matter). We may not know what the property value is until we measure it, but it is, in some sense, really there.

As we have seen with entangled states such as the two-qubit spin-1/2 state

$$|\psi\rangle = \frac{1}{\sqrt{2}}\left(|\uparrow\uparrow\rangle + |\downarrow\downarrow\rangle\right), \tag{17.2}$$

when we observe the orientation of the spin of one of the qubits, we might get $\uparrow$. Then when we measure the orientation of the other qubit's spin, we will get $\uparrow$. Such states are often called Bell states (recall Chapter 10) because of their use in illustrating Bell's theorem.

The "realistic" or "deterministic" account is that the two qubits both had those spin orientations before the measurements; all we did was to observe what is already there. These qubit properties determine how the qubit interacts with the measurement device and which light gets lit. The lights appear to be red or green randomly only because we don't know which set of properties is launched each time the source emits the qubits. We prefer to use the term deterministic account because "real," "reality," and "realism" are philosophically loaded terms; it would take us a whole other book to dig into the meaning of these words.

Now let's talk about "local." Local means that the outcome of the measurements depends only on the specific measurement device, its settings, and the properties of the entity the device interacts with. The results do not depend on what is going on at the other measurement device, at the source, at Simone's Café, or on some exo-planet on the other side of the galaxy. There are two parts to that requirement. One is practical: We have to be sure we exclude effects, such as the vibrations of the table on which the equipment is set up, "hidden" signals that might travel through the electrical power lines, and so on. We might also worry about whether choosing the measurement device settings affects the source or somehow affects the measurement devices themselves from a distance. We can even arrange to choose the device settings so they go into place *after* the source has emitted the entities. This rules out the possibility that the measurement device settings influence what the source sends out.

All these kinds of "hidden signal" effects can be ruled out by having the detectors far enough apart so that no signal can travel from one to the other during the time interval between when Alice's measurement device carries out its measurement and Bob's measurement device does its measurement. It is easy to make sure that there is not enough time for a hidden signal to travel from one detector to another since all signals have an upper speed limit: $\approx 3 \times 10^8$ m/s. This is the speed of light and the speed of gravitational waves and, as far as we know, is nature's fundamental speed limit. So, if we make sure that the detectors are far enough apart that a signal can't get from one to the other between two measurements, then we can rule out this kind of influence.

Taking all these possibilities into account, we find that the actual measurements agree with the quantum predictions, not the hidden signals' deterministic prediction. The bottom line of this argument is that having a local, deterministic model, although it may conform both to our intuition about how systems interact (the deterministic part) and to our knowledge about the limits of sending signals from one part of the world to another, it does not conform to the way the world works.

Bell's theorem generalizes this argument for different measurement setups and different questions about the correlations among the measurement results. In all cases for which experiments have been carried out, the experimental results agree with the predictions of quantum mechanics. The conclusion is that these experiments rule out any reasonable local, deterministic alternative to quantum mechanics. We wrote "reasonable" because you could always come up with some contrived theory that gives a local, deterministic account of the experiments. For example, you might propose that there is a mysterious matrix that surrounds us all and the entire universe, and the effects of the matrix "cause" what we call quantum probabilities. But we have no evidence that such a matrix exists, and we don't like to invent special things that explain just one phenomenon and don't have any other effects.

Another conclusion from these experiments is that the strange nonlocal correlations between measurements of entangled states is a fundamental part of the way nature works. Einstein called it "Spooky action at a distance." (Einstein was really good at catchy aphorisms!) We observed examples of that in Chapter 9, where the result of Bob's observation on one qubit seemed to affect the probabilities for what Alice would observe on another qubit, if the two-qubit state was entangled. These effects may be "spooky" when looked at from a deterministic viewpoint, but they seem to be an integral part of the quantum world.

The scientific community struggled to understand these effects for many years because they seem so counterintuitive. We have finally accepted them by remembering they only show up for conditional probabilities, for which, we must admit, our intuition is almost always wrong, even for classical probabilities.

Let's explain these effects. When we say that Bob's measurement probabilities depend on Alice's observation results, we are really saying that *if* Alice observes her qubit state and gets spin-up (for example), *then* Bob's measurement probabilities will be different from those when Alice observes spin-down. So, Bob's probabilities are conditioned (that's the *if–then* part) upon Alice's observations. Of course, all the measurement probabilities are conditioned upon the state in which the two qubits are prepared. The muddle occurs when we try to give a mechanistic ("deterministic") account of these correlations. In some sense, we just need to "go with the flow" and accept these strange quantum results as part of the way the world works.

We can turn the arguments around and ask: If quantum states have these strange effects, why don't we see them with everyday objects? Part of the answer has to do with the fact that everyday objects (which range from biological cells up to the entire universe) are built out of many, many qubits. This sounds straightforward but providing a detailed account of how that explains local determinism for everyday objects has turned out to be rather difficult. In particular, it has been challenging to specify a border between quantum land and classical, Newtonian land.

You might be interested in looking at the results of the experiment, as mentioned in Chapter 9, carried out by a team of Chinese scientists (Yin et al., 2017). They used a source of entangled photons emitted by a laser system in an Earth-orbiting satellite. The photons were linearly polarized and emitted in pairs in an entangled two-photon state. One of the photons was directed to a receiving station on Earth and the other went to another Earth-based receiving station 1200 km away. The receivers each had measurement devices that could select different linear polarization measurement directions. The large distances between the two receivers and the satellite ruled out secret signals that might explain the observed correlations. The results agreed with the predictions of quantum mechanics and ruled out local, deterministic models (á la Bell's theorem) to a high degree of precision

The degree of precision is specified by the standard deviation as a quantitative measure of uncertainties. If the standard deviation, a measure of the uncertainty in the experimental results, is small enough, we can clearly decide between the two sets of predictions and see how well they match the experimental results. In most areas of science having the difference between the predictions of one of the theories and the data larger than five times the standard deviation (sometimes called a "5-sigma" result) is viewed as providing very strong evidence for one set of predictions over the other. In the Chinese experiment, the differences were almost 300-sigma. It means there is no practical way the differences between the local, deterministic predictions and the experimental results could have happened "by accident."

It looks like we are stuck with quantum mechanics. But let's put a positive spin on this: All the machinery of entangled states seems to be a valid description of nature; therefore, we can be confident that the QIS and QC methods that we build from entangled states and other quantum theoretical components will provide a solid underpinning for our work.

There is another way of testing Bell's theorem: We replace the Bell state by what is called a Greenberger–Horne–Zeilinger (GHZ) state, named after three physicists Daniel Greenberger, Michael Horne, and Anton Zeilinger, who introduced these states in the late 1980s. GHZ states are superpositions of the basis states of a three-qubit system. An example of a GHZ state is

$$|\psi\rangle_{GHZ} = \frac{1}{2}\left[|000\rangle - |111\rangle\right]. \tag{17.3}$$

You can use the results of Chapter 9 (Section 9.8) to check that this is an entangled state. We won't go through the details of the arguments, but GHZ states provide stronger tests of Bell's theorem by showing that the quantum results satisfy certain equalities rather than just inequalities that we have seen for the two-qubit tests. Details of the argument can be found in Chapter 6 of Mermin's *Quantum Computer Science* (see Further Reading). Zeilinger won the Nobel Prize for his later work showing the experimental confirmation of these results and other Bell theorem tests.

17.3 The Measurement Problem

In Chapter 5, where we introduced quantum measurements, we mentioned what is called the "measurement problem." The difficulty is trying to find a theoretical description, ideally within the framework of standard quantum mechanics, for what happens when a qubit (or any other quantum entity) interacts with a measurement device.

As shown before, the probability interpretation of quantum mechanics (the Born rule) comes into play with measurements. Although in the book we haven't yet said much about how quantum states evolve in time, we have implicitly assumed that if we produce a qubit in a superposition state and launch the qubit to Bob, the state stays the same unless some interaction with other qubits intervenes. If these interactions do occur, we describe the changes in a state via matrices that represent reversible processes. The problem is that measurements, as normally construed, are irreversible. The *measurement principle* asserts that the appropriate state for the qubit after a measurement has been made is the basis state associated with the observed value of that property—spin-up or spin-down, for example. But the principle does not describe a mechanism by which that happens. How do we account for the irreversibility? This is one aspect of the measurement problem. The measurement problem continues to generate spirited debates among scientists and philosophers, at least those philosophers who specialize in the philosophy of science and find quantum mechanics to be a fertile field in which to plant the seeds of their ideas about reality and the nature of science.

At a practical level, we know how to make a measurement device work. For linearly polarized photons, for example, we use a polarizing beam splitter to separate the qubits depending on the measurement outcome and then we use a photodetector to indicate in which state the photon was found. The detection of the photon is an irreversible event. We can then store the result by turning on a light that goes with the appropriate channel or in a number that gets stored in a classical computer. So, the basic issue is how to reconcile the irreversibility of the measurement process with the normally reversible quantum processes among qubits.

The general strategy is to invoke arguments based on the number of qubits (or other quantum entities) that make up the measurement device. If the number of qubits is large enough, the interactions between the original qubit and the measurement qubits get spread out over so many processes that "for all practical purposes" (as John Bell wrote) the overall measurement process is irreversible even though each contributing individual process is reversible. We use the same kind of argument in describing thermal energy ("heat") flow from a hot object to a cool object. We know that for all practical purposes we never see thermal energy flow spontaneously from cool regions to hot regions. Individual particles in the cool region may gain or lose kinetic energy, but if more of them gain and fewer lose kinetic energy, the region's temperature goes up. Since energy is conserved, the other region's average energy must go down and consequently its temperature goes down. If we do want to move thermal energy from a cool region (the inside of your house during the summer) to a warmer region (the air outside your house in summer), we must use a gadget such as an air conditioner to move that energy in the desired direction.

For quantum mechanics, the difficulty has been in finding a good theoretical criterion to decide when a process is irreversible enough to count as a measurement. A measurement device is essentially a classical machine; so, we can say that the measurement problem is a problem in defining the boundary between the quantum world and the classical world.

Absent an agreed-upon solution to the measurement problem, we use our practical knowledge of what works for actual measurements, hold our noses, and proceed to build all sorts of interesting quantum gadgets, including quantum computers. In practical terms, we know how to make measurements and to check those measurement results with the theoretical predictions of quantum mechanics. This combination has worked extraordinarily well so far, which indicates that we are just not asking the right questions about measurements.

Some authors have decided that ordinary quantum mechanics is perfectly fine, but that we need to add additional ideas to account for what happens in a measurement. How to describe that reduction from many possible outcomes to one outcome is another form of the measurement problem in quantum mechanics. You may have heard of parallel universes, consistent histories, quantum Bayesians, and so on, which are the signatures of various attempts to figure out that connection. The Further Reading section has citations to books and articles where you can find further information about these issues.

17.4 Measurement and Decoherence

The measurement issues spill over into another aspect of quantum states—an aspect that is of major concern for QC. This problem is called *decoherence*. We design our measurement devices to interact with our qubits at specified times and locations. However, the qubits are not completely isolated from their surroundings—their environments. Most often the environment has electromagnetic fields (thermal radiation, light, and electric fields) and molecular interactions, for example, that affect our qubits. These interactions mean that we need to take the quantum states of the environment into account when describing the behavior of our qubits. In essence, the overall system (qubits plus the environment, which is of course nothing more than a vast collection of qubits itself) is described by an entangled state. We distinguish this situation from a measurement interaction because often (almost always) we have no control over the interaction of our qubits with the environment. As we discussed in the previous section, we build a measurement device so that only some of its states are correlated with our qubit's states.

If we want to take the environment into account, we could write the state of qubit plus the environment as

$$\underbrace{\left(a_0|A_0\rangle + a_1|A_1\rangle\right)}_{\text{qubit state}}\underbrace{\left\{c_a|e_a\rangle + c_b|e_b\rangle + \ldots..\right\}}_{\text{state of the environment}}, \tag{17.4}$$

where $|e_a\rangle$ is a state of the environment and c_a is the corresponding state amplitude. Eq. (17.4) is appropriate only if we assume that the qubit and the environment are independent. The more realistic case is an entangled state for the composite system qubit plus environment. But it is an entangled state in which the environment's contribution is generally unknown and in fact changing with time. It means that the relative phases in any qubit superposition state we prepare become washed out (they suffer decoherence) due to interactions with the environment. This is bad for quantum computing! One of the tricks-of-the-trade in building QCs is minimizing the effects of the environment on the computational qubits. Of course, we can't eliminate such effects completely and this is one reason a lot of effort goes into developing quantum error correction routines to detect and correct the effects due to unintended interactions with the environment.

Decoherence is an example of the automotive makers' old motto: "If you have a lemon, sell lemonade." There are ways of using these interactions to enhance the sensitivity of a qubit to its environment in a controlled way; so, the qubit acts as an enhanced sensor for the environmental conditions. For example, we might be able to use a qubit as an enhanced sensor to monitor magnetic fields or electromagnetic waves. In these situations, we focus on just one aspect of the environment, and in this case, Eq. (17.4) makes good sense.

As an example, let's look at an experiment (Lachance-Quiron et al., 2020) that shows a qubit can be used to detect the smallest possible magnetic excitation in a magnetic crystal. The excitation is called a "magnon" (in analogy with photon for electromagnetic radiation). The crystal was yttrium iron garnet, often used in microwave, optical, and data storage devices.

The experiment sets up an interaction between the magnon and the qubit via a microwave radiation device similar to the ones used in microwave ovens. That interaction leads to a shift in the energies associated with the qubit states depending on the number of magnons present in the crystal. (Under these experimental conditions, the qubit has two energy states $|E_0\rangle$ and $|E_1\rangle$.) The changes in the energy states then lead to changes in the results of measuring the qubit state. For example, if no magnon is present, you are more likely to measure the energy state E_1 while observing E_0 is correlated with having one or more magnons present. It turns out that we can measure these qubit energy changes with sufficient precision that we end up with enhanced sensitivity to the presence of the magnons compared to traditional magnetic field measurement methods. Enhanced quantum sensing might be the most practical use of QIS even if QCs never reach their full potential.

17.5 The Relationship between Classical and Quantum Computing

The discussion in the previous sections leads to the interesting question about the relationship between classical computing and quantum computing. The indistinct boundary between quantum states and quantum measurements carries over to this relationship. In Chapter 2, we showed that classical computing states—the high and low voltages in the typical modern computer—are persistent states: We can look at them to determine what gates should be doing, and we can connect them with wires to many gates at the same time without disturbing the state. In fact, with most computers today, you can turn off the computer, and when you turn it back on, it remembers what you had been doing just before you turned it off. The classical computing states are nothing if not persistent.

There is nothing magical about that persistence, at least from an engineering point of view. It has been known for well over a century that these states can be made persistent by building in feedback mechanisms that keep the circuitry in the desired state until you are ready to have it switch states. In principle, this could be mechanical feedback like an on–off electrical switch in your home. The lamp on a desk has pull-chain switches to turn the light bulbs on and off. If we pull on the chain, we hear a click indicating that some mechanical mechanism has turned the switch from the on-position to the off-position or vice versa. In electronic computers, engineers build in "flip-flops" and other switch-like mechanisms that play the same role for the high-voltage, low-voltage states of the circuits.

So, once again the notions of irreversibility and persistence come into play. The classical computer switches can be quite small with millions of them built into a single integrated circuit "chip." Such switches have even been built from single molecules that switch from one state to another by changing their shapes. The switching becomes irreversible in the sense that the molecule needs a "nudge" to go from one shape to the other.

From these perspectives, a classical computer is just a quantum computer that has been designed to have only two persistent states, which we traditionally call 0 and 1. We don't think that anyone is going to build a quantum computer and then add feedback so that it behaves like a classical computer. Today's classical computers are highly refined systems that do their jobs quickly and reliably, except when they fail a few hours before one of your important deadlines. But as a matter of principle, a classical computer is just a quantum device built from billions of qubits, organized to provide persistent computational states.

17.6 Entanglement and Quantum Measurements

Entangled states also show up, perhaps surprisingly, in some descriptions of quantum measurements. John von Neumann (1903–1957), one of the pioneers of quantum theory, proposed that measurements in quantum mechanics should be understood as creating entangled states between the quantum system state (a qubit's state in our language) and the measurement system (usually built out of a lot of qubits). At first sight, this formulation seems to contradict our mantra "Measurement Crushes Entanglement," as mentioned in Chapter 9 for two-qubit systems. Let's explore these ideas.

A measurement device is engineered, based on our knowledge of physics, so that it has a state $|M_0\rangle$ correlated with the Alice's basis state $|A_0\rangle$, for example, and another state $|M_1\rangle$ correlated with Alice's basis state $|A_1\rangle$: if we observe the measurement device in the state $|M_1\rangle$ after the measurement interaction, then we know the incoming state had a state component along $|A_1\rangle$. If you understand that sentence, you have come a long way in developing your understanding of QIS.

In von Neumann's picture, measurement systems are set up so that the output of a measurement device is expressed as

$$\frac{1}{\sqrt{2}}\{|A_0\rangle\,|M_0\rangle + |A_1\rangle\,|M_1\rangle\}. \tag{17.5}$$

Often, authors say, pointing to an expression like Eq. (17.5), that the measurement interaction leaves the measurement device in a superposition of the two states $|M_0\rangle$ and $|M_1\rangle$. Schrödinger's infamous cat, viewed as a measurement device for radioactive decay, is, in this view, both dead and alive. Based on our discussion of entangled states in previous chapters, this kind of statement is just plain wrong. We have seen that if a system of qubits is described by an entangled state, we cannot say anything about the states of the individual qubits. In fact, in some sense, the individual qubits don't "have" quantum states if the overall state is an entangled one. Then, the key question is: How does the combined object-measurement device system "collapse" to a specific measurement outcome? von Neumann's model does not address this issue.

If we want a piece of equipment to serve as a measurement device, its state after the measurement interaction must be persistent, as mentioned before, so we can record the result without changing the measurement device state. For example, in the situation described in Eq. (17.5), if we obtain the result $|M_0\rangle$ for the measurement apparatus, we know from the design of the apparatus that the qubit state must have had a component along $|A_0\rangle$. So, the measurement problem is then the problem of figuring out under what conditions a measurement device state is persistent and not significantly disturbed by itself being measured.

We know how to build such devices, and they work reliably. But how do we provide a theory that tells us, for example, how many qubits must the measurement device have to end up in a persistent state.

In the examples we showed earlier, after the measurement interaction the red light or the green light should be on and both Alice and Bob should be able to look at that state without disturbing it. As shown before, such persistence does not hold for generic quantum states. Measuring the measurement device (if it is described by an ordinary quantum state vector) would in general put the measurement device into a new state: one of the basis states of the device that is measuring the measurement device. What breaks this chain of measuring the previously measured measurement device? Eq. (17.5) certainly does not solve the measurement problem.

In practice, we think we know how to carry out a measurement on a quantum system using magnets, polarizers and light detectors, lasers, and so on. Once again, we see that the measurement problem is how to describe these results formally within the framework of quantum mechanics itself without bringing in notions such as "macroscopic," "large scale," and so on, which don't seem to be part of quantum mechanics.

17.7 Entanglement and Correlations of Measurement Outcomes

In Chapter 9, we observed that entangled quantum states have unusual properties. First, none of the qubits whose system state is entangled has its own state. Second, measurements on one part of the multiqubit entangled system affect the state we assign to the other parts. At a fundamental level, these issues are all tied to the concept of correlation. For an entangled state, outcomes of measurements carried out on different parts of the system are correlated.

Correlations are quantitative relationships between (or among) sets of data—an issue that is important in statistical analysis. However, for our purposes, we need only a few simple ideas about correlations. In this section, we show how correlations among measurement outcomes are related to entanglement. These results give us yet another way of determining if a quantum state is entangled. They also tell us that if we want to develop a quantum algorithm that includes correlations, we need to build in entangled states. Indeed, correlations occur in classical systems as well as in quantum systems. So, let's start by thinking about classical correlations. Imagine that we have two cubes: one red and one green. If Alice randomly sends one of the cubes to Bob and he observes that his cube is green, then he knows with certainty that Alice's cube is red and vice versa. In this case, there is a definite correlation between the colors that we observe. We might call this *deterministic* correlation because the individual observations are perfectly correlated.

There is another, and very important, type of classical correlation: correlations among stochastic (random) variables. Stochastic means that the results of repeated observations fluctuate from one measurement to the next. In some cases, the variables themselves, such as people's heights, vary among the population being studied. In other cases, for example in measuring outdoor temperatures, the temperature itself might be fluctuating due to changing weather conditions. In such cases, we can have a situation in which the individual observations do not exhibit obvious correlations, but the averages of the observations do. These correlations are the bread and butter of statistical analysis. For example, clinical studies of vaccine effectiveness look for correlations between the vaccine dose and (hopefully) a smaller probability of catching a disease.

However, statistical correlations can be misleading. In a famous 2012 paper in the New England Journal of Medicine, the author showed that there is a statistical correlation between the chocolate consumption per capita in a country and the number of scientists from that country who have won the Nobel Prize. This relationship does not hold for all countries but on average the correlation seems to be there. The author concluded that eating chocolate increases your cognitive capabilities.

Before you run out and spend all your money on chocolate, you need to remember that "correlation is not causation." Perhaps chocolate consumption has nothing directly to do with cognitive abilities but is just a reflection of the wealth of a country, which also leads to more investment in scientific research, which leads to more Nobel Prizes. Also, in this case the author mixed up large group data (the country's chocolate consumption as a whole) and small group data (the Nobel Prize winners). The author failed to check how much chocolate the Nobel Prize winners themselves actually consumed. This cautionary tale reminds us that we must be careful in drawing conclusions from statistical correlations

The correlations we need to describe quantum entanglement are relationships among *average* results of measurements. In statistics and probability theory, the average is also called the mean, the expected value, or the expectation value, though we could never figure out who is expecting what. We will stick with "average."

The history of the term "expectation value" is kind of interesting. In the 1600s, mathematicians became interested in predicting the outcome of different kinds of games of chance, basically various forms of gambling. They wanted to estimate how much they might expect to win or lose from the games with different rules and different odds. The word "expectation" later became associated with these calculations, and by the early 20th century, the letter E was used to express expected or expectation values.

How do averages give us a quantitative way of expressing correlations among measurements? There are many, but we focus on just one. Suppose we have two stochastic variables F and G—ones that have significant variation across a set of measurements. We make use of three averages: $\langle F\rangle$, $\langle G\rangle$, and $\langle FG\rangle$. The degree of correlation between F and G is often expressed by the so-called covariance:

$$\text{Covariance}(F, G) = \langle FG\rangle - \langle F\rangle\langle G\rangle . \tag{17.6}$$

If the covariance is equal to zero (or very close to zero compared to the averages), we say that the two variables are uncorrelated.

Try It 17.3 The following table gives the results of 10 sets of measurements of two stochastic variables F and G. Calculate the covariance of the results given in the table. Are the variables correlated or not? Explain your reasoning.

F	6.9	8.0	8.5	4.7	4.6	8.1	0.8	3.7	5.0	5.4
G	2.2	9.1	5.9	5.7	1.5	3.7	8.8	1.5	2.1	4.0

As shown before, quantum measurements involve probability and randomness; so, we might guess that covariance can be used to express correlations among quantum measurements. We use a two-qubit system to demonstrate explicitly quantum measurement correlations. We see that these correlations can often tell us whether the system state is entangled or not. In a sense, this method is a generalization of the entanglement criteria we introduced in Chapter 9.

We focus on a two-qubit system in which each of the qubits is a spin-1/2 object. In Chapter 7, we noted that for these qubits there are two possible states for the spin component along a spatial direction. We follow common practice and use the direction indicated by a z-axis. If we measure a qubit's spin along the z-axis, we get either "spin-up" with projection +1 (in suitable units) or "spin-down" with projection −1.

If we measure the z projection for a set of identically prepared qubits, each described by a superposition state

$$|\psi_1\rangle = a_\uparrow |\uparrow\rangle + a_\downarrow |\downarrow\rangle, \tag{17.7}$$

we get a random series of +1s and −1s. Since we are now expert quantum mechanicians, we know that the probability of getting +1 is $|a_\uparrow|^2$ and the probability for getting −1 is $|a_\downarrow|^2$, assuming as usual that $|a_\uparrow|^2 + |a_\downarrow|^2 = 1$.

For many situations, it is helpful to write the average in a slightly different way by making use of the probability (relative frequency) of each measurement outcome. For our spin-1/2 qubit, suppose we get spin-up $n_\uparrow$ times and spin-down $n_\downarrow$ with $n_\uparrow + n_\downarrow = N$. That means the measurement outcome is +1 for $n_\uparrow/N$ fraction of the measurements and −1 for $n_\downarrow/N$ fraction of the measurements. In other words, the relative frequency of getting +1 is $P_\uparrow = n_\uparrow/N$ and the relative frequency for −1 is $P_\downarrow = n_\downarrow/N$. If N is fairly large, we expect that these relative frequencies will be close to the probabilities we obtain from the superposition state amplitudes $P_\uparrow \approx |a_\uparrow|^2$ and $P_\downarrow \approx |a_\downarrow|^2$. So, we can write the average of the Pauli operator Z in terms of the probabilities for each measurement outcome and the outcome values (+1 and −1):

$$\begin{aligned}\langle Z\rangle &= (\text{probability of getting spin-up}) \times (+1 \text{ for spin-up}) \\ &\quad + (\text{probability of getting spin-down}) \times (-1 \text{ for spin-down}) \\ &= (+1)\,P_\uparrow + (-1)\,P_\downarrow = (+1)\,a_\uparrow^2 + (-1)\,a_\downarrow^2.\end{aligned} \tag{17.8}$$

In Chapter 22 (Eq. 22.5), we show you a different way at arriving at this result.

The advantage of writing the average this way is that it is easy to generalize: you add up the products of the probabilities for each of the measurement basis states and the value associated with Z in that state. It is important to note that the average will be different for different quantum states, which will have different amplitudes. If the probabilities for spin-up and spin-down are equal, the average $\langle Z \rangle = 0$, not equal to either of the two eigenvalues.

We are now ready to look at the two-qubit system. Let's label the qubits with our usual A and B. The quantities we want to calculate are the average of the Z operators $\langle Z_A \rangle$ and $\langle Z_B \rangle$ for each of the two qubits. The other quantity we need is $\langle Z_A Z_B \rangle$, which means that we measure Z for both qubits. It turns out that in general for an entangled two-qubit state $\langle Z_A Z_B \rangle \neq \langle Z_A \rangle \langle Z_B \rangle$ and for a simple product (nonentangled) state $\langle Z_A Z_B \rangle = \langle Z_A \rangle \langle Z_B \rangle$. The difference between $\langle Z_A Z_B \rangle$ and $\langle Z_A \rangle \langle Z_B \rangle$, the covariance, is a measure of the correlation in the qubits' system state:

$$\text{Covariance}\,(Z_A, Z_B) = \langle Z_A Z_B \rangle - \langle Z_A \rangle \langle Z_B \rangle \,. \tag{17.9}$$

As a specific example, suppose we have prepared the two qubits in the Bell state

$$|\psi_2\rangle = |\text{Bell}_{00}\rangle_{AB} = \frac{1}{\sqrt{2}} \left(|\uparrow_A\rangle\, |\uparrow_B\rangle + |\downarrow_A\rangle\, |\downarrow_B\rangle \right). \tag{17.10}$$

Since the spin-up and spin-down states are the measurement basis states for the Z operator, we know that

$$Z_A |\uparrow_A\rangle = +1 |\uparrow_A\rangle \quad \text{and} \quad Z_A |\downarrow_A\rangle = -1 |\downarrow_A\rangle \,. \tag{17.11}$$

When we apply Eq. (17.11) to the parts of the superposition state in Eq. (17.10), we obtain

$$\begin{aligned} Z_A |\uparrow_A\rangle\, |\uparrow_B\rangle &= +1 |\uparrow_A\rangle\, |\uparrow_B\rangle \\ Z_A |\downarrow_A\rangle\, |\downarrow_B\rangle &= -1 |\downarrow_A\rangle\, |\downarrow_B\rangle , \end{aligned} \tag{17.12}$$

with analogous results for measurements on B.

Using these results to calculate the average value of the Z measurements, we find

$$\langle Z_A \rangle = \underbrace{\frac{1}{2}}_{\text{probability for } \uparrow} \underbrace{(+1)}_{\text{result for } \uparrow} + \underbrace{\frac{1}{2}}_{\text{probability for } \downarrow} \underbrace{(-1)}_{\text{result for } \downarrow} = 0, \tag{17.13}$$

just as expected. Similarly, we obtain $\langle Z_B \rangle = 0$. We immediately conclude that $\langle Z_A \rangle \langle Z_B \rangle = 0$.

Now we need $\langle Z_A Z_B \rangle$. Let's write this out for each of the parts of Eq. (17.10):

$$\begin{aligned} Z_A Z_B |\uparrow_A\rangle |\uparrow_B\rangle &= (Z_A |\uparrow_A\rangle)(Z_B |\uparrow_B\rangle) = (+1)(+1) |\uparrow_A\rangle |\uparrow_B\rangle \\ Z_A Z_B |\downarrow_A\rangle |\downarrow_B\rangle &= (Z_A |\downarrow_A\rangle)(Z_B |\downarrow_B\rangle) = (-1)(-1) |\downarrow_A\rangle |\downarrow_B\rangle . \end{aligned} \tag{17.14}$$

From Eq. (17.14) and (17.8), we find

$$\langle Z_A Z_B \rangle = 1. \tag{17.15}$$

The covariance between these spin measurements is

$$\begin{aligned}\text{Covariance}\,(Z_A, Z_B) &= \langle Z_A Z_B\rangle - \langle Z_A\rangle\langle Z_B\rangle \\ &= 1 - 0 = 1.\end{aligned} \tag{17.16}$$

We conclude that the Z measurements for the two parts of the system described by the state in Eq. (17.10) are correlated and hence the state is an entangled state. In fact, this result tells us that the Z measurements for the Bell state in Eq. (17.10) are perfectly correlated.

Try It 17.4 Suppose the two-qubit state is the product state $|\psi_3\rangle = |\uparrow_A\rangle\,|\uparrow_B\rangle$. What is the value of the covariance of the Z_A and Z_B measurements? What about the state $|\psi_4\rangle = |\uparrow_A\rangle\,|\downarrow_B\rangle$? Explain the results.

Try It 17.5 Show that the covariance for Z measurements for the state $|\psi_3\rangle = |\text{Bell}_{11}\rangle = \frac{1}{\sqrt{2}}(|\uparrow_A\rangle|\downarrow_B\rangle - |\downarrow_A\rangle|\uparrow_B\rangle)$ equals -1. We say that for this state, the Z measurements are perfectly "anticorrelated." Whenever you get spin up for qubit A, you get spin down for qubit B and vice versa.

Let's sum up what we have learned: For a product (nonentangled) state, the covariance of two different measured variables equals 0. If the covariance is not 0, the state is an entangled one. However, it turns out that for a given entangled two-qubit state, you can always find two operators whose covariance is 0. It means that getting a covariance equal to 0 does not guarantee having a nonentangled state. If you get a 0 covariance for one pair of operators, calculating the covariance of another pair of operators will usually tell you if the state is entangled or not.

Measurements of correlations can also be used to test the "fidelity" of the implementation of quantum gates in real-life quantum communication devices and quantum computers. If the actual gates can maintain the correlations, the gates are said to have "high fidelity." It means that the errors introduced by the physical implementations of the gates are small.

17.8 Chapter Summary

- Bell's theorem provides theoretical predictions that can be used to test local, deterministic (hidden variable) models against the corresponding quantum models of correlations among measurement outcomes for multiqubit systems. Experiments rule out the local, deterministic models.

- The notion of measurement remains, in principle, somewhat problematic at the fundamental level. However, the "measurement problem" does not seem to provide any constraints in building actual quantum computers.
- The interaction of qubits with their environment leads to decoherence of the qubits' quantum states and thus requires quantum error correction in QIS and QCs.
- Structured decoherence can lead to enhanced quantum sensing devices, which in the short term may be more important than QCs.
- Correlations among measurement outcomes in multiqubit states can be used to test for entanglement.
- Measurements of correlations can also be used to test the "fidelity" of the implementation of quantum gates in real-life quantum communication devices and quantum computers.

17.9 Further Reading

Experiment Demonstrating Entanglement Distribution

[Yin et al., 2017]. J Yin, Y Cao, Y-H Li, S-K Liao, L Zhang, J-G Ren, W-Q Cai, W-Y Liu, B Li, H Dai, G-B Li, Q-M Lu, Y-H Gong, Y Xu, S-L Li, F-Z Li, Y-Y Yin, Z-Q Jiang, M Li, J-J Jia, G Ren, D He, Y-L Zhou, X–X Zhang, N Wang, X Chang, Z-C Zhu, N-L Liu, Y-A Chen, C-Y Lu, R Shu, C-Z Peng, J-Y Wang and J-W Pan, "Satellite-based entanglement distribution over 1200 kilometers," *Science* **356**, 1140–1144 (2017).

Using a Qubit as an Enhanced Quantum Sensor

[Lachance-Quiron et al., 2020]. D. Lachance-Quirion, S.P. Wolski, Y. Tabuchi, S. Kono, K. Usami, and Y. Nakamura, "Entanglement-based single-shot detection of a single magnon with a superconducting qubit," *Science* **367**, 425–428 (2020).

Bell's Theorem and Related Issues

Daniel F. Styer, *The Strange World of Quantum Mechanics* (Cambridge University Press, Cambridge, 2000). This book provides a nice discussion of experimental tests of Bell's theorem (pp. 41–47) without mathematics, but all the physics.

George Greenstein and Arthur Zajonc, *The Quantum Challenge: Modern Research on the Foundations of Quantum Mechanics*, 2nd Ed. (Jones and Barlett, Sudbury, MA 2006). This book provides a comprehensive and comprehensible treatment of experiments on Bell's theorem, delayed-choice measurements, and so on.

Louis Gilder's *The Age of Entanglement* (Vantage, New York, 2009) provides a nice summary of the discussions at the Amherst College 1990 Foundations of Quantum Mechanics Conference.

Tanya Bub and Jeffrey Bub, *Totally Random, Why Nobody Understands Quantum Mechanics, A Serious Comic on Entanglement* (Princeton University Press, Princeton 2018). This book provides a visual introduction to the mysteries of quantum mechanics, which is definitely not your grandmother's textbook on quantum mechanics. Bell's theorem (inequality) is described on pp. 36–41. More information is available at www.totallyrandom.info.

Chris Bernhardt, *Quantum Computing for Everyone* (MIT Press, Cambridge, MA, 2019). Bell's inequality is described on pp. 79–84.

David Mermin, *Quantum Computer Science* (Cambridge University Press, Cambridge, 2007). A nice discussion of using GHZ states in tests of Bell's theorem is given on pp. 154–158. Mermin also includes a thought-provoking discussion of several of the underlying issues such as "locality" and "elements of reality."

Robert Ross, "Computer simulation of Mermin's quantum device," *American Journal of Physics* **88**, 483–489 (2020). The article shows how to use a classical computer to simulate the quantum predictions for the Bell's theorem device described in this chapter.

Correlations

F. H. Messerli, "Chocolate consumption, cognitive function, and Nobel laureates," *New England Journal of Medicine* **367**, 1562–1564 (2012).

The conclusions about chocolate and Nobel Prizes have been critiqued many times. See, for example, Vladica Velickovic, "What everyone should know about statistical correlation," *American Scientist* **103**, 26 (2015). DOI: 10.1511/2015.112.26

David Mermin, "What is quantum mechanics trying to tell us," *American Journal of Physics* **66**, 753 (1988). The author argues that quantum mechanics can be freed of "the tyranny of measurements" by focusing on correlations.

Jed Brody and Gavin Guzman, "Calculating spin correlations with a quantum computer," *American Journal of Physics* **89**, 35–40 (2021). This article discusses correlations in quantum measurements and how to calculate them using a quantum computer accessible online.

Quantum State Time Dependence

18

Be ruled by time, the wisest counselor of all.

Plutarch, *Lives of the Noble Greeks and Romans, Pericles*

18.1 Time-Dependent State Vectors

We have seen how quantum state changes can be described by the action of quantum gates acting on quantum states, but the description has really been just a "black box" representation: We specify how the quantum gate causes an "input" state to become an "output" state without worrying about the details of how that change occurs. In essence, we assume that some clever scientists and engineers have designed and built the black boxes for the types of qubits we are using (photons, electrons, superconducting circuits, etc.). All we need to do is provide the input states and use the output states. This situation is like classical computing where most computer users can ignore the details of how semiconductors work in integrated circuits chips. If we think about what happens in a quantum algorithm, there are changes in the states over time. Nothing happens instantaneously.

In this chapter, we show how to describe mathematically the details of how quantum states change in time, which then gives us a way of understanding the manipulations that go on within quantum gates. Chapter 19 makes use of what we learn in this chapter to describe how quantum states change due to the interaction of spins (and their associated magnetic dipoles) with magnetic fields. We see that oscillating magnetic fields can be used to implement several types of quantum gates for these systems. The quantum states associated with atomic clocks (which make the global GPS system possible) are also described by those methods. In addition, time evolution plays a major role in quantum sensors, another rapidly growing area of quantum information science.

Of course, quantum states change with time in almost every physical, chemical, and biological system, not just the "artificial" ones we build in our labs. Change is necessary for life. In principle, the methods developed in this chapter can help us understand those changes as well.

As you might guess, we use operators (and their matrix representations) to describe how state vectors change in time. This approach fits in with our overall linear algebra approach to quantum physics, but we add a bit more mathematical flavor. We will show how the

Quantum Computing and Quantum Physics. Alice Flarend and Robert Hilborn, Oxford University Press.
 DOI: 10.1093/oso/9780197904930.003.0018

operator approach is related to the method that was used by Schrödinger and others about a hundred years ago (the 1920s). Their method used differential equations, which even today are the tools of choice in the sciences, engineering, economics, and many other fields concerned with providing a quantitative description of how things change with time (and with other variables).

Let's begin the discussion with our old friend, the general superposition state for a qubit (two-state) system:

$$|\psi(t)\rangle = a_1(t)|A_1\rangle + a_2(t)|A_2\rangle, \tag{18.1}$$

with the basis states $|A_1\rangle$ and $|A_2\rangle$ and the corresponding state amplitudes $a_1(t)$ and $a_2(t)$. Recall that, in general, the state amplitudes are complex numbers. Note the subtle change in notation. The key point is that when we say a state evolves in time, we really mean that the state amplitudes evolve in time; thus, we write them as explicit functions of time.

As mentioned above, our strategy is to introduce an operator $U_{\delta t}$ that acts on the state vector at a particular time t and generates the new state vector at a time $t + \delta t$. The symbol U is used for this operator to remind us that we want the operator to be a unitary operator.

Try It 18.1 Explain to a friend or colleague why we want to restrict ourselves to unitary operators.

The formal operation is expressed as

$$U_{\delta t}|\psi(t)\rangle = |\psi(t + \delta t)\rangle. \tag{18.2}$$

We first focus on small time steps δt because the operator U itself may change in time. This strategy also allows us to make connections with differential equations—yet another (useful) way to describe time evolution.

We introduce some notation that is commonly used for time evolution in quantum mechanics. First, we write the operator U as

$$U_{\delta t} = \mathrm{I} - i\frac{H}{\hbar}\delta t. \tag{18.3}$$

Let's explain each of the notational structures. First, I is the identity operator, so if $\delta t = 0$, no change occurs in the state vector, and we have $U_{\delta t=0}|\psi(t)\rangle = |\psi(t)\rangle$. Second, we expect the change in the state vector to be proportional to δt for small values of δt. In this limit, the operation result will be linear in δt. Third, by convention, we choose the operator H to be a Hermitian operator (that is, it is equal to its complex conjugate transpose $H = H^\dagger$ and consequently its eigenvalues will be real numbers). We discuss these issues once we have some specific examples.

H is called the Hamiltonian operator to recognize William Rowan Hamilton, a British scientist and mathematician, who, 100 years before quantum mechanics was developed, invented much of the mathematics that underlies quantum mechanics. In most cases, H represents a combination of kinetic and potential energies. $\hbar$ (pronounced "h bar") is Planck's constant (divided by 2π). In SI units, $\hbar = 1.05\ldots \times 10^{-34}$ Joule second and sets the energy and timescales for quantum physics. You may wonder why $i = \sqrt{-1}$ occurs in the operator. The simplest explanation is that having i multiply the operator H ensures that (to first order in δt) the overall operator will be unitary (its complex conjugate transpose is equal to its inverse) if H is Hermitian. The inverse operation simply replaces δt with $-\delta t$. (Recall that the complex conjugate operation replaces i with $-i$.)

We now use Eq. (18.3) in Eq. (18.2):

$$|\psi(t+\delta t\rangle = \left(\mathrm{I} - i\frac{H}{\hbar}\delta t\right)|\psi(t\rangle . \tag{18.4}$$

Expanding the right side of Eq. (18.4) gives us

$$|\psi(t+\delta t\rangle = |\psi(t)\rangle - i\frac{H}{\hbar}\delta t\,|\psi(t\rangle . \tag{18.5}$$

Eq. (18.5) shows how to calculate $|\psi(t+\delta t\rangle$ if we know $|\psi(t)\rangle$ and H. It is essentially an update rule to move forward in time by δt.

Try It 18.2 Go through the algebraic steps from Eq. (18.4) to Eq. (18.5). Explain what happens at each step.

Eq. (18.5) shows an abstract expression of the time evolution of a state vector. To carry out the calculation numerically, we want to shift our attention to the state amplitudes (complex numbers) and that means choosing a set of basis states. Recall that the state amplitudes in the A basis (see Eq. (18.1)) are given by $a_1(t) = \langle A_1|\,\psi(t)\rangle$ and $a_2(t) = \langle A_2|\,\psi(t)\rangle$; so, we may convert Eq. (18.5) into an equation for the state amplitudes by left multiplying all the terms with the left basis vector $\langle A_1|$

$$\langle A_1\,|\,\psi(t+\delta t)\rangle = \langle A_1\,|\,\psi(t)\rangle - i\frac{\delta t}{\hbar}\langle A_1|\,H\,|\psi(t)\rangle , \tag{18.6}$$

which is equivalent to

$$a_1(t+\delta t) = a_1(t) - i\frac{\delta t}{\hbar}\langle A_1|\,H\,|\psi(t)\rangle \tag{18.7}$$

and with analogous equations for $a_2(t)$.

Try It 18.3 Write out the equation analogous to Eq. (18.7) for $a_2(t)$.

These equations are numerical equations we can use to find how the state amplitudes evolve in time. But first we need to figure out what "quantum sandwich" terms like $\langle A_1|H|\psi(t)\rangle$ mean. Using Eq. (18.1) to write $|\psi(t)\rangle$ in the A basis, we find that

$$\begin{aligned} a_1(t+\delta t)-a_1(t) &= -i\frac{\delta t}{\hbar}\left[\langle A_1|H|A_1\rangle a_1(t) + \langle A_1|H|A_2\rangle a_2(t)\right] \\ a_2(t+\delta t)-a_2(t) &= -i\frac{\delta t}{\hbar}\left[\langle A_2|H|A_1\rangle a_1(t) + \langle A_2|H|A_2\rangle a_2(t)\right]. \end{aligned} \tag{18.8}$$

Try It 18.4 Write out the steps that lead from Eq. (18.7) to Eq. (18.8).

Recall that terms like $\langle A_1|H|A_2\rangle$ are called the matrix elements of the operator H in the A basis. For the Hamiltonian operator, these matrix elements are usually closely related to the energies associated with the qubit system. We may simplify the notation by using $H_{12} = \langle A_1|H|A_2\rangle$ and so on. Using that notation, we may write Eq. (18.8) as

$$\begin{aligned} a_1(t+\delta t)-a_1(t) &= -i\frac{\delta t}{\hbar}\left[H_{11}a_1(t) + H_{12}a_2(t)\right] \\ a_2(t+\delta t)-a_2(t) &= -i\frac{\delta t}{\hbar}\left[H_{21}a_1(t) + H_{22}a_2(t)\right]. \end{aligned} \tag{18.9}$$

These equations can be evaluated once the operator H is specified. The Hamiltonian operator will be different for different physical systems (electron spins versus superconducting circuits versus trapped ions, for example) and for different ways of manipulating those systems. In a subsequent section, we look at a few specific examples.

Math Aside The formalism using an operator (a linear transformation) with an "infinitesimal" parameter, such as δt, so the operator is "close to" the identity operator, is related to what is called a *Lie algebra*, named after Sophus Lie, a Norwegian mathematician (1842–1899). Associated with a Lie algebra is a Lie (mathematical) group of transformations. In physics, these transformations most often arise from symmetries of the system being studied. For example, for "time-translation symmetry," the equations of motion are unchanged if t is replaced by $t + T$ for all t and T. In our analysis, we used $T = \delta t$ to make the transformation operator just infinitesimally different from the identity operator. In addition to the time-translation transformations, there are similar transformations that depend on angular variables, which are related to rotation groups and, in quantum mechanics, to angular momentum operators. We also find the same formalism when we study spatial transformations $x \rightarrow x + \delta x$, which lead to operators associated with linear momentum.

Before moving on, let's look at Eq. (18.9) in more detail. First, both equations involve both state amplitudes, and we say they are "coupled" equations. Also note that once we

Two-state quantum time evolution

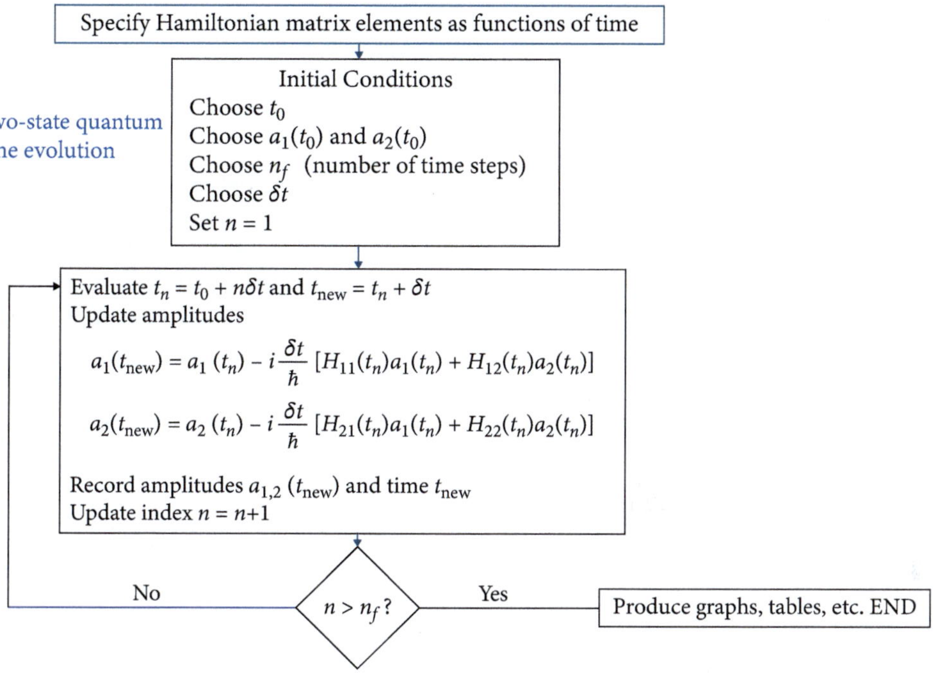

Fig. 18.1 The pseudocode for the numerical solution of Eq. (18.9).

have found $a_1(t + \delta t)$ and $a_2(t + \delta t)$, we can repeat the process to calculate $a_1(t + 2\delta t)$ and $a_2(t + 2\delta t)$, etc. This yields a straightforward iterative way of calculating how the state amplitudes (and hence the state) evolve in time. If H itself depends on time, we need to re-evaluate the matrix elements at each step. Figure 18.1 shows a pseudocode representation of the iteration steps.

Computational Aside: How do we know what value of δt to use in the iteration process? For quantum systems, the reduced Planck's constant $\hbar$ sets the scale of energy multiplied by time, as mentioned before. So, we want to choose δt such that $\delta t \left|H_{jk}^{(max)}\right| /\hbar$ is small compared to 1, where $\left|H_{jk}^{(max)}\right|$ is the largest (in absolute value) Hamiltonian matrix element for the particular system being studied. Often, it is better to define the Hamiltonian matrix elements in units of $\hbar$ to avoid having to deal with small numbers. We describe these kinds of details when we look at some specific examples of time evolution.

Next, let's look at a special case in which two of the matrix elements are zero: $H_{12} = 0 = H_{21}$. In this case, the equations are uncoupled:

$$\begin{aligned} a_1(t+\delta t) - a_1(t) &= -i\frac{\delta t}{\hbar}H_{11}a_1(t) \\ a_2(t+\delta t) - a_2(t) &= -i\frac{\delta t}{\hbar}H_{22}a_2(t). \end{aligned} \tag{18.10}$$

If the Hamiltonian operator does *not* depend on time, we show that the solutions to Eq. (18.10) are complex exponential functions of time:

$$a_1(t) = a_1(0)e^{-iH_{11}t/\hbar} \text{ and } a_2(t) = a_2(0)e^{-iH_{22}t/\hbar}. \tag{18.11}$$

How can we show that is true? There are two methods: (1) iterate Eq. (18.10) starting from $t = 0$ and ending at t and (2) turn Eq. (18.10) into differential equations, which are relatively easy to solve. If you would like to see the "standard" method (2), you may skip ahead to the Differential Equations Method section. The *iteration method* requires no knowledge of differential equations, but it is not commonly taught in quantum mechanics. If you are not interested in either of the derivation methods, you may skip ahead to the Comments on the Results section.

Iteration Method

The *iteration method* is analogous to the analysis of compound interest in finance. In fact, the exponential e was developed by Jacob Bernoulli in 1683 while solving a compound interest problem. The compound-interest-like analysis may be carried out in (at least) two ways (details available at https://wumbo.net/examples/derive-exponential-function-compound-interest/ and https://wumbo.net/examples/derive-exponential-function-compound-interest-alt/). We present the shorter of these ways.

Let's start by formulating the approximate iteration result for N steps ($t = N\delta t$) for the state amplitude $a_1(t)$:

$$\begin{aligned} a_1(t) &\approx \left[\underbrace{\left(1 - \frac{iH_{11}t}{\hbar N}\right)\cdots\left(1 - \frac{iH_{11}t}{\hbar N}\right)}_{N\text{ terms}}\right] a_1(0) \\ &= \left(1 - \frac{iH_{11}t}{\hbar N}\right)^N a_1(0) = \left(1 + \frac{rt}{N}\right)^N a_1(0), \end{aligned} \tag{18.12}$$

where we have introduced $r = -iH_{11}/\hbar$ to simplify the notation. Next, we define $m = N/(rt)$ to help match the Bernoulli form below and use that definition in Eq. (18.12) to obtain

$$a_1(t) \approx \left(1 + \frac{1}{m}\right)^{mrt} a_1(0). \tag{18.13}$$

To get the exact solution, we take the limit $m \to \infty$ in Eq. (18.13):

$$\lim_{m\to\infty}\left(1 + \frac{1}{m}\right)^{mrt} = \left[\lim_{m\to\infty}\left(1 + \frac{1}{m}\right)^m\right]^{rt}. \tag{18.14}$$

The limit within the square bracket in Eq. (18.14) is just what Bernoulli discovered:

$$\lim_{m\to\infty}\left(1 + \frac{1}{m}\right)^m = 2.71828... = e. \tag{18.15}$$

Try It 18.5 Use a spreadsheet or simple computer program to verify Eq. (18.15), starting with $m = 1$. How many steps do you need to get within 1% of the known value of e.

Using these results in Eq. (18.12) yields

$$a_1(\delta t) = e^{-iH_{11}\delta t/\hbar} a_1(0), \tag{18.16}$$

with an analogous equation for $a_2(t)$.

You may have noted that m, as we defined it, is an imaginary number, so you may legitimately worry about Eq. (18.15) with imaginary m. It is still correct, but proving so requires knowledge of complex variables beyond the prerequisites for this book. The next section provides a derivation that avoids that difficulty.

Differential Equations Method

The second method converts Eq. (18.10) into two relatively simple differential equations. We first divide both sides of Eq. (18.10) by δt and find that

$$\begin{aligned} \frac{a_1(t+\delta t) - a_1(t)}{\delta t} &= -i\frac{H_{11}a_1(t)}{\hbar} \\ \frac{a_2(t+\delta t) - a_2(t)}{\delta t} &= -i\frac{H_{22}a_2(t)}{\hbar}. \end{aligned} \tag{18.17}$$

If we think back to the definition of derivative in elementary calculus, we see that the left sides of the previous equations, in the limit $\delta t \to 0$, are the derivatives (rates of change) of the state amplitudes with respect to time. Hence, we may write Eqs. (18.17) as "differential equations" for the amplitudes:

$$\begin{aligned} \frac{da_1(t)}{dt} &= -i\frac{H_{11}a_1(t)}{\hbar} \\ \frac{da_2(t)}{dt} &= -i\frac{H_{22}a_2(t)}{\hbar}. \end{aligned} \tag{18.18}$$

In calculus, a differential equation is an equation that links the derivatives of functions to other parameters and other functions. The differential equations in Eq. (18.18) can be easily solved by recalling that there is only one function in mathematics whose derivative is proportional to the function itself, namely the exponential function. Using $\frac{de^{kx}}{dx} = ke^{kx}$, where k is a constant, it is easy to show that the solutions to Eqs. (18.18) are those in Eq. (18.11).

Try It 18.6 Verify that the results in Eq. (18.11) satisfy the differential equations in Eq. (18.18). Hint: Use the expression for the derivative of an exponential function.

Comments on the Results

Note that in this special (but important) case of a time-independent H with $H_{12} = H_{21} = 0$, the probabilities of finding the qubit in each of the A basis states does not depend on time. For example, by the standard Born rule the probability of finding the qubit in the basis state $|A_1\rangle$ is

$$|a_1(t)|^2 = a_1^*(t)a_1(t) = a_1^*(0)e^{iH_{11}t/\hbar}a_1(0)e^{-iH_{11}t/\hbar} = |a_1(0)|^2. \tag{18.19}$$

These states are called "energy eigenstates" because the matrix representation of the energy operator H acting on the basis state $|A_1\rangle$ in the column vector form is

$$\begin{pmatrix} H_{11} & 0 \\ 0 & H_{22} \end{pmatrix}\begin{pmatrix} 1 \\ 0 \end{pmatrix} = H_{11}\begin{pmatrix} 1 \\ 0 \end{pmatrix}. \tag{18.20}$$

So, we conclude that H_{11} is the eigenvalue of the operator H associated with the eigenvector $\begin{pmatrix} 1 \\ 0 \end{pmatrix}$, which is the column vector form of the basis state $|A_1\rangle$. Most traditional courses in quantum mechanics spend a lot of time finding energy eigenvalues and eigenvectors for specific physical systems like hydrogen atoms.

Try It 18.7 Write the equivalent of Eq. (18.20) for the basis state $|A_2\rangle$. What is the corresponding eigenvalue?

In more general situations in which all the matrix elements of H are nonzero and when $\hat{H}$ itself may vary in time, the appropriate differential equations are

$$\begin{aligned} \frac{da_1(t)}{dt} &= -\frac{i}{\hbar}\left[H_{11}(t)a_1(t) + H_{12}(t)a_2(t)\right] \\ \frac{da_2(t)}{dt} &= -\frac{i}{\hbar}\left[H_{21}(t)a_1(t) + H_{22}(t)a_2(t)\right]. \end{aligned} \tag{18.21}$$

These equations may be difficult to solve analytically (paper and pencil solutions), but computers easily find the solutions numerically. In fact, most computational packages (e.g. Python, Mathematica, MATLAB, etc.) have built-in functions that will do all the work for you.

18.2 The Schrödinger Equation

The abstract Eq. (18.5) for the state vector may also be expressed using a time derivative. The usual form for that (abstract) differential equation is

$$i\hbar\frac{d\,|\psi(t)\rangle}{dt} = H\,|\psi(t)\rangle, \tag{18.22}$$

and is called the *Schrödinger equation*, named after Erwin Schrödinger (1887–1961), Nobel laureate and a pioneer developer of quantum mechanics. Note that the Schrödinger equation suggests that we think of the time derivative $\frac{d}{dt}$ as an operator (linear transformation) acting on the state vector $|\psi(t)\rangle$. Of course, to solve the Schrödinger equation, we need to choose basis states and a Hamiltonian operator to convert Eq. (18.22) into a differential equation for numerical quantities.

In Chapter 19, we show how the formalism developed here plays out for a qubit system subject to an oscillating perturbation. Such a model is useful in describing the physics behind many quantum gate operations and it also provides a simple description of magnetic resonance imaging—a critical biomedical application.

18.3 Heisenberg Picture—Time-Dependent Operators

In the first part of this chapter, we focused our attention on the time evolution of quantum state vectors via the time dependence of the state amplitudes. This approach is called the Schrödinger picture of time dependence. There are other approaches to quantum time evolution. As an example, we discuss the so-called Heisenberg picture for time dependence, which allows the quantum operators rather than the quantum states to carry the time dependence. To motivate the Heisenberg picture, let's start with the expression for the average value of some quantum operator A for a time-dependent quantum state:

$$\langle\psi(t)|\, A \,|\psi(t)\rangle = \langle\psi(0)|\, e^{iHt/\hbar} A e^{-iHt/\hbar} \,|\psi(0)\rangle\,, \tag{18.23}$$

where we used the time evolution factors from Eqs. (18.11). We may view the right side of Eq. (18.23) as expressing the time dependence of the operator $A(t)$:

$$A(t) = e^{iHt/\hbar} A(0) e^{-iHt/\hbar}. \tag{18.24}$$

Unfortunately, Eq. (18.24) is difficult to evaluate if the operator A does not commute with the Hamiltonian H. However, we can find a differential equation for the operator that is simpler to use in many situations. Let's begin by using Eq. (18.24) for a small time step δt:

$$\begin{aligned} A(\delta t) &= e^{iH\delta t/\hbar} A(0) e^{-iH\delta t/\hbar} \approx (\mathrm{I} + iH\delta t/\hbar)\, A(0)(\mathrm{I} - iH\delta t/\hbar) \\ &= A(0) + iHA\delta t/\hbar - iHA\delta t/\hbar + (H(\delta t)/\hbar)^2 \\ &\approx A(0) + \frac{1}{i\hbar}\,[A, H]\,\delta t. \end{aligned} \tag{18.25}$$

In the first line, we used the power series expression for the exponential function, namely, $e^x \approx 1 + x$ for $|x| << 1$. In progressing from the second line of Eq. (18.25) to the third line, we moved i to the denominator (to have our final result agree with the conventional way of writing Eq. (18.26)) and assumed that $H\delta t/\hbar$ is small enough such that terms involving that quantity squared can be ignored compared to the linear terms. This removed the last term.

We also invoked the symbol for the commutator of two operators $[A, H] = AH - HA$. We like to call that commutator the "Ah-ha" rule to help us remember the order of the operators.

Next, we rearrange Eq. (18.25) in the form of a differential equation:

$$\frac{A(\delta t) - A(0)}{\delta t} \approx \frac{dA}{dt} = \frac{1}{i\hbar}[A, H]. \tag{18.26}$$

The last equality expresses the Heisenberg equation for the time evolution of an operator. If the operator A commutes with the Hamiltonian, then its derivative with respect to time is zero, which means that A is a constant. In physics language, we say that the expectation value of A (and in fact any function of the operator A) is "conserved" during the time evolution described by H. In general, the average value of A is calculated from $\langle\psi(0)|\, A(t)|\psi(0)\rangle$. In many practical applications, Eq. (18.26) is written in terms of the matrix elements (complex numbers), which can then be used in standard computer packages for finding solutions of differential equations. Let's represent the commutator as an operator $[A, H] = C$. Then, the Heisenberg time evolution equation may be written as

$$\frac{dA_{jk}}{dt} = \frac{1}{i\hbar} C_{jk} \tag{18.27}$$

using some convenient set of basis vectors.

In Chapter 20, Number Operators and Number States, we use the Heisenberg picture to see how those operators evolve with time. The Heisenberg picture is often used in theoretical calculations. It is also used to connect the quantum time evolution of the observable property represented by A to the time evolution of the corresponding classical property.

18.4 Chapter Summary

- In this chapter, we use a unitary operator U to describe how a quantum state vector evolves from time t to time $t + \delta t$ as shown in Eq. (18.2). By convention, that operator is expressed in terms of the identity operator I and a Hermitian operator H, called the Hamiltonian operator or just Hamiltonian for short. Using those operators and their matrix representations, we show how to find how the state's amplitudes change in time by iteratively applying Eq. (18.8).
- An alternative and more commonly used approach is to convert Eq. (18.8) to a set of coupled differential equations Eq. (18.21). This approach allows us to use the many mathematical and computational techniques for solving differential equations.
- The key "takeaway" message is that the Hamiltonian operator H allows us to calculate the time evolution of a quantum state. Eq. (18.22), the Schrödinger equation, encapsulates that principle in abstract form.

- As an alternative, we may use the Heisenberg picture to focus on the time evolution of quantum operators. In that picture, an operator A's time evolution is described by the Heisenberg equation $dA/dt = [A, H]/i\hbar$. The time dependence of an expectation (average) value is then calculated by using a quantum sandwich with the initial-time state functions and the time-dependent operator

$$\langle A(t)\rangle = \langle \psi(0)|\, A(t)|\psi(0)\rangle \,. \tag{18.28}$$

18.5 Further Reading

Most college-level textbooks on quantum mechanics have sections on time dependence and the Schrödinger equation. The Heisenberg equation shows up in slightly more advanced books. A good reference for both forms of time dependence at the level of this book is Claude Cohen-Tannoudji, Bernard Diu, and Frank Laloë, *Quantum Mechanics* (Wiley, New York, 1977).

19 Magnetic Precession and Resonance for a Spin-1/2 System

While this magnetic, peripatetic Lover,
He lived to learn, by no endeavor can magnets ever attract a Silver Churn.

William S. Gilbert, *Patience*

19.1 Spin and Magnetic Dipoles

In this chapter, we explore techniques that are used in "magnetic resonance." We bring to bear not only the linear algebra we have developed but also the physics of electricity and magnetism along with the lighthouse quantum ideas of Heisenberg uncertainty and the Schrödinger equation.

The name magnetic resonance implies that we observe the magnetic properties of a system by using oscillating magnetic fields, and we look for oscillation frequencies that give the strongest signal—a process called resonance: when the oscillation frequency of an applied external magnetic field matches an internal oscillation frequency of the system, a signal with a large amplitude occurs. We consider a spin-1/2 system, such as an electron, proton, neutron, or a neutral atom like sodium or silver, and subject it to various static and oscillating magnetic fields. All these systems possess a property called a magnetic dipole, which acts like a tiny bar magnet and interacts with those other magnetic fields. The resulting techniques are widely used in magnetic resonance imaging in medicine, in high precision atomic clocks that provide today's time standards, and in the manipulation of qubits in many quantum computing and quantum information systems. The high sensitivity and high precision of the techniques also make them useful in "quantum sensors" widely used in physics and the life sciences to monitor physical processes in both nonliving and living systems.

The analysis is based on the Schrödinger equation introduced in Chapter 18:

$$i\hbar \frac{d\,|\psi\rangle}{dt} = H\,|\psi\rangle, \tag{19.1}$$

where $\hbar$ ("h bar") is the reduced Planck's constant and $|\psi\rangle$ is the quantum state vector for the system under study. As usual, we specify the physics of the situation by providing the appropriate Hamiltonian operator H. For magnetic resonance, the Hamiltonian

Quantum Computing and Quantum Physics. Alice Flarend and Robert Hilborn, Oxford University Press.
 DOI: 10.1093/oso/9780197904930.003.0019

operator representing the potential energy of the magnetic dipole in a magnetic field is the dot product

$$H = -\vec{\mu} \cdot \vec{B}, \tag{19.2}$$

where $\vec{B}$ is the magnetic field at the location of the qubit, whose magnetic dipole is represented by an operator $\vec{\mu}$, which, as the arrow indicates, is a vector quantity (with x, y, and z components). The negative sign in Eq. (19.2) indicates that the object's magnetic energy is lowest (most negative) when the magnetic dipole vector and the magnetic field vector point in the same direction ($\cos 0° = 1$) and is highest (most positive) when the dipole vector and the magnetic field vector are in opposite directions ($\cos 180° = -1$). You might think of the magnetic dipole vector as a compass needle that tends to point in the direction of the magnetic field, thereby achieving its lowest energy state.

For the spin-1/2 systems we study, the magnetic dipole operator is proportional to the system's spin angular momentum operator $\vec{S}$. (We introduced the spin operator when we first discussed spin systems in Chapter 7. In this chapter, we use S_z in place of the operator Z to remind us of spin; similarly for the x and y components.) The connection between the magnetic dipole and the spin is given by

$$\vec{\mu} = \gamma \vec{S}. \tag{19.3}$$

The object's gyromagnetic ratio γ is positive for a particle whose magnetic dipole vector is parallel to its spin angular momentum vector and negative when the magnetic dipole vector and the spin angular momentum vector point in opposite directions. Elementary particles with a positive electric charge (such as a proton) usually have a positive gyromagnetic ratio. It is negative for a particle with negative electric charge (such as an electron).

Many quantum mechanics textbooks use electrons as paradigmatic particles and assume that the gyromagnetic ratio is negative. In this book, we assume that the gyromagnetic ratio is positive because protons and their magnetic moments are the qubits for magnetic resonance imaging. This choice avoids the confusion that might occur for electrons and other negatively charged particles for which spin-up means magnetic dipole down relative to some chosen axis, usually the direction of the local magnetic field.

As an aside, we note that an electrically neutral particle such as a neutron or a neutral atom can have a magnetic dipole even though it has no net electric charge. In these cases, γ depends on the internal structure of the system (the neutron can be thought of as a collection of quarks and the atom as a collection of a nucleus and a cloud of electrons).

To introduce some notation and to explore some of the basic physics of a magnetic dipole in a magnetic field, we first look at a situation in which the external magnetic field $\vec{B}$ points in a fixed direction, which we choose to be the z-direction of a Cartesian xyz coordinate system. In this case, the Hamiltonian is

$$H = -\gamma S_z B_0, \tag{19.4}$$

where B_0 is the strength of the external magnetic field at the location of the magnetic dipole and S_z is the z component of the spin operator. Note that some authors, as mentioned above, assume we are dealing only with electrons and drop the minus sign on the right of Eq. (19.4). We keep the minus sign to be more democratic.

Recalling the matrix form of the operator S_z, we may write the Hamiltonian as

$$H \doteq -\frac{\hbar\gamma B_0}{2}\begin{pmatrix}1 & 0\\ 0 & -1\end{pmatrix} = -\frac{\hbar\omega_0}{2}\begin{pmatrix}1 & 0\\ 0 & -1\end{pmatrix}, \tag{19.5}$$

where we introduced the parameter $\omega_0 = \gamma B_0$, which, for reasons discussed shortly, is called the Larmor frequency. Note that $\hbar\omega_0$ has dimensions of energy.

Since the Hamiltonian matrix is diagonal, we know that the diagonal elements are the energies for this two-state system and its eigenvectors are the spin-up and spin-down states. Recall that in the column vector form the spin-up and spin-down states are given by

$$|\uparrow\rangle_z \doteq \begin{pmatrix}1\\ 0\end{pmatrix} \text{ and } |\downarrow\rangle_z \doteq \begin{pmatrix}0\\ 1\end{pmatrix}. \tag{19.6}$$

If we let the Hamiltonian operate on those states, we find

$$\begin{aligned} H|\uparrow\rangle_z &\doteq -\frac{\hbar\omega_0}{2}\begin{pmatrix}1 & 0\\ 0 & -1\end{pmatrix}\begin{pmatrix}1\\ 0\end{pmatrix} = -\frac{\hbar\omega_0}{2}\begin{pmatrix}1\\ 0\end{pmatrix}\\ H|\downarrow\rangle_z &\doteq -\frac{\hbar\omega_0}{2}\begin{pmatrix}1 & 0\\ 0 & -1\end{pmatrix}\begin{pmatrix}0\\ 1\end{pmatrix} = -\frac{\hbar\omega_0}{2}\begin{pmatrix}0\\ -1\end{pmatrix} = \frac{\hbar\omega_0}{2}\begin{pmatrix}0\\ 1\end{pmatrix}. \end{aligned} \tag{19.7}$$

So, we see that the spin-up state has the lower energy eigenvalue $E_\uparrow = -\hbar\omega_0/2$ while the spin-down state has the higher energy eigenvalue $E_\downarrow = \hbar\omega_0/2$. (If the qubit has a negative gyromagnetic ratio, the roles of spin-up and spin-down are reversed.) Applying the external magnetic field causes a splitting in the energy depending on the spin of the particles. (This is called the Zeeman effect, which was discovered in the late 1800s.) Figure 19.1 shows the energy ladder (energy level) diagram for this situation. Note that the energy separation between the states increases if the magnitude of the static magnetic field B_0 increases. Energy ladders are probably familiar to you if you know about spectroscopy

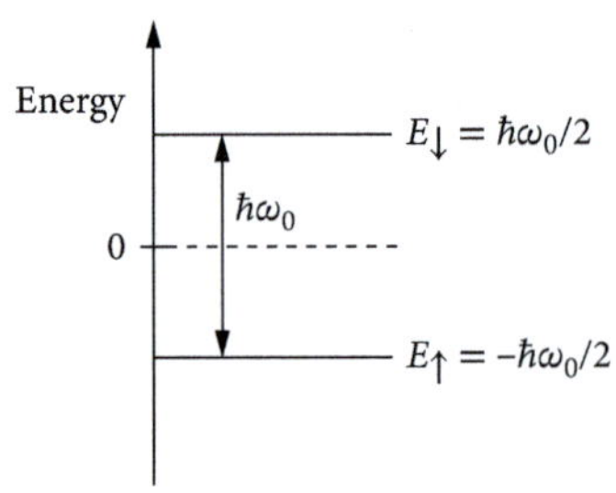

Fig. 19.1 The energy level diagram for a spin-1/2 qubit in a magnetic field in the z-direction. The parameter $\omega_0 = \gamma B_0$ is proportional to the strength B_0 of the magnetic field.

from astronomy or chemistry. They often represent a system that will be excited to a higher energy state and then releases a photon when it relaxes back to a lower energy state.

19.2 Magnetic Dipole Precession in a Magnetic Field

Applying the magnetic field to a spin-1/2 system with a magnetic dipole means we have a two-state system, and any state of the two-state system can be expressed as a linear combination (superposition) of the two energy eigenstates:

$$|\psi(t)\rangle = c_\uparrow(t)|\uparrow\rangle_z + c_\downarrow(t)|\downarrow\rangle_z. \tag{19.8}$$

You should recall from Chapter 18 that the time dependence of the coefficients of the energy eigenstates may be expressed as

$$c_\uparrow(t) = e^{-iE_\uparrow t/\hbar}c_\uparrow(0) \text{ and } c_\downarrow(t) = e^{-iE_\downarrow t/\hbar}c_\downarrow(0). \tag{19.9}$$

So, if we prepare the system in the spin-up state (for example), the probability of finding the system in that state is independent of time because

$$\begin{aligned} |c_\uparrow(t)|^2 &= c_\uparrow^*(t)c_\uparrow(t) \\ &= e^{+iE_\uparrow t/\hbar}c_\uparrow^*(0)e^{-iE_\uparrow t/\hbar}c_\uparrow(0) \\ &= c_\uparrow^*(0)c_\uparrow(0), \end{aligned} \tag{19.10}$$

with a similar result for $|c_\downarrow(t)|^2$.

Now let's look at a more general superposition state with state coefficients (amplitudes) given by

$$\begin{aligned} c_\uparrow(0) &= \cos(\theta/2) \\ c_\downarrow(0) &= \sin(\theta/2)\, e^{i\phi}. \end{aligned} \tag{19.11}$$

Recall from Chapter 7 that for a spin-1/2 system, the state space angle between basis state vectors is half the real space angle θ between the corresponding real space vectors—the actual magnetic dipole vector orientation, for example.

In the column vector form, the time-dependent state of the system is

$$\begin{aligned} |\psi(t)\rangle &\doteq \begin{pmatrix} e^{-iE_\uparrow t/\hbar}\cos(\theta/2) \\ e^{-iE_\downarrow t/\hbar}e^{i\phi}\sin(\theta/2) \end{pmatrix} \\ &= \begin{pmatrix} e^{i\omega_0 t/2}\cos(\theta/2) \\ e^{-i\omega_0 t/2}e^{i\phi}\sin(\theta/2) \end{pmatrix} = e^{i\omega_0 t/2}\begin{pmatrix} \cos(\theta/2) \\ e^{i(\phi-\omega_0 t)}\sin(\theta/2) \end{pmatrix}. \end{aligned} \tag{19.12}$$

From Eq. (19.12), we can easily read off that for a spin measurement along the z-direction the probability of observing spin-up along z is $\cos^2(\theta/2)$ while the probability of observing spin-down along z is $\sin^2(\theta/2)$.

Recall from Chapter 8 that we get more information about a quantum state by making measurements relative to other basis states. For example, we may want to find the probabilities for the outcomes of spin measurements along the x-direction. As an example, let's calculate the probability of getting spin-up along the x-direction using the row and column vector forms for state vectors. The row vector form of the spin-up state in the x basis is ${}_x\langle\uparrow| = \frac{1}{\sqrt{2}}(1 \quad 1)$. If the system is prepared in the state $|\psi(t)\rangle$ given by Eq. (19.12), the probability of getting spin-up with a measurement along the x-direction is

$$\begin{aligned} P_{\uparrow x} &= |{}_x\langle\uparrow|\psi(t)\rangle|^2 = \frac{1}{2}\left|(1 \quad 1)\begin{pmatrix} \cos(\theta/2) \\ e^{i(\phi-\omega_0 t)}\sin(\theta/2)\end{pmatrix}\right|^2 \\ &= \frac{1}{2}\left\{\cos^2(\theta/2) + \cos(\theta/2)\sin(\theta/2)\left[e^{i(\phi-\omega_0 t)} + e^{-i(\phi-\omega_0 t)}\right] + \sin^2(\theta/2)\right\} \\ &= \frac{1}{2}\{1 + \sin\theta\cos(\phi - \omega_0 t)\}. \end{aligned} \tag{19.13}$$

Try It 19.1 Work through the algebraic and trigonometric steps to check Eq. (19.13). Hints: Use $\cos^2\alpha + \sin^2\alpha = 1$, $\sin\alpha = 2\sin(\alpha/2)\cos(\alpha/2)$, and $e^{i\alpha} + e^{-i\alpha} = 2\cos\alpha$. Is the probability in Eq. (19.13) always positive?

Eq. (19.13) shows a notable feature: The probability for observing spin-up along x is time-dependent. It means that at least some of the average values of the spin vector components are time-dependent. Eq. (19.13) shows that the probability $P_{\uparrow x}$ varies in time like a cosine function with a frequency given by ω_0, the Larmor precession frequency, terminology borrowed from classical physics.

Try It 19.2 Calculate the following average (expectation) values of the components of the spin operators for the spin-up state along x and show that $\langle S_z\rangle = (\hbar/2)\cos\theta$, $\langle S_y\rangle = (\hbar/2)\sin\theta\sin(\phi - \omega_0 t)$, and $\langle S_x\rangle = (\hbar/2)\sin\theta\cos(\phi - \omega_0 t)$. Are these results consistent with the sketch of spin precession in Fig. 19.2?

The results of this Try It allow us to draw a geometric picture of the spin vector precessing, like a spinning top, around the z-direction. The z component is constant while the x and y components oscillate in time; we say that the spin angular momentum vector is precessing around the z-axis with an angular frequency ω_0. If the gyromagnetic ratio for the system is negative (e.g., as it is for an electron), the sense of precession is reversed.

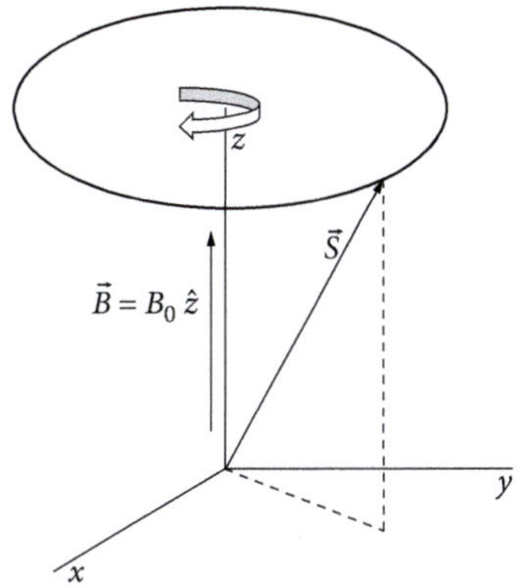

Fig. 19.2 The spin angular momentum vector $\vec{S}$ precessing around a magnetic field in the z-direction. The precession direction is appropriate for a qubit with a positive gyromagnetic ratio.

Figure 19.2 illustrates how the spin angular momentum vector for a qubit with a magnetic dipole precesses around the magnetic field in the z-direction. The projection of $\vec{S}$ along the magnetic field direction (that is, its z component) remains constant. The x component of $\vec{S}$ will vary in time as shown in Eq. (19.13). (The same is true of the y component.) The tip of $\vec{S}$ moves clockwise as viewed from above around the circular path centered on the z-axis.

19.3 Precession and the Heisenberg Uncertainty Relations

Now that we have seen spin precession in the superposition state Eq. (19.12), let's look at a simpler situation with the spin-up state $|\psi\rangle = |\uparrow\rangle_z$. In this case, we have $\langle S_z \rangle = \hbar/2$ and $\langle S_x \rangle = \langle S_y \rangle = 0$.

Try It 19.3 Use the results from the previous Try It to verify the statements in the previous paragraph. Hint: What is the appropriate value of θ for $|\uparrow\rangle_z$ in Eq. (19.12)?

We now use the *Heisenberg uncertainty relations* (Chapter 7, Section 7.7) to develop a pictorial representation for the spin vector behavior in this case. First, recall that for any two quantum operators A and C, the product of the standard deviations for the operators satisfies the Heisenberg inequality

$$\Delta A\, \Delta C \geq \frac{1}{2}\, |\langle [A, C] \rangle| \,. \tag{19.14}$$

Just as reminder, the standard deviation is $\Delta A = \sqrt{\langle A^2 \rangle - \langle A \rangle^2}$ and the right side of Eq. (19.14) is the absolute value of the mean value of the commutator between the two operators. In general, both sides of the equation depend on the precise state used to calculate the mean values and the standard deviations. The right side gives a minimum value for the product of the standard deviations. The inequality means the actual product for a particular state may be larger than the minimum.

We are interested in using Eq. (19.14) for the various spin-1/2 operators S_x, S_y, and S_z. We have already mentioned the mean values of these operators in the $|\uparrow\rangle_z$ state. In Chapter 7, we find that the squares of the spin-1/2 operators are equal to $(\hbar^2/4)\mathrm{I}$ and that the commutation relations are

$$[S_x, S_y] = i\hbar S_z, \tag{19.15}$$

with cyclic permutations of the component indices giving the other combinations of operators. So, we have in hand all the ingredients we need to apply Eq. (19.14) to this situation.

Since the squares of the spin-1/2 operator components are all multiples of the identity operator, we immediately find that $\langle S_x^2 \rangle = \hbar^2/4$ with identical results for the other two components. Let's work out Eq. (19.14) for the two operators S_x and S_y. The right side is just $\hbar^2/4$, so we have

$$(\Delta S_x)_{\uparrow_z} (\Delta S_y)_{\uparrow_z} \geq \hbar^2/4 \tag{19.16}$$

This result shows that in the $|\uparrow\rangle_z$ state both the x and y components of the spin operator have nonzero standard deviations and hence the results of measuring those components must lead to a spread in values.

In general, as mentioned previously, the standard deviations depend on the system's quantum state. For the $|\uparrow\rangle$ state, the two standard deviations turn out to be the same, each equal to $\hbar/2$. For example, for the x component, we have

$$(\Delta S_x)_\uparrow = \sqrt{\langle S_x^2 \rangle_\uparrow - \langle S_x \rangle_\uparrow^2} = \sqrt{\hbar^2/4 - 0} = \hbar/2. \tag{19.17}$$

To complete our picture of the spin vector's behavior, we need to look at the operator that represents the square of the length of the spin vector:

$$S^2 = S_x^2 + S_y^2 + S_z^2. \tag{19.18}$$

For a spin-1/2 system, we find

$$\langle S^2 \rangle = \langle S_x^2 \rangle + \langle S_y^2 \rangle + \langle S_z^2 \rangle = \frac{3}{4}\hbar^2, \tag{19.19}$$

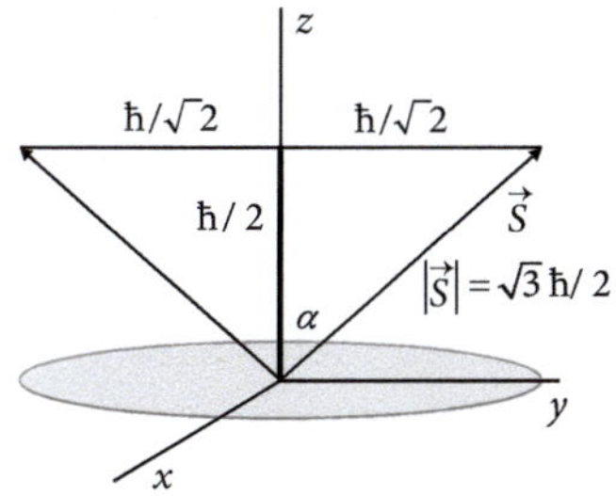

Fig. 19.3 A visualization of a spin-1/2 vector described by the state $|\uparrow\rangle_z$.

and the "length" (the square root of Eq. (19.19)) is $\sqrt{3}\,\hbar/2$. This result should surprise you. Here's why. Recall that the mean value of the z component of the spin vector in the $|\uparrow\rangle_z$ state is $\hbar/2$. Since the length of the spin vector we just found is longer than that, we learn that the spin vector is not completely aligned with the z-axis even for a spin-up state. If it were completely aligned with the z-axis, the x and y components would be zero and their standard deviations would be zero, in violation of the *Heisenberg uncertainty relation*. In other words, quantum mechanics shows that there are no quantum states for which a spin vector has definite (zero standard deviation) components along all three spatial directions. Can we find a visualization that is consistent with that result?

Figure 19.3 shows the geometry for a spin-1/2 vector in the spin-up state $|\uparrow\rangle_z$. The projection along the z-axis is given by $\langle S_z\rangle = \hbar/2$. The z component is a definite value and hence the standard deviation $\Delta S_z = 0$.

Try It 19.4 Use $\langle S_z^2\rangle_\uparrow = \hbar^2/4$ and $\langle S_z\rangle_\uparrow = \hbar/2$ to verify that $\Delta\hat{S}_z = 0$.

The gray circle in the xy plane has a radius $r = \sqrt{\langle S_x^2\rangle + \langle S_y^2\rangle} = \hbar/\sqrt{2}$. We should think of the x and y components of the spin vector as spread across the gray circle. This produces a distribution of measured values whose averages are zero and standard deviations are each $\hbar/2$. The combination of these effects suggest that we should think of the spin vector itself as distributed uniformly on a cone whose apex angle relative to the z-direction is α.

The angle between the spin vector and the z-axis is given by $\cos\alpha = 1/\sqrt{3}$. In this situation, as compared to the previous situation with a superposition of spin-up and spin-down, there is no precession around the z-axis.

Try It 19.5 Explain why $\cos\alpha = 1/\sqrt{3}$. What is the numerical value of α in both degrees and radians.

19.4 A Magnetic Field with Both z and x Components

In our next generalization, we look at the situation in which the magnetic field has both z and x components as shown in Fig. 19.4.

Following the notation commonly used in magnetic resonance work, we write the magnetic field vector as the sum of these two components: $\vec{B} = B_0\hat{z} + B_1\hat{x}$ (see Fig. 19.4). The easiest way to proceed would be to tilt the coordinate system so that the new z-direction is along $\vec{B}$ and then apply the results from the previous example. This certainly works for this case, but when we get to magnetic resonance with a static field in one direction and oscillating fields perpendicular to that direction, it will be easier to keep the field components separate.

The Hamiltonian for the system with a magnetic dipole subject to a magnetic field with components along the z- and x-directions is

$$\begin{aligned} H &= -\vec{\mu} \cdot \vec{B} = -\mu_z B_0 - \mu_x B_1 \\ &= -\gamma S_z B_0 - \gamma S_x B_1 \\ &= -\omega_0 S_z - \omega_1 S_x, \end{aligned} \tag{19.20}$$

where we introduced a new parameter $\omega_1 = \gamma B_1$ in analogy to $\omega_0 = \gamma B_0$. For this situation, we have $\tan\theta = B_1/B_0 = \omega_1/\omega_0$, $\sin\theta = \omega_1/\sqrt{\omega_0^2 + \omega_1^2}$, and $\cos\theta = \omega_0/\sqrt{\omega_0^2 + \omega_1^2}$.

In matrix form, the Hamiltonian operator is

$$H = -\omega_0 S_z - \omega_1 S_x \doteq -\frac{\hbar}{2}\begin{pmatrix} \omega_0 & \omega_1 \\ \omega_1 & -\omega_0 \end{pmatrix}. \tag{19.21}$$

A crucial point to note: In the z basis, the Hamiltonian matrix is *not* a diagonal matrix and hence the spin-up and spin-down basis vectors along the z-direction are no longer energy eigenstates.

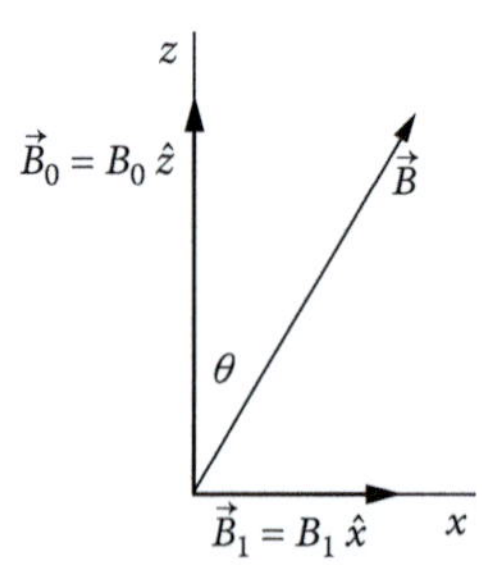

Fig. 19.4 $\vec{B}$ tilted at angle θ with respect to the z-axis. $\vec{B}$ has both a z component and an x component.

Try It 19.6 To show that $|\uparrow\rangle_z \doteq \begin{pmatrix}1\\0\end{pmatrix}$ is not an energy eigenstate, multiply the spin-up column vector by the matrix H in Eq. (19.21) to show that $H|\uparrow\rangle_z \neq$ (a number) $||\uparrow\rangle_z\rangle$.

How do we find out how the states vary in time in this situation? (We can no longer simply multiply the z state coefficients by $e^{-iE_{\uparrow,\downarrow}t/\hbar}$ to express the time evolution because these states are not energy eigenstates.) There are two ways to proceed. (1) Find the eigenvalues (label them E_A and E_B) and the eigenvectors of the Hamiltonian matrix in Eq. (19.21) and then express the desired measurement basis states (in this case spin-up and spin-down along the z-axis) as linear combinations of those eigenvectors, each of which is multiplied by the appropriate exponential $e^{-iE_{A,B}\,t/\hbar}$. (2) Use the matrix version of the Schrödinger equation to set up differential equations for the state coefficients in the measurement basis and then use standard differential equation techniques to find how the state coefficients evolve in time. We use method 1 first because it is good exercise for our state manipulation muscles. Then, we develop method 2 because it connects more smoothly to the magnetic resonance case in which the magnetic field in the x-direction oscillates in time.

We begin process 1 by finding the eigenvalues and eigenvectors associated with the Hamiltonian operator in Eq. (19.21). As usual, we find the eigenvalues E_A and E_B by setting the determinant of the characteristic matrix equal to zero:

$$-\frac{\hbar}{2}\begin{vmatrix}\omega_0-\lambda & \omega_1\\ \omega_1 & -\omega_0-\lambda\end{vmatrix}=0, \tag{19.22}$$

from which we easily find that $\lambda = \pm\sqrt{\omega_0^2+\omega_1^2} \equiv \pm\omega$ and the energy eigenvalues are

$$E_A = -\frac{\hbar\omega}{2} \qquad E_B = +\frac{\hbar\omega}{2}. \tag{19.23}$$

We chose the signs for the energies so that the energy eigenvalues $E_A \to E_{\uparrow_z} = -\hbar\omega_0/2$ and $E_B \to E_{\downarrow_z} = \hbar\omega_0/2$ as the x component of the magnetic field goes to zero (see Fig. 19.5), which shows how the energy eigenvalues vary with the parameter ω_1.

Try It 19.7 Use Eq. (19.23) to calculate the energy eigenvalues for a few values of ω_1 to check the values plotted in Fig. 19.5.

We label the corresponding energy eigenstates as $|E_A\rangle$ and $|E_B\rangle$. As mentioned previously, the time-dependent energy eigenstates can be written as

$$|\psi_{A,B}\rangle = e^{-iE_{A,B}t/\hbar}c_{A,B}(0)|E_{A,B}\rangle. \tag{19.24}$$

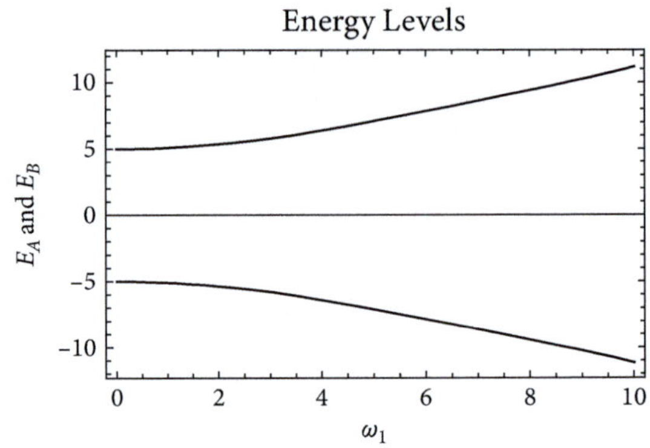

Fig. 19.5 A plot of the energy eigenvalues (divided by $\hbar$): E_A (lower curve) and E_B (upper curve) as a function of ω_1 for $\hbar\omega_0 = 5$.

So, if we can frame our problem in terms of energy eigenstates, the time evolution part is done. However, as mentioned previously, we often make measurements with basis states other than the energy eigenvectors. In particular, for magnetic resonance, we often use the spin-up and spin-down (relative to the z-axis) states as the measurement basis states and in general these basis states are not energy eigenstates.

We now express the energy eigenvectors associated with those energy eigenvalues as linear combinations of the $|\uparrow\rangle_z$ and $|\downarrow\rangle_z$ states, which we use as our measurement basis states:

$$\begin{aligned} |E_A\rangle &= a|\uparrow\rangle_z + b|\downarrow\rangle_z \\ |E_B\rangle &= c|\uparrow\rangle_z + d|\downarrow\rangle_z . \end{aligned} \tag{19.25}$$

For $|E_A\rangle$, the eigenvector equations become (dropping the common factor $\hbar/2$)

$$-\begin{pmatrix} \omega_0 & \omega_1 \\ \omega_1 & -\omega_0 \end{pmatrix}\begin{pmatrix} a \\ b \end{pmatrix} = -\omega\begin{pmatrix} a \\ b \end{pmatrix} \tag{19.26}$$

and for $|E_B\rangle$

$$-\begin{pmatrix} \omega_0 & \omega_1 \\ \omega_1 & -\omega_0 \end{pmatrix}\begin{pmatrix} c \\ d \end{pmatrix} = \omega\begin{pmatrix} c \\ d \end{pmatrix}. \tag{19.27}$$

Eq. (19.26) is equivalent to two linear equations

$$\begin{aligned} -\omega_0 a - \omega_1 b &= -\omega a \\ -\omega_1 a + \omega_0 b &= -\omega b. \end{aligned} \tag{19.28}$$

You can see that Eq. (19.28) determines only the ratio a/b of the coefficients. However, we have two physics conditions for the coefficients. The first is the normalization constraint

$|a|^2 + |b|^2 = 1$. The second is that an overall phase factor $e^{i\beta}$ in the coefficients does not affect probabilities. Hence, we may choose the coefficient a to be a positive real number so that $a \to 1$ as $\omega_1 \to 0$. In this limit, the state $|E_A\rangle$ becomes $|\uparrow\rangle_z$.

We use the lower line of Eq. (19.28) to write

$$\omega_1 a = b(\omega + \omega_0), \tag{19.29}$$

or equivalently,

$$\omega_1^2 a^2 = (\omega + \omega_0)^2 b^2. \tag{19.30}$$

The normalization condition gives

$$|a|^2 + |b|^2 = 1 \Rightarrow |a|^2 = 1 - |b|^2. \tag{19.31}$$

Using Eq. (19.29) in the normalization condition and recognizing that we may take both a and b as real numbers and then solving for b gives

$$b = \frac{\omega_1}{\sqrt{\omega_1^2 + (\omega + \omega_0)^2}}. \tag{19.32}$$

Using this result in Eq. (19.29), we find

$$a = \frac{(\omega + \omega_0)}{\sqrt{\omega_1^2 + (\omega + \omega_0)^2}}. \tag{19.33}$$

Try It 19.8 Use the normalization constraint to verify that the solutions in Eqs. (19.32) and (19.33) are correct and satisfy the normalization condition $|a|^2 + |b|^2 = 1$.

Try It 19.9 Verify that $a \to 1$ and $b \to 0$ as $\omega_1 \to 0$. Explain why those results make sense.

The analogous procedures applied to Eq. (19.27) give us the coefficients c and d for the $|E_B\rangle$ eigenvector:

$$c = \frac{\omega_1}{\sqrt{(\omega_0 + \omega)^2 + \omega_1^2}} \qquad d = \frac{\omega_0 + \omega}{\sqrt{(\omega_0 + \omega)^2 + \omega_1^2}}. \tag{19.34}$$

Try It 19.10 Challenge: Show that the state coefficients can be written as $a = \cos(\theta/2)$ and $b = -\sin(\theta/2)$, where $\cos\theta = B_0/B = \omega_0/\omega$ and $\sin\theta = B_1/B = \omega_1/\omega$ (see Fig. 19.4). This is a bit of a trigonometric challenge (hint: use half-angle formulae). Explain how the result is consistent with how state space angles are related to physical space angles for spin-1/2 systems as discussed in Chapter 7. Note that b is negative due the definition of θ in Fig. 19.4 compared to the state space angle between the E_A and spin-up state vectors.

We now work through an example to illustrate these procedures. Let's assume that we start with the system in the spin-up state $|\uparrow\rangle_z$, that is, $|\psi(0)\rangle = |\uparrow\rangle_z$, and we want to find the probability of observing the system in the spin-down state $|\downarrow\rangle_z$ as a function of time. This probability is expressed as usual by $P_{\uparrow\Rightarrow\downarrow}(t) = |_z\langle\downarrow|\psi(t)\rangle|^2$. We use Eq. (19.25) and the results in the previous Try It Challenge to express the state coefficients. From Eq. (19.25), we find that

$$\begin{aligned} |\psi(0)\rangle &= \frac{\omega+\omega_0}{\sqrt{\omega_1^2+(\omega+\omega_0)^2}}|E_A\rangle + \frac{\omega_1}{\sqrt{\omega_1^2+(\omega+\omega_0)^2}}|E_B\rangle \\ &= \cos(\theta/2)\,|E_A\rangle + \sin(\theta/2)\,|E_B\rangle = |\uparrow\rangle_z. \end{aligned} \tag{19.35}$$

Now it is straightforward to write $|\psi(t)\rangle$ in terms of the energy eigenstates:

$$|\psi(t)\rangle = e^{-iE_At/\hbar}\cos(\theta/2)\,|E_A\rangle + e^{-iE_Bt/\hbar}\sin(\theta/2)\,|E_B\rangle. \tag{19.36}$$

We then insert Eq. (19.36) into the probability amplitude $\langle\downarrow|\psi(t)\rangle$ and drop the z subscript for typological simplicity and find

$$\begin{aligned} \langle\downarrow|\psi(t)\rangle &= \langle\downarrow|\left[e^{-iE_At/\hbar}\cos(\theta/2)\,|E_A\rangle + e^{-iE_Bt/\hbar}\sin(\theta/2)\,|E_B\rangle\right] \\ &= e^{-iE_At/\hbar}\cos(\theta/2)\underbrace{\langle\downarrow|E_A\rangle}_{-\sin(\theta/2)} + e^{-iE_Bt/\hbar}\sin(\theta/2)\underbrace{\langle\downarrow|E_B\rangle}_{\cos(\theta/2)} \\ &= -\sin(\theta/2)\cos(\theta/2)\,e^{-iE_At/\hbar}\left[1 - e^{i(E_A-E_B)t/\hbar}\right]. \end{aligned} \tag{19.37}$$

Using $\sin 2\beta = 2\sin\beta\cos\beta$, $\sin^2\alpha = (1-\cos 2\alpha)/2$, and the Euler formula $e^{i\alpha} + e^{-i\alpha} = 2\cos\alpha$, we find

$$P_{\uparrow\Rightarrow\downarrow}(t) = |\langle\downarrow_z|\psi(t)\rangle| = \sin^2\theta\sin^2\left(\frac{(E_A-E_B)}{2\hbar}t\right) = \sin^2\theta\sin^2\left(\frac{\sqrt{\omega_0^2+\omega_1^2}\,t}{2}\right). \tag{19.38}$$

Try It 19.11 Check Eq. (19.38) for several simple cases: (1) We expect that $P_\downarrow(0) = 0$ because we start in the state $|\uparrow\rangle$. Explain the reasons. (2) Suppose the magnetic field lies entirely in the x-direction and hence $\theta = \pi/2$. Sketch a graph of $P_\downarrow(t)$ in this case. Explain what is going on.

Try It 19.12 Show that Eq. (19.38) may be written entirely in terms of ω_0 and ω_1:

$$P_{\uparrow\Rightarrow\downarrow}(t) = \frac{\omega_1^2}{\omega_0^2 + \omega_1^2}\sin^2\frac{\sqrt{\omega_0^2 + \omega_1^2}}{2}t \tag{19.39}$$

The combination $\omega = \sqrt{\omega_0^2 + \omega_1^2}$ is called the Rabi frequency and Eq. (19.39) is known as the Rabi formula, after the Nobel laureate physicist Isidor Rabi, who pioneered magnetic precession experiments on atoms and molecules in the 1930s. These experiments ultimately led Rabi and his colleagues to invent magnetic resonance in 1938.

19.5 Differential Equation Approach

In the previous section, we find the time dependence of probabilities by expressing the qubit state vector as a superposition of the energy eigenvectors for the system. As mentioned before, we could also find these results by using the Schrödinger equation to generate differential equations for the state coefficients. The matrix form of the Schrödinger equation in the terms of the coefficients for the states $|\uparrow\rangle_z$and $|\downarrow\rangle_z$ for a magnetic dipole in a static magnetic field is

$$i\hbar\frac{d|\psi\rangle}{dt} = H|\psi\rangle \doteq i\hbar\begin{pmatrix}\dot{c}_\uparrow(t)\\ \dot{c}_\downarrow(t)\end{pmatrix} = -\frac{\hbar}{2}\begin{pmatrix}\omega_0 & \omega_1\\ \omega_1 & -\omega_0\end{pmatrix}\begin{pmatrix}c_\uparrow(t)\\ c_\downarrow(t)\end{pmatrix}, \tag{19.40}$$

where the overdot means differentiation with respect to time $\dot{c} = dc/dt$. Canceling the $\hbar$s on both sides of the second equation in (19.40) and moving the i to the right side gives us two coupled differential equations:

$$\begin{aligned}\frac{dc_\uparrow(t)}{dt} &= \frac{i}{2}\left[\omega_0 c_\uparrow(t)+\omega_1 c_\downarrow(t)\right]\\ \frac{dc_\downarrow(t)}{dt} &= \frac{i}{2}\left[\omega_1 c_\uparrow(t)-\omega_0 c_\downarrow(t)\right].\end{aligned} \tag{19.41}$$

We can convert these into two *uncoupled* equations by differentiating both equations with respect to t and then using Eq. (19.41) to get rid of the first time derivatives. Taking the time derivative of the expressions in Eq. (19.41) gives us

$$\begin{aligned}\frac{d^2c_\uparrow(t)}{dt^2} &= \frac{i}{2}\left[\omega_0\frac{dc_\uparrow(t)}{dt} + \omega_1\frac{dc_\downarrow(t)}{dt}\right]\\ \frac{d^2c_\downarrow(t)}{dt^2} &= \frac{i}{2}\left[\omega_1\frac{dc_\uparrow(t)}{dt} - \omega_0\frac{dc_\downarrow(t)}{dt}\right].\end{aligned} \tag{19.42}$$

After substituting the expressions for the first time derivatives from Eq. (19.41) into Eq. (19.42) and using $\omega = \sqrt{\omega_0^2 + \omega_1^2}$, we find after a bit of algebra that

$$\begin{aligned} \frac{d^2 c_\uparrow(t)}{dt^2} &= -\frac{1}{4}\left(\omega_0^2 + \omega_1^2\right) c_\uparrow(t) = -\frac{1}{4}\omega^2 c_\uparrow(t) \\ \frac{d^2 c_\downarrow(t)}{dt^2} &= -\frac{1}{4}\left(\omega_0^2 + \omega_1^2\right) c_\downarrow(t) = -\frac{1}{4}\omega^2 c_\downarrow(t). \end{aligned} \tag{19.43}$$

It is easy to verify by direct substitution that the general solutions to the equations in Eq. (19.43) can be written as

$$\begin{aligned} c_\uparrow(t) &= k \cos\frac{\omega t}{2} + \ell \sin\frac{\omega t}{2} \\ c_\downarrow(t) &= m \cos\frac{\omega t}{2} + n \sin\frac{\omega t}{2}, \end{aligned} \tag{19.44}$$

where the four parameters k, ℓ, m, and n are determined by the initial conditions, $c_\uparrow(0)$ and $c_\downarrow(0)$.

A specific example will be helpful. Suppose we start (at $t = 0$) with the system in the spin-up state. Hence, the initial conditions are $c_\uparrow(0) = 1$ and $c_\downarrow(0) = 0$. Hence, we must have $k = 1$ and $m = 0$. To find ℓ and n, we use Eq. (19.41) evaluated at $t = 0$ and use $c_\uparrow(0) = 1$ and $c_\downarrow(0) = 0$. From this we find $\ell = i\omega_0/\omega$ and $n = i\omega_1/\omega$. Using these results, we have

$$\begin{aligned} c_\uparrow(t) &= \cos(\omega t/2) + i\frac{\omega_0}{\omega} \sin(\omega t/2) \\ c_\downarrow(t) &= i\frac{\omega_1}{\omega} \sin(\omega t/2). \end{aligned} \tag{19.45}$$

Try It 19.13 Follow through with the analysis of Eq. (19.44) and show that you end up with the results stated in Eq. (19.39) for the case $|\psi(0)\rangle = |\uparrow\rangle_z$. Hint: For Eq. (19.39), all we need is the probability of measuring spin-down, namely, $|c_\downarrow(t)|^2$. Also, recall that $\sin\theta = \omega_1/\omega$.

19.6 Magnetic Resonance with an Oscillating Magnetic Field

In magnetic resonance devices, the magnetic field in the xy plane is time-dependent (see Fig. 19.6).

$$\begin{aligned} \vec{B}_1 &= B_1 \left[\hat{x}\cos\omega t - \hat{y}\sin\omega t\right] \\ \vec{B}_0 &= B_0\hat{z}. \end{aligned} \tag{19.46}$$

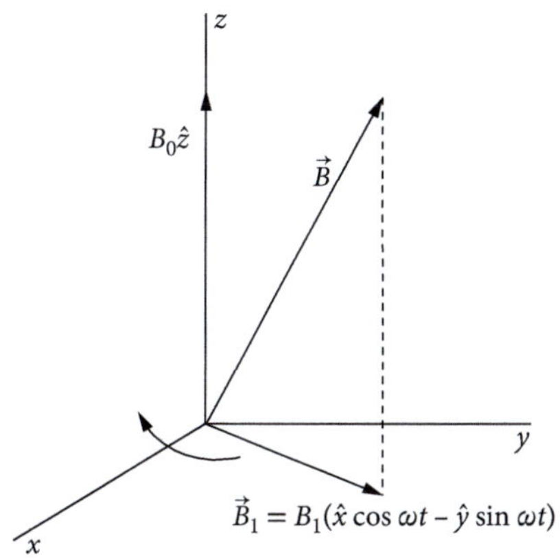

Fig. 19.6 The magnetic field $\vec{B}_1$ in the xy plane is time-dependent and rotates around the z-axis in the clockwise sense as viewed from above.

This is just like the previously described static magnetic field situation, but now the magnetic field has a component that rotates in the xy plane with an angular oscillation frequency ω, which we are free to adjust. (Note the change in the meaning of the symbol ω. It is now the angular oscillation frequency of the x and y components of the time-dependent part of the magnetic field.)

Try It 19.14 Use several values of t in Eq. (19.46) to see that the xy components of the magnetic field rotate in the clockwise sense indicated in Fig. 19.6. Hint: Try $t = 0$, $t = \pi/(2\omega)$, and $t = \pi/\omega$.

To find how the states evolve in time, as usual we need to write down the appropriate Hamiltonian operator. In analogy to Eq. (19.20), we have

$$\begin{aligned} H &= -\vec{\mu}\cdot\vec{B} = -\mu_z B_0 - \mu_x B_1\cos\omega t + \mu_y B_1 \sin\omega t \\ &= -\gamma S_z B_0 - \gamma S_x B_1 \cos\omega t + \gamma S_y \sin\omega t \\ &= -\omega_0 S_z - \omega_1 S_x \cos\omega t + \omega_1 S_y \sin\omega t. \end{aligned} \tag{19.47}$$

We shortly see why we have chosen the S_x and S_y terms with opposite signs[1]. Using the matrix representations of the spin components, the matrix form of the Hamiltonian

[1] Other textbooks, such as McIntyre, *Quantum Mechanics* (2023), assume that γ, the gyromagnetic ratio, is negative and choose plus signs for all the terms on the right in the first line of Eq. (19.47). That procedure is fine for electrons but not for protons. In the end, the results turn out to be essentially the same except that some frequencies are negative.

operator is

$$H \doteq -\frac{\hbar}{2}\begin{pmatrix} \omega_0 & \omega_1 e^{+i\omega t} \\ \omega_1 e^{-i\omega t} & -\omega_0 \end{pmatrix} \tag{19.48}$$

Try It 19.15 Use the matrix form of the spin operators to show that Eq. (19.47) leads to Eq. (19.48).

In principle, we could find the eigenvalues and eigenvectors of the matrix in Eq. (19.48) and proceed along the lines we worked out previously. However, the time dependence in the Hamiltonian operator makes that a bit tricky. Alternatively, we could (and will) simply use Eq. (19.48) in the Schrödinger equation and solve the resulting differential equations. For this approach, we deploy a mathematical "trick" based on the following idea: The magnetic field in the xy plane rotates around the z-axis. If we viewed that xy plane field from a reference frame that also rotates around the z-axis with the same angular velocity, the magnetic field in the xy plane would appear to be a static field. We could then use the results previously derived for the static magnetic field situation (see Eq. (19.45)).

In analogy to Eq. (19.41), the differential equations for the state coefficients associated with the Hamiltonian in Eq. (19.48) are (putting in the time dependence explicitly)

$$\begin{aligned} \frac{dc_\uparrow(t)}{dt} &= \frac{i}{2}\left[\omega_0 c_\uparrow(t)+\omega_1 e^{i\omega t} c_\downarrow(t)\right] \\ \frac{dc_\downarrow(t)}{dt} &= \frac{i}{2}\left[\omega_1 e^{-i\omega t} c_\uparrow(t)-\omega_0 c_\downarrow(t)\right]. \end{aligned} \tag{19.49}$$

The "trick" to remove the time-dependent exponentials is implemented by introducing new state coefficients. In particular, we define new coefficients $a_\uparrow$ and $a_\downarrow$

$$c_\uparrow(t) = e^{i\omega t/2} a_\uparrow(t) \qquad c_\downarrow(t) = e^{-i\omega t/2} a_\downarrow(t). \tag{19.50}$$

Next, we need to evaluate the time derivatives that appear on the left in Eq. (19.49) using the calculus rule for taking the derivative of a product function:

$$\begin{aligned} \dot{c}_\uparrow(t) &= \frac{i\omega}{2} e^{i\omega t/2} a_\uparrow(t)+e^{i\omega t/2} \dot{a}_\uparrow(t) \\ \dot{c}_\downarrow(t) &= \frac{-i\omega}{2} e^{-i\omega t/2} a_\downarrow(t)+e^{-i\omega t/2} \dot{a}_\downarrow(t). \end{aligned} \tag{19.51}$$

If we substitute Eq. (19.51) into Eq. (19.49) and introduce $\Delta\omega \equiv \omega_0 - \omega$, we find that the time-dependent exponentials drop out, just what we were hoping for, and we are left with

$$\begin{aligned} \dot{a}_\uparrow(t) &= \frac{i}{2}\left[(\Delta\omega)\ a_\uparrow(t)+\omega_1 a_\downarrow(t)\right] \\ \dot{a}_\downarrow(t) &= \frac{i}{2}\left[\omega_1\, a_\uparrow(t)-(\Delta\omega)\ a_\downarrow(t)\right]. \end{aligned} \tag{19.52}$$

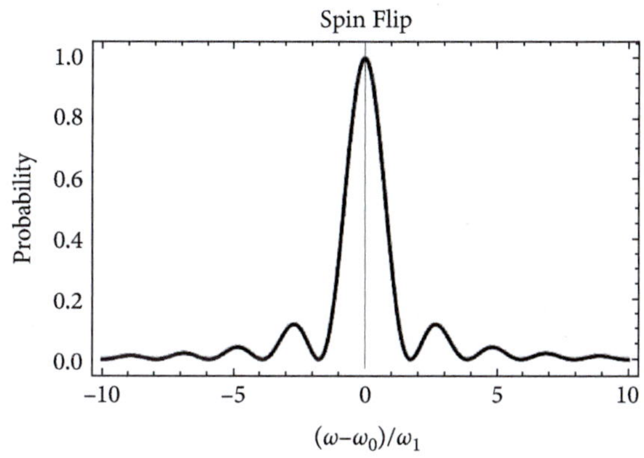

Fig. 19.7 The probability of measuring spin-down (starting from the spin-up state) as a function of the detuning (in units of ω_1) of the oscillating field frequency ω from the Larmor frequency ω_0. ω_1 is the Larmor frequency associated with the oscillating magnetic field strength. The field is applied for a time $T = \pi/\omega_1$.

These equations are mathematically equivalent to Eq. (19.41). $\Delta\omega \equiv \omega_0 - \omega$ is called the detuning: the difference between the precession frequency associated with B_z and the frequency of the oscillating magnetic field.

Let's now look at the situation in which the initial state is spin-up along the z-axis. We may, based on the analogy with Eq. (19.39), immediately write the probability of observing spin-down as

$$P_{\uparrow\Rightarrow\downarrow}(t)= \frac{\omega_1^2}{(\Delta\omega)^2+\omega_1^2}\sin^2\frac{\sqrt{(\Delta\omega)^2+\omega_1^2}}{2}t = \frac{\omega_1^2}{\Omega^2}\sin^2\frac{\Omega}{2}t, \tag{19.53}$$

which we call the generalized Rabi formula and $\Omega = \sqrt{(\Delta\omega)^2 + \omega_0^2}$ is the generalized Rabi frequency.

Figure 19.7 shows a plot of the probability of observing a spin-down state (assuming we start with a spin-up state) as a function of the detuning of the oscillating magnetic field frequency from the Larmor precession frequency ω_0 associated with the magnetic field B_0. The oscillating field was applied for a time $T = \pi/\omega_1$. The probability reaches a maximum when $\Delta\omega = 0$, a condition called "resonance." We see that resonance occurs when the oscillation frequency of the oscillating field matches the Larmor frequency. In a sense, the oscillating magnetic field follows the magnetic dipole vector, as it precesses at the Larmor frequency. In classical physics language, the associated torque causes the dipole to flip from the up direction to the down direction most effectively when $\Delta\omega = 0$.

Note that some textbooks (e.g., McIntyre, 2023) do not show the smaller side oscillations. These textbooks seem to have plotted just the prefactor multiplying the sine term in Eq. (19.53).

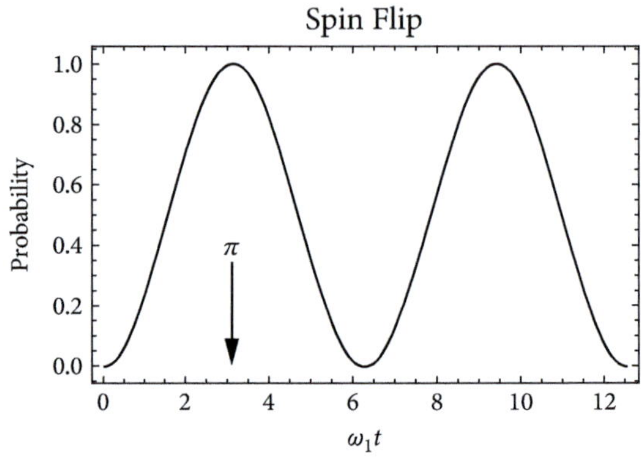

Fig. 19.8 A plot of the spin-flip probability as a function of time (in units of $1/\omega_1$) with the detuning $\Delta\omega = 0$. Note that a complete spin-flip occurs when $\omega_1 t = \pi$.

There are some interesting spin-flip probability features that are important for quantum computing, quantum information science, and quantum sensing. Figure 19.8 shows the spin-flip probability as a function of time when the detuning between the oscillating field frequency and the Larmor frequency is zero. Note that when $t = \pi/\omega_1$, the system is completely in the spin-down state. The net result is that we have changed the initial spin-up state of the qubit to a spin-down state. If the oscillating field is turned off quickly at that time, we have applied what is called a π pulse.

Try It 19.16 What is the state of the system if the oscillating field is turned off at $t = \pi/(2\omega_1)$, which is called a $\pi/2$ pulse?

Try It 19.17 What are the quantum gates that correspond to those two types of pulses? Hint: We know from Chapter 6 that the Hadamard gate acts on a single basis state to produce a linear superposition of the basis states: $\mathrm{H}\,|\uparrow\rangle = \left(1/\sqrt{2}\right)(|\uparrow\rangle + |\downarrow\rangle)$.

From Try It 19.17, we conclude that a $\pi/2$ pulse is a physical implementation of a Hadamard gate. Both π and π /2 pulses are widely used to implement quantum gates.

Let's look at more details of the probabilities. Somewhat surprisingly, a spin flip may occur no matter how weak the oscillating field amplitude B_1 is (as long as it is not exactly zero) (refer to Fig. 19.8). If ω_1 decreases (that is, B_1 decreases), then it takes a larger value of t for the product $\omega_1 t$ to reach a particular value, say, π. We just need to be patient and wait longer times for smaller values of B_1.

Let's now consider the effects of detuning. Figure 19.9 shows the spin-flip probability as a function of time when the detuning is *not* zero. In this case, the spin-flip probability does not reach 1, but nevertheless there is a nonzero probability for a spin-flip to occur. This result

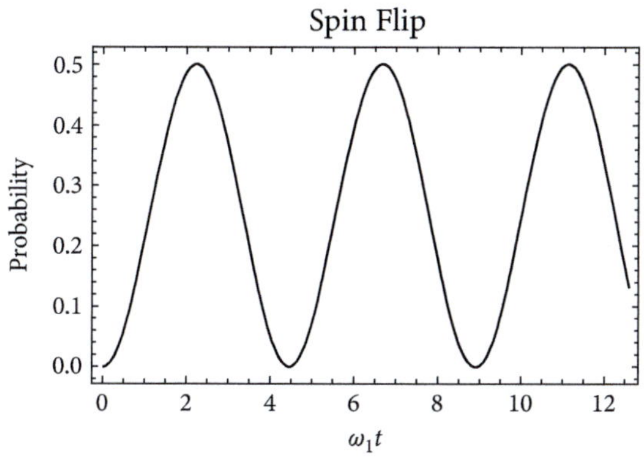

Fig. 19.9 A plot of the spin-flip probability when the detuning is $\Delta\omega = \omega_1$.

has important implications: a weak oscillating magnetic field in the system's environment might also occasionally produce some change in the qubit state in unwanted (and often undesirable) ways if its frequency spectrum has some overlap with the range of frequencies near the resonance frequency (see Fig. 19.7). For this reason, quantum computational and quantum sensing systems go to great lengths to reduce these kinds of "noisy" environmental fields.

We point out that other two-state quantum systems, such as those involving two energy levels in trapped ions or trapped neutral atoms or circulating currents in superconducting devices, exhibit similar dynamics with appropriate changes in the Hamiltonian operator and the types of oscillating fields used to induce transitions between states.

Magnetic Resonance Imaging (MRI)

MRI works by using sensitive electronics and spatially varying magnetic fields to locate clusters of magnetic dipoles (often protons or phosphorous-31 nuclei) inside materials both living and nonliving. The spatially varying magnetic fields means that only some of the magnetic dipoles will be in resonance for a given oscillating magnetic field frequency. The behavior of the interaction varies with the type of tissue (there is a different Hamiltonian for different substances) so different structures and organs can be imaged. By seeing how the strength of the MRI signal varies with position, medical personnel can locate tumors or other abnormal tissues and examine the active function of parts of the heart, for example. A simplified animated model of MRI can be found at https://phet.colorado.edu/en/simulations/mri. The PhET site also has some instructional materials that guide your use of the simulation.

Since MRI is usually looking at large collections of magnetic dipoles, the analysis of its magnetic resonance processes is more complicated than the analysis of resonance for a single qubit. In particular, the interactions among the dipoles themselves and with their chemical environment introduce phenomena that are absent for a single isolated qubit situation. These collective interactions cause two main effects. (1) If no oscillating magnetic fields are applied, the number of spins in the higher of the two energy states decreases while the number in the lower energy state increases. This "population" change is called population relaxation. The final populations primarily depend on the temperature of the system and its chemical environment as described by "thermal relaxation." Relaxation occurs exponentially with time, $\exp[-t/T_1]$, where the parameter T_1 is called the population relaxation time. (2) The second effect is a loss of the coherence between the components of the quantum superposition state for the magnetic dipole system. This is called "coherence relaxation" or "decoherence." This also occurs, on average, exponentially with time, $\exp[-t/T_2]$, with a parameter T_2. Since both of these characteristic times depend on the magnetic dipole's environment, measurements of those relaxation processes give us information about that environment. Because the environment is affected by the nature of the molecules in the environment, magnetic resonance (without the imaging part) is widely used in chemistry to help determine molecular structures and molecular composition.

When these relaxation processes are important, as they are in most MRI applications, the theoretical description of the processes becomes more complicated because it involves both the quantum mechanics important for single qubits and the thermal effects involving many atoms and molecules in the environment. The good news is that these theoretical descriptions are now highly developed and allow us to use magnetic resonance imaging in a wide variety of medical and materials applications.

19.7 Chapter Summary

- A system with a magnetic dipole precesses in the presence of a static magnetic field.
- If the static magnetic field is not along the spin measurement direction, the precession can lead to a "spin flip": If we start with a spin-up (along the measurement direction) state, there will be some probability of later finding the spin-down state. That probability oscillates in time with an oscillation frequency proportional to the strength of the magnetic field at the location of the qubit.
- When the qubit is subject to both a static magnetic field and an oscillating magnetic field, the system exhibits resonance when the oscillating field's frequency matches the Larmor precession frequency associated with the qubit's state. In this case, the system's state oscillates between spin-up and spin-down states.
- Even when the oscillation frequency does not exactly match the Larmor frequency, there is some probability of observing a spin flip.

19.8 Further Reading

David McIntyre, *Quantum Mechanics: A Paradigms Approach* (Cambridge University Press, Cambridge, 2023). This book presents a nice treatment of spin precession and magnetic resonance.

Wikipedia article "Magnetic resonance (quantum mechanics)" https://en.wikipedia.org/wiki/Magnetic_resonance_(quantum_mechanics)

I. Rabi, N. F. Ramsey, and J. Schwinger, "Use of rotating coordinates in magnetic resonance problems," *Reviews of Modern Physics* **26**(2), 167–171 (1954). This paper provides a nice description of using rotating coordinates in both classical and quantum magnetic resonance. The paper also includes a short description of the history of the use of rotating coordinates in magnetic resonance problems.

20 Number Operators and Number States

So teach us to number our days, that we may apply our hearts unto wisdom.

Psalm 90

20.1 Introduction

For most of our discussions of quantum gates, quantum operators, and quantum circuits for quantum information science (QIS) and quantum computing (QC), we have focused on multiqubit states with a fixed number of qubits. In this chapter, we explore the quantum mechanics of systems of particles in which the number of particles may change. We focus on a simple example—a system of photons in which photons may be emitted (increasing the number of photons) or absorbed (decreasing the number of photons). More sophisticated examples are systems of the so-called elementary particles that can be emitted and destroyed in interactions among the particles. The discovery of these processes was one of the major advances in physics in the 20th century. These processes are described by relativistic quantum field theories. The linear transformations we study for a simple system of particles play a major role in these advanced theories.

As shown later, operator algebra, which is ultimately just linear algebra, may be used to describe these many-particle systems in a compact way. This methodology is becoming increasingly important in QIS and QC. The same operator algebra is also widely used in condensed matter physics, nuclear physics, atomic physics, and elementary particle physics. So, it is good to have these methods in your quantum toolbox. In Chapter 21, we show that the tools we develop in this chapter for number states can also be applied to states that are functions of position and momentum, two properties that dominate classical mechanics.

20.2 Number States

You should recall that we described multiqubit systems by using tensor product vector spaces and tensor products of state vectors. However, as mentioned previously, we have assumed that the number of qubits (or particles more generally) remains fixed. If the number of particles can change, then the methods we used previously still work but they become

Quantum Computing and Quantum Physics. Alice Flarend and Robert Hilborn, Oxford University Press.
 DOI: 10.1093/oso/9780197904930.003.0020

rather cumbersome. The method we explore in these notes allows us to handle the change in the number of particles in a relatively straightforward way.

The quantum states we use can be labeled by a complete set of eigenvalues for the properties of the system (focusing just on the properties that may change, such a position, energy, and spin orientation). By "complete" we mean that the eigenvectors corresponding to these eigenvalues form a complete set of basis states for that system: Any quantum state of that system can be expressed as a linear combination of the basis states. Other system properties, such as mass, spin angular momentum magnitude, and electric charge, are assumed to remain fixed. Note that some of the properties, such as spin orientation, have discrete eigenvalues, while others, like position and energy, may have continuous variable eigenvalues.

A key feature in our methodology is to specify the number of particles described by each quantum state. We call each set of eigenvalues, which may be a mix of discrete and continuous values, a "mode." For example, modes for photons have a specified energy, direction of travel, and linear polarization (for example, vertical linear polarization). We then specify how many photons in our system have those state specifications. We often need to know "how many photons occupy a particular mode." As shown later, the system might have 0, 1, 2, ... up to very large numbers of such photons. In other words, the system we want to describe consists of state vectors, each one specifying the number of particles present in the various modes available to these particles.

This set of occupation number state vectors constitutes a vector space, called Fock space, named after Vladimir Fock (1898–1974), who first developed this methodology in 1932. We label the basis states with positive integers for the number of particles present in each of the modes: $|0\rangle$, $|1\rangle$, $|2\rangle$, $\dots |n\rangle$, ... These states are called Fock states—states with definite numbers of particles. Some authors call these "occupation numbers" (meaning something like how many particles "occupy" the specified state). When we first introduced state vectors, we noted that thinking about state vectors as something particles "have" or "occupy" can be misleading, particularly for entangled states, which are so important in QIS and QC. So, we will avoid the term occupation number.

To keep the notation relatively simple, we focus on just one mode and not specify the set of properties explicitly, indicating just the number of particles. Later we describe how to handle systems with more than one mode.

The Fock states are the eigenstates of a number operator N

$$N|n\rangle = n\,|n\rangle . \tag{20.1}$$

As usual, we use an uppercase letter to indicate a Hermitian operator. The corresponding lowercase letter represents the corresponding eigenvalue. In Eq. (20.1), n serves two roles: (1) it labels the state vectors and (2) it is the eigenvalue of the number operator N for that state. The state $|0\rangle$ with no particles is called the *vacuum state.*

As with other vector spaces, a general state vector may be written as a linear superposition of the number basis states:

$$|\psi\rangle = \sum_{n=0}^{\infty} c_n\,|n\rangle, \tag{20.2}$$

where c_n is the (complex) amplitude for the state $|n\rangle$. In other words, the number states are basis states for the Fock vector space.

There are two distinct and important number state features to note. First, since there is no limit on the number of particles in the system (at least in this theoretical model), the Fock state vectors constitute an infinite-dimensional vector space. Second, a general state in this state space does not have a definite number of particles associated with it. For example, in the superposition state $|\phi\rangle = c_2\,|2\rangle + c_3\,|3\rangle$, there is a part with two particles and another part with three particles. Needless to say, it is a challenge to develop a mental model of such a state. But we can easily calculate the average number of particles:

$$\langle N\rangle = 2|c_2|^2 + 3|c_3|^2. \tag{20.3}$$

Try It 20.1 Explain why Eq. (20.3) gives the average number of particles. What is the average number if $c_2 = 1/\sqrt{3}$ and $c_2 = \sqrt{2/3}$?

20.3 Raising and Lowering Operators

It turns out to be useful to write the number operator as a product of two other operators a and $a^\dagger$, where the superscript † indicates the Hermitian conjugate:

$$N = a^\dagger a. \tag{20.4}$$

$a^\dagger$ is read as "a dagger" or "a adjoint." Even though a and $a^\dagger$ are not Hermitian operators (hence, they cannot represent measurable properties), their product is Hermitian, which is what we need because the number of particles in the system is a measurable property.

Try It 20.2 Show that $a^\dagger a$ is a Hermitian operator. What about $aa^\dagger$?

We assume that number states are orthonormal. The orthogonality condition $\langle n \mid n'\rangle = 0$ if $n \neq n'$ holds because of the general orthogonality property of eigenvectors of Hermitian operators. As shown below, we can always adjust the amplitudes multiplying the state vectors to make sure they have unit length.

We now explore the results of acting on the number states with these operators. These operations result in an increase or a decrease in the number of particles in the system. For many other quantum systems, we expressed the state right vectors as column vectors and the operators by matrices. But, as mentioned previously, the number states live in an infinite-dimensional vector space, so each column vector would have an infinite number of entries and the matrices would be $(\infty \times \infty)$-dimensional arrays—forgive us for not using column vectors and matrices. It turns out that all we need is what is called operator algebra:

manipulating the abstract expressions for the operators following the rules of linear algebra. We simply write down the results of the operations on the number states and see how they make sense. Once we have the operations on number states, we can use those results on any state vector written as a linear superposition of the number states.

First, we specify what happens when the operator a acts on a number state:

$$a\,|n\rangle = \sqrt{n}\,|n-1\rangle \tag{20.5}$$

for $n \geq 1$.

Try It 20.3 Why do we exclude $n = 0$?

We see that the a operator acting on a number state gives us a state whose number is reduced by 1. Consequently, the a operator is called a lowering operator, an annihilation operator, or a destruction operator because it describes the removal of one particle from the system.

The operation of the $a^{\dagger}$ operator acting on a number state is

$$a^{\dagger}\,|n\rangle = \sqrt{n+1}\,|n+1\rangle\,. \tag{20.6}$$

The $a^{\dagger}$ operator acting on a number state results in a state with one additional particle. Hence, $a^{\dagger}$ is called a raising operator or a creation operator because its application to a Fock state results in a state with the particle number increased by 1. The two operators together are called ladder operators.

We may think of the raising and lowering operators as moving the state up and down the ladder as shown in Fig. 20.1. If we use the vertical position to indicate the energy of the state, the spacing between any pair of adjacent states is ΔE. For our photon example, increasing the number of photons by one increases the energy of the system by E_{photon}.

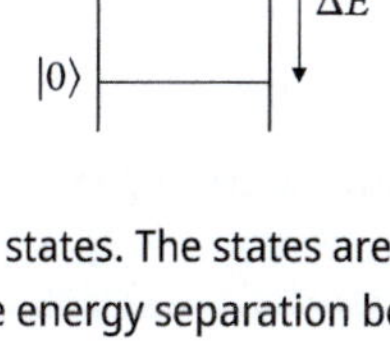

Fig. 20.1 The ladder diagram for number states. The states are represented by the horizontal lines. When the states are also energy eigenstates, the energy separation between adjacent states is indicated as ΔE.

You may have wondered where these square root factors in Eqs. (20.5) and (20.6) came from. They are present so that Eq. (20.1) is satisfied if N is replaced by the product of the ladder operators. Let's see how that works out:

$$a^\dagger a\,|n\rangle = a^\dagger\left(\sqrt{n}\,|n-1\rangle\right) = \sqrt{n}\left(a^\dagger\,|n-1\rangle\right) = \sqrt{n}\sqrt{n}\,|n\rangle = n\,|n\rangle\,. \tag{20.7}$$

Try It 20.4 Explain what happens for each step in Eq. (20.7) and pay particular attention to the origin of the second $\sqrt{n}$ factor.

As we have discovered in other parts of quantum mechanics, it is important to see what happens when a pair of operators acts on a state but in opposite orders. We know that the order matters because $AB\,|\psi\rangle$ may not equal $BA\,|\psi\rangle$, that is, the operators may not commute. So, let's see what happens when we reverse the order of the two ladder operators:

$$aa^\dagger\,|n\rangle = a\left(a^\dagger\,|n\rangle\right) = \sqrt{n+1}\,(a\,|n+1\rangle) = \sqrt{n+1}\sqrt{n+1}\,|n\rangle = (n+1)\,|n\rangle\,, \tag{20.8}$$

which is different from the result in Eq. (20.7). Combining the results from Eqs. (20.7) and (20.8), we find that

$$\left(aa^\dagger - a^\dagger a\right)|n\rangle = |n\rangle \tag{20.9}$$

and hence we have

$$aa^\dagger - a^\dagger a = \left[a, a^\dagger\right] = \mathrm{I}, \tag{20.10}$$

where, as usual, I is the identity operator and $\left[a, a^\dagger\right]$ is the commutator of the two operators. The two operators a and $a^\dagger$ do not commute, and consequently, the order of operation is important.

Normalization

Assume that $\langle 1|1\rangle = 1$ so that the state vector is normalized. We want to find the factor that will make $\langle 2|2\rangle = 1$. Here is how this works: We can use $a^\dagger$ to produce $|2\rangle$ from $|1\rangle$ and then evaluate $||2\rangle|^2 = |a^\dagger|1\rangle|^2 = \langle 1|aa^\dagger|1\rangle = \langle 1|a^\dagger a + \mathrm{I}|1\rangle = \langle 1|a^\dagger a|1\rangle + 1 = 1 + 1 = 2,$ where we used the commutator relation in Eq. (20.10). So, we need $\left(1/\sqrt{2}\right)|2\rangle$ to have the state normalized. It is easy to see that the state with n particles is normalized if we use the factor $1/\sqrt{n}$. Note the general result agrees with our initial assumption for $n = 1$. (Always good to check!)

Try It 20.5 Show by the same kind of reasoning that the vacuum state is normalized $\langle 0 | 0 \rangle = 1$.

We can also think about generating the state $|n\rangle$ by repeatedly applying the raising operator to the vacuum state $|0\rangle$:

$$|n\rangle = \frac{\left(a^{\dagger}\right)^{n}}{\sqrt{n!}} |0\rangle . \qquad (20.11)$$

The factorial function is $n! = n \times (n-1) \times (n-2) \cdots 2 \times 1$.

Try It 20.6 Why do we need the factorial function in Eq. (20.11)?

Connections with Energy

If we describe a system of photons each with the same frequency, then we know from quantum physics that each photon carries an energy $E_{\text{photon}} = hf = \hbar\omega$, where h is Planck's constant and f is the frequency associated with the classical wave model of light. Instead of f (the number of wave cycles per unit time), authors often use $\omega = 2\pi f$ and the reduced Planck constant $\hbar = h/(2\pi)$ to avoid factors of 2π that would appear in calculations had they used f. In this case, we write $E_{\text{photon}} = \hbar\omega$.

The ladder diagram in Fig. 20.1 can be thought of as an energy level diagram. The spacing between each rung on the energy ladder is the energy added by a single photon. It also means that the energy operator (the Hamiltonian operator) for the system is given by the number operator multiplied by the single photon energy

$$H = E_{\text{photon}} a^{\dagger} a. \qquad (20.12)$$

It turns out that there is energy associated with the vacuum state $|0\rangle$ of the system even though there are no photons associated with that state. This energy is called the vacuum energy and turns out to be $E_{\text{vacuum}} = \hbar\omega/2$, i.e., just half the energy carried by one photon. We explore this idea below. If we include the vacuum energy, the Hamiltonian is

$$H = E_{\text{photon}} \left(a^{\dagger} a + \frac{1}{2} \right). \qquad (20.13)$$

Time Dependence of the Ladder Operators

In many situations, including understanding vacuum energy, it is useful to find the time dependence of quantities involving the ladder operators by putting the time dependence in the operators themselves rather than in the amplitudes of the state vectors. To do that, we invoke the Heisenberg picture (see Chapter 18). In the Heisenberg picture, the ladder operators' time dependence is given by the following differential equations:

$$\frac{da}{dt} = \frac{i}{\hbar}[a, H] \text{ and } \frac{da^\dagger}{dt} = \frac{i}{\hbar}\left[a^\dagger, H\right]. \tag{20.14}$$

To solve those equations, our first task is to evaluate the commutators. For a system of photons, the commutator of the Hamiltonian and the lowering operator is

$$\begin{aligned}[a, H,] &= E_{\text{photon}}\left[a, a^\dagger a\right] = E_{\text{photon}}\left(aa^\dagger a - a^\dagger aa\right) \\ &= E_{\text{photon}}\left(aa^\dagger - a^\dagger a\right)a = E_{\text{photon}}a\end{aligned} \tag{20.15}$$

In the last line of Eq. (20.15), we used the commutator in Eq. (20.10).

Try It 20.7 Work through the steps in Eq. (20.15) and explain what is happening. Why were we able to leave out the +1/2 term from the Hamiltonian in Eq. (20.13)?

The differential equation satisfied by the lowering operator is then

$$\frac{da}{dt} = i\frac{E_{\text{photon}}}{\hbar}a \tag{20.16}$$

We have seen this type of differential equation before with the derivative proportional to the function itself and (hopefully) recognize that the solution is

$$a(t) = a(0)e^{iE_{\text{photon}}t/\hbar}. \tag{20.17}$$

The equation for $a^\dagger$ is then "obviously"

$$a^\dagger(t) = a^\dagger(0)e^{-iE_{\text{photon}}t/\hbar}. \tag{20.18}$$

Try It 20.8 Explain why Eq. (20.18) is the obvious solution for the raising operator. If you are skeptical, work through the equivalent of Eq. (20.15) and use this result to find the equivalent of Eq. (20.16).

20.4 Vacuum Energy

We now put together an argument to explain why there is energy associated with the vacuum (no particle) state. There are two Hermitian operators—the sum of the raising and lower operators and $i = \sqrt{-1}$ times the difference of the raising and lower operators—which have been waiting patiently on the sidelines. It is now time to bring them onto the playing field. Since the sum and differences (with the $\sqrt{-1}$ factor) are Hermitian, they possibly represent measurable quantities. Spoiler alert: We show that for photons the difference of the raising and lowering operators is related to the oscillating electric field associated with light. (For those familiar with classical electromagnetic theory, you may know that the sum of raising and lowering operators is related to the electromagnetic vector potential function.)

We focus on the average value of various operators with respect to number states since these averages represent the results of measurements. We use the difference of the raising and lowering operators and leave the sum as an exercise for the reader.

The operator for the electric field $\boldsymbol{\mathcal{E}}$ associated with a photon is proportional to the difference between the raising and lowering operators:

$$\boldsymbol{\mathcal{E}} \propto i(a - a^\dagger). \tag{20.19}$$

We show you the proportionality factor shortly.

Try It 20.9 Show that $i(a - a^\dagger)$ is a Hermitian operator. Explain your reasoning.

The average value of the difference of the raising and lowering operators is easily seen to be zero for a number state:

$$i\langle n|\, a - a^\dagger \,|n\rangle = i\langle n|an\rangle - i\langle n|a^\dagger n\rangle = i\sqrt{n}\,\langle n|n-1\rangle - i\sqrt{n+1}\,\langle n|n+1\rangle = 0 \tag{20.20}$$

because of the orthogonality of the number states. So, whatever the property associated with the difference of the operators is, it has a zero average value for number states.

Since we know that the energy density associated with a classical electric field is proportional to the square of the field value, we can get some hints of what is going on by looking at the average value of the *square* of the difference of the raising and lowering operators:

$$\langle n|\, i^2\left(a - a^\dagger\right)^2 |n\rangle = -1\left[\langle n|\, a^2\,|n\rangle - \langle n|\, aa^\dagger + a^\dagger a\,|n\rangle + \langle n|\left(a^\dagger\right)^2|n\rangle\right]. \tag{20.21}$$

The first and last parts of the right side of the previous equation are zero because of the orthogonality of the number states. In the middle term, we use the commutator in Eq. (20.10) to replace $aa^\dagger$ with $a^\dagger a + \mathrm{I}$ from which we find

$$\langle n|\left(2a^\dagger a + \mathrm{I}\right)|n\rangle = 2n + 1. \tag{20.22}$$

Thus, even in the vacuum state (with $n = 0$), the average value of the square of the difference of the raising and lowering operators is not zero.

Try It 20.10 Carry out the same analysis for the sum of the raising and lowering operators $a + a^\dagger$.

Let's expand on the notion that electrical energy density is proportional to the electric field squared. We can interpret the operator in physics terms by thinking about the energy associated with the state in terms of the electric field in the classical picture of an electromagnetic wave. For almost all types of waves, we can associate the square of some quantity with the energy carried by the wave. For an electromagnetic wave, the quantity is, as just mentioned, the electric field. For sound waves in air, the energy is proportional to the square of the pressure variations in the air. For water waves, the energy is proportional to the square of the height of the wave disturbance measured from the quiet water level.

For the electric field case, we know from introductory physics and engineering classes that the energy expression for the average energy associated with the electric field in a volume V is $\varepsilon_0 \mathcal{E}^2 V/2$. The electric field magnitude is indicated by $\mathcal{E}$, and ε_0 is an electromagnetic constant used to get SI units (volts/meter) for the field. The ½ comes in because we are averaging the oscillating field over at least one period of oscillation. (The field itself oscillates as a cosine function such as $\cos \omega t$ and the average of $\cos^2 \omega t$ over one period is ½.) These ideas suggest that the electric field operator for photons with frequency ω may be expressed as[1]

$$\mathcal{E} = \left(\frac{2\hbar\omega}{\varepsilon_0 V}\right)^{1/2} i\left(a - a^\dagger\right). \tag{20.23}$$

For the electric field squared associated with a single photon, we use that definition and the energy expression to find that

$$\frac{1}{2}\varepsilon_0 \mathcal{E}^2 V = \frac{1}{2}\varepsilon_0 \frac{2\hbar\omega}{\varepsilon_0 V} V = \hbar\omega. \tag{20.24}$$

If we use Eqs. (20.22)–(20.24), we find that

$$\langle n| \frac{1}{2}\varepsilon_0 \mathcal{E}^2 V |n\rangle = \frac{1}{2}\hbar\omega\,(2n + 1) = \hbar\omega\left(n + \frac{1}{2}\right). \tag{20.25}$$

Eq. (20.25) shows that the $n = 0$ state (the vacuum state) has electromagnetic energy associated with it. In quantum mechanics language, the average electric field in any number state is zero, but the average of the field squared has a nonzero value, even in the state with no photons present. We talk about "quantum fluctuations" of the electric field. In fact, experiments have been carried out to show that these fluctuations have actual measurable effects.

[1] Various authors place the factor of 2 in different places in these expressions. These differences are based on different versions of the definition of field energy.

Let's pause for a methodological comment. We arrived at these interesting physical conclusions by using abstract operator algebra informed by what we know from classical physics. These operator algebra methods are part of linear algebra involving the manipulation of matrices since operators may be represented by matrices. However, for number states, the matrices are infinite dimensional, which makes their manipulations more difficult. So, we use the operator algebra method.

Multiple Modes

In many situations, the physical object we are considering has several physical properties that may change because of interactions. For example, for photons, the properties include type of polarization, direction of travel, and energy (proportional to the corresponding wave frequency). We can define raising and lowering operators labeled by the specific values of these properties. For example, we might define $a^{\dagger}_{\text{vlp},\vec{r},\omega}$ as the raising operator that acts on the vacuum state to add one photon traveling in the $\vec{r}$ direction with vlp polarization and energy $\hbar\omega$.

The number basis states for each photon are then labeled by both the photon number n and the list of properties associated with that mode: $|n\rangle_{\text{vlp}.\vec{r},\omega}$. The "power" of quantum processes arises because we can, at least in principle, prepare photon states that are linear combinations of states associated with different modes. Specialized equipment is needed to produce those superposition states. For example, a partially reflecting/partially transmitting mirror (beamsplitters) can produce superposition states of different travel directions. Polarization devices, such as linear polarizers, quarter-wave plates, and so on, can produce polarization superpositions as we have seen many times in this book. There are also nonlinear devices, such as "parametric down-converters," that can remove a higher frequency (higher energy) photon and add two photons of lower frequency (lower energy) to the system. The field of "quantum optics" has employed many of these kinds of superposition states for use in QIS and QC, as well as tests of fundamental quantum principles. See Further Reading for more information about quantum optics.

20.5 Uncertainty Relation for Number Operators and Electric Field Operators

In this section, we look at the *uncertainty relation* as applied to number states. The general *Heisenberg uncertainty relation* shows that the product of the standard deviations for the two operators is related to the average value of the commutator:

$$(\Delta N)_{\psi}(\Delta \mathcal{E})_{\psi} \geq \frac{1}{2}\left|\langle [N, \mathcal{E}] \rangle_{\psi}\right|. \tag{20.26}$$

Eq. (20.26) shows the uncertainty relationship for the number operator and the electric field operator (dropping constants for now) involves the commutator $[N, a - a^{\dagger}]$. (We may also drop the i factor because the *uncertainty relation* involves the absolute value of the commutator.) First, let's look at the commutator of those two operators. We evaluate the two parts individually:

$$\left[a^{\dagger}a, a\right] = a^{\dagger}aa - aa^{\dagger}a = -\left(aa^{\dagger} - a^{\dagger}a\right)a = -\left[a, a^{\dagger}\right]a = -a, \tag{20.27}$$

where we used Eq. (20.10). The other part gives

$$\left[a^{\dagger}a, a^{\dagger}\right] = a^{\dagger}aa^{\dagger} - a^{\dagger}a^{\dagger}a = a^{\dagger}\left(aa^{\dagger} - a^{\dagger}a\right) = \left[a, a^{\dagger}\right]a^{\dagger} = a^{\dagger}. \tag{20.28}$$

From these two equations, we see that the number operator and the electric field operator do not commute:

$$\left[a^{\dagger}a, (a - a^{\dagger})\right] = -(a^{\dagger} + a). \tag{20.29}$$

Try It 20.11 For a photon number basis state $|\psi\rangle = |n\rangle$, show that $(\Delta N)_{\psi} = 0$. Is that consistent with the right side of Eq. (20.26)?

In principle, we could use Eq. (20.29) in Eq. (20.26) to find the lower limit for the product of standard deviations for superposition states. However, the calculation is moderately complicated and depends on the form of the number state superposition; so, we just provide a qualitative argument. For a state $|\psi\rangle$ expressed as a superposition of number states, there will be some "spread" in the number of photons. This spread is expressed as the standard deviation ΔN. Similarly for the spread in electric field values. If we let ΔN increase (wider range of photon number), the spread in electric field values can be made smaller and still have the product of standard deviations satisfy the *uncertainty relation*. This particular application of the *Heisenberg uncertainty principle* is an example of the complementarity between particle and wave pictures in quantum mechanics: There is no quantum state in which the number of photons (a particle property) and the associated electric field (a wave property) are both precisely known (standard deviations are zero).

20.6 Applications in QC and QIS

These ideas are used in a wide range of applications in QC and QIS. In some situations, photons are used to manipulate qubits to prepare quantum states and to manipulate these states via quantum gates. In other situations, the photons themselves are the qubits and their states can be prepared and manipulated in ways that are described by combinations of raising and lowering operators.

Here is an example of using raising and lowering operators in the description of a two-dimensional lattice of qubits. In this case, the qubits are superconducting microwave cavities with a resonant frequency ω. The number of photons in a cavity is a measure of its "excitation." The lattice sites are labeled by indices i and j. The Hamiltonian (energy) operator for the system is written as[2]

$$H_{\text{lattice}} = \hbar\omega \sum_i a_i^\dagger a_i - \hbar t \sum_{(i,j)} \left(a_i^\dagger a_j + a_j^\dagger a_i\right) \tag{20.30}$$

We recognize the first term on the right as the product of the number of sites and the energy $\hbar\omega$ per site. The second term has a sum over all pairs of sites. The term $a_i^\dagger a_j$ removes an excitation from site labeled j and creates an additional one at site labeled i. $\hbar t$ is the energy associated with that transfer. The properties of the states and their evolution in time can be calculated using generalizations of the operator techniques introduced in this chapter.

20.7 Fermion Number States and Ladder Operators

The analysis in the previous sections focused on systems for which the number of particles associated with a specific mode could range from 0 to infinity. In physics, these kinds of particles are called bosons, after the Indian physicist S. Bose, who, in the 1920s, pioneered the theory of systems of such particles. However, nature has a surprise in store for us. There is another family of particles for which the occupation number is either 0 or 1. There are only two possibilities: no particles in a particular mode or just 1. Such particles are called fermions, named after Enrico Fermi.

What distinguishes bosons from fermions? Another surprise: The distinguishing feature is their intrinsic spin angular momentum. Recall that the fundamental unit for spin angular momentum is $\hbar/2$. It turns out that bosons have an even number $(0, 2, 4, \ldots)$ units of spin angular momentum while fermions have an odd number $(1, 3, 5, \ldots)$ of such units. Among the elementary particles, photons have spin angular momentum $\hbar$ and the infamous Higgs boson has 0 units, and hence both are bosons. On the other hand, particles such as electrons, protons, neutrons, quarks, and neutrinos have spin angular momentum $\hbar/2$ and are fermions. Although there remains some controversy about *why* these values of spin angular momentum are associated with fermions and bosons, there is overwhelming empirical evidence for that distinction.

Why do we care about bosons versus fermions? Since the building blocks of ordinary matter are electrons, protons, and neutrons, the limitation on occupation numbers to just 0 and 1 for fermions has a significant impact on the structure of nuclei, atoms, and molecules. You may have heard of the *Pauli exclusion principle* in introductory physics or chemistry courses. The *exclusion principle* states that no more than one fermion (think electron) can occupy a given quantum state. That restriction then leads to the orbital shell structure of atoms and the periodic table of the elements. Without the *exclusion principle*, the chemistry of the elements would all be similar and life as we know it would not exist.

[2] Alicia J. Kollár, Mattias Fitzpatrick, and Andrew A. Houch, "Hyperbolic lattices in circuit quantum electrodynamics," *Nature* **571** (7763), 45–50 (2019).

Let us now turn to raising and lowering operators for fermions. We label the fermion raising and lowering operators as $c^\dagger$ and c. These operators act on the two possible fermion Fock states:

$$c^\dagger |n = 0\rangle = |n = 1\rangle \qquad c\,|n = 1\rangle = |n = 0\rangle . \tag{20.31}$$

We have labeled the states by indicating the number of particles associated with the states to avoid confusion with the $|0\rangle$ and $|1\rangle$ computational basis states. We also have

$$c^\dagger |n = 1\rangle = 0 \qquad c\,|n = 0\rangle = 0. \tag{20.32}$$

The first equation shows that we can't put one more fermion into an already occupied state. The second reminds us that the lowest number of particles is 0. Combining Eqs. (20.31) and (20.32), we find that $c^\dagger c^\dagger = 0$ and $cc = 0$.

Based on our discussion of boson number operators, we may define a fermion number operator

$$N = c^\dagger c. \tag{20.33}$$

Let's see what happens when N acts on the $|n = 1\rangle$ Fock state for fermions:

$$c^\dagger c\,|n = 1\rangle = c^\dagger\,|n = 0\rangle = |n = 1\rangle . \tag{20.34}$$

Eq. (20.34) shows that the eigenvalue of the number operator associated with the $|n = 1\rangle$ state is 1. (That's why we labeled the states that way.) For the $|n = 0\rangle$ state, we obtain

$$c^\dagger c\,|n = 0\rangle = 0\,|n = 0\rangle , \tag{20.35}$$

which shows that the number operator eigenvalue associated with the $|n = 0\rangle$ state is 0.

As we observed with the boson ladder operators, knowing what happens when the operators act in the opposite order is an important part of the operator algebra. It turns out that the fermion ladder operators satisfy what is called the anticommutator expression

$$cc^\dagger + c^\dagger c = \left\{c, c^\dagger\right\} = \mathrm{I}. \tag{20.36}$$

Because the left side of Eq. (20.36) uses a plus sign, while the commutator uses a minus sign, the form with the plus sign is called the anticommutator. The curly braces { } are traditionally used to denote the anticommutator. The anticommutator relates the two orders in which the operators may act: $cc^\dagger$ versus $c^\dagger c$.

Let's check that the operator algebra for fermions gives us a consistent set of results. We begin by evaluating the average value of the operator product $N(\mathrm{I} - N)$ for the Fock states:

$$\begin{aligned}\langle n|\, N(\mathrm{I} - N)|n\rangle &= \langle n|\, c^\dagger c(\mathrm{I} - c^\dagger c)|n\rangle = \langle n|\, c^\dagger c - c^\dagger c c^\dagger c\,|n\rangle \\ &= \langle n|\, c^\dagger c - c^\dagger c(\mathrm{I} - cc^\dagger\,|n\rangle = \langle n|\, c^\dagger c c c^\dagger\,|n\rangle = 0.\end{aligned} \tag{20.37}$$

In progressing from the first term on the right of Eq. (20.37) to the second term, we used the anticommutator relation, Eq. (20.36). The left side of Eq. (20.37) is just $n(1-n)$. So, we end up with $n(1-n) = 0$ for the eigenvalues of the fermion number operator. Hence, the two possible solutions are $n = 0$ and $n = 1$, in agreement with what we found before.

We may also use some operator algebra to find matrix representations for the fermion raising and lowering operators. Since there are only two eigenvalues (and two corresponding eigenstates), we deal with a two-dimensional state space. We know from our discussions of spin-1/2 systems that the Pauli operators $\sigma_x, \sigma_y, \sigma_z$ (or X, Y, Z) and the identity operator form an operator basis for two-state systems and any operators in a two-dimensional state space can be written as a linear combination of these operators. For a two-dimensional Fock space, we write the ladder operators as

$$c^{\dagger} = \frac{1}{2}\left(\sigma_x + i\sigma_y\right) \qquad c = \frac{1}{2}\left(\sigma_x - i\sigma_y\right). \tag{20.38}$$

We recall that the Pauli operators satisfy the following relations:

$$\sigma_x^2 = \sigma_y^2 = I \qquad \left[\sigma_x, \sigma_y\right] = 2i\sigma_z. \tag{20.39}$$

Let's use Eqs. (20.38) and (20.39) in the fermion number operator

$$\begin{aligned} N = c^{\dagger}c &= \frac{1}{2}\left(\sigma_x + i\sigma_y\right)\frac{1}{2}\left(\sigma_x - i\sigma_y\right) \\ &= \frac{1}{2}\mathrm{I} + \frac{1}{2}\sigma_z. \end{aligned} \tag{20.40}$$

Try It 20.12 Write out the details in Eq. (20.40).

Since the eigenvalues of σ_z are ± 1, we see that the eigenvalues of N are 0 and 1, confirming what we found before.

Multiple Fermion Modes

There is one unexpected complication when dealing with fermions in different modes. One might expect that fermion operators associated with different modes would commute with each other since the modes are independent of one another. However, to have a consistent operator algebra for fermions, we need to have the operators associated with different modes *anticommute*. Let's add a subscript to the fermion operators to label the modes: c_j and $c_j^{\dagger}$. So, for two different modes, the operators satisfy the anticommutation relations $\left\{c_j, c_k^{\dagger}\right\} = 0$, $\left\{c_j, c_k\right\} = 0$, and $\left\{c_j^{\dagger}, c_k^{\dagger}\right\} = 0$. Since $\left\{c_j, c_k\right\} = 0$ means $c_j c_k = -c_k c_j$, we will have a lot of minus signs to keep track of and that makes calculations involving fermion operators more complex than the corresponding calculations for boson operators.

20.8 Chapter Summary

- Raising and lowering operators $a^\dagger$ and a may be used to define number states, which are eigenstates of the number operator $N = a^\dagger a$. For bosons, the number eigenvalues are 0, 1, 2, ... with no upper bound.
- The raising and lowering operators change the number state by $+1$ or -1, respectively.
- The i times the difference of the lowering and raising operators is a Hermitian operator that can be interpreted as representing the electric field associated with photons.
- Because the number operator and the electric field operator do not commute, the *Heisenberg uncertainty relation* shows that we cannot have a quantum state in which both the number of photons and the electric field are both precisely known (standard deviations both equal 0).
- Fermion raising and lowering operators are described by anticommutation relations.
- The fermion number eigenvalues for a specific mode are 0 and 1.
- Fermion raising and lowering operators for different modes anticommute.

20.9 Further Reading

Many undergraduate quantum mechanics texts introduce ladder operators when discussing the simple harmonic oscillator (mass on a spring) model (see Chapter 21).

Most textbooks on quantum optics and photonics have introductions to raising and lowering operators. Here are two suggestions:

Christopher C. Gerry and Peter L. Knight, *Introductory Quantum Optics* (Cambridge University Press, 2005).

Serge Haroche and Jean-Michel Raimond, *Exploring the Quantum: Atoms, Cavities, and Photons* (Oxford University Press, Oxford, 2006).

Momentum and Position States and the Simple Harmonic Oscillator

21

Life is a constant oscillation between the sharp horns of dilemmas.

H. L. Mencken

21.1 Momentum and Position

In classical mechanics, position and linear momentum (mass times velocity) are often the primary physical variables used for describing trajectories of everything from planets to rockets to small particles. (Henceforth, we drop the adjective "linear" for momentum, but it will be implied.) They are equally important in quantum mechanics, but there is a complexity at the quantum level. Position and momentum, as shown later, are complementary properties in the sense of the *uncertainty relation*: It is impossible to have a quantum state in which position and momentum are both precisely known properties. Recall that "precisely known" means that their standard deviations are zero. This may not seem to be a big deal unless we remember that in classical mechanics you need precise values of both position and momentum to predict the trajectory of an object. So, in some sense, quantum mechanics has pulled the rug out from under one of the most important principles in classical mechanics.

To show how quantum mechanics handles position and momentum, including modeling the simple harmonic motion that can be used to represent many systems such as molecular bonds, we need to introduce operators for these properties. We also need basis state vectors. To keep the notation relatively simple, we restrict ourselves to one-dimensional motion, along, say, the x-axis of some coordinate system. The operator for the x component of the momentum is indicated by P_x (the symbol traditionally used in physics for momentum), and the position operator is indicated by X. (Note that we denote the x position operator by an italic symbol to distinguish it from the Pauli X operator.) The eigenvalues for these operators are written as p_x and x, respectively. The eigenvectors are, as usual, labeled by the eigenvalues:

$$P_x|p_x\rangle = p_x|p_x\rangle\,. \tag{21.1}$$

Quantum Computing and Quantum Physics. Alice Flarend and Robert Hilborn, Oxford University Press.
 DOI: 10.1093/oso/9780197904930.003.0021

These eigenvectors constitute the momentum basis states. For the corresponding coordinate (position) basis states, we have

$$X|x\rangle = x|x\rangle . \tag{21.2}$$

For a three-dimensional situation, we would include coordinate basis states for y and z and for the y and z components of the momentum.

Most introductory treatments of quantum mechanics focus on quantum states in the coordinate rather than momentum basis, that is, a general quantum state may be written as a superposition of the coordinate basis state vectors $|x\rangle$. However, since the x coordinate is a continuous (rather than discrete) property, the superposition must be written as an integral rather than just a simple sum:

$$|\psi\rangle = \int \langle x|\psi\rangle|x\rangle \, dx = \int |x\rangle \langle x|\psi\rangle \, dx. \tag{21.3}$$

As usual, the Dirac bracket $\langle x|\psi \rangle$ is a number representing the projection of the state vector $|\psi\rangle$ along the coordinate state basis vector $|x\rangle$. In the third term in Eq. (21.3), we moved the Dirac bracket to the right of the state vector to indicate that the integral of the coordinate projection operator $|x\rangle \langle x|$ is equal to the identity operator $\int |x\rangle \langle x| \, dx = \mathrm{I}$. For historical reasons, the coordinate Dirac bracket is called the wave function:

$$\langle x|\psi \rangle \equiv \psi(x). \tag{21.4}$$

Since $\psi(x)$ is a quantum state amplitude, we expect that its square (strictly speaking, its absolute-value-squared, because it may be a complex number) has something to do with probability. For a continuous variable like x, $|\psi(x)|^2$ is actually the probability "density," specifically the probability per unit distance along the x-axis. In other words, $|\psi(x)|^2 dx$ is the probability of finding the quantum object in the range $[x, x + dx]$ along the x-axis.

Average Values in the Coordinate Basis

Since we now have operators with continuous parameters, when we calculate average values, we need to integrate over x rather than just sum over basis states as we had done for spin systems, for example. For an operator A, the average value of measurements of the property associated with A when the system is prepared in a state described by the wave function $\psi(x)$ is given by

$$\langle A\rangle = \langle\psi| A|\psi\rangle = \int dx \; \psi^*(x)A\psi(x). \tag{21.5}$$

We have not yet specified the form of the operator A, but in most cases, it is given as a function of x. In general, we integrate over all values of x from $-\infty$ to $+\infty$.

21.2 Spatial Displacement and Momentum Operators

Our first goal is to find equations that allow us to find $\psi(x)$ for a specified physical situation. We follow the line of reasoning used in Chapter 18 on time dependence, but now we focus on spatial displacements rather than time displacements. We start by introducing a unitary operator that shifts the x coordinate by an amount δx:

$$|\psi(x+\delta x)\rangle = U_{\delta x}\,|\psi(x)\rangle\,. \tag{21.6}$$

Based on our discussion of time displacements in Chapter 18, we anticipate that $U_{\delta x}$ will have the form

$$U_{\delta x} = \mathrm{I} + i\frac{D_x}{\hbar}\delta x \tag{21.7}$$

for small enough δx. D_x, the infinitesimal spatial displacement operator, will be a Hermitian operator. The term infinitesimal means that the operator $U_{\delta x}$ shifts the x coordinate by a very small amount δx.

Math Aside Our choice of a plus sign in Eq. (21.7) is due to the need to have the sign choices for the time displacement operator and for the position displacement operator give a result in agreement with a convention used in the description of waves. (Wave functions in quantum mechanics share some properties with the functions that describe waves in classical mechanics.) In particular, the result in Eq. (21.28) has the argument $(kx - \omega t)$, which describes a wave traveling in the positive x-direction (assuming that k and ω are positive parameters). In other words, as t increases (becomes more positive), x needs to become more positive to have the same value for the argument $(kx - \omega t)$. This illustrates why we want to have opposite signs for the spatial and temporal displacement operators.

Using Eq. (21.7) in Eq. (21.6) yields

$$|\psi(x+\delta x\rangle = |\psi(x)\rangle + \frac{i}{\hbar}D_x\,|\psi(x)\rangle\,\delta x \tag{21.8}$$

or with some rearrangement

$$\frac{|\psi(x+dx\rangle - |\psi(x)\rangle}{\delta x} = \frac{d\,|\psi(x)\rangle}{dx} = \frac{i}{\hbar}D_x\,|\psi(x)\rangle = \frac{i}{\hbar}P_x\,|\psi(x)\rangle\,, \tag{21.9}$$

where we used the standard definition of derivative in calculus. We have also introduced the symbol P_x for the momentum operator. Solving for the operator D_x, we find

$$D_x\,|\psi(x)\rangle = P_x\,|\psi(x)\rangle = -i\hbar\frac{d\,|\psi(x)\rangle}{dx}. \tag{21.10}$$

We say that infinitesimal spatial displacement operator (or equivalently, the momentum operator) in the coordinate basis is proportional to the derivative with respect to that coordinate. (Note that it is perfectly legitimate to think of a derivative as a linear transformation. See the Wikipedia article "Operational Calculus.")

Math Aside Is P_x a Hermitian operator? If it is to represent an observable quantity such as momentum, it should be. Let's find out. First, recall that if P_x is a Hermitian operator it must satisfy

$$\langle\phi| P_x |\psi\rangle\rangle = \langle\psi| P_x|\phi\rangle^*. \tag{21.11}$$

We now use P_x in its coordinate basis in Eq. (21.10) to obtain

$$\langle\phi| P_x |\psi\rangle = \langle\phi| - ih\frac{d}{dx} |\psi\rangle = -i\hbar \int dx\phi^*(x)\frac{d}{dx}\psi(x). \tag{21.12}$$

We now use everyone's favorite (we are being ironic) mathematical tool: integration by parts. Recall the product rule for differentiation

$$\frac{d}{dx}(fg) = g\,df + f\,dg. \tag{21.13}$$

The integral in Eq. (21.12) has an integrand that is equivalent to $f\,dg$ if we associate f with $\phi^*(x)$ and g with $\psi(x)$. (This is equivalent to the standard "integration by parts" formula $\int f dg = fg - \int g df$.) Using these associations, we may rewrite Eq. (21.12) as

$$\begin{aligned} -i\hbar \int dx\phi^*(x)\frac{d}{dx}\psi(x) &= -i\hbar\left\{\int dx\frac{d}{dx}[\phi^*(x)\psi(x)] - \int dx\psi(x)\frac{d}{dx}\phi^*(x)\right\} \\ &= -i\hbar\left\{[\phi^*(x)\psi(x)]_{-\infty}^{+\infty} - \int dx\psi(x)\frac{d}{dx}\phi^*(x)\right\} \\ &= \langle\psi| P_x|\phi\rangle^*, \end{aligned} \tag{21.14}$$

which shows that P_x is a Hermitian operator. Let's explain the details. In progressing from the last term in the first line of Eq. (21.14) to the second line, we used the *fundamental theorem of calculus* $\int_a^b dx\,(dF(x)/dx) = [F(x)]_a^b$. We then invoked the assumption that wave functions for any reasonable quantum system need to go to zero as $x \to \pm\infty$. In particular, we use $[\phi^*(x)\psi(x)]_{-\infty}^{+\infty} = 0$. (We will mention an exception to this rule later in this chapter.) Finally, we used the complex conjugation operation applied to the Dirac bracket. This example illustrates an important mathematical point: Whether or not a particular operator is Hermitian depends on the mathematical properties of the states on which the operator acts.

Try It 21.1 Check the algebra in progressing from the middle line of Eq. (21.14) to the last line.

Challenge: If you are willing to accept that P_x represents linear momentum, you may skip this box. But if you are (appropriately) skeptical, how do we know that representation is correct? To answer this question, we need to go beyond Eq. (21.10) and invoke other results from quantum mechanics: If we find that the results of having P_x act on quantum states yields properties that we expect to find for momentum, then we may legitimately claim that P_x represents momentum.

First, as shown later, the expectation value $\langle\psi|\, P_x\, |\psi\rangle$ is independent of time if $|\psi\rangle$ is a state describing a situation in which there are no external interactions with the system. In classical mechanics, this corresponds to the statement that if there are no (net) external forces acting on a system (free particle), its linear momentum is constant (independent of time). Second, for a wave packet (a linear superposition of several momentum basis states), the average of P_x is equal to the mass of the object times the group velocity of the packet. In classical mechanics, this corresponds to the fact that momentum is given by the product of the mass and its velocity. The details of the second connection require knowledge about wave group velocity, which is beyond the scope of this book (see Further Reading).

A third, and rather subtle, argument is based on symmetry. The mathematician Emmy Noether (1893–1935) showed that if a system's potential energy function has a continuous symmetry (for example, being unchanged under spatial displacements), then the system's behavior is described by a conservation law. In the case of spatial displacements, the law is the conservation of linear momentum.

21.3 **Coordinate Basis States and Momentum Eigenstates**

Now we want to find the momentum eigenstates as a function of the coordinate x. Using the eigenvalue equation for linear momentum, we have $P_x|p_x\rangle = p_x|p_x\rangle$ in Eq. (21.10) with $|p_x\rangle$ replacing the generic $|\psi(x)\rangle$. We can find the momentum eigenvector expressed by the coordinate basis states by left multiplying by $\langle x|$, which gives us a differential equation for the Dirac bracket $\langle x|p_x\rangle$:

$$-i\hbar\frac{d}{dx}\langle x|p_x\rangle = p_x\langle x|p_x\rangle, \tag{21.15}$$

whose solution is

$$\langle x|p_x\rangle = a\,e^{ip_x x/\hbar} = a[\cos(p_x x) + i\sin(p_x x)]. \tag{21.16}$$

In analogy with Eq. (21.4), we call $\langle x|p_x\rangle$ the momentum wave function.

Let's check that $\langle x|p_x\rangle = ae^{ip_x x/\hbar}$ is a solution of Eq. (21.15). First, we evaluate the left side of Eq. (21.15) by taking the derivative with respect to x: $\frac{d}{dx}a\,e^{ip_x x/\hbar} = a(ip_x/\hbar)e^{ip_x x/\hbar}$. We then multiply that result by $-i\hbar$ and find that the left side is equal to the right side. We have found the solution.

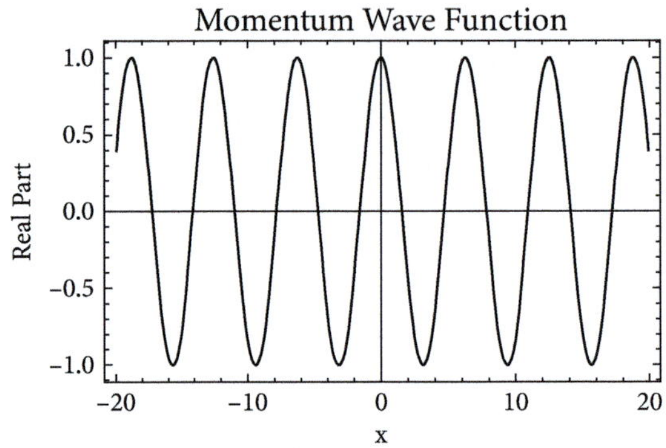

Fig. 21.1 The real part of the amplitude of $\langle x|p_x\rangle$ from Eq. (21.16) with $a = 1$.

In Eq. (21.16), the parameter a is a normalization constant. In the last term, we have separated the result into its real and imaginary parts. We see that the terms oscillate with x and support the choice of the name "wave" function. Figure 21.1 shows a plot of the real part of the momentum wave function as a function of x. The distance between successive maxima of the momentum wave function is called the de Broglie wavelength, after Louis de Broglie, who proposed in 1924 that any particle with a momentum p has associated with it a wave of wavelength $\lambda = h/p$, where h is Planck's constant. The de Broglie wavelength is often misinterpreted as a particle like a marble oscillating up and down analogous to a yo-yo or a paddleball on an elastic string. Instead, the de Broglie wavelength describes the wave nature in the wave–particle duality nature of matter.

Let's calculate the expectation value of the operator $P_x = -i\hbar\, d/dx$ for the momentum wave function given in Eq. (21.16). The expectation value is given by the quantum sandwich

$$\begin{aligned}\langle P_x\rangle = \langle\psi|\, P_x\, |\psi\rangle &= |a|^2 \int_{-\infty}^{\infty} dx\, e^{-ip_x/\hbar}\left(-i\hbar\frac{d}{dx}\right) e^{ip_x/\hbar} \\ &= |a|^2 \int_{-\infty}^{\infty} dx\, e^{-ip_x/\hbar} p_x e^{ip_x/\hbar} = p_x|a|^2 \int_{-\infty}^{\infty} dx\, e^{-ip_x/\hbar} e^{ip_x/\hbar} = p_x.\end{aligned} \tag{21.17}$$

We see that the expectation value of P_x in this state is just the parameter p_x, which is independent of time as expected for the free (no force) particle case. The last equality in Eq. (21.17) follows from the normalization of the momentum wave function.

The normalization of the momentum wave function is a bit tricky. If we calculate the absolute-value-squared of the Dirac bracket, we have $|\langle x|p_x\rangle|^2 = |a|^2$, which is independent of position. When we integrate this result over all x, we get infinity. (Note that the wave function in Eq. (21.16) violates the previous Math Aside's assumption about wave functions going to zero at very large distances.) A bit of an embarrassment because we want all state vectors to be normalized. We won't go into the details, but it turns out that if we use $a = 1/\sqrt{2\pi\hbar}$ and employ the momentum wave functions only as ingredients in forming coordinate superposition states that can be normalized, we won't get into any mathematical difficulties.

We can also express wave functions as a superposition of the momentum basis vectors:

$$\langle x|\psi\rangle = \int \langle x|p_x\rangle \langle p_x|\psi\rangle \, dp_x = \int e^{ip_x x/\hbar}\psi(p_x)dp_x, \tag{21.18}$$

where $\psi(p_x)$ is the state function in the momentum basis. The momentum basis is useful in describing scattering problems in atomic, nuclear, and particle physics. Here, scattering means two or more objects collide with each other, and we want to relate the properties of the incoming objects to the properties of the outgoing objects.

Before going into more details of the momentum and position basis states, we summarize the basic properties of state vectors, basis states, inner products, and operators for a quantum system with continuous variables. Consider a function $F(x)$ of the continuous variable x. State vectors may be represented abstractly by a Dirac right vector as $|F\rangle$ or more concretely as the bracket $\langle x|F\rangle = F(x)$, which in essence is the state coefficients of F in terms of the x basis states. If we use basis vectors $h(x) = \langle x|h\rangle$, we may write the inner product of F and h as

$$\langle F|h\rangle = \int F^*(x)h(x)dx, \tag{21.19}$$

which allows us to express $F(x)$ as a superposition (integral) of the $h(x)$ basis functions:

$$F(x) = \int \langle F|h\rangle\langle h|x\rangle \, dx. \tag{21.20}$$

Note that Eq. (21.19) involves the complex conjugate of the function $F(x)$.

Measurable properties of the system are represented by Hermitian operators, which are functions of x (and perhaps other variables). If the quantum state is described by the wave function $\psi(x)$, then the average value of the operator, say A, is found from the quantum sandwich

$$\langle A\rangle = \int \langle\psi|A|\psi\rangle \, dx. \tag{21.21}$$

21.4 Generalized Schrödinger Equation

Next, we recall from Chapter 18 that the state vector's time evolution is described by the Schrödinger equation

$$i\hbar\frac{d\psi}{dt} = H|\psi\rangle, \tag{21.22}$$

where H is the Hamiltonian operator, which we assume is the system's energy operator. To be specific, let's assume that for an object with mass m, its energy is a sum of kinetic energy $\frac{1}{2}mv_x^2 = p_x^2/(2m)$ and potential energy $V(x)$, again restricting ourselves to one-dimensional motion along an x-axis. We assume that the potential energy can be described by a function of position alone, a common situation in classical physics like when a mass raised to a higher height has a larger gravitational potential energy. Since momentum in the x-direction is given by $mv_x = p_x$, we may write the Hamiltonian operator in the coordinate representation as

$$H = \frac{1}{2m}P_x^2 + V(x) = -\frac{\hbar^2}{2m}\frac{d^2}{dx^2} + V(x). \tag{21.23}$$

Try It 21.2 Check that calculating P_x^2 from Eq. (21.10) gives Eq. (21.23).

We use Eq. (21.23) in Eq. (21.22) to find a generalized Schrödinger equation that takes into account both the spatial and temporal dependence of the wave function. Since we have two independent variables x and t, we use partial derivatives in place of ordinary derivatives:

$$i\hbar\frac{\partial\,|\psi\rangle}{\partial t} = -\frac{\hbar^2}{2m}\frac{\partial^2\,|\psi\rangle}{\partial x^2} + V(x)|\psi\rangle\,. \tag{21.24}$$

To find an equation for the wave function $\psi(x, t)$, we left multiply both sides of Eq. (21.24) by the coordinate basis left vector $\langle x|$. Since the spatial derivatives are with respect to x and because $V(x)$ depends only on x (by assumption), we obtain

$$i\hbar\frac{\partial\,\langle x|\psi\rangle}{\partial t} = -\frac{\hbar^2}{2m}\frac{\partial^2\,\langle x\,|\psi\rangle}{\partial x^2} + V(x)\langle x\,|\psi\rangle\,, \tag{21.25}$$

which in terms of the wave function is

$$i\hbar\frac{\partial\psi(x,t)}{\partial t} = -\frac{\hbar^2}{2m}\frac{\partial^2\psi(x,t)}{\partial x^2} + V(x)\psi(x,t). \tag{21.26}$$

Eq. (21.26) represents the Schrödinger equation in the coordinate basis. We can, in principle, solve the Schrödinger equation using analytic or numerical techniques once we have specified the potential energy function $V(x)$.

The Free Particle Case: A Connection with Classical Waves

Let's look at a situation in which $V(x) = 0$, that is, there are no interactions (no potential energy functions), and we say that the quantum object is a "free particle." The Schrödinger equation becomes

$$i\hbar\frac{\partial\psi(x,t)}{\partial t} = -\frac{\hbar^2}{2m}\frac{\partial^2\psi(x,t)}{\partial x^2}. \tag{21.27}$$

This might seem difficult to solve—after all, it is a partial differential equation—but it turns out that the solution is relatively simple:

$$\psi(x, t) = ae^{i(kx-\omega t)}, \tag{21.28}$$

where a is a normalization constant. To understand how we attach physical meaning to the other constants k and ω, solve Try It 21.3.

Try It 21.3 Verify that Eq. (21.28) is a solution to Eq. (21.27) by evaluating the derivatives. It turns out that we may give the parameters k and ω physical interpretations: $\hbar k = p_x$ and $\hbar\omega = p_x^2 / (2m)$, the energy of a free particle. To get the first interpretation, evaluate the solution at time $t = 0$.

Try It 21.4 Use the momentum operator in Eq. (21.10) to verify that Eq. (21.28) is a state with momentum p_x.

We note that the free particle wave function suffers from the same normalization difficulty we observed with the momentum eigenstate in the coordinate basis.

Math Aside The exponential function that appears in Eq. (21.28) also appears in the Fourier transform, which we introduced in Chapter 15. You may have also seen it in physics, engineering, or linear algebra classes. For example, suppose we have an oscillating wave that contains a range of frequencies described by a function $g(\omega)$. The corresponding function of time is called the Fourier transform $\tilde{g}(t)$, which is given by

$$\tilde{g}(t) = \frac{1}{\sqrt{2\pi}} \int_{-\infty}^{+\infty} g(\omega)e^{-i\omega t} d\omega. \tag{21.29}$$

The inverse Fourier transform is

$$g(\omega) = \frac{1}{\sqrt{2\pi}} \int_{-\infty}^{+\infty} \tilde{g}(t)e^{i\omega t} dt. \tag{21.30}$$

Similar spatial Fourier transform pairs can be defined with $f(x)$, $\tilde{f}(k)$, and $e^{\pm ikx}$. In other words, the connection between the coordinate basis and the momentum basis is the same as the connection between a function and its Fourier transform (see Eq. (21.18)). Note that different authors reverse the signs in the exponentials in Eqs. (21.29) and (21.30); they may also use different normalization factors.

The quantum version of the Fourier transform is treated in detail in Chapter 15.

21.5 Position and Momentum Commutation Relations and the Simple Harmonic Oscillator

As shown in Chapter 20, the commutation relation between two operators leads to essentially the entire operator algebra for these operators. So, let's take a look at the commutator between the X and P_x operators by seeing how they act on a coordinate basis wave function:

$$\begin{aligned}(XP_x - P_xX)\,\psi(x) &= x\left(-i\hbar\frac{d\psi(x)}{dx}\right) + i\hbar\frac{d}{dx}\,(x\psi(x)) \\ &= -i\hbar x\left(x\frac{d\psi(x)}{dx}\right) + i\hbar\psi(x) + + i\hbar x\frac{d\psi(x)}{dx} = i\hbar\psi(x).\end{aligned} \tag{21.31}$$

In progressing from the first line of the previous equation to the second line, we used the standard product rule for derivatives $d(ab)/dx = a\ (db/dx) + b\,(da/dx)$. In the abstract operator form, Eq. (21.31) is expressed as

$$[X, P_x] = i\hbar \mathrm{I}. \tag{21.32}$$

Math Aside If we extend position and momentum states to three dimensions, we need to know the commutation relations for six operators: X, Y, Z and P_x, P_y, and P_z. We assume that motions in the X–, Y–, and z-directions are independent; so, all three of X, Y, and Z commute with each other. Similarly, all three momentum operators commute with each other. However, the position and momentum operators for a particular direction do not commute. So, analogous to Eq. (21.32), we have $[Y, P_y] = i\hbar\mathrm{I}$ and $[Z, P_z] = i\hbar\mathrm{I}$. Also, the "mixed pairs" of operators such as Y and P_z commute.

You should recall that we have observed a similar commutation relation with the ladder operators a and $a^\dagger$ (See Chapter 20, Eq. 20.10). The only difference is the factor of $\hbar$, which reflects the units used for position and momentum. Taking these units into account, we define operators

$$a = \sqrt{\frac{m\omega}{2\hbar}}\left(X + i\frac{P_x}{m\omega}\right) \qquad a^\dagger = \sqrt{\frac{m\omega}{2\hbar}}\left(X - i\frac{P_x}{m\omega}\right). \tag{21.33}$$

For a concrete example, consider a mass m oscillating on a spring. X and P_x are the quantum operators for the mass's position and linear momentum along the x-axis. ω is a parameter whose significance will be revealed shortly. We see that these definitions connect the operator algebra of the ladder operators to a specific position–momentum quantum system—one which is widely used in many applications of quantum mechanics.

It is straightforward to use operator algebra to show that these operators satisfy $[a, a^\dagger] = \mathrm{I}$, exactly the commutation relation for the ladder operators. We then know that the operators in Eq. (21.33) serve as ladder operators for the position–momentum quantum system.

Try It 21.5 Use operator algebra to show that $[a, a^\dagger] = \mathrm{I}$ and pay careful attention to the order in which the operators act.

We can now use the operator algebra to find the Hamiltonian for the physical system described by the operators in Eq. (21.33) simply by evaluating the equivalent of the number operator for ladder operators. After a bit of operator algebra, we find

$$a^\dagger a = \frac{1}{\hbar}\left(\frac{P_x^2}{2m} + \frac{1}{2}m\omega^2 X^2\right) + \frac{1}{2}. \tag{21.34}$$

Try It 21.6 Use the operators in Eq. (21.33) to derive Eq. (21.34).

Let's interpret the terms in parentheses in Eq. (21.34). The first term is an operator for the kinetic energy of the system. The second term plays the role of potential energy. You may recall from introductory physics or engineering that a potential energy function proportional to the square of the position (relative to some origin) describes an oscillating system such as a mass on a spring. Such a system is called a simple harmonic oscillator (SHO)—"simple" because the system oscillates with a single frequency ω. The same model is often used to describe molecular bonds, electromagnetic waves with a single frequency, and many other physical systems.

The last term in Eq. (21.34) represents an additional energy present even when the sum of kinetic and potential energies are zero. We introduced this "vacuum energy" in our discussion of ladder operators in Chapter 20.

We can express the Hamiltonian operator for the system as

$$H = \hbar\omega\left(a^\dagger a + \mathrm{I}/2\right). \tag{21.35}$$

From the algebra of the ladder operators, we know that the energies for this system are spaced by multiples of $\hbar\omega$. The lowest three energy eigenvalues are

$$\begin{aligned} E_0 &= \hbar\omega/2 \\ E_1 &= \hbar\omega + \hbar\omega/2 = (3/2)\hbar\omega \\ E_2 &= (5/2)\hbar\omega. \end{aligned} \tag{21.36}$$

So, we see that the energy eigenstates of the SHO are also number operator eigenstates. Figure 21.2 shows the energies in an energy ladder diagram.

It is interesting to note that several hardware platforms used for quantum computers, notably trapped ions and trapped neutral atoms, use the SHO model. The mechanisms for trapping the ions or atoms produce interactions that are modeled reasonably well by

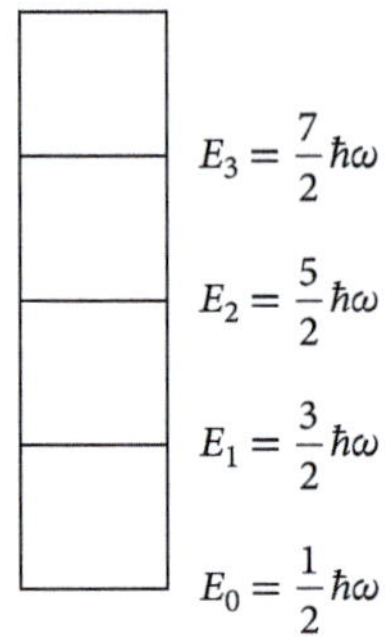

Fig. 21.2 An energy ladder diagram for the simple harmonic oscillator.

a potential energy function proportional to x^2, the ion's or atom's displacement from its equilibrium trap location. Moving the ions or atoms among various SHO energy states is used in several quantum computational algorithms.

21.6 More Simple Harmonic Oscillator Properties

To illustrate the power of operator algebra, we show how to find the wave functions for the SHO. First, we note that the lowering operator acting on the vacuum state gives 0: $a\,|0\rangle = 0$. In the coordinate representation, this condition gives a differential equation for the lowest energy state wave function $\langle x|0\rangle = \psi_0(x)$. Here is how this works: Using the definition of the lowering operator and expressing P_x in its coordinate representation, we find

$$\left(x + \frac{\hbar}{m\omega}\frac{d}{dx}\right)\psi_0(x) = 0, \tag{21.37}$$

which is equivalent to

$$\frac{d}{dx}\psi_0(x) = -\frac{m\omega}{\hbar}x\,\psi_0(x). \tag{21.38}$$

This differential equation is similar to the ones that have an exponential solution except for the extra factor of x on the right. This suggests constructing a solution by putting a factor of x in the argument of the exponential function The proposed solution to Eq. (21.38) is

$$\psi_0(x) = \left(\frac{m\omega}{\pi\hbar}\right)^{1/4} e^{-m\omega x^2/(2\hbar)}\,. \tag{21.39}$$

Since these wave functions are energy eigenfunctions, it is easy to write down their time dependence since that involves the exponential function. For example, for the $n = 0$ state we have

$$\psi_0(x,t) = \psi_0(x,0)\,e^{-iE_0 t/\hbar}. \tag{21.40}$$

Math Aside The procedure used to find the solution to Eq. (21.38) was not exactly rigorous mathematics, but since we can check that the solution works and since the solutions of linear differential equations are unique, we can be sure we have a correct solution. Of course, we could always add a constant (independent of x) to the right side of Eq. (21.39) and it would still satisfy Eq. (21.38). However, to have the wave function normalized, this constant needs to be zero. We also point out that the Schrödinger equation (21.26) could be used directly to find the wave functions and the energy eigenvalues. However, to carry out that program you need to be proficient in partial differential equations and special functions such as Hermite polynomials.

Figure 21.3 shows the plot of the wave function and position probability density function for the $n = 0$ state of the SHO. We can see that the position probability density is largest near $x = 0$, but there is some spread in that probability density.

Figure 21.4 shows the plot of the wave function and probability density function for the $n = 1$ state of the SHO. (Note that neither plot is normalized.) We can see that the probability density is zero at $x = 0$ and largest at some distance away from $x = 0$. In this case, the SHO particle is more likely to be found away from $x = 0$. In a homework problem posted on the book's website, we show how to derive the $n = 1$ wave function.

Note that the $n = 1$ wave function crosses the x-axis once (at a so-called node) while the $n = 0$ wave function has no nodes. This is a general feature of quantum mechanics: Higher energy wave functions have more nodes in a given spatial interval than lower energy wave functions. Why is that? More nodes in a given spatial interval means that the wave function must have regions of higher curvature in that interval. By Schrodinger's equation, the curvature (represented by the second derivative with respect to x) means higher energy.

We could find all the other wave functions for the SHO by using the raising operator to generate more differential equations that are relatively easy to solve for these wave functions. If you are interested in seeing how that is solved, see the books listed in Further Reading.

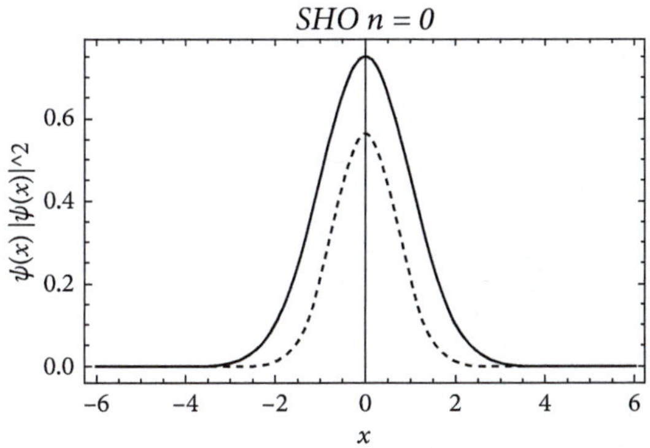

Fig. 21.3 A plot of the simple harmonic oscillator lowest ($n = 0$) energy state wave function (solid curve) and position probability density (dashed) versus position. The wave function and probability density have been normalized with $m\omega/\hbar = 1$.

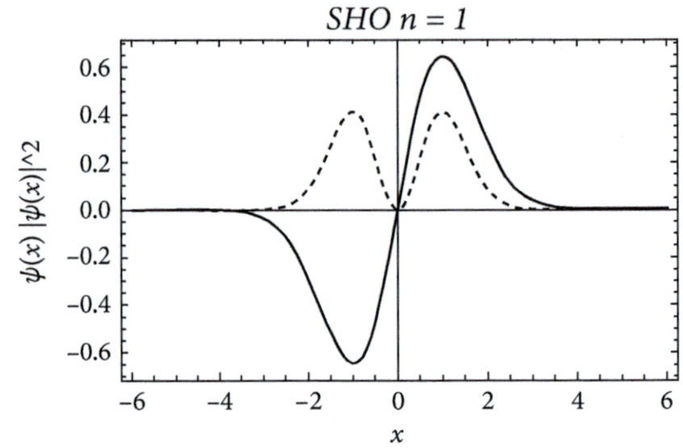

Fig. 21.4 A plot of the simple harmonic oscillator $n = 1$ energy state wave function (solid curve) and position probability density (dashed) versus position. The wave function and position probability density have been normalized with $m\omega/\hbar = 1$.

21.7 Uncertainty Relation for the SHO

It is instructive to look at the *uncertainty relation* for position and momentum for the SHO. The *uncertainty relation* we want to examine is

$$(\Delta X)_\psi (\Delta P_x)_\psi = \frac{1}{2}\left|\langle [X, P_x]\rangle_\psi\right| = \frac{\hbar}{2} \tag{21.41}$$

Note that for the SHO, the minimum product of the standard deviations is independent of the state used for the calculation.

Try It 21.7 Use Eq. (21.32) to calculate the middle term in Eq. (21.41). Verify that the result is $\hbar/2$.

Try It 21.8 Challenge: Calculate the standard deviations ΔX and ΔP_x for the lowest energy state of the SHO by two methods. (1) Express the standard deviations in terms of the ladder operators and use the operator algebra to find their values for the lowest energy state. (2) Use calculus to find $(\Delta x)^2 = \int x^2 \psi_0^* \psi_0 dx$. You can see from Fig. 21.2 that the average value of x is zero. You can find $(\Delta P_x)^2$ by using the coordinate representation of the momentum operator and the fact that the average value of the momentum is zero.

The *uncertainty relation* for position and momentum for the SHO is an example of the famous dictum of quantum mechanics: It is not possible to have a quantum state whose position and momentum are both precisely known (zero standard deviations). We also say that the momentum wave functions illustrate the wave properties of the system while the coordinate parts focus on the particle aspects. In other words, for a particle the position probability density is concentrated in some relatively small spatial region and for a wave the probability density is spread out over a wide range of positions, just as a wave is typically spread out in space. These features illustrate what is called wave–particle duality: Some quantum states for the SHO are closer to what we expect for a particle; others are more like classical waves.

Try It 21.9 Use the *uncertainty relation* for position and momentum to argue that for a state with a small standard deviation in momentum ($\Delta p_x \to 0$) requires that the state be spread over a large spatial region ($\Delta x \to \infty$).

21.8 Chapter Summary

- Using a linear transformation (operator) that produces a spatially displaced state vector, we find that the operator representing momentum is $-i\hbar\frac{d}{dx}$, a derivative with respect to position.
- The operator P_x for momentum in the x-direction and position operator X do not commute. Their commutator is given by $XP_x - P_xX = [X, P_x] = i\hbar\mathrm{I}$.
- The Hamiltonian (energy) operator for a simple harmonic oscillator system is given by $H = \frac{P_x^2}{2m} + \frac{1}{2}m\omega^2X^2$.
- From X and P_x, we may construct two non-Hermitian operators that act as raising and lowering operators for the simple harmonic oscillator system.
- The *uncertainty relation* applied to the simple harmonic oscillator system indicates that there is no quantum state for which the position and the momentum are both known precisely (the standard deviations are zero).

21.9 Further Reading

D. McIntyre *Quantum Mechanics, A Paradigms Approach* (Cambridge University Press, Cambridge, 2023) (see Chapter 9, The Harmonic Oscillator).

Leonard Susskind and Art Friedman, *Quantum Mechanics, The Theoretical Minimum* (Basic Books, New York, 2014). Chapter 10 covers raising and lowering operators and the simple harmonic oscillator.

22 Density Operators and Density Matrices

The world in all doth but two nations bear,
The good, the bad, and those mixed everywhere.

Andrew Marvell, "The Loyal Scot", 1650–1652

22.1 Limitations of Pure-State Vectors

Up to now, we have implicitly assumed that we describe qubit quantum states by what are traditionally called *pure* quantum state vectors. It means that the qubit state vector $|\psi\rangle$ can be written as a linear combination (superposition) of a set of basis vectors, for example, the computational (standard) basis vectors $|0\rangle$ and $|1\rangle$:

$$|\psi\rangle = c_0 |0\rangle + c_1 |1\rangle . \tag{22.1}$$

As shown before, the same idea can be applied to multiqubit states. In this analysis, we have assumed that we know the numerical values of the state amplitudes c_0 and c_1 or that they could, at least in principle, be determined up to an overall phase through measurements.

There are, however, practical considerations that force us to generalize the formal description of quantum states. For example, sometimes we know that our apparatus is preparing a qubit in a pure quantum state, but we don't know exactly which state; that is, we don't know the exact state amplitudes, say c_0 and c_1 in Eq. (22.1). However, we may know (or can measure) the distribution of possible amplitudes by carrying out the state preparations many times and making measurements relative to appropriate basis states. In such situations, we may know the probabilities of observing the possible amplitudes. Note that this is an additional type of probability resulting from the actual physical situation rather than the inherent probabilistic nature of the quantum realm. This additional probability is not related to the measurement probabilities given by $|c_0|^2$, $|c_1|^2$, and the Born rule. You might call this additional type of probability "mixture probability." We explain the term mixture in the following section.

Quantum Computing and Quantum Physics. Alice Flarend and Robert Hilborn, Oxford University Press.
 DOI: 10.1093/oso/9780197904930.003.0022

A concrete example might be helpful. Suppose we are preparing the state of a qubit (a spin-1/2 system) that has a magnetic dipole. A pure state describing the qubit can always be written as a superposition of spin-up and spin-down basis states:

$$|\psi\rangle = c_0 |\uparrow\rangle + c_1 |\downarrow\rangle . \tag{22.2}$$

But we may not be able to control exactly which values of c_0 and c_1 our apparatus produces. For example, the magnetic field direction that determines the up–down direction in the apparatus that produces the initial state may fluctuate due to changing (and uncontrolled) mechanical instabilities. In essence, the source is producing an ensemble (collection) of quantum states. How do we describe the state of the system in this case? More importantly, how do those additional probabilities affect the predictions for measurements performed on that qubit?

The second practical issue arises if the qubit has significant interactions with its environment. This occurs, for example, if the qubit is part of a multiqubit state described by an entangled composite state vector. The interactions with other qubits may wipe out the phase differences between the state amplitudes in a process called *decoherence*. In this situation, the standard (pure-state) superposition of basis states is no longer a good description of the qubit. Since we know that the phase difference between the amplitudes affects what we observe with measurements, this is an important issue.

To summarize, we need to use more general ways of describing quantum state in two cases: (1) in practical cases when the system is actually a statistical ensemble of systems, each with its own pure state, and (2) the more fundamental case in which we have a multiqubit entangled composite system, and we make measurements on only one of the qubits. Unfortunately, in both of these situations, the quantum pure-state formalism cannot directly deal with these issues. We need a more general way of constructing quantum states. The surprise is that there is a formalism that allows us to deal with both these situations in a systematic way. The "price" we must pay is learning how to use a "density operator" to describe a quantum state in a more general way—moving beyond the pure states we have come to love and cherish. The upside is that this formalism provides a method of getting state information about subsets within a composite system.

We shall see that quantum entanglement plays an important role in describing these more general states. Our analysis often focuses on the matrix representation of these density operators, so you may want to review matrix linear algebra as you read this chapter. The density operator has applications in many areas of physics and in quantum information and quantum computing. Since the physics of many of those applications involves concepts beyond the scope of this book, we focus on using density operators (and their associated matrices) to characterize quantum states and only briefly mention applications in this chapter.

22.2 Density Operators and Density Matrices

The appropriate generalization is based on the notion of a projection operator, which we introduced previously in implementing the transformations from one set of basis states to another in Chapter 8. For a pure state, the projection operator is written as $|\psi\rangle\langle\psi|$, a product of a right vector and a left vector. Recall that this product is different from the scalar (inner) product $\langle\psi|\,\psi\rangle$.

For a two-state system, the projection operator, just like any other operator, can be represented by a 2×2 matrix. Let's see how this works. Suppose we write the right and left state vectors as column and row vectors as we have done before:

$$|\psi\rangle \doteq \begin{pmatrix} c_0 \\ c_1 \end{pmatrix} \qquad \langle\psi| \doteq \begin{pmatrix} c_0^* & c_1^* \end{pmatrix}. \tag{22.3}$$

Then, the projection operator is represented by

$$|\psi\rangle\langle\psi| \doteq \begin{pmatrix} c_0 \\ c_1 \end{pmatrix}\begin{pmatrix} c_0^* & c_1^* \end{pmatrix} = \begin{pmatrix} |c_0|^2 & c_0 c_1^* \\ c_0^* c_1 & |c_1|^2 \end{pmatrix}. \tag{22.4}$$

Note that multiplying a row vector by a column vector is an *outer* product and produces a matrix as a result. In Eq. (22.4), the matrix is a representation of the projection operator.

For historical reasons, $\rho = |\psi\rangle\langle\psi|$ is called the *density operator* (of a pure state) and the corresponding matrix is called the *density matrix*. The adjective *density* indicates a connection to a probability density (or distribution). If we were inventing terminology from scratch, we would prefer to call $|\psi\rangle\langle\psi|$ the *pure-state projection operator* (PSPO—because we must have an acronym). The proposed name reminds us that we are dealing with state projections, not mass per unit volume—the usual definition of density. But the terms density operator and density matrix are deeply embedded in the jargon used by quantum mechanicians. As users of quantum mechanics, we need to speak that language.

Math Aside It might seem that we are making a major change from our previous way of describing quantum states as vectors in a vector space. However, the projection operators themselves can be thought of as belonging to a matrix space—a generalization of the notion of vector space. The matrix space contains superpositions of matrices and allows for multiplying matrices by scalars, just the two operations we need to define a "vector" space. We can also talk about basis matrices and the dimensions of the matrix space and define an inner product based on the trace operation. So, in that sense, the density operator and density matrix are a straightforward generalization of the vector space description of quantum states. For a spin-1/2 system, the Pauli matrices along with the identity matrix act as basis matrices in the sense that any operator, including the density operator, for the two-state system can be written as a linear superposition of the basis matrices.

The density operator is a Hermitian operator and its corresponding matrix is a Hermitian matrix: $\rho = (\rho^T)^*$, i.e., the density matrix is equal to the complex conjugate of its transpose. Consequently, the eigenvalues of the density matrix are real.

We know that the eigenvalues of a matrix play an important role in quantum mechanics. What are the eigenvalues of the density matrix in Eq. (22.4)? Since this matrix is a 2×2 matrix, it will have two eigenvalues λ_1 and λ_2, which satisfy $\lambda_1 + \lambda_2 = \mathrm{Tr}\rho$ and $\lambda_1\lambda_2 = \det\rho$. In this case, $\lambda_1 + \lambda_2 = 1$ and $\lambda_1\lambda_2 = 0$. Hence, the two eigenvalues are 0 and 1. A matrix whose eigenvalues are positive or zero is called a positive semidefinite matrix.

Before digging deeper into the use of density operators and density matrices, let's look at the calculation of average (expectation) values and see how that calculation sets the stage for density operators. As discussed before, the average value of an operator A for a system described by a pure state $|\psi\rangle = c_0\,|0\rangle + c_1\,|1\rangle$ is expressed as a quantum sandwich:

$$\begin{aligned}\langle A\rangle_\psi &= \langle\psi|\,A\,|\psi\rangle = \left(c_0^*\,\langle 0| + c_1^*\,\langle 1|\right)\hat{A}\left(c_0\,|0\rangle + c_1\,|1\rangle\right) \\ &= c_0^* c_0 A_{00} + c_0^* c_1 A_{01} + c_1^* c_0 A_{10} + c_1^* c_1 A_{11},\end{aligned} \tag{22.5}$$

where $A_{jk} = \langle j|A|k\rangle$ with $j, k = 0, 1$ are the matrix elements of the operator A with respect to the computational basis state vectors.

We see that the last line of Eq. (22.5) looks like a number generated from the product of two matrices. One matrix has the elements A_{jk}. The other matrix has elements that may be labeled (aforethought) as ρ_{kj} so that

$$\rho \doteq \begin{pmatrix} c_0 c_0^* & c_0 c_1^* \\ c_1 c_0^* & c_1 c_1^* \end{pmatrix}. \tag{22.6}$$

Recall that the elements of the resulting product are the scalar products of the row vectors of the left matrix and the column vectors of the matrix on the right. Writing out the matrix product in detail gives

$$\rho A \doteq \begin{pmatrix} c_0 c_0^* & c_0 c_1^* \\ c_1 c_0^* & c_1 c_1^* \end{pmatrix}\begin{pmatrix} A_{00} & A_{01} \\ A_{10} & A_{11} \end{pmatrix} = \begin{pmatrix} c_0^* c_0 A_{00} + c_0 c_1^* A_{10} & c_0^* c_0 A_{01} + c_0 c_1^* A_{11} \\ c_1 c_0^* A_{00} + c_1 c_1^* A_{10} & c_1 c_0^* A_{01} + c_1^* c_1 A_{11} \end{pmatrix}. \tag{22.7}$$

We recognize that the last line in Eq. (22.5) is just the sum of the principal (upper-left to lower-right) diagonal elements of the matrix on the right side of Eq. (22.7). This sum is called the *trace* of the matrix and is usually written as $\mathrm{Tr}(\rho A)$. The important conclusion is that if we express the projection operator (and its generalization) as a matrix and the operator whose average value we want as a matrix, the average value is simply the trace of the matrix product of the two matrices:

$$\langle A\rangle = \mathrm{Tr}\,(\rho A)\,. \tag{22.8}$$

Since we use the equivalent of Eq. (22.8) in most of what follows, let's explicitly write $\mathrm{Tr}(BC)$ for any two operators B and C in terms of their matrix elements in a basis set of state vectors $|u_n\rangle$. Here are the details:

$$\begin{aligned}\mathrm{Tr}(BC) &= \underbrace{\sum_n \langle u_n| BC |u_n\rangle}_{\text{sum of diagnonal matrix elements}} = \sum_n \langle u_n| B \underbrace{\sum_m |u_m\rangle\langle u_m|}_{\mathrm{I}} C |u_n\rangle \\ &= \sum_{m,n} B_{nm} C_{mn},\end{aligned} \tag{22.9}$$

where the second equality used the "completeness relationship" for the basis vectors $\sum_m |u_m\rangle\langle u_m| = \mathrm{I}$.

Try It 22.1 Check that Eq. (22.9) gives the result in Eq. (22.7) for $B \to \rho$ and $C \to A$ for a qubit.

Statistical Ensemble Case

Now, let's return to our generalization arguments. In the ensemble case, we build a density operator when the probabilities p_j of the various possible states $|\psi^{(j)}\rangle$ are known. The label j indicates the elements of the set of state amplitudes. For example, in the computational basis, the jth state is written as

$$|\psi^{(j)}\rangle = c_0^{(j)}|0\rangle + c_1^{(j)}|1\rangle. \tag{22.10}$$

We then write the density operator as

$$\rho = \sum_{j=1}^{k} p_j |\psi^{(j)}\rangle\langle\psi^{(j)}| = \sum_{j=1}^{k} p_j \rho^{(j)}, \tag{22.11}$$

where k is the number of pairs of state amplitudes in the distribution and $\rho^{(j)} = |\psi^{(j)}\rangle\langle\psi^{(j)}|$ is the density operator for the jth state. In this situation, we say that the system is described by a *mixed* state. The operator is a sum of projection operators weighted by the appropriate *probabilities.* (If we wanted to continue our invention of new terminology, the operator in Eq. (22.11) might be called a mixed-state projection operator.) Each of the density matrices may be written as

$$\rho^{(j)} \doteq \begin{pmatrix} \left|c_0^{(j)}\right|^2 & c_0^{(j)} c_1^{*(j)} \\ c_0^{*(j)} c_1^{(j)} & \left|c_1^{(j)}\right|^2 \end{pmatrix}. \tag{22.12}$$

Note that the density operator for a mixed state uses probabilities, not amplitudes, when combining the individual density operators. In that sense, it is quite different from a superposition state. If only one of the p_js is nonzero, then $p_j = 1$ and we are left with a pure-state density operator.

We now want to find the average (expectation) value of an operator D if the system is described by an ensemble of pure states. From general probability theory, the average value is

$$\langle D\rangle_\rho = p_1\langle D\rangle_{\psi_1} + p_2\langle D\rangle_{\psi_2} + \ldots + p_k\langle D\rangle_{\psi_k}, \tag{22.13}$$

where we have used a subscript ρ to remind us that we are considering a mixed state characterized by a density operator. On the right of Eq. (22.13), the subscripts on the average values indicate the state used to calculate the individual average values. Compare this expression to the one developed in Chapter 17 (Section 17.7).

Try It 22.2 This Try It is a reminder that we are using just ordinary probability theory for calculating averages. If you are familiar with these methods, you may skip this Try It. Suppose you give a quiz in your class and 10% of the students score 40, 20% score 55, and 70% score 83. What is the average grade for the class? Interpret this example using Eq. (22.13) and explain your reasoning.

If you know the probabilities and the elements of the two matrices, we may write Eq. (22.13) as

$$\langle D\rangle_\rho = \mathrm{Tr}\,(\rho D) = \sum_j p_j \sum_{mn} \rho_{nm}^{(j)} D_{mn}. \tag{22.14}$$

Spin-1/2 Example

Suppose we prepare a spin-1/2 qubit in a state described in the spin-up, spin-down basis such that in a series of repeated state preparation procedures half the time the apparatus produces $|\uparrow\rangle$ and half the time the state $|\downarrow\rangle$. What is the appropriate density operator and density matrix for this case? First, we identify the probabilities $p_\uparrow = 1/2$ and $p_\downarrow = 1/2$. Thus, we may write the density operator as

$$\rho_{\text{mixed}} = \frac{1}{2}\,|\uparrow\rangle\langle\uparrow| \;+\; \frac{1}{2}\,|\downarrow\rangle\langle\downarrow|. \tag{22.15}$$

We know that Eq. (22.15) represents a mixed state because it contains the weighted sum of two PSPOs. In the matrix form, we obtain

$$\rho_{\text{mixed}} \doteq \begin{pmatrix} 1/2 & 0 \\ 0 & 1/2 \end{pmatrix}. \tag{22.16}$$

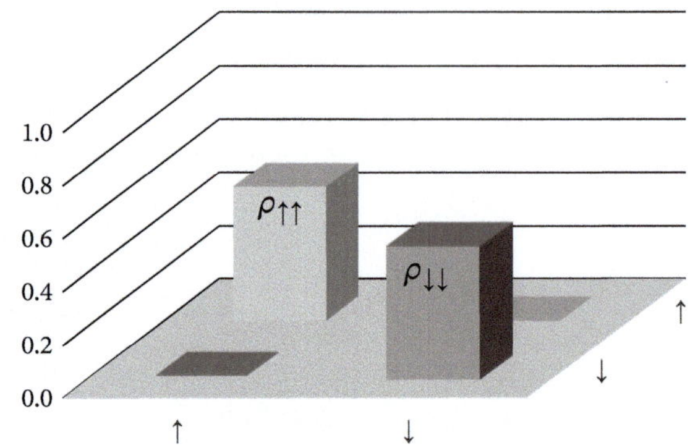

Fig. 22.1 A column plot of the mixed-state density matrix elements in Eq. (22.16). If the matrix elements were complex, we would display separate plots for the real and imaginary parts.

Figure 22.1 shows the density matrix elements as four vertical columns rising from a horizontal plane. The height of a column is proportional to the magnitude of the matrix element. The mixed-state density matrix has no off-diagonal elements (the column height is zero) and has, in a sense, "lost" the off-diagonal terms that express the "coherence" between the two basis state vectors.

Let's use the density operator method to evaluate the average value for a measurement of S_x, the x component of the spin angular momentum. The matrix representation of that operator is $\frac{\hbar}{2}\begin{pmatrix} 0 & 1 \\ 1 & 0 \end{pmatrix}$. For the Eq. (22.16) density matrix, the average value is

$$\begin{aligned} \langle S_x \rangle &= \mathrm{Tr}(\rho_{\text{mixed}} S_x) = \mathrm{Tr}\left[\frac{\hbar}{2}\begin{pmatrix} 1/2 & 0 \\ 0 & 1/2 \end{pmatrix}\begin{pmatrix} 0 & 1 \\ 1 & 0 \end{pmatrix}\right] \\ &= \frac{\hbar}{2}\mathrm{Tr}\begin{pmatrix} 0 & 1/2 \\ 1/2 & 0 \end{pmatrix} = 0. \end{aligned} \tag{22.17}$$

To re-enforce the importance of coherence, let's write out the *pure*-state density operator and matrix for a superposition state that seems to be closely related to the one in Eq. (22.15):

$$|\psi\rangle = \frac{1}{\sqrt{2}}|\uparrow\rangle + \frac{1}{\sqrt{2}}|\downarrow\rangle. \tag{22.18}$$

The density operator and density matrix for this pure state are

$$\begin{aligned} \rho_{\text{pure}} &= |\psi\rangle\langle\psi| = \frac{1}{2}\left[|\uparrow\rangle\langle\uparrow| + |\uparrow\rangle\langle\downarrow| + |\downarrow\rangle\langle\uparrow| + |\downarrow\rangle\langle\downarrow|\right] \\ &\doteq \frac{1}{2}\begin{pmatrix} 1 & 1 \\ 1 & 1 \end{pmatrix}. \end{aligned} \tag{22.19}$$

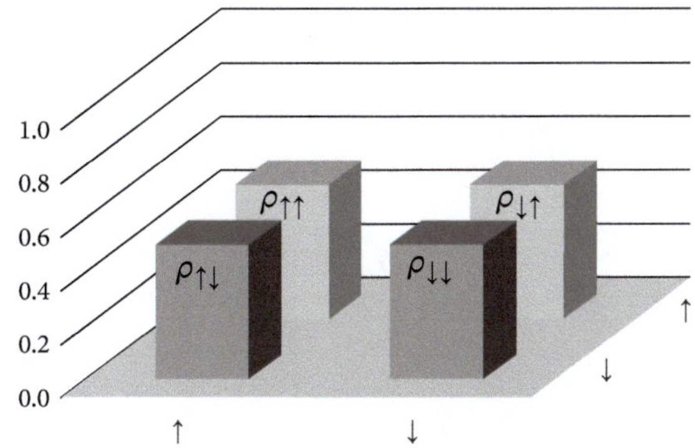

Fig. 22.2 A column plot of the pure-state density matrix elements in Eq. (22.19).

In the pure-state case, the density matrix has the critical off-diagonal matrix elements needed to account for the coherence between the basis state components of the state vector. Figure 22.2 shows the pure-state density matrix elements as four vertical columns rising from a horizontal plane. The height of a column is proportional to the magnitude of the matrix element. In this example, the pure-state density matrix elements are of equal magnitude.

Let's calculate the average value of S_x for the system described by the pure-state density matrix from Eq. (22.19):

$$\begin{aligned}\langle S_x \rangle &= \mathrm{Tr}(\rho_{\mathrm{pure}} S_x) = \mathrm{Tr}\left[\frac{\hbar}{4}\begin{pmatrix}1 & 1\\ 1 & 1\end{pmatrix}\begin{pmatrix}0 & 1\\ 1 & 0\end{pmatrix}\right] \\ &= \frac{\hbar}{4}\mathrm{Tr}\begin{pmatrix}1 & 1\\ 1 & 1\end{pmatrix} = \frac{\hbar}{2}.\end{aligned} \tag{22.20}$$

We see that the apparently similar states, one pure and one mixed, give significantly different measurement predictions.

Try It 22.3 Suppose we asked for the average value of S_z, the z component of the spin angular momentum. The operator for S_z is $(\hbar/2)\begin{pmatrix}1 & 0\\ 0 & -1\end{pmatrix}$. Use the density matrices in Eqs. (22.16) and (22.19) to find the average values of S_z for both the mixed and pure states given above.

The takeaway lesson from Try It 22.3 is that the pure and mixed states may give the same average value for some operator(s) but different average values for others. Recall that we need a series of state preparations and measurements to find these averages.

Try It 22.4 Can you find some other pure state for which the average value of S_z is zero and the average value of S_x is also zero? If so, what is the average value of S_y? Is it always possible to distinguish a mixed state from a pure state by some series of measurements?

Let's write out the most general density matrix for a two-state system. You may recall from our discussion of the Bloch sphere in Chapter 7 that the most general spin-1/2 state vector may be written as

$$|\psi\rangle = \cos(\theta/2)\,|\uparrow\rangle + e^{i\phi}\sin(\theta/2)\,|\downarrow\rangle, \tag{22.21}$$

where θ is the colatitude of the state vector in the Bloch sphere picture and ϕ is the longitude in this picture. The density matrix for this system is then

$$\rho \doteq \begin{pmatrix} \cos^2(\theta/2)) & e^{i\phi}\sin(\theta/2)\cos(\theta/2) \\ e^{-i\phi}\sin(\theta/2)\cos(\theta/2) & \sin^2(\theta/2) \end{pmatrix}. \tag{22.22}$$

22.3 Mathematical Properties of Density Operators and Density Matrices

Density operators and their matrix representations have some interesting mathematical properties that are useful in checking calculations. First, we note that the density operator is a Hermitian operator; so, its eigenvalues must be real numbers.

Try It 22.5 Show that the previous statement is true both for the pure and mixed states. What does that tell us about the eigenvalues and eigenvectors of the density operators? We discuss the eigenvalues and eigenvectors in Section 22.7.

The next feature is the trace of a density matrix, which, in any basis, is equal to 1. To show that, we calculate the Dirac brackets of the projection operators with respect to basis states $|\phi_k\rangle$:

$$\begin{aligned} \mathrm{Tr}\,\rho &= \mathrm{Tr}\left(\sum_j\sum_k p_j \langle\phi_k|\,\psi_j\rangle\langle\psi_j|\,\phi_k\rangle\right) = \sum_j p_j \mathrm{Tr}\left[\langle\phi_k|\,\psi_j\rangle\langle\psi_j|\,\phi_k\rangle\right] \\ &= \sum_j p_j \sum_k \langle\psi_j|\,\phi_k\rangle\langle\phi_k|\,\psi_j\rangle = \sum_j p_j = 1. \end{aligned} \tag{22.23}$$

In the last step in Eq. (22.23), we assumed that the basis state vectors are orthonormalized.

The next important property is that the expectation value of the density operator in any state is itself positive semidefinite. That last phrase is mathematical language for "greater than or equal to 0." Here is the proof:

$$\langle\Phi|\,\rho\,|\Phi\rangle = \sum_j p_j \langle\Phi|\,\psi_j\rangle \langle\psi_j|\,\Phi\rangle = \sum_j p_j \left|\langle\psi_j|\,\Phi\rangle\right|^2 \geq 0. \tag{22.24}$$

The final property we look at is the square of the density operator. The general result is that $\rho = \rho^2$ for a pure-state density operator but $\rho \neq \rho^2$ for a mixed-state density operator. For a mixed state, the trace of the *square* of the density operator is less than 1, that is, $\mathrm{Tr}[\rho_{\mathrm{mixed}}]^2 < 1$. Thus, we may use the trace to provide a numerical measure of the degree of "mixture." Rather than giving a formal proof, let's simply check the results for the two density matrices in Eqs. (22.16) and (22.19). First, for the mixed state, we have

$$\rho^2_{\mathrm{mixed}} \doteq \begin{pmatrix} 1/2 & 0 \\ 0 & 1/2 \end{pmatrix}\begin{pmatrix} 1/2 & 0 \\ 0 & 1/2 \end{pmatrix} = \begin{pmatrix} 1/4 & 0 \\ 0 & 1/4 \end{pmatrix} \neq \rho_{\mathrm{mixed}} \doteq \begin{pmatrix} 1/2 & 0 \\ 0 & 1/2 \end{pmatrix}. \tag{22.25}$$

For the pure state, we obtain

$$\rho^2_{\mathrm{pure}} \doteq \frac{1}{4}\begin{pmatrix} 1 & 1 \\ 1 & 1 \end{pmatrix}\begin{pmatrix} 1 & 1 \\ 1 & 1 \end{pmatrix} = \frac{1}{2}\begin{pmatrix} 1 & 1 \\ 1 & 1 \end{pmatrix} = \rho_{\mathrm{pure}}. \tag{22.26}$$

We can also recognize a mixed-state density matrix because its off-diagonal elements are zero or smaller in magnitude than those in the corresponding pure-state density matrix. If all the off-diagonal elements are zero, we say that the state is maximally mixed. Using this criterion, we see that the mixed-state density matrix in Eq. (22.25) is maximally mixed.

22.4 Density Operators for Multiqubit Systems: The Effects of Entanglement

We now turn our attention to multiqubit systems and see how we can express their states using the density operator approach. This analysis leads to some fundamental and interesting results that appear in many quantum information science applications. We use multiqubit entangled states as models to show how averaging over all the other qubits leads to a density matrix for a single qubit. By way of contrast, recall from Chapter 9 that an entangled multiqubit system does not provide a state vector for the individual qubits. In this section, we focus on two-qubit systems. Three-qubit systems are treated in the following section.

Two-qubit systems will be sufficient to understand the basic principles. Spoiler alert: For a pure two-qubit entangled state, the density operator for a single qubit of a multiqubit state, obtained by averaging the trace of the system's density operator over the other qubit's basis states, is a mixed-state, not a pure-state, density operator. In essence, this averaging is a method to account for the correlations between one qubit and the other qubits in the system. We discuss the importance of this result later.

As described in Chapter 9, the states for a two-qubit system live in a vector space that is the tensor product of two individual vector spaces, each associated with one of the qubits. We choose $|u\rangle_A$ to be the basis states in vector space A (Alice's) and $|v\rangle_B$ to be the basis states in vector space B (Bob's). (In what follows, we use the computational basis states in the examples. In a product state, Alice's basis state will be on the left and Bob's will be on the right.)

Any pure state for the composite two-qubit system can be written as a linear superposition of products of these two sets of computational basis states:

$$|\Psi\rangle = \sum_{n,p=0,1} c_{np} |u_A\rangle |v_B\rangle. \tag{22.27}$$

As shown before, most linear combinations of the product states describe an entangled overall system state, and consequently, we cannot assign a (pure) quantum state to either of the two qubits individually. However, as shown later, we can use a density operator to describe each qubit individually.

The density operator for the overall system is defined as

$$\rho = |\Psi\rangle \langle\Psi| . \tag{22.28}$$

Using the computational basis states, we may write Eq. (22.27) as

$$|\Psi\rangle = c_{00} |0_A\rangle |0_B\rangle + c_{01} |0_A\rangle |1_B\rangle + c_{10} |1_A\rangle |0_B\rangle + c_{11} |1_A\rangle |1_B\rangle . \tag{22.29}$$

In the matrix form, Eq. (22.28) becomes

$$\rho \doteq \begin{pmatrix} |c_{00}|^2 & c_{00}c_{01}^* & c_{00}c_{10}^* & c_{00}c_{11}^* \\ c_{01}c_{00}^* & |c_{01}|^2 & c_{01}c_{10}^* & c_{01}c_{11}^* \\ c_{10}c_{00}^* & c_{10}c_{01}^* & |c_{10}|^2 & c_{10}c_{11}^* \\ c_{11}c_{00}^* & c_{11}c_{01}^* & c_{11}c_{10}^* & |c_{11}|^2 \end{pmatrix}. \tag{22.30}$$

The question we want to address is: What is the appropriate density operator if we measure only qubit A's properties, averaged over Bob's unmeasured part of the system? Let's denote the density operator for the A subspace by $\rho(A)$ and its matrix elements in the computational basis as

$$\rho_{nm}(A) = \langle n| \rho(A)|m\rangle , \tag{22.31}$$

with $m, n = 0_A, 1_A$. We may also define $\rho_{nm}(B)$ for subspace B. The key idea is that we find $\rho(A)$ by taking the *partial* trace (summing over just the qubit B states) of the full density matrix for ρ:

$$\rho(A) = \mathrm{Tr}_B \rho. \tag{22.32}$$

To simplify the notation, we use $|0_A\rangle$ and $|1_A\rangle$ for the A subspace basis vectors and $|0_B\rangle$ and $|1_B\rangle$ for the B subspace. Then, Eq. (22.32) becomes

$$\rho(A) = \mathrm{Tr}_B\rho = \langle 0_B|\,\rho\,|0_B\rangle + \langle 1_B|\,\rho\,|1_B\rangle. \tag{22.33}$$

The general case (see Try It 22.8) is a bit complex, so let's restrict ourselves to the situation in which the composite system is described by a pure state with only two of the basis states

$$|\Phi\rangle = c_{00}\,|0_A\rangle\,|0_B\rangle + c_{11}\,|1_A\rangle\,|1_B\rangle = c_{00}\,|0_A0_B\rangle + c_{11}\,|1_A1_B\rangle, \tag{22.34}$$

where we used the compact notation $|0_A\rangle\,|0_B\rangle = |0_A0_B\rangle$ for the product states. You should recall that such a state (for general values of the amplitudes) is an entangled state. The density operator for this state is

$$\begin{aligned}\rho_\Phi &= |\Phi\rangle\langle\Phi| \\ &= c_{00}c_{00}^*\,|0_A0_B\rangle\langle 0_A0_B| + c_{00}c_{11}^*\,|0_A0_B\rangle\langle 1_A1_B| \\ &\quad + c_{11}c_{00}^*\,|1_A1_B\rangle\langle 0_A0_B| + c_{11}c_{11}^*\,|1_A1_B\rangle\langle 1_A1_B|.\end{aligned} \tag{22.35}$$

The subscript Φ reminds us that the resulting density operator depends on the composite-system state $|\Phi\rangle$. The coefficients of the diagonal terms express the probabilities of measuring the system in basis states $P(0_A0_B) = |c_{oo}|^2$ and $P(1_A1_B) = |c_{11}|^2$ while the coefficients of the off-diagonal terms express the coherence (interference) between the two basis states.

We now apply Eq. (22.33) to find the density operator for the subspace A:

$$\rho_\Phi(A) = \mathrm{Tr}_B\rho_\Phi = \langle 0_B|\,\rho_\Phi\,|0_B\rangle + \langle 1_B|\,\rho_\Phi\,|1_B\rangle = c_{00}^*c_{00}\,|0_A\rangle\langle 0_A| + c_{11}^*c_{11}\,|1_A\rangle\langle 1_A|. \tag{22.36}$$

In other words, Eq. (22.36) expresses averaging the composite-system density operator over the B part of the system. The result for the reduced density operator in the matrix form is

$$\rho_\Phi(A) \doteq \begin{pmatrix} |c_{00}|^2 & 0 \\ 0 & |c_{11}|^2 \end{pmatrix}. \tag{22.37}$$

We see that the off-diagonal (coherence) terms are zero and hence this is a maximally mixed-state density matrix.

While progressing to Eq. (22.37), we used $\langle 0_B|\,0_A0_B\rangle = \langle 0_B|\,0_B\rangle\,|0_A\rangle = |0_A\rangle$ and $\langle 0_B|\,0_A1_B\rangle = \langle 0_B|\,1_B\rangle\,|0_A\rangle = 0$ with analogous expressions for the other inner-product-like terms. We see that this "reduced" density operator is exactly what we expect for a *mixed* state if we identify the probabilities as $p_0 = c_{00}^*c_{00}$ and $p_1 = c_{11}^*c_{11}$:

$$\rho_\Phi(A) = p_0\,|0_A\rangle\langle 0_A| + p_1\,|1_A\rangle\langle 1_A|, \tag{22.38}$$

which we can use to calculate average values for subspace A. Note that if we had used a different composite state in place of the one in Eq. (22.34), we would end up with a different density operator for $\rho(A)$, that is, the reduced density operator for A depends on the state of the composite system.

Note that the partial trace operation allows us to construct a mixed-state density matrix (or operator) description for just one part of the two-qubit system. This contrasts with the statement that in general we cannot assign a (*pure*) quantum state to any subpart of an entangled-state system.

In the case when $|c_{00}|^2 = |c_{11}|^2$, we find that $\rho_\Phi(A) = \rho_\Phi(B)$ and are proportional to the identity operator. We say that the individual qubits A and B are maximally disordered.

The key point is that if we have an entangled system but make measurements with an operator that acts on only one of the constituents of that system and we average over the other part of the system, the results for expectation values will be just those we would obtain if the subsystem being measured were described by a mixed-state density operator.

Why is this important? The mixed-state density operator can be used as a model of the results of the interaction between the system and its environment. In the case we have looked at, the "environment" for part A of the system is the B part. In more general cases, the B part might be a many, many-qubit system, perhaps even a measurement device.

Many authors (see Further Reading, Kuzyk 2023) claim that these results demonstrate that a qubit's interaction with its environment (here, the other qubit in the two-qubit entangled state) is what "causes" the appearance of mixed states. According to these authors, there is no need to invoke ensembles to generate mixed states. We argue that the situation is more complicated and that the nature of the reduced density matrix after the trace operation depends on both the composite-system state vector, which expresses the correlations among the qubits in the composite-system entangled state, and on the number of qubits in the system (see Section 22.5 for three-qubit examples).

Try It 22.6 Repeat the analysis above to find the reduced density matrix $\rho_\Phi(B)$ associated with Bob's qubit.

So far, we have shown how to reduce the density operator for the composite system to reduced density operators for each part of the system. Can we work in the other direction? Given $\rho_\Phi(A)$ and $\rho_\Phi(B)$, can we reconstruct $\rho_\Phi(AB)$? The general answer is no. In carrying out the partial traces, we have lost information about the correlation of the two qubits.

Product States

Suppose the composite two-qubit system is prepared in a mixture of simple product states. In this case, the density matrix for the composite state is

$$\rho(AB) = \sum_{j,k} p_k q_j \left(|\Phi^{(k)}\rangle \otimes |\Psi^{(j)}\rangle \right) \left(\langle \Phi^{(k)}| \otimes \langle \Psi^{(j)}| \right), \tag{22.39}$$

and the density matrices for the A and B qubits individually are

$$\rho(A) = \sum_k p_k \left|\Phi^{(k)}\right\rangle \left\langle\Phi^{(k)}\right| \text{ and } \rho(B) = \sum_j q_j \left|\Psi^{(j)}\right\rangle \left\langle\Psi^{(j)}\right|. \tag{22.40}$$

For that product state, the density matrix for composite state is the Kronecker product of the individual density matrices:

$$\rho\,(AB) = \rho(A) \otimes \rho(B), \tag{22.41}$$

where $\otimes$ indicates the Kronecker product between the two reduced density operators.

To prove this result, we use the density matrices in Eq. (22.40) to write

$$\rho(A) \otimes \rho(B) = \sum_{j,k} p_k q_j \left(\left|\Phi^{(k)}\right\rangle \left\langle\Phi^{(k)}\right|\right) \otimes \left(\left|\Psi^{(j)}\right\rangle \left\langle\Psi^{(j)}\right|\right), \tag{22.42}$$

where p_k and q_j are the appropriate probabilities. From linear algebra, the Kronecker product and the outer product are interchangeable in the sense that

$$\left(\left|\Phi^{(k)}\right\rangle \left\langle\Phi^{(k)}\right|\right) \otimes \left(\left|\Psi^{(j)}\right\rangle \left\langle\Psi^{(j)}\right|\right) = \left(\left|\Phi^{(k)}\right\rangle \otimes \left|\Psi^{(j)}\right\rangle\right)\left(\left\langle\Phi^{(k)}\right| \left\langle\Psi^{(j)}\right|\right), \tag{22.43}$$

from which we see that Eq. (22.41) is satisfied.

Note that if $\rho(AB)$ is the density operator for a pure product state, then $\rho(A)$ and $\rho(B)$ are also density operators for pure states. If $\rho(AB)$ is a density operator for a mixed state, we have shown that the reduced density matrices are also for mixed states.

More on Entangled States

In Chapter 9, we showed that a general two-qubit state could be written as a linear superposition of product basis states in the computational basis:

$$|\psi\rangle = c_{00}\,|0_A 0_B\rangle + c_{01}\,|0_A 1_B\rangle + c_{10}\,|1_A 0_B\rangle + c_{11}\,|1_A 1_B\rangle, \tag{22.44}$$

and that if the coefficients satisfied the relationship

$$s = 2\,|c_{00}c_{11} - c_{01}c_{10}| = 0, \tag{22.45}$$

then the state is *not* an entangled state. Recall that the parameter s can range from 0 to 1. When $s = 1$ we say that the state has maximal entanglement. For example, if $c_{01} = 0 = c_{10}$ and $c_{00} = c_{11} = 1/\sqrt{2}$, then we have $s = 1$. In this case, Eq. (22.44) is

$$|\psi\rangle = \frac{1}{\sqrt{2}}\,|0_A 0_B\rangle + \frac{1}{\sqrt{2}}\,|1_A 1_B\rangle, \tag{22.46}$$

one of the Bell states (see Eqs. (10.14) and (10.15)). As described in Chapter 9, the parameter s can be used as a measure of the "degree of entanglement."

Try It 22.7 Suppose we have a two-qubit system that can be described by either of two pure spin-up, spin-down states:

$$|\Phi\rangle = \frac{1}{2}|\uparrow_A\uparrow_B\rangle + \frac{\sqrt{3}}{2}|\downarrow_A\downarrow_B\rangle \tag{22.47}$$

or

$$|\psi\rangle = \frac{\sqrt{3}}{2}|\uparrow_A\uparrow_B\rangle - \frac{1}{2}|\downarrow_A\downarrow_B\rangle . \tag{22.48}$$

Show that the reduced density matrices for subsystem B obtained by tracing over A is

$$\begin{aligned}\rho_\Phi(B) &= \mathrm{Tr}_A\,|\Phi\rangle\langle\Phi| = \frac{1}{4}|\uparrow_B\rangle\langle\uparrow_B| + \frac{3}{4}|\downarrow_B\rangle\langle\downarrow_B| \\ \rho_\psi(B) &= \mathrm{Tr}_A\,|\psi\rangle\langle\psi| = \frac{3}{4}|\uparrow_B\rangle\langle\uparrow_B| + \frac{1}{4}|\downarrow_B\rangle\langle\downarrow_B| .\end{aligned} \tag{22.49}$$

These results indicate that the two reduced density matrices predict different measurement results for the B subsystem. What is the parameter s (Eq. (22.45)) for this two-qubit system? Are these states maximally entangled states?

Try It 22.8 Challenge: For the general two-qubit system state in Eq. (22.29), use the procedures outlined above to find the reduced density operator for subspace A, namely, $\rho(A) = \mathrm{Tr}_B\rho = \langle 0_B|\rho|0_B\rangle + \langle 1_B|\rho|1_B\rangle$. To be completely general, assume that the state amplitudes are complex variables.

(a) Show that the reduced density operator for subspace A is

$$\begin{aligned}\rho(A) = &\left(|c_{00}|^2 + |c_{01}|^2\right)|0_A\rangle\langle 0_A| + \left(c_{00}c_{10}^* + c_{01}c_{11}^*\right)|0_A\rangle\langle 1_A| \\ &+ \left(c_{10}c_{00}^* + c_{11}c_{01}{}^*\right)|1_A\rangle\langle 0_A| + \left(|c_{01}|^2 + |c_{11}|^2\right)|1_A\rangle\langle 1_A| .\end{aligned} \tag{22.50}$$

(b) Compare this result with those for the probabilities in Sections 9.5 and 9.6. Explain why they agree or disagree. Note that for the general case, the off-diagonal matrix elements of the reduced density operator are not equal to zero.

(c) If the two-qubit system state is just a simple product state, the reduced density operator for the A subspace should be that of a pure state and we should have $\mathrm{Tr}\rho(A) = \mathrm{Tr}\left[\rho(A)^2\right]$. Show that Eq. (22.50) leads to the conditions for c_{00}, c_{01}, c_{10}, and c_{11} stated in Chapter 9 (Eq. (9.21)).

22.5 Connecting Reduced Density Matrices, Probabilities, and the Generalized Born Rule

We have developed several ways to predict the results of measurements on a multiqubit system. Section 9.4 shows how to extract the probabilities for state measurement outcomes directly from the coefficients associated with the (pure) state for the composite system. Section 9.5 shows that the "collapsed state after measurement" results are the same as those calculated directly from conditional probabilities. How are the reduced density matrix results related to those other methods? We would expect that the results are the same if we deal with the same composite multiqubit system state because we are "simply" using different calculational techniques. Let's check this notion with a specific example.

We use the two-qubit composite state given in Eq. (22.34). In particular, we calculate the probability of observing the state $|0_A\rangle$ via three methods: (1) the probabilities found directly from the composite-system state vector (as we did in Chapter 9), (2) the reduced density matrix for subsystem A, and (3) the full density matrix for the composite system.

Table 9.4, with an appropriate change in notation from spin-up, spin-down basis states to computational basis states, shows that the probability of observing the state $|0_A\rangle$ is $P(0_A) = |c_{00}|^2$ since for the example's state $c_{01} = 0$. How about the density operator approach? The density operator for the AB composite system is

$$\begin{aligned} \rho_\Phi(AB) &= |\Phi\rangle\langle\Phi| \\ &= c_{00}c_{00}^* |0_A 0_B\rangle\langle 0_A 0_B| + c_{00}c_{11}^* |0_A 0_B\rangle\langle 1_A 1_B| \\ &\quad + c_{11}c_{00}^* |1_A 1_B\rangle\langle 0_A 0_B| + c_{11}c_{11}^* |1_A 1_B\rangle\langle 1_A 1_B| . \end{aligned} \tag{22.51}$$

The density operator for the A subsystem is given by Eq. (22.37). The operators for measurements targeting $|0_A\rangle$ and $|1_A\rangle$ are

$$M(0_A) \doteq \begin{pmatrix} 1 & 0 \\ 0 & 0 \end{pmatrix} \qquad M(1_A) \doteq \begin{pmatrix} 0 & 0 \\ 0 & 1 \end{pmatrix}. \tag{22.52}$$

So, the probabilities of obtaining 0_A and 1_A as measurement outcomes are given by

$$\begin{aligned} \langle M(0_A)\rangle &= \mathrm{Tr}\left[\rho_\Phi(A) M(0_A)\right] = |c_{00}|^2 \\ \langle M(1_A)\rangle &= \mathrm{Tr}\left[\rho_\Phi(A) M(1_A)\right] = |c_{11}|^2 \end{aligned} \tag{22.53}$$

in agreement with the direct probability approach of Chapter 9.

We could also use the *composite*-system density operator in Eq. (22.51) along with the Kronecker product for the composite-system operator

$$M(0_A) \otimes \mathrm{I}_B \doteq \begin{pmatrix} 1 & 0 \\ 0 & 0 \end{pmatrix} \otimes \begin{pmatrix} 1 & 0 \\ 0 & 1 \end{pmatrix} = \begin{pmatrix} 1 & 0 & 0 & 0 \\ 0 & 1 & 0 & 0 \\ 0 & 0 & 0 & 0 \\ 0 & 0 & 0 & 0 \end{pmatrix}. \tag{22.54}$$

This product describes measuring for $|0_A\rangle$ in the A subsystem and doing nothing (the identity operator) in the B subsystem. The result is

$$P(0_A) = \mathrm{Tr}\left[\rho_\Phi(AB)\{M(0_A) \otimes \mathrm{I}_B\}\right] = |c_{00}|^2. \tag{22.55}$$

We could also use the generalized Born rule to calculate that probability

$$P(0_A) = |\langle 0_A | \Phi\rangle|^2 = |c_{00}|^2 \langle 0_B \mid 0_b\rangle = |c_{00}|^2. \tag{22.56}$$

We see that all these methods give the same result for the state probability measurements. Although the methods give the same results for the pure-state case, the density operator methods have the advantage of also applying to mixed states for which the direct probability methods and the generalized Born rule do not work.

Conditional Density Matrices

In Chapter 9, we observed that conditional probabilities such as $P(0_A|1_B)$ could be deployed to calculate the results obtained using Lüder's rule for the state of a system after a measurement had been made on part of that system. For systems described by density operators, we may define a conditional density matrix. To see how we might find such an operator, we begin with a two-qubit system example developed in Chapter 9. The general state for the qubit system is given by Eq. (22.49) and its density matrix is given by Eq. (22.30). If we make a measurement on qubit B and find the result 0_B, we know by Lüder's rule that the state of the composite system after the measurement is a product state

$$|\psi_{\text{after}}\rangle \propto |0_B\rangle \left[c_{00}\,|0_A\rangle + c_{10}\,|1_A\rangle\right]. \tag{22.57}$$

To ensure that the A part of the state is normalized, we need to divide the right side of Eq. (22.57) by $d = \sqrt{|c_{00}|^2 + |c_{10}|^2}$. Thus, after the measurement, qubit A is described by the state

$$|\psi_{\text{after}}(A)\rangle = \frac{1}{d}\left[c_{00}\,|0_A\rangle + c_{10}\,|1_A\rangle\right]. \tag{22.58}$$

We define a conditional density operator built from the state in Eq. (22.58):

$$\rho\,(A|0_B) = |\psi_{\text{after}}(A)\rangle\,\langle\psi_{\text{after}}(A)| = \frac{1}{d^2}\begin{pmatrix} |c_{00}|^2 & c_{00}c_{10}^* \\ c_{10}c_{00}^* & |c_{11}|^2 \end{pmatrix}. \tag{22.59}$$

Note that this conditional density operator is Hermitian and has a trace equal to 1. You should check to see if $\rho(A|0_B)^2 = \rho\,(A|0_B)$ and determine if the operator corresponds to a pure state or a mixed state. Note that the conditional density operator is distinctly different from the subsystem A reduced density operator given in Try It 22.8.

Obviously, the next issue is to determine a procedure to go directly from the composite state density operator to the conditional density operator. For a two-qubit system in a composite pure state, we outline a procedure that works, but for more general cases, the issue is an open research question. Here is the two-qubit system argument.

We assume that we start with the two-qubit system in a composite pure state described by the density operator ρ_ψ. We want to find the conditional density operator $\rho\,(A|B)$, where

B stands for either 0_B or 1_B. Introducing a projection operator $P_{B^k} = |b_k\rangle\langle b_k|$ with $k = 0, 1$ for the B measurement, we may express the probability of getting 0_B or 1_B as

$$p_k = \text{Tr}\left(\rho_\psi P_{B^k}\right), \tag{22.60}$$

which involves a trace over the full matrix product. With $|\alpha_k\rangle$ indicating the state describing the A qubit after the B measurement yields b_k, the state of the composite system can now be written as

$$|\psi\rangle = \sum_k \sqrt{p_k}\,|\alpha_k\rangle \otimes |b_k\rangle. \tag{22.61}$$

Note that $|\alpha_k\rangle$ will generally be a linear combination of the A basis states: $|\alpha_k\rangle = a_0\,|0_A\rangle + a_1\,|1_A\rangle$. Then, we will have $p_k = |a_0|^2 + |a_1|^2$.

The conditional density operator for the A subsystem given that b_k was the result of measuring B is

$$\rho\left(A|b_k\right) = \frac{1}{p_k}\text{Tr}_B\left(\rho_\psi P_{B^k}\right). \tag{22.62}$$

Note that the trace in Eq. (22.62) is the trace over just the B basis states.

Let's use this procedure on the example used in the first part of this section that used $|0_B\rangle$ as the B state after measurement. The appropriate projection operator is

$$P_{B^0} = |0_B\rangle\langle 0_B| \doteq \begin{pmatrix} 1 & 0 \\ 0 & 0 \end{pmatrix}, \tag{22.63}$$

and ρ_ψ is given by Eq. (22.30). First, we calculate the probabilities p_k and obtain

$$\begin{aligned} p_0 &= \text{Tr}\left[(\rho P_{B^0})\right] \doteq \text{Tr}\left[\begin{pmatrix} |c_{00}|^2 & c_{00}c_{10}^* \\ c_{00}^* & |c_{11}|^2 \end{pmatrix}\begin{pmatrix} 1 & 0 \\ 0 & 0 \end{pmatrix} + \begin{pmatrix} |c_{00}|^2 & c_{00}c_{10}^* \\ c_{00}^* & |c_{11}|^2 \end{pmatrix}\begin{pmatrix} 0 & 0 \\ 0 & 1 \end{pmatrix}\right] \\ &= |c_{00}|^2 + |c_{01}|^2. \end{aligned} \tag{22.64}$$

The conditional density matrix is then given by Eq. (22.62):

$$\begin{aligned} \rho(A|0_B) &= \frac{1}{p_{00}}\left[\langle 0_B|\,\rho_\psi P_{B^0}\,|0_B\rangle + \langle 1_B|\,\rho_\psi P_{B^1}\,|1_B\rangle\right] \\ &\doteq \frac{1}{|c_{00}|^2 + |c_{10}|^2}\begin{pmatrix} |c_{00}|^2 & c_{00}c_{10}^* \\ c_{10}c_{00}^* & |c_{10}|^2 \end{pmatrix}, \end{aligned} \tag{22.65}$$

just what we found in Eq. (22.59).

22.6 Three-Qubit Density Matrices

Let's work out our matrix muscles on a system with three qubits. Although this section is moderately demanding because of the increased size of the matrices, it is worth at least a quick reading to see how issues of entanglement for mixed states are connected to decoherence, measurements, and quantum error correction. We see that having an additional qubit increases both the types of reduced density matrices available to describe the two-qubit subsystems and the number of reduced density matrices we can construct. For example, if we label the three qubits as A, B, and C, we will have reduced density matrices for the AB, AC, and BC subsystems. Each of these subsystems can then be used to find reduced density matrices for two of the qubits. Obvious questions arise: Will the reduced density matrix for qubit A be the same if it is deduced from the AB subsystem and the AC subsystem? Do the subsystem density matrices always involve mixed states? We explore these kinds of questions by looking at specific three-qubit systems.

Eq. (9.26) and the following material show how a pure state for such a system could be formulated. In the abbreviated form, the state vector is

$$|\psi\rangle = c_{000}\,|000\rangle + c_{001}\,|001\rangle + \cdots, \tag{22.66}$$

which in general will have eight terms in the superposition state. The system density operator is given by a 64-term projection operator

$$\rho = |c_{000}|^2\,|000\rangle\,\langle 000| + c_{000}c^*_{001}\,|000\rangle\,\langle 001| + \cdots. \tag{22.67}$$

The corresponding density matrix will be an 8×8 matrix

$$\rho \doteq \begin{pmatrix} |c_{000}|^2 & c_{000}c^*_{001} & c_{000}c^*_{010} & \cdots \\ c_{001}c^*_{000} & |c_{001}|^2 & \cdots & \\ \vdots & \vdots & |c_{010}|^2 & \cdots \end{pmatrix}. \tag{22.68}$$

We want to find reduced density matrices that arise from averaging over parts of the composite state. Let's do this in stages. We label the subspaces as A, B, and C. We can construct reduced density operators for the AB, AC, and BC subsystems, and from each of these, we can construct individual density operators. To minimize the complexity of the calculations, let's assume that the state vector for the composite system is the so-called Greenberger–Horne–Zeilinger (GHZ) state, the three-qubit extension of a Bell state,

$$|\psi_{\text{GHZ}}\rangle = \frac{1}{\sqrt{2}}\left[|0_A 0_B 0_C\rangle + |1_A 1_B 1_C\rangle\right]. \tag{22.69}$$

The corresponding density operator is

$$\begin{aligned}\rho_{\text{GHZ}} = |\psi_{\text{GHZ}}\rangle\langle\psi_{\text{GHZ}}| &= \frac{1}{2}\left[|0_A 0_B 0_C\rangle\langle 0_A 0_B 0_C| + |0_A 0_B 0_C\rangle\langle 1_A 1_B 1_C|\right] \\ &+ \frac{1}{2}\left[|1_A 1_B 1_C\rangle\langle 0_A 0_B 0_C| + |1_A 1_B 1_C\rangle\langle 1_A 1_B 1_C|\right].\end{aligned} \tag{22.70}$$

Now let's take the trace over C and find what happens to the AB subsystem: Is it described by a mixed-state or pure-state density operator?

The AB subsystem density operator is given by

$$\rho_{\text{GHZ}}(AB) = \text{Tr}_C \rho_{\text{GHZ}} = \langle 0_C|\,\rho_{\text{GHZ}}\,|0_C\rangle + \langle 1_C|\,\rho_{\text{GHZ}}\,|1_C\rangle. \tag{22.71}$$

In evaluating the trace, we use $\langle 0_C|\, 0_A 0_B 0_C\rangle = |0_A 0_B\rangle$, $\langle 0_C|\, 0_A 0_B 1_C\rangle = |0_A 0_B\rangle\langle 0_C|\, 1_C\rangle = 0$, and other analogous inner products. We obtain

$$\langle 0_C|\,\rho_{\text{GHZ}}\,|0_C\rangle = \frac{1}{2}\,|0_A 0_B\rangle\langle 0_A 0_B| \tag{22.72}$$

and

$$\langle 1_C|\,\rho_{\text{GHZ}}\,|1_C\rangle = \frac{1}{2}\,|1_A 1_B\rangle\langle 1_A 1_B|. \tag{22.73}$$

Using Eqs. (22.72) and (22.73) in Eq. (22.71) yields

$$\rho_{\text{GHZ}}(AB) = \frac{1}{2}\,|0_A 0_B\rangle\langle 0_A 0_B| + \frac{1}{2}\,|1_A 1_B\rangle\langle 1_A 1_B| \doteq \begin{pmatrix} 1/2 & 0 & 0 & 0 \\ 0 & 0 & 0 & 0 \\ 0 & 0 & 0 & 0 \\ 0 & 0 & 0 & 1/2 \end{pmatrix}. \tag{22.74}$$

Since there are no nonzero off-diagonal matrix elements, this is a mixed-state density matrix. Note that the first line of Eq. (22.74) looks exactly like the density matrix for a statistical ensemble with probabilities = ½ for each part of the ensemble.

If we now take the trace of $\rho_{\text{GHZ}}(AB)$ with respect to qubit B's states, we find

$$\rho_{\text{GHZ}}(A) = \text{Tr}_B \rho_{\text{GHZ}}(AB) = \begin{pmatrix} 1/2 & 0 \\ 0 & 1/2 \end{pmatrix}, \tag{22.75}$$

another mixed-state density matrix. The reduced density matrix $\rho_{\text{GHZ}}(A)$ is said to be maximally disordered because it is a multiple of the identity matrix.

Try It 22.9 Carry out the partial trace over A of $\rho_{\text{GHZ}}(AB)$ in Eq. (22.74) to get $\rho_{\text{GHZ}}(B)$. Are the two single-qubit density operators the same?

Try It 22.10 Carry out the same calculations for the AC and BC subsystems. Explain similarities and differences.

Let's try the trace operations with a different composite-system state vector:

$$|\Phi\rangle = \frac{1}{2}\left[|0_A 0_B 0_C\rangle + |0_A 1_B 0_C\rangle + |0_A 0_B 1_C\rangle + |1_A 1_B 0_C\rangle\right]. \tag{22.76}$$

Try It 22.11 Is this an entangled state? Explain your reasoning.

The density operator for this state is

$$\begin{aligned}
\rho_\Phi &= |\Phi\rangle\langle\Phi| \\
&= \frac{1}{4}\left[|0_A 0_B 0_C\rangle\langle 0_A 0_B 0_C| + |0_A 0_B 0_C\rangle\langle 0_A 1_B 0_C| + |0_A 0_B 0_C\rangle\langle 0_A 0_B 1_C|\right. \\
&\quad \left. + |0_A 0_B 0_C\rangle\langle 1_A 1_B 0_C|\right] + \frac{1}{4}\left[|0_A 1_B 0_C\rangle\langle 0_A 0_B 0_C| + |0_A 1_B 0_C\rangle\langle 0_A 1_B 0_C|\right. \\
&\quad \left. + |0_A 1_B 0_C\rangle\langle 0_A 0_B 1_C| + |0_A 1_B 0_C\rangle\langle 1_A 1_B 0_C|\right] + \frac{1}{4}\left[|0_A 0_B 1_C\rangle\langle 0_A 0_B 0_C|\right. \\
&\quad \left. + |0_A 0_B 1_C\rangle\langle 0_A 1_B 0_C| + |0_A 0_B 1_C\rangle\langle 0_A 0_B 1_C| + |0_A 0_B 1_C\rangle\langle 1_A 1_B 0_C|\right] \\
&\quad + \frac{1}{4}\left[|1_A 1_B 0_C\rangle\langle 0_A 0_B 0_C| + |1_A 1_B 0_C\rangle\langle 0_A 1_B 0_C| + |1_A 1_B 0_C\rangle\langle 0_A 0_B 1_C|\right. \\
&\quad \left. + |1_A 1_B 0_C\rangle\langle 1_A 1_B 0_C|\right].
\end{aligned} \tag{22.77}$$

First, we carry out the trace over C:

$$\rho_\Phi(AB) = \mathrm{Tr}_C \rho_\Phi = \langle 0_C|\rho_\Phi|0_C\rangle + \langle 1_C|\rho_\Phi|1_C\rangle. \tag{22.78}$$

We evaluate the two terms on the far right of Eq. (22.78) separately. For $\langle 0_C|\rho_\Phi|0_C\rangle$, we get zero unless the term in Eq. (22.77) has 0_C on the far right of both the right vector and left parts, for example $|1_A 0_B 0_C\rangle\langle 1_A 1_B 0_C|$. The results are

$$\begin{aligned}
\langle 0_C|\rho_\Phi|0_C\rangle &= \frac{1}{4}\left[|0_A 0_B\rangle\langle 0_A 0_B| + |0_A 0_B\rangle\langle 0_A 1_B| + |0_A 0_B\rangle\langle 1_A 1_B|\right] \\
&\quad + \frac{1}{4}\left[|0_A 1_B\rangle\langle 0_A 0_B| + |0_A 1_B\rangle\langle 0_A 1_B| + |0_A 1_B\rangle\langle 1_A 1_B|\right] \\
&\quad + \frac{1}{4}\left[|1_A 1_B\rangle\langle 0_A 0_B| + |1_A 1_B\rangle\langle 0_A 1_B| + |1_A 1_B\rangle\langle 1_A 1_B|\right].
\end{aligned} \tag{22.79}$$

The other part of the trace over C is given by

$$\langle 1_C| \rho_\Phi |1_C\rangle = \frac{1}{4}\left[|0_A 0_B\rangle\langle 0_A 0_B|\right]. \tag{22.80}$$

Eq. (22.80) has only one term because there is only one term in Eq. (22.77) that has 1_C in both C parts of the projection operator terms.

The resulting reduced density matrix is

$$\rho_\Phi(AB) \doteq \begin{pmatrix} 1/2 & 1/4 & 0 & 1/4 \\ 1/4 & 1/4 & 0 & 1/4 \\ 0 & 0 & 0 & 0 \\ 1/4 & 1/4 & 0 & 1/4 \end{pmatrix}. \tag{22.81}$$

Does Eq. (22.81) represent a pure state or a mixed state? The easiest way to check is to calculate $[\rho_\Phi(AB)]^2$. For a mixed state, $\rho^2 \neq \rho$ and $\text{Tr}\rho^2 < 1$. Is the AB complex described by a product state or an entangled state?

Try It 22.12 Carry out the matrix calculations to show that $\rho_\Phi(AB)$ is a mixed-state density matrix for the AB two-qubit subsystem, but it is not a maximally disordered density matrix. Does it represent an entangled state? Explain your reasoning.

We now carry out the trace over B to obtain the reduced density matrix for qubit A:

$$\rho_\Phi(A) = \text{Tr}_B \rho_\Phi(AB) \doteq \begin{pmatrix} 3/4 & 1/4 \\ 0 & 1/4 \end{pmatrix}. \tag{22.82}$$

This is a valid density matrix since its trace = 1. It is a mixed-state density matrix because $\rho^2 \neq \rho$ and $\text{Tr}\rho^2 < 1$. To find the reduced density matrix $\rho_\Phi(B)$ for subsystem B, we carry out the trace of $\rho_\Phi(AB)$ in Eq. (22.81) over A. The result is

$$\rho_\Phi(B) \doteq \begin{pmatrix} 1/2 & 0 \\ 0 & 1/2 \end{pmatrix}. \tag{22.83}$$

In this example, the reduced density matrices for A and B are different while for the GHZ state they are identical. The reason is that this example's state vector does not treat the three qubits symmetrically, while they are symmetric in the GHZ state.

Try It 22.13 Compare $\rho_\Phi(AB)$ with $\rho_\Phi(A) \otimes \rho_\Phi(B)$ to find if the AB complex is an entangled subsystem or a product-state subsystem.

To see how general the previous results are, we now consider a three-qubit system described by a so-called W state:

$$|\psi_W\rangle = \frac{1}{\sqrt{3}}\left[|001\rangle + |010\rangle + |100\rangle\right], \tag{22.84}$$

which is also an entangled state. The state is named after Wolfgang Dür[1], who with two colleagues, first discussed this state in 2000. It has the unusual feature of giving an entangled-state reduced matrix for two of the qubits if it is averaged over the states of one of the qubits. Here are the details.

The density matrix for the composite system is given by

$$\begin{aligned} 3\rho_W = 3\,|\psi_W\rangle\langle\psi_W| &= (|001\rangle + |010\rangle + |100\rangle)\,(\langle 001| + \langle 010| + \langle 100|) \\ &= |001\rangle\langle 001| + |001\rangle\langle 010| + |001\rangle\langle 100| \\ &\quad + |010\rangle\langle 001| + |010\rangle\langle 010| + |010\rangle\langle 100| \\ &\quad + |100\rangle\langle 010| + |100\rangle\langle 010| + |100\rangle\langle 100|\,. \end{aligned} \tag{22.85}$$

We compute the reduced matrix $\rho_W(AB)$ from

$$\begin{aligned} \rho_W(AB) = \mathrm{Tr}_C\rho_W &= \langle 0_C|\,\rho_W\,|0_C\rangle + \langle 1_C|\,\rho_W\,|1_C\rangle \\ &= \frac{1}{3}\left[|00\rangle\langle 00| + |01\rangle\langle 01| + |01\rangle\langle 10| + |10\rangle\langle 01| + |10\rangle\langle 10|\right] \\ &\doteq \frac{1}{3}\begin{pmatrix} 1 & 0 & 0 & 0 \\ 0 & 1 & 1 & 0 \\ 0 & 1 & 1 & 0 \\ 0 & 0 & 0 & 0 \end{pmatrix}. \end{aligned} \tag{22.86}$$

The matrix form indicates that the AB reduced density operator is a mixed-state operator.

We now calculate the reduced density operators $\rho_W(A)$ and $\rho_W(B)$. The results are

$$\begin{aligned} \rho_W(A) &= \langle 0_B|\,\rho_W(AB)|0_B\rangle + \langle 1_B|\,\rho_W(AB)|1_B\rangle \\ &= \frac{2}{3}\,|0_A\rangle\langle 0_A| + \frac{1}{3}\,|1_A\rangle\langle 1_A| \\ &\doteq \begin{pmatrix} 2/3 & 0 \\ 0 & 1/3 \end{pmatrix} \end{aligned} \tag{22.87}$$

[1] https://en.wikipedia.org/wiki/Wolfgang_D%C3%BCr

and

$$\begin{aligned}\rho_W(B) &= \langle 0_A|\,\rho_W(AB)|0_A\rangle + \langle 1_A|\,\rho_W(AB)|1_A\rangle \\ &= \frac{2}{3}\,|0_B\rangle\langle 0_B| + \frac{1}{3}\,|1_B\rangle\langle 1_B| \\ &\doteq \begin{pmatrix} 2/3 & 0 \\ 0 & 1/3 \end{pmatrix}.\end{aligned} \tag{22.88}$$

The two reduced density matrices are the same even though the composite state system is not maximally entangled and may not seem to treat the three qubits symmetrically. However, the composite state system is cyclic, in the sense that if you make the replacements $A \to B \to C \to A$, you end up with the same state and that leads to identical reduced density matrices for A and B.

The Kronecker product of $\rho_W(A)$ and $\rho_W(B)$ in Eqs. (22.87) and (22.88) is used to see if the product is equal to $\rho_W(AB)$ in which case the latter is the product state density matrix. We obtain

$$\begin{aligned}\rho_W(A) \otimes \rho_W(B) &\doteq \begin{pmatrix} 2/3 & 0 \\ 0 & 1 \end{pmatrix} \otimes \begin{pmatrix} 2/3 & 0 \\ 0 & 1 \end{pmatrix} \\ &= \begin{pmatrix} 4/9 & 0 & 0 & 0 \\ 0 & 2/9 & 0 & 0 \\ 0 & 0 & 2/9 & 0 \\ 0 & 0 & 0 & 1/9 \end{pmatrix}.\end{aligned} \tag{22.89}$$

Comparing Eqs. (22.89) to (22.86) shows that they are not the same. We conclude that $\rho_W(AB)$ is not the density matrix for a product state. The corresponding state must be an entangled state.

Let's summarize what we have discussed about three-qubit states and their density matrices. If we carry out the trace over one of the qubit's basis states (say, C), the reduced density matrix describes a mixed state for the composite system AB. If that reduced density matrix is traced over the B qubit's basis states, then the reduced density matrix for A alone is a mixed-state density matrix.

The classification of states for a three-qubit entangled state is modestly complicated. It has been shown that three-qubit states can be classified into four categories: separable states (the qubits have no correlations with each other), bipartite entangled states, W, and GHZ states.

We have shown that averaging the density matrix over the states except those of one of the qubits leads to states described by a mixed-state density matrix for that remaining qubit. Since these mixed states have reduced or zero off-diagonal matrix elements, we are tempted to say that the remaining qubit has experienced decoherence due to the averaging over the states of the other qubits with which it is entangled. That statement is misleading because the qubit was never described by a pure state from which the coherence might disappear. It is the composite system that is described by a pure state. Nevertheless, the reduced density matrix result provides a model for the effects of a qubit interacting with its environment. Finding ways to undo that decoherence lies at the core of much of quantum error correction.

Given a mixed-state density matrix, is it possible to find the particular pure state that led to the mixed-state density matrix by the trace operations? The general answer is "no," but a process called "purification" identifies classes of pure states that lead to a particular mixed state.

As mentioned previously, reduced density matrices are widely used in quantum information science and in quantum error correction methods. The challenge is to "engineer" the entangled composite state so that it will have the properties needed for your purposes.

22.7 Spin-1/2 Systems' Density Matrices and the Pauli Matrices

As mentioned in Chapter 6, for a two-state (spin-1/2) system, all operators can be expressed as linear combinations of the identity operator I and the Pauli operators σ_k (or their matrix representations). How does this work out for the density operator? It turns out that the density operator can be expressed as

$$\rho = \frac{1}{2}\left(\mathrm{I} + \vec{V}^B \cdot \vec{\sigma}\right), \tag{22.90}$$

where $\vec{V}^B$ is the so-called Bloch vector. (Some authors use $\vec{s}$ for the Bloch vector, but this may cause confusion due to the use of $\vec{S}$ for the spin angular momentum vector.) The components of the Bloch vector are equal to the expectation values of the corresponding Pauli operators. For example, $V_x^B = \langle \sigma_x \rangle$.

Let's convert Eq. (22.90) into its matrix equivalent:

$$\rho \doteq \begin{pmatrix} \rho_{11} & \rho_{12} \\ \rho_{21} & \rho_{22} \end{pmatrix} = \frac{1}{2}\begin{pmatrix} 1 + V_z^B & V_x^B + iV_y^B \\ V_x^B - iV_y^B & 1 - V_z^B \end{pmatrix}. \tag{22.91}$$

Try It 22.14 Use the matrix representation of Eq. (22.90) to derive Eq. (22.91).

Try It 22.15 Use the Pauli matrices and the density matrix in Eq. (22.91) to show that $V_x^B = \langle \sigma_x \rangle$ with analogous results for the y and z components.

For a mixed state, the magnitude of the Bloch vector is less than 1. The density matrix has two eigenvalues $g_1 = 1/2\left[1 + \left|\vec{V}^B\right|\right]$ and $g_2 = 1/2\left[1 - \left|\vec{V}^B\right|\right]$ and two eigenvectors

$\vec{u}_{+,-} = \pm\vec{V}^B / \left|\vec{V}^B\right|$. With this information, we can represent the mixed state for a single qubit by vectors in a Bloch Sphere diagram where $\vec{V}^B$ points along the direction of the expectation value of the spin angular momentum and the two eigenvectors of the density matrix point parallel and antiparallel to $\vec{V}^B$. Normalized pure states are represented by points on the surface of the Bloch sphere, whose radius equals 1. Mixed states are represented by points inside the sphere because the trace of the square of a mixed-state density matrix is less than one.

It is interesting to point out that an analogous representation for light polarization was developed by Henri Poincaré in the late 1800s, some 25 years before the development of quantum mechanics. Poincaré provided a geometric representation, now called the Poincaré sphere, motivated by the polarized light Stokes parameters first introduced in 1852. A mixed state in quantum mechanics is analogous to a state of partial polarization for light waves. Pure polarization states are represented by points on the surface of the Poincaré sphere while partially polarized states are represented by points inside the Poincaré sphere.

22.8 Density Matrices and Measurements

If the system state is described by a density operator and we carry out a measurement of some property represented by an operator A, what is the appropriate density operator *for the composite system* when the measurement yields the eigenvalue a_j? (Note that this is a different question from the ones we asked about creating reduced density matrices by averaging over other qubits' basis states.) The easiest way to answer this question is to write the operator A as a projection operator P_j involving its eigenvectors $\left|a_j\right\rangle$ labeled by the corresponding eigenvalue:

$$A = \sum_j a_j \left|a_j\right\rangle \left\langle a_j\right| \equiv \sum_j a_j P_j . \tag{22.92}$$

In the usual account of quantum measurements based on the Lüders rule, we say that after the measurement, the appropriate state vector for the measured qubit is the eigenvector associated with the observed eigenvalue. It means that the density operator after the measurement is the original density matrix projected on the eigenvector associated with the observation's outcome a_j:

$$\rho^{\text{after}} = \frac{P_j \rho P_j}{Tr\left(\rho P_j\right)}. \tag{22.93}$$

Try It 22.16 What do you expect Eq. (22.93) to give if the system is in a pure state? Work through the details to see if you obtain the expected result.

22.9 Time-Dependent Density Operators

How does a density operator evolve in time? We know this must happen because quantum states evolve in time. Let's begin with a mixed-state density operator

$$\rho(t) = \sum_j p_j \left|\psi^{(j)}(t)\right\rangle \left\langle\psi^{(j)}(t)\right| . \tag{22.94}$$

We assume that the mixture probabilities p_j do not depend on time. We know from Chapter 18 that the right vectors satisfy Schrödinger's equation

$$i\hbar \frac{d\left|\psi^{(j)}(t)\right\rangle}{dt} = H\left|\psi^{(j)}(t)\right\rangle . \tag{22.95}$$

The corresponding left vector satisfies

$$-i\hbar \frac{d\left\langle\left|\psi^{(j)}(t)\right\rangle\right|}{dt} = \left\langle\psi^{(j)}(t)\right| H, \tag{22.96}$$

which is essentially the complex conjugate of Eq. (22.95) if the Hamiltonian operator H is a Hermitian operator.

If we take the derivative of Eq. (22.94) with respect to time, we obtain

$$\frac{\partial \rho}{\partial t} = \sum_j p_j \left[\frac{\partial \left|\psi^{(j)}(t)\right\rangle}{\partial t} \left\langle\psi^{(j)}(t)\right| + \left|\psi^{(j)}(t)\right\rangle \frac{\partial \left\langle\psi^{(j)}(t)\right|}{\partial t} \right]. \tag{22.97}$$

Substituting Eqs. (22.95) and (22.96) into Eq. (22.97) yields

$$i\hbar \frac{\partial \rho}{\partial t} = [H, \rho] , \tag{22.98}$$

where, as usual, $[H, \rho] = H\rho - \rho H$ is the commutator between the two operators.

Try It 22.17 Work through the algebra in progressing from Eq. (22.97) to Eq. (22.98).

You may recall seeing a similar differential equation in Section 18.3 for the time dependence of an operator in the Heisenberg picture, but there is a sign difference in Eq. (22.98). This difference occurs because we are viewing the time dependence for the density operator in the Schrödinger picture. (We used the Schrödinger equation in Eq. (22.97).) The main point of this section is that we can find the time dependence of the density operator using operator techniques such as the commutator.

However, if the system is described by a mixed state, the situation is more complicated. In particular, the appropriate time evolution operator is not generally a Hermitian operator. These techniques are beyond the scope of this book. If you are interested in learning more about those techniques, search for information on quantum statistical mechanics.

Energy Eigenstates

If we use energy eigenstates (eigenstates of the Hamiltonian operator H) as the basis states, the resulting density matrix has some interesting properties. First, the diagonal elements, which represent the populations in the corresponding energy eigenstates, are independent of time. To see where that result comes from, form a $\langle E_n| \, |E_n\rangle$ quantum sandwich of both sides of Eq. (22.98) and show that

$$\langle E_n| H\rho - \rho H |E_n\rangle = E_n \langle E_n| \rho |E_n\rangle - \langle E_n| \rho |E_n\rangle E_n = 0, \tag{22.99}$$

where we used the eigenvalue equations for the left and right vector versions of eigenvectors $\langle E_n| H = \langle E_n| E_n$ and $H |E_n\rangle = E_n |E_n\rangle$. Defining $\rho_{nn} = \langle E_n| \rho |E_n\rangle$, we conclude that

$$i\hbar \frac{d\rho_{nn}(t)}{dt} = 0. \tag{22.100}$$

The time dependence of the off-diagonal elements is described by a relatively simple differential equation:

$$i\hbar \frac{d\rho_{mn}(t)}{dt} = (E_m - E_n)\rho_{mn}(t). \tag{22.101}$$

The solution to Eq. (22.101) is

$$\rho_{mn}(t) = e^{-i(E_n - E_m)t/\hbar} \rho_{mn}(0), \tag{22.102}$$

which you can easily check by substituting Eq. (22.102) into Eq. (22.101). We see that the off-diagonal terms ($m \neq n$), representing the coherences, oscillate with complex exponential time dependence (like cosines and sines). The oscillation frequency depends on the differences between the corresponding energy eigenvalues.

Thermodynamic Equilibrium

As mentioned at the beginning of this chapter, the density operator allows us to describe the states of a system influenced by interactions with its environment. If the quantum system is in thermal equilibrium with its environment that has an absolute temperature T, the appropriate density operator is

$$\rho = \frac{1}{Z} e^{-H/k_B T}, \tag{22.103}$$

where H is the relevant Hamiltonian operator, k_B is called Boltzmann's constant, $k_B = 1.38... \times 10^{-23}$J/K (J indicates energy in Joules and K means degrees Kelvin), and Z is the so-called partition function from classical statistical mechanics: $Z = \mathrm{Tr}\left(e^{-H/k_B T}\right)$. For our purposes, Z is just a number whose value does not influence our general conclusions.

Let's evaluate the density matrix in Eq. (22.103) using the energy eigenvectors $|E_n\rangle$ associated with the Hamiltonian (energy) operator. We find that the diagonal elements of the matrix are

$$\rho_{nn} = \frac{1}{Z}\langle E_n|\, e^{-H/k_B T}\, |E_n\rangle = \frac{1}{Z} e^{-E_n/k_B T}, \tag{22.104}$$

where we used the result that a function of an operator acting on an eigenstate of that operator, for example $f(\hat{A})|a_n\rangle = f(a_n)\,|a_n\rangle$, gives the function of the eigenvalue times the eigenstate. The last exponential term in Eq. (22.104) is called the Boltzmann factor and shows that in thermal equilibrium the population of the energy eigenstates falls off exponentially with the energy associated with the states. That is, the population piles up in the set of states with the lowest energies and is smaller for higher energy states. The product $k_B T$ sets the energy scale for the Boltzmann factor. (If several quantum states have the same energy, you need to take into account the "degeneracy" number: the number of states with that energy.)

Let's look at the off-diagonal density matrix elements. For thermal equilibrium, we obtain

$$\rho_{mn} = \frac{1}{Z}\langle E_m|\, e^{-H/k_B T}\, |E_n\rangle = \frac{1}{Z} e^{-E_n/k_B T}\langle E_m|\, E_n\rangle = 0. \tag{22.105}$$

The last equality comes from the orthogonality of the energy eigenstates. Since all the off-diagonal density matrix elements are zero, the system in the thermal equilibrium case has no coherence whatsoever. These thermal equilibrium quantum states are used in nongate quantum computers using what is called quantum annealing.

22.10 Chapter Summary

- When a quantum system has a mix (an ensemble) of pure quantum states, pure-state vectors can no longer describe the properties of this system. However, density operators and density matrices can predict the results of measurements on mixed-state systems.
- For two-qubit pure states, averaging the system's density operator over one subsystem using the partial trace method leaves the other subsystem described by a mixed-state density operator.
- If a two-qubit system is described by a maximally entangled state ($s = 1$), the reduced density operators for the two qubits are the same.
- For two-qubit states that are not maximally entangled, the reduced density operators are not the same and measurements can distinguish between them.
- Mixed-state density operators may be used as models of the effects of the interaction between the quantum system and its environment.

- For a single qubit spin-1/2 system, a mixed state may be represented by a point that falls inside the Bloch sphere picture for this system.
- Three-qubit systems have more complicated density matrices and several types of mixed states. Nevertheless, starting with a pure state for the composite system, averaging over the states of two of the qubits, leaves the third qubit described by a mixed-state density matrix. The mixed-state density matrix depends on the details of the composite-system pure state.
- The Hamiltonian (energy-conserving) time evolution of a density matrix is described by a differential equation involving the density operator and the Hamiltonian operator for the system.
- Using a partition function, we may describe the density operator for a state in thermal equilibrium with an environment at temperature T. That corresponding matrix has all off-diagonal equal to zero.

22.11 Further Reading

Almost all upper-level undergraduate quantum mechanics texts discuss the density operator methods but few, if any, discuss the methods in terms of quantum information science.

Kuzyk (2023) suggests several ways of introducing density operators to make the connections with classical statistical ensembles more transparent. M. G. Kuzyk, "The emergence of classical mixtures from an entangled quantum state." *American Journal of Physics*, **91**(10), 858–860 (2023). https://doi.org/10.1119/5.0063636

Chang et al. (2024) describe several variations on the Bloch disk and the Bloch sphere to provide visualizations of the distinctions between states with different degrees of entanglement and between different kinds of mixed states. However, the results are applicable only to state vectors with real number amplitudes. L.-H. H Chang, S. Roccaforte, X. Zu, and P. Cadden-Zimansky, "Geometric visualizations of single and entangled qubits," *American Journal of Physics*, **92**(7), 528–537 (2024). https://doi.org/10.1119/5.0137901

A treatment of density operators and density matrices in a quantum computing and quantum information context can be found in P. Y. Lee, H. Ji, and Ran Cheng, *Quantum Computing & Information: A Scaffolding Approach* (Polaris QCI Publishing, Middletown, NY, 2024).

23 Quantum Perturbation Theory

A great perturbation in nature, to receive at once the benefit of sleep, and do the effects of watching.

Shakespeare, *Macbeth*

23.1 Introduction

In quantum mechanics, just as in every other field of science and engineering, the number of problems we can solve exactly is quite limited. In introductory physics, for example, to make a problem solvable by elementary calculus often requires we ignore effects such as air resistance and friction. We can make some progress in solving more complicated problems by using computational numerical methods, but these methods have their own limitations as well.

Rather than just give up on more complex problems, a better strategy is to start with a problem you can solve exactly and then use a general method called perturbation theory to make reasonable estimates of effects not included in the exactly solvable problem. For example, in calculating the motion of the planets in our solar system, we might first calculate the motion of each of the planets in its orbit around the Sun using Newton's laws of motion and Newton's law of gravity, which expresses the gravitational force that the Sun exerts on the planets. In this process, we first ignore the (hopefully, small) gravitational interactions between various planets. We then take into account these interplanetary gravitational interactions as small corrections to the orbits calculated without them. For our solar system, this method works quite well, and we are able to predict the motion of the planets far into the future (or in the past, to account for the alignment of planets that were noticed historically, for example).

Quantum mechanics uses a similar strategy. It turns out that it is relatively easy to calculate the energy eigenvalues for a hydrogen atom (one electron and one proton) using Coulomb's law for the electrical attraction between the two particles and the Schrödinger equation from quantum mechanics. But that calculation ignores the interactions involving the magnetic dipoles of the electron and the proton and the effects of Einstein's special theory of relativity, both of which cause small, but measurable, changes in those eigenvalues. To take these effects into account, we use what is called quantum perturbation theory, which is analogous to classical perturbation theory used for planetary motion.

Quantum Computing and Quantum Physics. Alice Flarend and Robert Hilborn, Oxford University Press.
 DOI: 10.1093/oso/9780197904930.003.0023

In this chapter, we use some simple examples to illustrate how perturbation theory works in quantum mechanics. The full theory is rather complicated and treated mostly in upper undergraduate- and graduate-level quantum mechanics courses. To get the flavor of quantum perturbation theory, we look at two cases: (1) a situation in which a time-independent interaction causes a shift in energy eigenvalues and (2) a case in which time-dependent effects cause transitions between energy eigenstates. Both situations have applications in quantum computing (QC) and quantum information science (QIS).

Perturbation theory mainly appears in describing the physics of the qubits that make up the QC or QIS system of interest. It rarely, if ever, enters the construction of quantum algorithms. However, perturbation theory is common in broader discussions of QC and QIS; so, it is important to be familiar with the basic concepts.

23.2 Time-Independent Quantum Perturbation Theory

We first develop the theory of perturbations that do not depend on time. We start with a Hamiltonian operator $H^{(0)}$ whose energy eigenvalues and eigenvectors we know. The superscript (0) reminds us that this operator has eigenvalues and eigenvectors considered to be the "zeroth" approximation to the problem we want to solve. We focus on energy eigenvectors and eigenvalues because finding corrections to those eigenvectors and eigenvalues is the primary goal of most quantum perturbation applications. Once we have these corrections to the energy eigenvectors, we can use the corrected eigenvectors as basis states to see their effects on other quantum properties of the system.

In the standard form of quantum perturbation theory (see, for example, McIntyre's *Quantum Mechanics* in Further Reading), the operator, the known eigenvectors, labeled $\left|\phi_n^{(0)}\right\rangle$, and the eigenvalues $E_n^{(0)}$ are related by the time-independent Schrödinger equation

$$H^{(0)}\left|\phi_n^{(0)}\right\rangle = E_n^{(0)}\left|\phi_n^{(0)}\right\rangle. \tag{23.1}$$

The subscript n is a label for the energy eigenvalues with the convention that $E_m^{(0)} < E_n^{(0)}$ if $m < n$.

The more complete problem is specified by a "perturbation" Hamiltonian operator $H^{(1)}$. Consequently, the Schrödinger equation we want to solve is

$$\left(H^{(0)} + \lambda H^{(1)}\right)|\phi_n\rangle = E_n|\phi_n\rangle. \tag{23.2}$$

The superscript (1) turns out to be a useful bookkeeping choice as discussed shortly. The parameter λ is another useful bookkeeping tool, which disappears from the final results.

Unfortunately, in most cases Eq. (23.2) cannot be solved exactly. That is where quantum perturbation theory comes in. The strategy is to assume that the energy eigenvalues in Eq. (23.2) can be written as a sum of corrections to the 0th order energy eigenvalues:

$$E_n = E_n^{(0)} + \lambda E_n^{(1)} + \lambda^2 E_n^{(2)} + \cdots. \tag{23.3}$$

The superscripts help us keep track of the "order" of the corrections. In other words, $E_n^{(1)}$ is called the first-order correction to the energy eigenvalue, $E_n^{(2)}$ is the second-order correction, and so on. Similarly, we assume that the eigenvectors in Eq. (23.2) can be written as a sum of corrections to the original (known) eigenvectors

$$|\phi_n\rangle = \left|\phi_n^{(0)}\right\rangle + \lambda \left|\phi_n^{(1)}\right\rangle + \cdots. \tag{23.4}$$

We now display the results of quantum perturbation theory and provide some physical interpretation before working through the derivation of these results in the following section. The first-order correction to the energy eigenvalue turns out to be

$$E_n^{(1)} = \left\langle \phi_n^{(0)} \right| H^{(1)} \left|\phi_n^{(0)}\right\rangle \equiv H_{nn}^{(1)}, \tag{23.5}$$

the average (expectation) value of the perturbation Hamiltonian operator with respect to the *original* energy eigenvectors.

The first-order correction to the state vector is

$$\left|\phi_n^{(1)}\right\rangle = \sum_{m \neq n} \frac{\left\langle \phi_m^{(0)} \right| H^{(1)} \left|\phi_n^{(0)}\right\rangle}{E_n^{(0)} - E_m^{(0)}} \left|\phi_m^{(0)}\right\rangle. \tag{23.6}$$

The second-order correction to the energy eigenvalue is

$$E_n^{(2)} = \sum_{m \neq n} \frac{\left|\left\langle \phi_m^{(0)} \right| H^{(1)} \left|\phi_n^{(0)}\right\rangle\right|^2}{E_n^{(0)} - E_m^{(0)}}. \tag{23.7}$$

Let's give simple physical explanations of Eqs. (23.5)–(23.7). Note that Eq. (23.5) shows that the first-order correction to the energy eigenvalue for the nth state is just the average (expectation) value of the perturbation Hamiltonian operator, which expresses an interaction of the quantum system that results in this energy shift. The "corrected" energy is then just $E_n \approx E_n^{(0)} + E_n^{(1)} = E_n^{(0)} + H_{nn}^{(1)}$.

In a similar way, Eq. (23.6) shows that the correction to the energy eigenvector is expressed as a weighted sum of all the original eigenvectors (except for the one associated with the nth state). The weighting factor is the matrix element $H_{mn}^{(1)}$ of the perturbing Hamiltonian connecting the mth eigenstate to the nth eigenstate divided by the original energy difference between these two states. Generally, the states that are closer in energy to $E_n^{(0)}$ are weighted more heavily than states with larger energy separations. Using Eq. (23.6) in Eq. (23.4) gives the corrected energy eigenvector for the nth state as $|\phi_n\rangle \approx \left|\phi_n^{(0)}\right\rangle + \left|\phi_n^{(1)}\right\rangle$.

Finally, the second-order energy eigenvalue correction is a weighted sum of the absolute-value-squared of the matrix elements of the perturbing Hamiltonian that connects the nth state to all the other energy eigenstates, divided by the energy differences between these states. Note that the numerator of the terms in Eq. (23.7) has the dimensions of energy-squared while the denominator's dimensions are energy, thus giving terms with dimensions of energy—just what we would expect for energy corrections. Further note that if $E_m^{(0)} < E_n^{(0)}$, the mth contribution to the energy correction will be positive and will increase the energy eigenvalue, while the reverse is true if $E_m^{(0)} > E_n^{(0)}$.

As shown later, it is straightforward, but algebraically tedious, to develop yet higher-order corrections. However, for many practical purposes, the results in Eqs. (23.5)–(23.7) are sufficient.

23.3 Derivation of the Time-Independent Quantum Perturbation Theory Expressions

We now turn to the derivation of the results presented in the previous section. We begin by using Eqs. (23.3) and (23.4) in Eq. (23.2) to obtain

$$\begin{aligned}&\left[H^{(0)} + \lambda H^{(1)}\right]\left[\left|\phi_n^{(0)}\right\rangle + \lambda\left|\phi_n^{(1)}\right\rangle + \cdots\right] \\ &= \left[E_n^{(0)} + \lambda E_n^{(1)} + \lambda^2 E_n^{(2)} + \cdots\right]\left[\left|\phi_n^{(0)}\right\rangle + \lambda\left|\phi_n^{(1)}\right\rangle + \cdots\right].\end{aligned} \tag{23.8}$$

Next, we expand the products in Eq. (23.8) and group terms according to the powers of λ. Since λ is an adjustable parameter, the terms involving the same power of λ on the left and right of Eq. (23.8) must be equal. For example, for terms that are independent of λ, we obtain

$$H^{(0)}\left|\phi_n^{(0)}\right\rangle = E_n^{(0)}\left|\phi_n^{(0)}\right\rangle, \tag{23.9}$$

which is just the original Schrödinger equation (Eq. (23.1)).

For those terms that involve λ to the first power, we have

$$H^{(1)}\left|\phi_n^{(0)}\right\rangle + H^{(0)}\left|\phi_n^{(1)}\right\rangle = E_n^{(0)}\left|\phi_n^{(1)}\right\rangle + E_n^{(1)}\left|\phi_n^{(0)}\right\rangle. \tag{23.10}$$

Note that superscripts in each term sum to 1, a nice mnemonic. The terms involving λ^2 give

$$H^{(0)}\left|\phi_n^{(2)}\right\rangle + H^{(1)}\left|\phi_n^{(1)}\right\rangle = E_n^{(0)}\left|\phi_n^{(2)}\right\rangle + E_n^{(1)}\left|\phi_n^{(1)}\right\rangle + E_n^{(2)}\left|\phi_n^{(0)}\right\rangle. \tag{23.11}$$

Note that the superscripts in each term sum to 2.

We now introduce an important assumption: We assume that each first-order correction to the energy eigenvectors can be written as a linear superposition of the original energy eigenvectors

$$\left|\phi_n^{(1)}\right\rangle = \sum_{m\neq n} c_{nm}\left|\phi_m^{(0)}\right\rangle. \tag{23.12}$$

The coefficients c_{nm} will be determined shortly. In linear algebra language, we are assuming that the original energy eigenvectors continue to span the state space, even with the presence of the perturbation Hamiltonian operator $H^{(1)}$. Note that the sum excludes the basis state $\left|\phi_n^{(0)}\right\rangle$ because it is helpful to make the first-order correction $\left|\phi_n^{(1)}\right\rangle$ to the state vector orthogonal to the original energy eigenvector $\left|\phi_n^{(0)}\right\rangle$. Later, we do the same for the second-order correction to the state vector.

Using Eq. (23.12) in Eq. (23.10) and rearranging terms yields

$$\left[H^{(0)} - E_n^{(0)}\right]\sum_m c_{nm}\left|\phi_m^{(0)}\right\rangle = \left[E_n^{(1)} - H^{(1)}\right]\left|\phi_n^{(0)}\right\rangle. \tag{23.13}$$

On the left of Eq. (23.13), we note that $H^{(0)}$ acts on each term in the sum to give the corresponding eigenvalue multiplying its corresponding eigenvector. Using this fact, we find that Eq. (23.13) becomes

$$\left[H^{(0)} - E_n^{(0)}\right]\sum_m c_{nm}\left|\phi_m^{(0)}\right\rangle = \sum_m c_{nm}\left(E_m^{(0)} - E_n^{(0)}\right)\left|\phi_m^{(0)}\right\rangle = \left[E_n^{(1)} - H^{(1)}\right]\left|\phi_n^{(0)}\right\rangle. \tag{23.14}$$

Try It 23.1 Go through the algebra that leads from Eq. (23.10) to Eq. (23.14).

We left multiply both sides of Eq. (23.14) by $\left\langle\phi_j^{(0)}\right|$ and recall that the original eigenvectors corresponding to different eigenvalues are orthogonal $\left\langle\phi_j^{(0)}\middle|\phi_m^{(0)}\right\rangle = \delta_{jm}$. Only one term of the sum on the left survives, the one with $j = m$. Then, Eq. (23.14) becomes

$$c_{nj}\left(E_j^{(0)} - E_n^{(0)}\right) = E_n^{(1)}\delta_{jn} - \left\langle\phi_j^{(0)}\right|H^{(1)}\left|\phi_n^{(0)}\right\rangle. \tag{23.15}$$

In drawing conclusions from Eq. (23.15), there are two cases to consider: (1) with $j = n$ and (2) with $j \neq n$. For the first case, Eq. (23.15) becomes

$$E_n^{(1)} = \left\langle\phi_n^{(0)}\right|H^{(1)}\left|\phi_n^{(0)}\right\rangle, \tag{23.16}$$

which is the result presented in Eq. (23.5): The first-order correction to the energy eigenvalue is simply the average (expectation) value of the perturbing Hamiltonian operator with respect to the original energy eigenstates.

The second case ($j \neq n$) gives us the first-order correction to the energy eigenvector. In this case, Eq. (23.15) becomes

$$c_{nj} = \frac{\left\langle\phi_j^{(0)}\right|H^{(1)}\left|\phi_n^{(0)}\right\rangle}{\left(E_n^{(0)} - E_j^{(0)}\right)}. \tag{23.17}$$

We now set $j = m$ and use Eq. (23.17) in Eq. (23.12) to obtain

$$\left|\phi_n^{(1)}\right\rangle = \sum_{m\neq n}\frac{\left\langle\phi_m^{(0)}\right|H^{(1)}\left|\phi_n^{(0)}\right\rangle}{\left(E_n^{(0)} - E_m^{(0)}\right)}\left|\phi_m^{(0)}\right\rangle, \tag{23.18}$$

which is the result stated previously in Eq. (23.6).

Try It 23.2 Show how Eq. (23.18) follows form Eq. (23.16).

Eq. (23.18) shows why perturbation theory can be difficult to apply. Not only do we need to know the original eigenvectors and eigenvalues, we must also calculate the matrix elements of the perturbing Hamiltonian $H^{(1)}$. Furthermore, we need to carry out the sum over all the original eigenvectors. Often, the terms in the sum decrease in magnitude because of the increase in the difference between the energy eigenvalues that appear in the denominator; so, we may not need to include more than a few terms in the sum to get reasonable numerical accuracy. It is hard to make generalizations because the details of the matrix elements and energy differences depend on the specific physical system being described.

Note that we need to be careful about what happens when $E_n^{(0)} = E_m^{(0)}$ because that would lead to a zero in the denominator of Eq. (23.18)—definitely, a mathematical "no–no." When the energy eigenvalue differences are zero, they are said to be "degenerate." Fortunately, it is not too difficult to work out what is called "degenerate perturbation theory" to cover these cases. If you encounter a degenerate eigenvalue case, consult any standard upper-level textbook on quantum mechanics to find out what to do. The essential idea is that you identify the two states whose 0th-order energy eigenvalues are equal and treat them as a two-level system, whose eigenvalues and eigenvectors you want to find with the perturbing Hamiltonian included.

We now return to the remainder of the derivation. The parts that involve λ^2 are found in Eq. (23.11). We define the second-order correction to the energy eigenvectors as a superposition of the original energy eigenvectors:

$$\left|\phi_n^{(2)}\right\rangle = \sum_{\ell \neq n} d_{n\ell} \left|\phi_\ell^{(0)}\right\rangle. \tag{23.19}$$

Just like our treatment of the first-order eigenvector correction, we exclude the eigenvector $\left|\phi_n^{(0)}\right\rangle$ from the sum to make the second-order correction orthogonal to the original eigenvector.

We rearrange Eq. (23.11) according to the order of the corrections to the eigenvectors to obtain

$$\left(H^{(0)} - E_n^{(0)}\right)\left|\phi_n^{(2)}\right\rangle = \left(E_n^{(1)} - H^{(1)}\right)\left|\phi_n^{(1)}\right\rangle + E_n^{(2)}\left|\phi_n^{(0)}\right\rangle. \tag{23.20}$$

Using Eqs. (23.12) and (23.19) in Eq. (23.20) yields

$$\begin{aligned} H^{(0)} \sum_{\ell \neq n} d_{n\ell} \left|\phi_\ell^{(0)}\right\rangle - E_n^{(0)} \sum_{\ell \neq n} d_{n\ell} \left|\phi_\ell^{(0)}\right\rangle &= E_n^{(1)} \sum_{m \neq n} c_{nm} \left|\phi_m^{(0)}\right\rangle - H^{(1)} \sum_{m \neq n} c_{nm} \left|\phi_m^{(0)}\right\rangle \\ &\quad + E_n^{(2)}\left|\phi_n^{(0)}\right\rangle. \end{aligned} \tag{23.21}$$

Since $H^{(0)}\left|\phi_\ell^{(0)}\right\rangle = E_\ell^{(0)}\left|\phi_\ell^{(0)}\right\rangle$, we may write Eq. (23.21) as

$$\sum_{\ell\neq n} d_{n\ell}E_\ell^{(0)}\left|\phi_\ell^{(0)}\right\rangle - \sum_{\ell\neq n} d_{n\ell}E_n^{(0)}\left|\phi_\ell^{(0)}\right\rangle = E_n^{(1)}\sum_{m\neq n} c_{nm}\left|\phi_m^{(0)}\right\rangle - H^{(1)}\sum_{m\neq n} c_{nm}\left|\phi_m^{(0)}\right\rangle + E_n^{(2)}\left|\phi_n^{(0)}\right\rangle. \tag{23.22}$$

Left multiply both sides of Eq. (23.22) by $\left\langle\phi_n^{(0)}\right|$. We see that the left side of Eq. (23.22) will be zero because $\left\langle\phi_n^{(0)}\middle|\phi_\ell^{(0)}\right\rangle = 0$ since $\ell \neq n$ in the sum. Similarly, the first term on the right of Eq. (23.22) will be zero for the same reason. Using c_{mn} from Eq. (23.17) leaves us with

$$E_n^{(2)} = \sum_{m\neq n} c_{nm}\left\langle\phi_n^{(0)}\right|H^{(1)}\left|\phi_m^{(0)}\right\rangle = \sum_{m\neq n}\frac{\left|H_{mn}^{(1)}\right|^2}{E_n^{(0)} - E_m^{(0)}}, \tag{23.23}$$

in agreement with the previously stated result in Eq. (23.7). Again, we see that the degenerate eigenvalue case will require special attention. The previous remarks about dealing with the sum of matrix elements and energy differences apply here as well.

Try It 23.3 Derive Eq. (23.23) from Eq. (23.20). Explain your reasoning.

23.4 Time-Independent Perturbation Theory Example

To show how we make use of quantum perturbation theory, we consider a perturbation of the simple harmonic oscillator (SHO) system described in Chapter 21. This example, like all uses of perturbation theory, involves evaluating the matrix elements that occur in Eqs. (23.5)–(23.7) and then carrying out the sums in those equations.

We evaluate the effects of the following perturbation Hamiltonian:

$$H^{(1)} = kx^4, \tag{23.24}$$

where k is a parameter whose dimensions are energy/(distance)^4. This perturbation Hamiltonian is often called a "quartic perturbation" because it involves the coordinate x to the 4th power. It is commonly used in the theory of elastic materials and quantum field theory (elementary particle theory).

In the coordinate basis, the operator x is replaced by the coordinate x. Figure 23.1 shows a plot of the SHO potential energy as a function of the coordinate x without the perturbation term and with the perturbing Hamiltonian included. Because the perturbation Hamiltonian is proportional to x^4, it will have larger effects as x increases.

To find the first-order energy eigenvalue correction, we evaluate the average value of the perturbation Hamiltonian with respect to the $n = 0$ (ground) state of the SHO. We perform

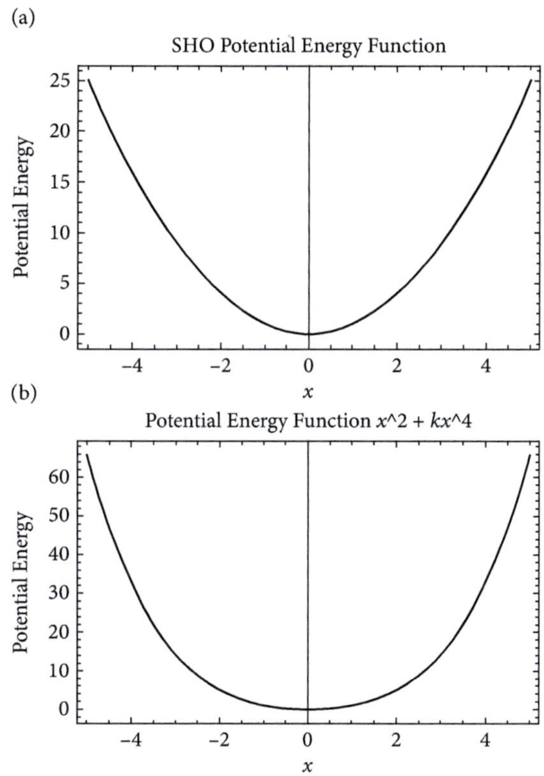

Fig. 23.1 In the graph part (a), we plot the potential energy function for the simple harmonic oscillator (SHO) with no perturbation. In the graph part (b), we plot the combined potential energy function for the SHO perturbed by the Hamiltonian in Eq. (23.24) with $k = 0.065$.

the calculation by two ways. First, we use the SHO ground-state wave function expressed as a function of the coordinate x. Then, we carry out the calculation for other SHO states using the ladder operators from Chapter 21.

For the wave function method, we need to evaluate the integral associated with the Dirac bracket quantum sandwich:

$$E^{(1)}_{n=0} = \left\langle \phi^{(0)}_{n=0} \right| kx^4 \left| \phi^{(0)}_{n=0} \right\rangle = \int_{-\infty}^{+\infty} dx\, kx^4 \left(\phi^{(0)}_{n=0}(x) \right)^2 . \tag{23.25}$$

From Chapter 21, we know that the normalized ground state wave function in the coordinate representation for the unperturbed SHO is

$$\phi^{(0)}_{n=0} = \left(\frac{m\omega}{\pi\hbar} \right)^{1/4} \exp\left[-\frac{m\omega x^2}{2\hbar} \right], \tag{23.26}$$

where m is the oscillator's mass, ω is its oscillation frequency, and $\hbar$ is the usual reduced Planck's constant. Using Eq. (23.26) in Eq. (23.25) yields

$$E^{(1)}_{n=0} = \left(\frac{m\omega}{\pi\hbar}\right)^{1/2} \int_{-\infty}^{+\infty} dx\, kx^4 \exp\left[-\frac{m\omega x^2}{\hbar}\right]. \tag{23.27}$$

This integral may be intimidating, but you can find a related integral in math tables on the web $\int_{-\infty}^{+\infty} dx\, x^4 \exp\left[-bx^2\right] = \frac{3}{4}\sqrt{\frac{\pi}{b^5}}$. Setting $b = m\omega/\hbar$, we find

$$E^{(1)}_{n=0} = \frac{3k}{4}\left(\frac{\hbar}{m\omega}\right)^2. \tag{23.28}$$

To see if this result makes sense, note that $\hbar/(m\omega)$ has dimensions of distance-squared and recall that k has dimensions of energy/(distance^4). The overall results should have dimensions of energy. Note that the first-order energy correction is proportional to k.

Math Aside There is a nice "trick" that allows you to calculate integrals like the one in Eq. (23.27) fairly easily. The idea is to start with the well-known integral $\int_{-\infty}^{+\infty} e^{-bx^2} dx = \sqrt{\pi/b}$ for $b > 0$. (What happens if $b \le 0$?) The trick is to differentiate both sides of that result with respect to the parameter b. For example,

$$\frac{d}{db}\int_{-\infty}^{+\infty} e^{-bx^2} dx = \int_{-\infty}^{+\infty} (-x^2)e^{-bx^2} dx = \frac{d}{db}\sqrt{\pi/b} = -\frac{\sqrt{\pi}}{2}b^{-3/2}$$

From which we have $\int_{-\infty}^{+\infty} x^2 e^{-bx^2} dx = (1/2)\sqrt{\pi/b^3}$. Successive derivatives give integrals of increasing powers of x^2. Why don't we worry about these types of integrals with the exponential multiplied by an odd power of x?

The second calculational method uses the ladder operators. From Chapter 21, we know that the x operator can be expressed as a sum of the lowering and raising operators. In what follows, we use the ^ symbol to indicate operators to distinguish the operator (e.g., $\hat{x}$) from the coordinate value (e.g., x). The x operator is

$$\hat{x} = \left(\frac{\hbar}{2m\omega}\right)^{1/2} (\hat{a} + \hat{a}^\dagger) \tag{23.29}$$

and consequently

$$\hat{x}^4 = \frac{1}{4}\left(\frac{\hbar}{m\omega}\right)^2 \left(\hat{a} + \hat{a}^\dagger\right)^4. \tag{23.30}$$

To evaluate the expectation value of $H^{(1)} = k\hat{x}^4$, we need to calculate

$$E^{(1)}_n = \frac{k}{4}\left(\frac{\hbar}{m\omega}\right)^2 \langle n| \left(\hat{a} + \hat{a}^\dagger\right)^4 |n\rangle. \tag{23.31}$$

This calculation is not as bad as it may seem because it turns out that many of the quantum sandwich contributions are zero. Moreover, we can find the result for any SHO energy eigenstate (not just the ground state) without knowing the wave functions for these states.

We begin by evaluating the sum of the ladder operators raised to the 4th power. Since these operators do not commute, we need to be careful about the order in which the operators appear. Here is the calculation:

$$\begin{aligned}(\hat{a} + \hat{a}^\dagger)^4 &= (\hat{a} + \hat{a}^\dagger)^2(\hat{a} + \hat{a}^\dagger)^2 \\ &= (\hat{a}^2 + \hat{a}\hat{a}^\dagger + \hat{a}^\dagger\hat{a} + \hat{a}^{\dagger 2})(\hat{a}^2 + \hat{a}\hat{a}^\dagger + \hat{a}^\dagger\hat{a} + \hat{a}^{\dagger 2}) \\ &= \hat{a}^4 + \hat{a}^3\hat{a}^\dagger + \hat{a}^2\hat{a}^\dagger\hat{a} + \hat{a}^2\hat{a}^{\dagger 2} \\ &\quad + \hat{a}\hat{a}^\dagger\hat{a}^2 + \hat{a}\hat{a}^\dagger\hat{a}\hat{a}^\dagger + \hat{a}\hat{a}^\dagger\hat{a}^\dagger\hat{a} + \hat{a}^\dagger\hat{a}\hat{a}^{\dagger 2} \\ &\quad + \hat{a}^\dagger\hat{a}^3 + \hat{a}^\dagger\hat{a}\hat{a}\hat{a}^\dagger + \hat{a}^\dagger\hat{a}\hat{a}^\dagger\hat{a} + \hat{a}^\dagger\hat{a}\hat{a}^{\dagger 2} \\ &\quad + \hat{a}^{\dagger 2}\hat{a}^2 + \hat{a}^{\dagger 2}\hat{a}\hat{a}^\dagger + \hat{a}^{\dagger 3}\hat{a} + \hat{a}^{\dagger 4}.\end{aligned} \tag{23.32}$$

Note that each of the 16 terms on the right of Eq. (23.32) is comprised of four operator factors.

We now evaluate the average value of this sum of operators. To get a nonzero contribution, we must have an equal number of *a*s and $a^\dagger$s. For example, $\langle n = 0|\hat{a}^{\dagger 4}|n = 0\rangle \propto \langle n = 0|n = 4\rangle = 0$ because of the orthogonality of the SHO energy eigenvectors. There are just 6 (out of 16) terms that satisfy this criterion. These six matrix elements can be evaluated by the repeated use of $\hat{a}|n\rangle = \sqrt{n}|n-1\rangle$ and $\hat{a}^\dagger|n\rangle = \sqrt{n+1}|n+1\rangle$. For example,

$$\begin{aligned}\langle n|\hat{a}\hat{a}\hat{a}^\dagger\hat{a}^\dagger|n\rangle &= \sqrt{n+1}\langle n|\hat{a}\hat{a}\hat{a}^\dagger|n+1\rangle = \sqrt{n+1}\sqrt{n+2}\langle n|\hat{a}\hat{a}|n+2\rangle \\ &= \sqrt{n+1}\sqrt{n+2}\sqrt{n+2}\langle n|\hat{a}|n+1\rangle \\ &= \sqrt{n+1}\sqrt{n+2}\sqrt{n+2}\sqrt{n+1}\langle n|n\rangle = (n+1)(n+2).\end{aligned} \tag{23.33}$$

Figure 23.2 shows a nice graphical method to keep track of the evaluation of the matrix elements. The evaluation starts on the right with an operator acting on the state vector $|n\rangle$. The numerical factor that results from this operation is placed next to the line going to the state produced by that operation. The resulting states (e.g., $|n+1\rangle$) are indicated next to the corresponding filled circle. The overall numerical factor for the matrix element is the *product* of the individual factors indicated in the diagram. In particular, for Fig. 23.2, the factors are $\sqrt{n+1}\sqrt{n+2}\sqrt{n+2}\sqrt{n+1} = (n+1)(n+2)$.

Table 23.1 lists the six operator products that give nonzero results for $\langle n|H^{(1)}|n\rangle$.

To check the results in Table 23.1, we evaluate the total for $n = 0$ (the ground state of the system), which we can compare to the results of the wave function contribution. From the table, we get the total contribution for $n = 0$ as 3, which when used in Eq. (23.31) yields

$$E^{(1)}_{n=0} = \frac{3}{4}k\left(\frac{\hbar}{m\omega}\right)^2 \tag{23.34}$$

in agreement with Eq. (23.28) from the wave function approach.

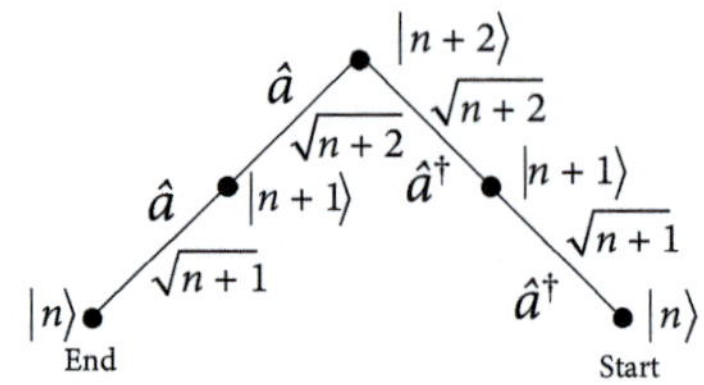

Fig. 23.2 A graphical representation of the set of operations embodied in Eq. (23.33). The filled circles indicate the states. The evaluation starts on the right and ends on the left. The numerical terms between the states and to the right of the lines indicate the result of applying the operator to the filled circle state. The operators are shown to the left of the states on which they operate.

Table 23.1 Six operator products that provide nonzero results.

Operator product	$\langle n \vert H^{(1)} \vert n \rangle$ contribution
$\hat{a}^2\hat{a}^{\dagger 2}$	$(n+1)(n+2)$
$\hat{a}^{\dagger 2}\hat{a}^2$	$n(n-1)$
$\hat{a}\hat{a}^\dagger\hat{a}\hat{a}^\dagger$	$(n+1)^2$
$\hat{a}^\dagger\hat{a}\hat{a}^\dagger\hat{a}$	n^2
$\hat{a}^\dagger\hat{a}\hat{a}\hat{a}^\dagger$	$n(n+1)$
$\hat{a}\hat{a}^\dagger\hat{a}^\dagger\hat{a}$	$n(n+1)$
Total of the contributions	$6n^2 + 6n + 3$

Try It 23.4 Evaluate all the other operator products in Table 23.1 and verify the entries in the second column.

To see how the perturbation term affects the energy eigenvalues, we use

$$E_n \approx E_n^{(0)} + E_n^{(1)} = n\hbar\omega + \frac{3}{4}k\left(\frac{\hbar}{m\omega}\right)^2 \left(6n^2 + 6n + 3\right). \tag{23.35}$$

Since $k(\hbar/m\omega)^2$ is an energy, let's replace it by $\varepsilon\hbar\omega$, where ε is a dimensionless number. We then write the energy eigenvalues as

$$E_n \approx E_n^{(0)} + E_n^{(1)} = \left[n + \frac{3}{4}\varepsilon\left(6n^2 + 6n + 3\right)\right]\hbar\omega. \tag{23.36}$$

We see that including the x^4 perturbation term means that the energy eigenvalues are no longer equally spaced. The system is now an "anharmonic" oscillator. As n increases, the separation between adjacent energy states increases.

Try It 23.5 Use Eq. (23.36) to calculate $E_{n+1} - E_n$. When this difference becomes much larger than $\hbar\omega$, the first-order correction to the energy eigenvalues is no longer an adequate approximation and we would need to include the second-order energy correction using Eq. (23.23).

We could also use the ladder operator method to find the first-order correction to the unperturbed state vectors and the second-order correction for the energy. For these results, we need to evaluate "nondiagonal" matrix elements $\langle m| H^{(1)} |n\rangle$ to which all 16 operator products in Eq. (23.32) contribute—a straightforward but tedious calculation. These results then need to be used in Eqs. (23.6) and (23.7). The energy differences in the denominators are simply $(n - m)\hbar\omega$. We would still need to evaluate the sums. We leave the details as exercises for the reader.

In the state vector correction equation, the numerators' largest term will be proportional to n^2 while the denominators have only a factor of n. Mathematically, this shows that the sum does not converge. This is a fairly common result in perturbation theory, indicating a limit to its applicability. The situation is even worse for the second-order energy corrections that have n^4 terms in the numerator and terms proportional to n in the denominator. However, if the perturbation Hamiltonian is small enough (due to the smallness of the parameter k or its equivalent), then the results may still give us reasonable approximations to various correction terms.

23.5 Time-Dependent Perturbation Theory

We now consider how we carry out perturbation theory if the perturbing part of the Hamiltonian depends on time:

$$H = H^{(0)} + \lambda H^{(1)}(t). \tag{23.37}$$

We assume that the unperturbed system's quantum states are the energy eigenvectors of $H^{(0)}$:

$$H^{(0)} \left|\phi_n^{(0)}\right\rangle = E_n^{(0)} \left|\phi_n^{(0)}\right\rangle \tag{23.38}$$

with the energy eigenvalues $E_n^{(0)}$. As before, we assume that the states with the perturbation present can be written as a linear superposition of the original eigenvectors

$$|\psi(t)\rangle = \sum_n c_n(t) \left|\phi_n^{(0)}\right\rangle. \tag{23.39}$$

We now use Eq. (23.39) in the time-dependent Schrödinger equation (Chapter 18, Eq. (18.22))

$$i\hbar\frac{d|\psi\rangle}{dt} = H|\psi\rangle$$

$$i\hbar\frac{d}{dt}|\psi(t)\rangle = i\hbar\sum_n\left[\dot{c}_n(t)\left|\phi_n^{(0)}(t)\right\rangle + c_n(t)\frac{d}{dt}\left|\phi_n^{(0)}\right\rangle\right] \tag{23.40}$$

$$= H^{(0)}\sum_n c_n(t)\left|\phi_n^{(0)}(t)\right\rangle + \lambda H_n^{(1)}(t)\sum_n c_n(t)\left|\phi_n^{(0)}(t)\right\rangle.$$

The original energy eigenvectors have the time dependence

$$\left|\phi_n^{(0)}(t)\right\rangle = e^{-iE_n^{(0)}t/\hbar}\left|\phi_n^{(0)}(0)\right\rangle. \tag{23.41}$$

So, the time derivative in the second line of Eq. (23.40) is

$$i\hbar\frac{d}{dt}\left|\phi_n^{(0)}(t)\right\rangle = E_n^{(0)}\left|\phi_n^{(0)}(t)\right\rangle. \tag{23.42}$$

When used in Eq. (23.40), these terms just cancel the first terms in the second line of Eq. (23.40). We now use the Dirac trick and left multiply both sides of Eq. (23.40) by $\left\langle\phi_k^{(0)}\right|$ and write

$$i\hbar\dot{c}_k(t) = \sum_n c_n(t)\left\langle\phi_n^{(0)}(t)\right|\lambda H^{(1)}(t)\left|\phi_n^{(0)}(t)\right\rangle. \tag{23.43}$$

These are a set of coupled first-order differential equations. So far, no approximations have been made.

In a typical application of time-dependent perturbation theory, we assume that the system starts off in one of the original energy eigenstates at time $t = 0$. Let's suppose that is the jth state, in which case we have $c_j(0) = 1$ and $c_{n\neq j}(0) = 0$. Of course, other initial states are possible, but this choice is relatively simple and illustrates the basic principles of time-dependent perturbation theory while minimizing the algebraic complexity. This choice of initial conditions means that for times close to $t = 0$ we may write

$$i\hbar\dot{c}_j(t) \approx c_j(0)\left\langle\phi_k^{(0)}(t)\right|\lambda H^{(1)}(t)\left|\phi_j^{(0)}(t)\right\rangle \tag{23.44}$$

for $k \neq j$.

We now assume that the perturbing Hamiltonian can be factored into two parts—one that depends on time and the other that may depend on position coordinates $\vec{r}$:

$$H^{(1)}(\vec{r}, t) = g^{(1)}(\vec{r})f(t). \tag{23.45}$$

This is similar to the Hamiltonian we used for magnetic resonance in Chapter 19 where $H = -\gamma S_z B_z \cos \omega t$. We now write the matrix elements on the right side of Eq. (23.44) as

$$\begin{aligned}\left\langle \phi_k^{(0)}(t)\right| \lambda H^{(1)}(t)\left|\phi_j^{(0)}(t)\right\rangle &= e^{iE_k^{(0)}t/\hbar}e^{-iE_j^{(0)}t/\hbar} f(t)\left\langle \phi_k^{(0)}(0)\right| g \left|\phi_j^{(0)}(0)\right\rangle \\ &= g_{kj}e^{i\omega_{kj}t} f(t),\end{aligned} \tag{23.46}$$

where we have introduced $\omega_{kj} = (E_k^{(0)} - E_j^{(0)})/\hbar$ and the matrix elements $g_{kj} = \left\langle \phi_k^{(0)}(0)\right| g \left|\phi_j^{(0)}(0)\right\rangle$.

Using these results in Eq. (23.44) and integrating with respect to time yields

$$c_k(t) \approx \frac{g_{kj}}{i\hbar}\int_0^t dt'\, e^{i\omega_{kj}t'} f(t') \tag{23.47}$$

for t relatively small so that the c coefficients do not change significantly. (This is where the "small perturbation" part comes in.)

We want to find the probability that if the system started in the state $\left|\phi_j^{(0)}\right\rangle$ at time $t = 0$, it will be found in the state $\left|\phi_k\right\rangle$ at time t. The Born rule shows that probability is given by

$$\text{Prob}\,(j \to k) = \left|c_k(t)\right|^2 = \frac{\left|g_{kj}\right|^2}{\hbar^2}\left|\int_0^t dt' f(t') e^{i\omega_{kj}t'}\right|^2. \tag{23.48}$$

This shows that the time-dependent perturbation can cause transitions among the original energy eigenstates.

We need to specify the time part of the perturbing Hamiltonian to solve Eq. (23.48). In many situations of interest to QC and QIS, the perturbation oscillates in time with a cosine time dependence

$$f(t) = \cos \omega t = \frac{1}{2}\left[e^{i\omega t} + e^{-i\omega t}\right], \tag{23.49}$$

where we used the Euler formula to replace the cosine with two complex exponentials. (The reason for this substitution will become apparent in the next step of the calculation.) Using Eq. (23.49) in Eq. (23.47), we find

$$\begin{aligned}c_k(t) &\approx \frac{g_{kj}}{i\hbar}\int_0^t dt'\, e^{i\omega_{kj}t'}\frac{1}{2}\left[e^{i\omega t'} + e^{-i\omega t'}\right] \\ &= \frac{g_{kj}}{2i\hbar}\left[\frac{e^{i(\omega_{kj}+\omega)t} - 1}{\omega_{kj}+\omega} + \frac{e^{i(\omega_{kj}-\omega)t} - 1}{\omega_{kj}-\omega}\right].\end{aligned} \tag{23.50}$$

Rather than finding the absolute-value-squared of Eq. (23.50), let's look at two cases: (1) $E_j^{(0)} < E_k^{(0)} \Rightarrow \omega_{kj} > 0$ and (2) $E_j^{(0)} > E_k^{(0)} \Rightarrow \omega_{kj} < 0$. In the first case, the first term inside the square brackets in the second line of Eq. (23.50) will be much smaller than the second term if $\omega \approx \omega_{kj}$ (taking $\omega > 0$ as usual). It means that the probability to go from a lower energy state to a higher energy state is large compared to the opposite. The quantum system is absorbing energy from the perturbation interaction. For case (2), the signs are reversed and the interaction with the perturbing Hamiltonian favors a transition from a higher energy state to a lower energy state. So, the quantum system is giving up energy to the perturbing interaction. We say that the perturbation causes (or induces) emission.

Let's work through the absorption case. Using Eq. (23.50) (with only the second term in square brackets) in Eq. (23.48) we find

$$\begin{aligned} \text{Prob}_{j\to k} = |c_k(t)|^2 &= \frac{|g_{kj}|^2}{\hbar^2} \frac{1 - e^{i(\omega_{kj}-\omega)t} - e^{-i(\omega_{kj}-\omega)t} + 1}{(\omega_{kj} - \omega)^2} \\ &= \frac{|g_{kj}|^2}{\hbar^2} \frac{4\sin^2\left((\omega_{kj} - \omega)t/2\right)}{(\omega_{kj} - \omega)^2}. \end{aligned} \tag{23.51}$$

We see that the probability is proportional to the absolute-value-squared of the matrix element of the spatial part of the perturbing Hamiltonian. It means that some states will be favored over others due to the state dependence of that matrix element, which provides a "selection rule" for possible transitions.

The time-dependent part shows that the probability oscillates in time just as we observed with magnetic resonance in Chapter 19. The denominator in Eq. (23.51) shows that the transition probability will be large if the perturbation frequency ω is close to the frequency ω_{kj} associated with the energy difference between the two states: a resonance effect just like the one we found for magnetic resonance.

Try It 23.6 Work through the algebra in Eq. (23.51). Explain your reasoning.

Try It 23.7 Go through the same calculation but for the emission case (when $E_j^{(0)} > E_k^{(0)} \Rightarrow \omega_{kj} < 0$). Explain your reasoning.

23.6 Chapter Summary

- We have shown that the low-order corrections to energy eigenvalues and eigenstates predicted by time-independent perturbation theory are given by the following three equations:

$$E_n^{(1)} = \langle \phi_n^{(0)} | H' | \phi_n^{(0)} \rangle = H_{nn}^{(1)}$$

$$\left|\phi_n^{(1)}\right\rangle = \sum_{m \neq n} \frac{\left\langle \phi_m^{(0)} \right| H^{(1)} \left| \phi_n^{(0)} \right\rangle}{E_n^{(0)} - E_m^{(0)}} \left|\phi_m^{(0)}\right\rangle$$

$$E_n^{(2)} = \sum_{m \neq n} \frac{\left|\left\langle \phi_m^{(0)} \right| H^{(1)} \left| \phi_n^{(0)} \right\rangle\right|^2}{E_n^{(0)} - E_m^{(0)}}$$

- We showed how to calculate the results of perturbation theory for a quartic x^4 perturbation of the simple harmonic operator potential energy function both by using an integral involving the SHO ground state and raising and lowering operators that gave us the results for any SHO energy state.
- Time-dependent quantum perturbation theory indicates that such perturbations can cause transitions among the system's quantum states. These transitions caused by the perturbation can be interpreted as absorption and emission of energy by the original system.

23.7 Further Reading

David McIntyre, *Quantum Mechanics: A Paradigms Approach* (Cambridge University Press, Cambridge, 2023)

C. Cohen-Tannoudji, B. Diu, and F. Laloë, *Quantum Mechanics* (Wiley, New York, 1977)

PART V

Looking to the Future

Present and Future QIS and QC

24

It's tough to make predictions, especially about the future.

Apparently, an old Danish proverb, but also attributed to Yogi Berra and Niels Bohr.

24.1 QC and QIS Overview

We want to congratulate you, our dedicated readers, on your persistence through a lot of challenging material. You should be proud of your work. Both Alice and Bob hope that you will continue your studies of quantum information science (QIS) and quantum computing (QC). In this chapter, we suggest several "next steps" you might take to deepen your understanding of QIS and QC. We also discuss programming quantum computers and developments such as the quantum internet. A quick survey of current QC and QIS technology will give you a sense of the broad range of skills deployed in those fields. We will also rub the quantum crystal ball to suggest areas of QIS and QC to keep an eye on over the next few years and some ethical issues that are important in QIS and QC. However, we need to remind ourselves that the fields of QC and QIS are evolving rapidly. Consequently, some of the detailed information provided in this chapter will soon be out-of-date.

QIS and QC are still in their beginning stages of development. Scientific journals are publishing new results in practically every issue. Governments, especially in the United States, Europe, and China, are pouring research funds into QIS and QC. Major technology companies, such as Microsoft, Google, and IBM, are also active in these areas. As of this writing (2026), Google, IonQ, IBM, and several other companies have QCs with 25–100 qubits. There is obviously both intellectual excitement in these developments and many possibilities for rewarding careers. QC and QIS are truly multidisciplinary fields requiring experts in physics, computer science, mathematics, engineering, materials science, electronics technology, and of course in business management to get the resources to make all this work and to use QC and QIS in projects that will benefit the world.

Let's get an overview of QC before discussing future directions for the field. A good place to start is the diagram of the components of a quantum computer system shown in Fig. 24.1. It starts with software and ends with hardware.

Level A in the diagram includes the development of quantum algorithms and quantum programming languages that users can employ to implement algorithms without worrying

Quantum Computing and Quantum Physics. Alice Flarend and Robert Hilborn, Oxford University Press.
 DOI: 10.1093/oso/9780197904930.003.0024

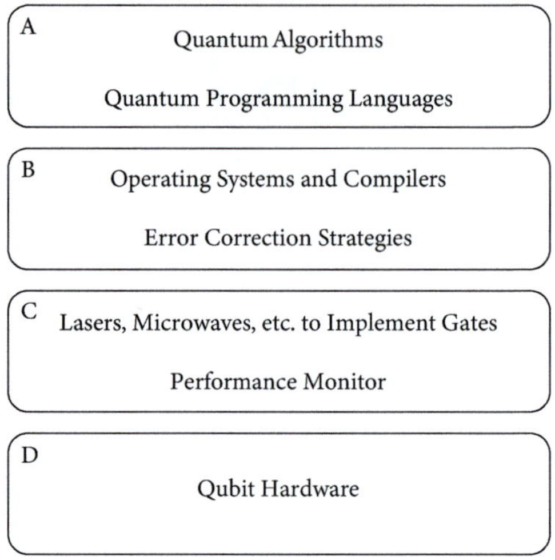

Fig. 24.1 A diagram of a quantum computer system.

about the details of the qubit hardware. This is similar to the situation with classical computers, where we use spreadsheets, word processors, and programming languages, such as Python, R, or C#, that are (we hope) understandable to humans, without needing to know anything about the actual computer hardware.

Layer B refers to the operating systems that coordinate the operations of the computers' physical components and the translation of the code written in the programming language to the code that the computer hardware itself understands. For classical computers, we have operating systems such as Unix, Linux, Windows, and iOS, which translate the high-level language commands into what the hardware understands: assembly language or machine language. In QC, layer B includes translating the code into qubit gate operations. We observed several examples of that translation in Chapters 12–16.

Layer C contains hardware that is set up more or less permanently to organize gate operations and carry out measurements. This hardware is called quantum firmware and is used to improve stability and reduce various qubit errors. There is a tremendous amount of engineering work required to develop firmware. That layer also carries out the error correction protocols like those mentioned in Chapter 13.

The bottom layer D is the actual qubit hardware—the physical machinery to prepare, process, and measure the qubits. More information about this is provided in Section 24.2.

The takeaway message is that making a working (and useful) QC requires many kinds of skills and many ways of thinking. Most QC users will probably focus on the top layer. However, a quantum algorithm or a QC program is useless if there is no machine on which to implement the program. To construct a QC, we need people to build the components, physically install the electrical wires, etc. It means that there are many QC and QIS employment opportunities in the trades too.

A way to gain experience with QCs is to use a classical computer to simulate a QC. The basic idea is that the gates acting on quantum states can be represented, as we have shown

Table 24.1 Examples of commercial companies producing quantum computers.

Company	Type of quantum computer
Alice & Bob	Cat qubit technology
Amazon	Superconducting circuits
D-Wave Systems	Quantum annealing
Google	Superconducting circuit
IBM	Superconducting circuit
Intel	Electron spins in silicon
IonQ	Trapped ions
Microsoft	Topological superconducting circuit
Quantum Computing, Inc.	Integrated photonics
Quantinuum	Trapped ions
Rigetti Computing	Superconducting circuits
Xanadu	Nanophotonics

For additional information, see J. Roundy in Further Reading.

again and again in this book, by matrices acting on column and row vectors. Most classical programming languages handle these calculations easily. The only tricky part is simulating quantum measurements. This can be performed by using computer-generated random numbers to produce measurement outcomes consistent with the probabilities encoded in the quantum states. Of course, this method does require some knowledge of computer programming, but it is not overly complicated. The Further Reading section shows citations to papers that illustrate how these simulations work. However, the simulations become very time-consuming as the number of qubits grows, and they will not be able to replace a QC when large numbers of qubits are required. After all, if they could, we would not need to build QCs. Nevertheless, the simulations will help test algorithms and build your intuition about how QCs operate (Table 24.1).

Web-Based Access to Quantum Computers

Amazon Braket https://aws.amazon.com/braket/

Google Cloud https://quantumai.google/cirq/google/concepts

IBM Quantum Composer https://quantum.ibm.com/composer/files/new

Intel Quantum SDK https://www.intel.com/content/www/us/en/developer/tools/quantum-sdk/overview.html

Microsoft Azure Quantum azure.microsoft.com

24.2 Physics of Qubits

In this section, we briefly describe the types of physical qubits being deployed in today's (as of 2026) quantum computers. The discussion is primarily qualitative because understanding the details requires substantial background in physics, materials science, and engineering. With the rapidly developing technology in QC, what we write here will most likely be out-of-date within a year or two. Moreover, as mentioned before, most QC users will not need to know these details, any more than the typical classical computer user needs to understand the details of the semiconductor devices used in their laptops and smartphones. Of course, the details are important if you want to build a QC. In fact, there are many fascinating and challenging issues to be addressed to build fully functioning quantum computers and, as mentioned before, a lot of great jobs in the labs and businesses that are designing and building these machines.

Before looking at specific systems, let's discuss what is needed in a QC qubit. It is truly amazing that we can controllably manipulate individual atoms or other atomic-like particles. Transistors on silicon wafers in classical computers are small, but the atomic world is even smaller. Of course, a qubit must be a quantum object whose state can be manipulated into superposition states and changed via gates that are external, controllable stimuli. These stimuli can be laser light or other electromagnetic fields applied at different places and times in the circuit. These quantum objects must be available in large enough quantities and have nearly identical properties to make a useful system. As mentioned earlier, the largest general-purpose QC systems at this time have about 50–100 qubits, but this number is likely to increase dramatically over the next few years.

The qubits must be able to be measured. In most practical QCs, the measurement bases stay fixed, and we apply various gates to transform the states of the qubits. As shown in Chapter 8, using different basis states is a pathway to extracting the maximum information from a quantum system. Importantly, the objects must be robustly immune to unwanted outside influences to limit changes in these states, known as maintaining coherence. While we can use error correction protocols as discussed in Chapter 13, there are limits to the fixes, and it is more efficient, when possible, to reduce the effects due to undesirable outside influences.

The two types of qubits currently used in many general-purpose quantum computers are superconducting circuits and trapped atomic ions. In both these systems, the qubits are acted upon and measured through interactions with electromagnetic fields, using the tremendous advances in lasers and microwave devices developed over the past 50 years or so.

Superconducting Qubits

Let's start with qubits made from superconducting materials. Superconductors are materials that allow electrical currents to flow through them without electrical resistance. It means that if you get the current started, it will flow forever. Superconductivity comes about through quantum effects discovered during the first quantum revolution that cause the materials' charge carriers, usually electrons, to pair up (in the so-called Cooper pairs) in such a way that electrical resistance disappears. However, with the materials available today, the materials must be cooled to very low temperatures (typically 1 K—one degree above absolute zero—or lower) and isolated from their warmer surrounding environment to operate properly as superconducting qubits.

Superconducting qubits use superconducting materials separated by a thin insulating layer. Such a structure is called a Josephson junction device, named after Nobel laureate Brian Josephson, who predicted their behavior theoretically in the 1960s. Normally, the insulator would stop the electrons from flowing from one material to another, but, thanks to another quantum phenomenon called tunneling, the Cooper pairs can penetrate that layer. When the Josephson junction is placed in an external electrical circuit, the combination has a preferred oscillation frequency, usually in the microwave frequency range. Measuring the oscillation frequency allows you to observe properties of the qubit's quantum state. Conversely, by having the qubit interact with externally controlled microwaves you can manipulate the quantum state and perform gate operations to run an algorithm. There are many different flavors of superconducting qubits, but the version that is used in several types of QCs are called transmons.

Why are they called transmons? Physicists like to name atomic-sized or smaller objects with words that end with "on." You are familiar with photon, electron, proton, neutron, and ion. "Transmon" is an abbreviation for "transmission-line shunted plasma oscillation," which is quite a mouthful. In simple terms, a transmon is a circuit built from a Josephson junction device and other electrical components, including a region that traps microwaves, much like your microwave oven but on a much smaller size scale, typically a few millimeters on a side.

Although a transmon is much larger than an electron or ion, its electrical current behavior can be described as a quantum two-state system with different amounts of current in the two states. Measurement, control, and coupling between the transmons is performed by observing the characteristics of the microwaves emitted by the transmons or shining microwaves on them to shift them from one quantum state to another.

Since transmons are manufactured devices, we can't guarantee that they will all be exactly the same, and the QC must be designed to tolerate slight differences among the qubits. However, the technology of Josephson devices is well established, and we know a lot about how to control and measure microwaves, thanks to decades of work in microwave communications devices and their use in both commercial and basic science applications.

Trapped Ions

Several companies have developed quantum computers using atomic ions as the qubits. An ion is an atom or molecule that has lost or gained one or more electrons, leaving it with a net electrical charge. This charge makes it relatively easy to use electric and magnetic fields to push the ions around and to trap them individually in separate spatial regions. In current QCs, the total area occupied by the trapped ions is about 0.1 mm × 0.4 mm, which is small compared to a human but very large compared to a typical atom's diameter—about 0.1 nm $\left(10^{-10}\mathrm{m}\right)$.

Using techniques that have been developed to a high level of sophistication over the past five or six decades, you can operate on those states with light from lasers. You can create superposition states and entangled states and can make measurements of various kinds.

Ion qubits don't need super-cold temperatures, and all the qubits are exactly the same because they are atoms with no additional manufacturing needed. Since the qubits are manipulated by laser light and not wires, every qubit in the system can interact with any other qubit in the system, giving what is called complete connectivity. The main challenge is getting the laser beams that control the qubit states connected uniformly to all the qubits and synchronized in a controlled way. Many of these laser light manipulations make use of π pulses and $\pi/2$ pulses mentioned in Chapter 19.

The company IonQ uses ytterbium (Yb) positively charged ions in its quantum computer. Yb is used because its atomic energy states are convenient for manipulation by laser light, and it is relatively easy to trap the ions in specific spatial locations with electric fields. The collective motion of the ions in the trap region provides another quantum degree of freedom; so, each ion contributes to an energy qubit and is part of the quantized collective motion, which can be modeled like the simple harmonic oscillator in Chapter 21. Because Yb has been used in many sensitive atomic clock experiments, the technology for manipulation Yb ions and their states is well understood.

Let's now look at some other physical systems that are being developed for use as qubits in QIS and QC applications.

Nitrogen Vacancies in Diamond

Another interesting qubit is a "vacancy" in diamond crystals next to an "impurity" atom substituting for one of the carbon atoms (Fig. 24.2). Diamond is a crystalline form of carbon. A local gap in the crystal structure is called a vacancy. Suppose a nitrogen atom

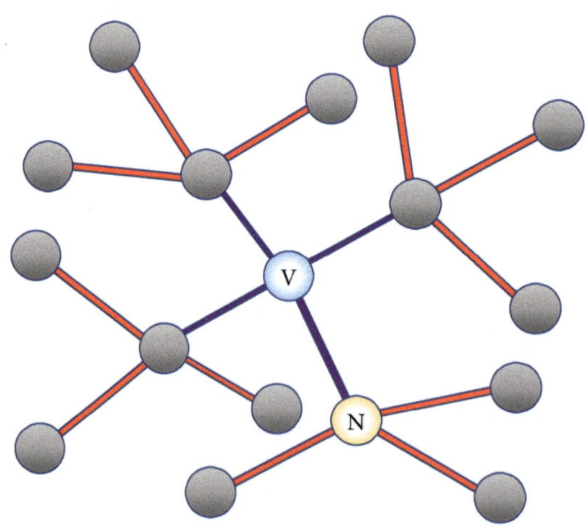

Fig. 24.2 A schematic drawing of carbon atoms (gray) in a diamond lattice, a vacancy V (blue), and a substitute nitrogen atom N (yellow). The colored lines represent chemical bonds. Only a few of the bonds are shown. The nitrogen atom and the vacancy constitute a quantum object that can be used as a qubit.

replaces one of the carbon atoms in the crystal next to the vacancy. The nitrogen atom plus vacancy system has energy states that lead to the absorption and emission of visible light; so, the vacancy plus nitrogen atom is called a *color center*. These energy states are easily manipulated with light from lasers. Furthermore, the electrons' spin can be manipulated with magnetic fields (via their magnetic dipoles as mentioned in Chapter 19), giving us another "handle" to operate on the qubit states. As of this writing, an Australian company, Quantum Brilliance, is building a QC based on these nitrogen vacancies in diamond. Another qubit possibility is to use the spins associated with the nuclei of the atoms in a nitrogen vacancy. This possibility has already been implemented in laboratory situations.

In addition to nitrogen substitutions, scientists are also exploring the use of tin atoms in diamond. The tin atom version has the advantage that it is relatively easy to build into nanophotonic structures that combine the vacancies with material shapes that trap microwaves, which can be used to manipulate and read the quantum states. In addition, the tin version can operate at somewhat higher temperatures than the nitrogen vacancies, thus simplifying the cooling systems needed to build the QC. Although tin vacancies have yet to be used in QC and QIS devices, their advantages seem likely to drive the research and engineering needed to make that happen.

Photon Qubits

You may wonder why we don't use photons for qubits in QCs. We discussed how to use their polarization states for quantum cryptography. Photons are easy to produce and detect. There are two problems with using photons as qubits in a QC. First, photons travel fast—at the speed of light. So, they might travel through the QC faster than we can act on them with gates. You can, however, use long optical fibers to provide time delays if needed. The second problem is that we need to have the photons "made to order" and ready to be manipulated at specific times. In most common photon sources (flashlights, light-emitting diodes, candle flames, lasers, and the like), the photons are produced randomly in time. There are some laboratory devices that can produce photons in pairs. Then, the detection of one of the photons (the "herald") shows that the other photon is on its way. But these devices produce the pairs at random times. However, there are other laboratory techniques that produce single photons at controllable times. One example is placing atoms between pairs of mirrors ("optical cavities"). This setup can produce large interactions between the atoms and photons in the cavity in such a way that a photon can be emitted upon demand. This technique is still at the laboratory development stage and has not yet been used in a practical QC.

Another strategy is to sort the photons into time bins (rather than polarization modes or path modes). Two different time bins are equivalent to a two-state qubit. Three-state qubits ("qutrits"), four-state qubits ("ququads"), and so on are also relatively easy to produce this way. However, QC work with photons is just in its developmental stage and it remains to be seen if a reliable QC with a large enough number of qubits can be constructed using photons.

Neutral Atom Qubits

Neutral (no net electric charge) atoms are another possible set of qubits. With the so-called laser-cooling techniques, it is possible to trap atoms in a vacuum in regular spatial arrays set up by intersecting light beams. The interactions between the atoms can be controlled relatively easily and the atoms' states can be manipulated and measured by other light beams. The techniques for manipulating and measuring the trapped atoms are highly developed and are similar to the trapped-ion methods mentioned previously. Our guess is that QCs based on trapped neutral atoms will be operational over the next few years.

Quantum Dots

A quantum dot is a material in which electrons are confined to a small (few nm = few $\times 10^{-9}$ m) region. This is larger than an atom, but the dot behaves in some ways like a single atom. "Dot" simply means that the confinement region is small, and when observed, it looks like a dot. The colors of light absorbed and emitted by the confined electron depend on the size of the dot rather than the chemical composition. By way of contrast, different colors that appear in fireworks are due to different chemical compounds. Larger quantum dots tend to emit and absorb red light (longer wavelengths), while smaller dots absorb and emit blue light (shorter wavelengths) (see Fig. 24.3).

There are two common ways to produce quantum dots: (1) put small particles, usually semiconductor materials, in liquid solutions and (2) use electric and magnetic fields to trap the electrons near the surface of some material. The latter method allows greater control over the position of the trap region, and many quantum dots can be created near one another, leading to a compact set of qubits. By the way, quantum dots may also be used in solar cells to enhance the efficiency of the cells in converting solar light energy into electrical energy. They are already being used to make more vibrant televisions.

You can buy solutions of quantum dots that emit a range of colors when stimulated by ultraviolet light. You can even make the solutions at home. Quantum dots available in commercial solutions are often small clusters of cadmium, selenium, or other semiconductor materials. But because the dots are in solution, they are not easily controllable for QC purposes.

The quantum dots that are being explored for possible use in QCs are fabricated in thin layers of solid semiconductors and other materials with the electrons trapped by applied electric and magnetic fields. These are sometimes called "gate-confined" states. Here, gate

Fig. 24.3 Quantum dots fluoresce (emit light) when illuminated by ultraviolet light. The smaller dots on the left emit violet and blue light, while the larger ones, on the right, emit yellow and red light.

does not mean logic gates but rather materials to which electric voltages are applied to produce the electric fields that trap the charged particles, just like a physical gate in a fence can be used to help corral horses. Quantum dot qubits have the advantage that they can be fabricated in dense arrays using techniques similar to those widely used in integrated-circuit construction. However, many technical issues, such as uniformity and removal of disturbances from other charges in the system, need to be addressed before quantum dots can be widely used in QCs.

Topological Qubits

There is another possibility for qubits that is very different from the others. Scientists have been searching for many years for materials that could be used to produce "topological" qubit states. These kinds of quantum states depend on the topological properties (edges, holes, etc.) of the device. Since the states depend on these geometric properties of the device, it is believed that they will be less susceptible to local effects such as material impurities and stray electric and magnetic fields. Defects in semiconductor nanowires might provide the appropriate devices, but much basic research remains to be done.

In all these current and future qubits, improving the properties of the materials and reducing interactions between the qubit and its environment may reduce the need for error correction and sophisticated quantum control systems that make QCs quite complex. Research in materials science will play a large role in making otherwise fragile qubits more robust.

24.3 Quantum Information Science and the Quantum Internet

Throughout this book we have used QIS and QC as if they were friendly siblings, always being careful to give them equal time. Indeed, they are closely related, but in some sense, QIS is more general because it includes both the transmission of information and computations carried out on this information.

Just as QC can be viewed as a generalization of classical computing, QIS can be viewed as a generalization of classical information science. Both QIS and classical information science are concerned with the overall theory of information, not just computer algorithms. QIS, of course, takes advantage of the constraints and opportunities provided by quantum mechanics. This often leads to a deep level of abstraction that makes the concepts difficult to fathom.

When the authors first learned about information theory, it seemed to have little overlap with what we worry about when sending information via email, X (formerly known as Twitter), or a postcard. We finally realized that information theory is really not about information in the everyday sense, but more about the abstract structure of what is being sent back and forth between Bob and Alice, for example, and the advantages and disadvantages of various kinds of information structures. Information science includes how to store and retrieve information, how much information can be accurately sent, and how to make corrections for errors.

Let's add learning more about information theory and QIS to our list of topics to study and understand the broader principles that underly QC.

Quantum Internet

One area in which QIS is likely to make a big splash within the next few years is the so-called quantum internet. This does not mean faster streaming services for your favorite cat videos. Today's internet does a great job in producing the fast transfer of information from point A to point B, regardless of where A and B are. However, we know that "bad actors" can often access that information and use it for malevolent purposes. A quantum internet has the potential to provide super-high security data transmission over long distances. Polarized photons appear to be the practical choice for the qubits because they can travel extended distances at high speeds over special polarization-maintaining optical fibers, thus allowing their quantum states to be undisturbed as they travel.

There are two approaches to quantum-enabled secure communications. The first is called postquantum cryptography that uses standard public cryptography keys that are believed to be sufficiently resistant to current quantum algorithms. The second uses quantum mechanics to distribute the cryptographic key and is called *quantum key distribution* (QKD). We described a simple version of QKD in Chapter 8. *Quantum state teleportation* (Chapter 10) is part of QKD as well. Commercial quantum communication packages are now available that operate over city-sized distances (see Further Reading).

For longer distance communications, devices called repeaters are needed because the probability that a photon will be lost due to absorption or scattering increases with the length of the optical fiber. Such repeaters are used in classical transmission lines to amplify the signal, which is inevitably degraded by various "losses" in the electrical cables or, nowadays, in optical fibers.

If we are using photons and their quantum states to transmit information, how do we build a quantum repeater? This would require detecting and then rebuilding the quantum state. The infamous *no-cloning* theorem states that it is not possible in general. If that idea jumped into your mind before you read it, then you are thinking quantum mechanically.

To get around the *no-cloning* theorem problem, Bob and Alice can make use of a combination of quantum state teleportation and entanglement swapping (Chapter 10). They each entangle a pair of photons at the ends of the communication channel, and they each send one of the photons toward the repeater unit. When the repeater successfully receives a photon from either Bob or Alice, it stores it in quantum memory and waits for a photon from the other person. When the second photon is received, the repeater performs a Bell state measurement on the two photons and the entanglement is swapped (as discussed in Chapter 10) between the two memory photons. They are then launched on their respective ways toward the receivers.

A classical computer memory stores a pattern of 0s and 1s and allows us to read out that pattern later. A quantum memory stores a quantum state and allows us to produce a qubit described by that quantum state. Again, quantum memories are just beginning to be developed, but the idea is that a photon qubit could interact with other qubits (perhaps trapped atoms or ions). These other qubits cannot be easily transported, but they can store the state temporarily until they are commanded to emit a photon whose state is the same as that of the incoming photon.

Although a fully functioning quantum repeater has yet to be built, research labs are testing various repeater components. In addition, some researchers are using classical computer simulations of quantum repeater networks. Such simulations will help engineers come up with a design protocol for the repeater and allow them to study the complex dynamics of a repeater system. For more information about the *quantum internet*, refer to the references in the Further Reading section.

24.4 Quantum Computing, Machine Learning, and Artificial Intelligence

Two of the most rapidly growing areas in computer science are machine learning and artificial intelligence (widely known as AI). In simple terms, machine learning means training a computer to recognize patterns—either spatial patterns or temporal patterns or a combination of the two. More generally, machine learning is a procedure by which a computer can extract information from a database. The training consists of giving the computer some input data and the desired output and having the software work out mathematical connections that allow the computer to recognize similar patterns in input data on which it has not been trained. For example, you might give the computer several images of different kinds of dogs. The computer analyzes these images and comes up with some rules that allow it to produce an output "dog" when similar images are presented to it later. Much of machine learning involves solving systems of linear equations. There are quantum algorithms that can carry out these solutions more efficiently than classical computers.

Related to image recognition is image processing. As a concrete example, cameras in satellites in the Earth orbit send images that are useful for weather prediction, monitoring environmental changes (for example, erosion and glacier deterioration), and military reconnaissance. If we could put fast image-processing computers on satellites, we could substantially reduce the data transmission and reception requirements (Fan et al. 2017) because the entire image would not need to be transmitted. Once QCs become more portable, image processing is likely to be one of the first space applications of QCs. Until then, quantum machine learning is likely to focus on solving problems related to quantum mechanics, such as quantum simulation.

AI is similar to but broader than machine learning. The computer is trained to sense its environment (the World Wide Web) and learn what happens when it responds to this environment in various ways as it tries to achieve some specified goal. AI uses the environment to make decisions whereas machine learning mostly just classifies data. For example, AI allows a computer to learn to be competitive in strategic games, such as Go and chess, to control an autonomously operating automobile, to provide intelligent routing of messages in content delivery networks, and to run sophisticated military simulations. With AI, the computer is supposed to be able to learn all these things without human intervention.

Generative AI can, at least in principle, use what it has learned to generate prose answers to questions, create a requested image, provide a medical diagnosis, or compose a piece of music in a specified style. However, current AI is far from perfect and is known to provide false information to an individual prompt. It often seems to hallucinate, as evidenced by wrong reasoning and inclusion of irrelevant information. It has become so mainstream that the authors of this book feel compelled to attest that AI was not used in the generation of this manuscript.

What is the role of QC in machine learning and AI? The complete answer is not yet known, but current work focuses on using quantum computer algorithms to speed up mathematical operations that are part of classical machine learning and AI. These operations include solving systems of linear equations, using what is called principal-component analysis to focus on the major features of large data sets, and finding eigenvectors and eigenvalues of matrices.

24.5 Alternative Quantum Computer Architectures

Thinking ahead about what is coming in QC, we should mention that there are different ways of designing QCs. So far, we have described QCs based on quantum gates interacting with qubits. Quantum gate architecture is much like the architecture in classical computers. But there are other ways qubits can be manipulated and measured for computational purposes. These alternative structures may turn out to be faster or more easily implemented than the gate-based structures that have been used so far. Moreover, alternative architectures may help circumvent noise, loss, and decoherence effects.

As one alternative, we have already mentioned briefly in Chapter 4 and earlier in this chapter the possibility of using quantum systems with more than two states. It turns out there are some advantages of using these multistate systems in terms of energy efficiency, coding flexibility, and a reduction in susceptibility to noise and other decoherence effects. But so far, it is qubit (two-state) technology and algorithms based on qubit manipulation that have dominated QIS and QC. Developing technologies to produce qutrits (and, more general qudits) in specified states and getting them involved in entangled states is still in the development stage. But we believe this is an aspect of QIS and QC that may take off in the next decade or so.

Quantum Annealing

A quite different architecture uses qubits arranged in arrays with various connections between the qubits. Instead of using gate operations on one or two qubits, these systems have the qubits undergo a type of time evolution called quantum annealing. You may have heard about annealing in your high school shop class, where you annealed some pieces of glass and some metal welding joints. Quantum annealing is similar to this process. You computationally "heat up" a system of qubits and then let them "cool" slowly. The slow cooling allows the system, be it blown glass, a metal weld, or the qubits, to relax slowly into its lowest energy state. The opposite process—quenching—cools the system quickly, freezing in defects and resulting in a state of higher energy.

In quantum annealing, you prepare the qubits individually in some state and then use controllable coupling between the qubits to form an entangled state. You next use external fields, often magnetic fields, to control the speed of relaxation to the lowest energy state. To make quantum annealing useful for computation, you need to formulate your problem in terms of minimizing the system's energy or, more generally, optimizing some property of the system. This turns out to be possible in many cases, including search problems, factoring problems similar to those addressed by the Shor algorithm, and molecular configuration problems.

The Canadian company D-Wave has been selling quantum annealing devices based on superconducting qubits. The advantage of this method is that it is relatively easy to include many qubits. However, you usually have only a limited number of connections among the qubits. The company's website (see Further Reading) has useful information about the quantum annealing architecture and the types of problems that can be formulated to take advantage of this technique. Some authors characterize quantum annealing as an example of "hardware acceleration," in which the specially designed qubit circuitry carries out a fixed task efficiently but lacks the flexibility of a general-purpose computer processor.

Measurement-Based Quantum Computing on Cluster States

Yet another architecture uses clusters of qubits and implements quantum algorithms by making state-dependent measurements on a sequence of clusters. For example, you might have your qubits set up in an array as shown in Fig. 24.4. The array can be three dimensional.

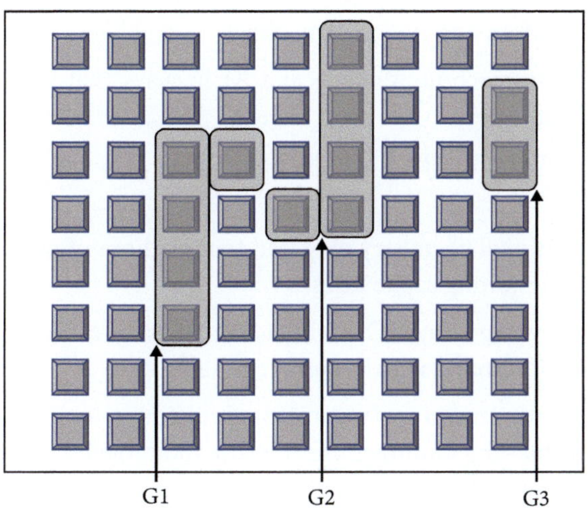

Fig. 24.4 General scheme of a quantum computer with cluster states. The qubit clusters G1, G2, and G3 (the beveled rectangles with gray shading in the background) are measured one after the other. The results of earlier measurements determine the measurement bases of later ones.

You divide the qubits into clusters (the shaded groups shown in the figure) where the size and shape of the clusters have been chosen to implement a particular quantum algorithm. After preparing each cluster in an entangled state, you then make a measurement of a single qubit in cluster 1. The result of that measurement affects what you do with cluster 2. Eventually, the measurement outcomes give you the sought-for answer. The whole computation is nothing but a series of single-qubit measurements. Alas, as we know, the measurements destroy the entangled states; so, they can be used only once. To run the algorithm again (as part of a computational loop, for example), you need to reset the entangled states.

24.6 Ethical Issues

Ethical issues might not be foremost in your mind if you are interested in QC and QIS. Your thoughts probably tend to focus on qubits, superposition, entanglement, and measurement—rightfully so. But because information technology has become so integrated with our lives, both professionally and personally, we know that its benefits are accompanied by side effects—sometimes serious and dangerous side effects. We are all familiar with identity theft, malicious hacking, ransomware, distribution of misinformation—the list goes on. We should be worried that QIS and QC might enhance the capabilities to break security codes, to increase the gathering of private information, and to deploy more sophisticated ways to disrupt the enemy's (competitor's) information technology systems.

These issues are already present in our society. The use of QIS and QC will likely accelerate their rate of growth, making each more pervasive and more difficult to deal with. For example,

- QC and QIS may make it easier to access your private information unless you are using a sophisticated quantum encryption algorithm.
- They may accelerate machine learning and AI in ways that we as a society are not prepared to deal with.
- Hackers and information technology disrupters may gain enhanced capabilities with the use of QCs.
- Quantum computational chemistry may lead to a plethora of new drugs, which might overwhelm our drug-testing capabilities.
- If new drugs are developed by QC methods, who gets access to these new drugs? Just those in the country whose scientists designed the drugs?
- QC chemistry may also lead to new ways of manipulating genetic codes—those of plants, animals, and humans—by finding more efficient and targeted gene-editing techniques such as CRISPR.

Other ethical issues are more focused on social justice rather than technology. We should be asking hard questions such as

- Will QC and QIS widen the gap between the haves and the have-nots, both within a country and among countries? Will only those companies and countries that have access to QCs be able to enjoy their benefits?
- How do we ensure diversity, equity, and inclusion in the QC and QIS workforce?
- How do we adapt our education systems to meet the new workforce needs generated by QC and QIS?

We encourage you to take these issues seriously and to talk about them with friends, colleagues, and those who make technology and educational policy.

24.7 Preparing for Your Quantum Career

This is a good time to give our dedicated readers some advice about next steps. Although there are many specific jobs that need to be done in building and operating a QC, each requiring specialized knowledge, there is some common background needed by almost everyone who wants to build or use QCs and other quantum-information-processing devices.

In 2020, the US White House Office of Science and Technology Policy released a report *Key Concepts for Future QIS Learnings* (see Further Reading). The report was the product of a conference attended by secondary school and university educators, industry researchers, and representatives from educational and professional organizations. Because the learning and excitement of quantum does not start in adulthood, there are also high school frameworks that could and should be incorporated into the secondary physics, computer science, and chemistry curricula. Here is the list of key concepts from the report with brief explanatory material:

1. *Quantum information science* exploits quantum principles to transform how information is acquired, encoded, manipulated, and applied. QIS encompasses quantum computing, quantum communication, and quantum sensing, and spurs other advances in science and technology.
2. *A quantum state* is a mathematical representation of a physical system, such as an atom, and provides the basis for processing quantum information.
3. Quantum applications are designed to carefully manipulate fragile quantum systems without observation to increase the probability that the final *measurement* will provide the intended result.
4. *The quantum bit, or qubit,* is the fundamental unit of quantum information and is encoded in a physical system, such as polarization states of light, energy states of an atom, or spin states of an electron.
5. *Entanglement,* a correlation among multiple qubits, is a key property of quantum systems necessary for obtaining a quantum advantage in most QIS applications.

6. For quantum information applications to be successfully completed, fragile quantum states must be preserved or kept *coherent.*
7. *Quantum computers*, which use qubits and quantum operations, will solve certain complex computational problems more efficiently than classical computers.
8. *Quantum communication* uses entanglement sent over a transmission channel, such as optical fiber, to transfer quantum information between different locations.
9. *Quantum sensing* uses quantum states to detect and measure physical properties with the highest precision allowed by quantum mechanics.

We have covered all the concepts in this book, with the exception of quantum sensing. This area requires more detailed knowledge of physics, chemistry, materials science, and engineering.

Let's discuss the mathematical knowledge needed for your future work in QIS and QC. In this book, we have kept the math requirements to a minimum, but we also showed that we need extended mathematical reasoning to understand how quantum algorithms work. To dig deeper into QC and QIS, you will need to be on good-speaking terms with complex variables and, by implication, functions of complex variables. You should certainly learn more about linear algebra, which is the math underlying right and left vectors, state space, quantum gates (operators), and so on. You don't necessarily need a full course in linear algebra because a lot of information about linear algebra is available on the web. However, a good course or a good textbook will provide a systematic overview, which will clarify the mathematical context of those parts of linear algebra needed in QC and QIS.

It is also important to learn more about computer science since fundamentals such as the theory of computation, algorithms, data structures, computer security, and cryptography apply equally well to quantum computing and classical computing. If you are interested in building and operating actual quantum computers, you will need to learn more about physical qubits such as polarized photons, atoms and molecules with spin, trapped ions, and superconducting circuits—the devices used currently in QC and QIS. Courses in this area are just beginning to be offered by colleges and universities.

We should also mention that many of the jobs in QIS and QC are "engineering" positions: developing and building devices based on the underlying quantum and materials principles, including electrical, cryogenics, vacuum, and nanofabrication systems. These can be exciting and well-paying positions with the opportunities to work on many kinds of QIS and QC systems (Asfaw et al. 2021).

Brief biographical sketches of 52 women scientists and engineers working in QIS and QC are available in Quantum Daily (2021) (see Further Reading).

24.8 Grand Finale

We appreciate your persistence and devotion to learning more about QC and QIS. We confidently predict that over the next few decades the advances in QC and QIS will dramatically alter almost every area of science and technology. We believe that you, devoted readers, are now well positioned to understand these advances and, if you so decide, to join the QC and QIS workforce that will make this happen. Whether this is your first journey into the world of quantum or your last, you have chosen to be part of what is now being called the *second quantum revolution*. Just as the first quantum revolution in the 20th century brought us lasers, semiconductors, classical computers, solar cells, nuclear power, radiation treatments, and magnetic resonance imaging in medicine, the second quantum revolution has the promise of changing lives in dramatic and unexpected ways. We wish you the best as you continue your QC and QIS journeys.

By the way, April 14 is the Annual World Quantum Day, so mark your calendars. And look for local activities that you might join. Why April 14? Well, another way of stating this date is 4.14—still wondering? Take a look at Planck's constant in units of eV s!

When we began writing this book, we were at first worried that it might not be possible to develop a presentation that is both accessible to a broad audience and faithful to the concepts of quantum mechanics and their applications in quantum algorithms. We trust that we were reasonably successful in achieving these goals. We also hope that you have learned as much about QIS and QC as we did in writing the book.

We anticipate that some of our readers will contribute to the development of new quantum algorithms, quantum programming languages, and the building of actual QCs. The ground is fertile; the seeds are ready to be planted; we just need a lot of people to tend the plants and harvest the fruits of our labors.

Let's wrap up with some quantum humor. We know that it sounds oxymoronic, especially to those not versed in quantum, but here it goes:

There are three types of people in this world:
Those who understand quantum computing;
Those who do not understand quantum computing;
And those who simultaneously do and do not understand quantum computing.

24.9 Chapter Summary

- Further work in QIS and QC requires understanding many diverse fields. Obviously, quantum physics, complex variables, probability, linear algebra, and other mathematical tools are needed. In addition, an understanding of the nature of information and how it is shared is also an important underlying foundation for manipulating the quantum world.
- Exciting applications of QC to machine learning, AI, and a quantum internet are currently being explored. There will most likely be a pairing of classical and quantum computing to increase the efficiency and therefore the power of these tools.

24.10 Further Reading

Commercial Quantum Computing Companies

J. Roundy, TechTarget, March 28, 2025. https://www.techtarget.com/searchdatacenter/feature/Companies-building-quantum-computers#:~:text=IBM%20has%20a%20strong%20history,to%20support%20future%20quantum%20processors

The Quantum Insider hosts articles about quantum companies, policies, and technical developments in QC. https://thequantuminsider.com/

Open-Access Resources

The journal *Physical Review X Quantum* is an open-access journal. Most of the articles are moderately technical, but we have listed several that give overviews of QIS and QC. The journal also has occasional tutorials on basic quantum topics.

www.arXiv.org is an online repository of preprint articles, freely accessible.

Wikipedia has many articles dealing with QC and QIS principles and techniques. https://en.wikipedia.org/wiki/Main_Page

QIS and QC Overview

National Academies of Sciences, Engineering, and Medicine 2019. *Quantum Computing: Progress and Prospects* (The National Academies Press, Washington, DC 2019). Free download available at https://www.nap.edu/read/25196/chapter/1

Ivan H. Deutsch, "Harnessing the power of the second quantum revolution," *Physical Review X Quantum* **1**, 020202 (2020). This article presents a broad overview of QIS and QC and their relationship to classical computing.

Harrison Ball, Michael J. Biercuk, and Michael R. Hush, "Quantum firmware and the quantum computing stack," *Physics Today* **74**, 29 (2021). This article presents an expanded discussion of the organization of quantum computing development as illustrated in our Fig. 24.1.

The US federal website https://www.quantum.gov/ has an extensive collection of reports on all aspects of QC and QIS.

Yuri Alexee et al., "Quantum computer systems for scientific discovery," *Physical Review X Quantum* **2**, 017001 (2021). This is a wide-ranging article about current and future QCs and should be accessible to anyone who has finished this book.

Physics of Qubits

Superconducting qubits

The following website has good introductory-level information on using superconducting qubits with microwave resonators.

https://www.jst.go.jp/erato/nakamura_mqm/english/project_overview/p01.html

Trapped ions

IonQ. https://ionq.com

The website of the company IonQ has a nice introduction to QC with trapped ions.

Nitrogen vacancies in diamond

V. Sewani, H Vallabhapurapu, Y. Yang, et al., "Coherent control of NV^- centers in diamond in quantum teaching lab," *American Journal of Physics* **88**, 1156–1169 (2020). This article provides a nice introduction to the energy states associated with a nitrogen vacancy in diamond and gives details of an experimental setup to prepare and measure the appropriate quantum states.

Evangelia Takou and Sophia E. Economou, "Tin qubits give diamond a new shine," *Physics* **14**, 105 (2021) (www.physics.aps.org/articles/v14/105). Using tin atoms in place of nitrogen atoms in diamonds has several advantages.

Neutral atom qubits

David S. Weiss and Mark Saffman, "Quantum computing with neutral atoms," *Physics Today* **70**(7), 44 (2017).

Semiconductor spins

Lieven M. K. Vandersypen and Mark Eriksson, "Quantum computing with semiconductor spins," *Physics Today* **72**(8), 38 (2019).

Photons as qubits for QC

Peter B. R. Nisbet-Jones, Jerome Dilley, Annemarie Holleczek, Oliver Barter, and Axel Kuhn, "Photonic, qubits, qutrits, and ququads accurately prepared and delivered on demand," arXiv:1203.5614v2 (2012).

PsiQuantum (Palo Alto, CA) is developing a photonic QC. https://psiquantum.com/

Nathalie P. de Leon et al., "Materials challenges and opportunities for quantum computing hardware," *Science* **372**, 253 (2021). This article discusses the materials science research needed to improve superconducting qubits, gate-defined quantum dots, color centers, ion traps, and topological qubits in semiconductor nanowires.

Examples of Quantum Computing Companies

Alice & Bob. https://alice-bob.com/

IonQ. https://ionq.com. Trapped-ion quantum computer.

Honeywell Quantum Solutions. https://www.honeywell.com/us/en/company/quantum. Trapped-ion quantum computer.

Rigetti Quantum Computer. https://www.rigetti.com/. Superconducting circuit QC.

IBM Qiskit. https://www.ibm.com/quantum/qiskit. It provides access to IBM superconducting circuit QCs and simulators as well as many tutorials on all aspects of QC.

Google. https://quantumai.google/. Superconducting qubit gate-based computer system.

Microsoft Azure Quantum. https://azure.microsoft.com/en-us/solutions/quantum-computing/

Microsoft is attempting to build a quantum computer based on Majorana (topological) qubits. That is still a work in progress as of this writing (2025). Microsoft has available a quantum development kit for the Q# quantum programming language and Azure Quantum projects.

D-Wave. https://www.dwavesys.com/. A quantum annealing system rather than a general-purpose quantum computer.

Quantum internet

Marcos Curty, Koji Azuma, and Hoi-Kwong Lo, "A quantum leap in security," *Physics Today* **74**, 37 (2021). This article shows how the use of one- and two-photon interference can lead to enhanced cryptographic protocols.

R. Valivarthi et al., "Teleportation systems towards a quantum internet," arXiv:2007.11157v2 (2020).

Commercially available quantum information networks. https://www.aliroquantum.com/

QuISP is a simulation package for quantum repeater networks. https://aqua.sfc.wide.ad.jp/quisp_website/

A nice survey of current quantum key distribution systems. https://www.nature.com/articles/s41534-019-0221-4#Tab1

The US National Security Agency currently does not recommend the use of QKD systems. https://www.nsa.gov/what-we-do/cybersecurity/quantum-key-distribution-qkd-and-quantum-cryptography-qc/

Quantum Computing, Machine Learning, and AI

"Machine Learning" Wikipedia

"Artificial Intelligence" Wikipedia

Xanadu. https://pennylane.ai/. This provides a lot of information about quantum machine learning.

Harrigan Johnson and Gimeno-Segovia, *Programming Quantum Computers* (O'Reilly Media Inc., Sebastopol, CA, 2019). For "Quantum Machine Learning," see Chapter 13.

"Quantum Image Processing," Wikipedia

Fan et al. 2017: F. Fan, A. M. Iliyasu, and P.Q, Le, "Quantum image processing: a review of advances in its security technologies," *International Journal of Quantum Information* **15**(3), 1730001–1730044 (2017). doi:10.1142/S0219749917300017 (open access).

Alternative QC Architectures

Ehud Altman et al., "Quantum simulators: architectures and opportunities," *Physical Review X Quantum* **2**, 017003 (2021). This article provides a wide-ranging survey of current and future QC qubits and architectures with generally accessible descriptions and no math.

Qutrits

Qutrits can be used in quantum teleportation. https://www.scientificamerican.com/article/qutrit-experiments-are-a-first-in-quantum-teleportation/

Quantum annealing

"D-Wave Systems," Wikipedia

Measurement-based QCs

Robert Raussendorf, Daniel E. Browne, and Hans J. Briegel, "Measurement-based quantum computing on cluster states," *Physical Review A* **68**, 022312 (2003).

Sara Bartolucci et al., "Fusion-based quantum computation," arXiv:2101.09310 (2021). This article describes a QC architecture based on entangling measurements and resource states.

David Hunger, "Quantum logic at a distance," *Science* **371**, 576 (2021). This article describes work demonstrating that QCs can be scaled by breaking up the system into smaller sets of neutral atom qubits connected by optical fibers.

M. Pompili et al., "Realization of a multimode quantum network of remote solid-state qubits," *Science* **372**, 259 (2021). This article uses entanglement among three nitrogen vacancies in diamond, including one memory qubit.

Ethical Issues

Ronald de Wolf, "The potential impact of quantum computers on society," arXiv:1712.05380 (2017). An essay focusing on cryptography, optimization, and simulation of quantum systems.

Cathy O'Neil, *Weapons of Math Destruction* (Broadway Books, New York, 2017). This book provides a candid look at the downsides of machine learning and big data.

Quantum Education

The report *Key Concepts for Future QIS Learners* is available at https://qis-learners.research.illinois.edu/

David Matthews, "How to get started in quantum computing," *Nature* **591**, 166–167 (2021). Available online at https://www.nature.com/articles/d41586-021-00533-x

Abraham Asfaw et al., "Building a quantum engineering program," arXiv:2108.01311 (2021). The title is a bit misleading; in addition to designing quantum engineering curricula, the authors write about a wide range of QIS and QC careers and the topics you should study to become involved in these careers.

Resources for teaching about QC and QIS can be found at https://q12education.org/learning-materials-framework

Qubit-by-qubit. https://www.qubitbyqubit.org/. The organization offers a year-long high school course on QC and QIS. It also runs summer workshops (1–2 weeks and 4 weeks) for both high school and college students.

Quantum.country. https://quantum.country/. The site provides a series of "essays" about QC and QIS. The essay "Quantum computing for the very curious" introduces quantum states and quantum gates but assumes some familiarity with linear algebra and complex numbers. The essay might serve as a review of Chapters 1–5 and 9 of this book. The pace is fast, but the writing is lucid and informative.

Quantum for all: K-12 quantum education materials. https://quantumforall.org/

National Q-12 Education Partnership. https://q12education.org/. The site has links to many quantum educational materials. https://q12education.org/learning-materials

PhETs. The site provides freely accessible simulations on a variety of STEM topics: search for "quantum phenomena." https://phet.colorado.edu/

ACEQM. Online materials for an introductory quantum mechanics course. https://www.physport.org/curricula/ACEQM/ (instructors only).

Qubitekk. www.qubitekk.com. It sells equipment for quantum education and research using photons, which include single-photon detectors, entangled photon sources, and so on.

EPiQC. Enabling Practical Scale Quantum Computing. https://www.epiqc.cs.uchicago.edu/. The site includes QC and QIS educational materials at various levels.

D. Candela, "Undergraduate computational physics projects on quantum computing," *American Journal of Physics* **83**, 688 (2015). This article shows how to simulate a quantum computer to illustrate gate operations on qubit states and to simulate the Grover and Shor algorithms. The article's exercises provide nice extensions of the methods described in this book.

Mark Beck, *Quantum Mechanics: Theory and Experiment* (Oxford University Press, Oxford, 2012). Chapters 1–8 and 17 provide a somewhat higher-level coverage of Chapters 1–11 of this book. This is an excellent resource for physics and engineering majors.

People in QIS and QC

https://q12education.org/. The site hosts short bios of a diverse set of QC and QIS workers.

Short bios of 52 women scientists and engineers working in QC and QIS. https://thequantuminsider.com/2021/08/24/52-wonder-women-working-in-industry-as-quantum-scientists-engineers/

APPENDIX

Quantum Toolkit

How I need a drink, alcoholic of course, after all these problems involving quantum mechanics.

(The number of letters in each word gives π to 15 significant figures.)
Applied Optics **6**, 234 (1967)

In this appendix, we provide a summary, without proofs or derivations, of mathematical expressions that occur frequently in quantum information science (QIS) and quantum computing (QC). We also collect terminology dealing with complex numbers, matrices, and vectors.

A.1 Complex Numbers

Complex numbers are generalizations of real numbers. Each complex number has two parts, historically called the real part and the imaginary part. A complex number z can be written in several equivalent forms:

$$z = x + iy = (x, y), \tag{A.1}$$

where x is the real part and y is the imaginary part. (We assume that x and y are real numbers.) $i = \sqrt{-1}$ is the imaginary unit vector. Note that engineers often use $j = \sqrt{-1}$.

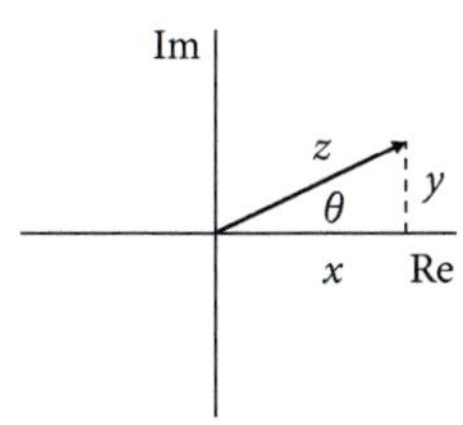

Fig. A.1 Argand diagram. A complex number $z = x + iy$ may be represented graphically on an Argand diagram with the real part on the horizontal axis and the imaginary part on the vertical axis.

A. **Magnitude, Addition, and Complex Conjugation**
The *magnitude* of a complex number, written as $|z|$, is given by

$$|z| = \sqrt{x^2 + y^2} \tag{A.2}$$

The *argument* (phase) of a complex number is defined as

$$\arg z = \arctan \frac{y}{x}. \tag{A.3}$$

The angle θ in Fig. A.1 is $\arg z$.

The *complex conjugate* of a complex number z is

$$z^* = x - iy, \tag{A.4}$$

that is, you simply reverse the sign of the imaginary part of z. The magnitude (absolute value) of the complex number z can then be written as

$$|z| = (z^* z)^{1/2}. \tag{A.5}$$

Complex exponential functions $e^z = e^{x+iy} = e^x e^{iy}$ and $|e^z| = e^x$. The form e^{iy} appears in many different parts of quantum mechanics.

B. **Euler Formula**
The Euler formula is

$$e^{iy} = \cos y + i \sin y, \tag{A.6}$$

with

$$|e^{iy}| = 1 \tag{A.7}$$

for any real number y. The Euler formula combines exponential functions, trigonometric functions, and complex numbers.

C. **Polar Form of a Complex Number**
A complex number may be written as $z = |z|\, e^{i\theta} = re^{i\theta}$, where $\theta = \arctan(y/x)$.

D. **Rotations in Argand Diagrams**

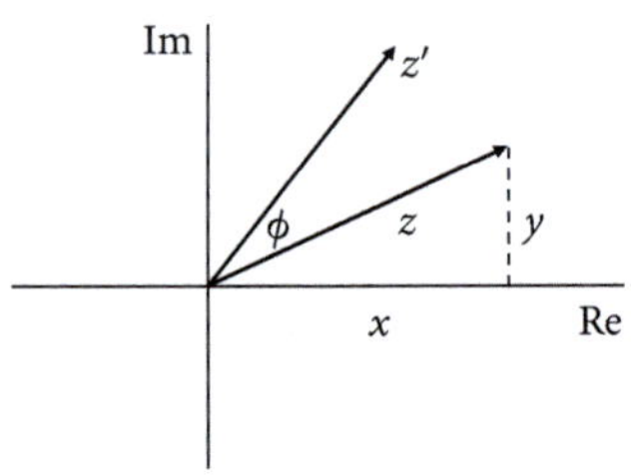

Fig. A.2 The complex number (vector) z is rotated counterclockwise through the angle ϕ to form a new complex number (vector) z'.

A rotation in the Argand diagram can be expressed as

$$re^{i\theta}e^{i\phi} = re^{i(\theta+\phi)}. \tag{A.8}$$

ϕ is often called a phase or phase shift (Fig. A.2).

E. **Multiplication of Complex Numbers**
For two complex numbers $z_1 = x_1 + iy_1$ and $z_2 = x_2 + iy_2$, the product of the two is given by $z_1 z_2 = (x_1 + iy_1)(x_2 + iy_2) = (x_1x_2 - y_1y_2) + i(x_1y_2 + x_2y_1)$. Alternatively, the polar form of a product of two complex numbers is $z_1 z_2 = r_1 r_2 e^{i(\theta_1+\theta_2)}$.

A.2 Linear Algebra and Quantum Mechanics

The underlying mathematics for most of quantum mechanics is just linear algebra. The primary exception is the use of differential equations to find wave functions, in particular, the energy eigenvectors of Hamiltonian (energy) operators. In the following table, we indicate the correspondence between quantum mechanics and linear algebra. Details are provided in the following sections and throughout the main text of this book.

Linear algebra	Quantum mechanics
Vector space	Hilbert space (a vector space with complex vector amplitudes and a definition of a vector length)
Vectors	States or state vectors
Column vector	Right vector, ket vector $\vert\psi\rangle$, and column vector
Row vector	Left vector, bra vector $\langle\psi\vert$, and row vector
Linear combination of vectors	Superposition states
Linear transformations	Operators
Eigenvectors	Eigenvectors and eigenstates
Eigenvalues	Eigenvalues
Inner product	Inner product, scalar product, and dot product
Tensor product	Tensor product
Successive linear transformation	Successive operators and successive matrices
Outer product of vectors	$\vert A\rangle\langle B\vert$ = operator

A.3 Quantum States

A. **Superposition State**
A superposition state is a linear combination of basis states weighted by state amplitudes. For example, a superposition of computational basis states $|0\rangle$ and $|1\rangle$ for a two-state (qubit) system is

$$|\psi\rangle = a_0|0\rangle + a_1|1\rangle, \tag{A.9}$$

where a_0 and a_1 are the state amplitudes.

B. **Normalization Condition**
$a_0^2 + a_1^2 = 1$ for real amplitudes and $a_0^*a_0 + a_1^*a_1 = 1$ for complex amplitudes.

C. **Probability Interpretation (Born Rule)**
The probability of getting $|0\rangle$ as a measurement outcome for a system described by the state $|\psi\rangle$ is $|\langle 0|\,\psi\rangle|^2 = |a_0|^2$ while $|\langle 1|\,\psi\rangle|^2 = |a_1|^2$ is the probability of observing $|1\rangle$.

D. **Right and Left Vectors (Column and Row Vectors)**
Right vectors and column vectors:

$$|\psi\rangle = a_0|A_0\rangle + a_1|A_1\rangle \doteq \begin{pmatrix} a_0 \\ a_1 \end{pmatrix}. \tag{A.10}$$

We use the symbol $\doteq$ to indicate "represented by." A column vector is a representation of a right vector.

A left vector is represented by a row vector:

$$\langle\psi| = a_0^* \langle A_0| + a_1^* \langle A_1| \doteq \begin{pmatrix} a_0^* & a_1^* \end{pmatrix}. \tag{A.11}$$

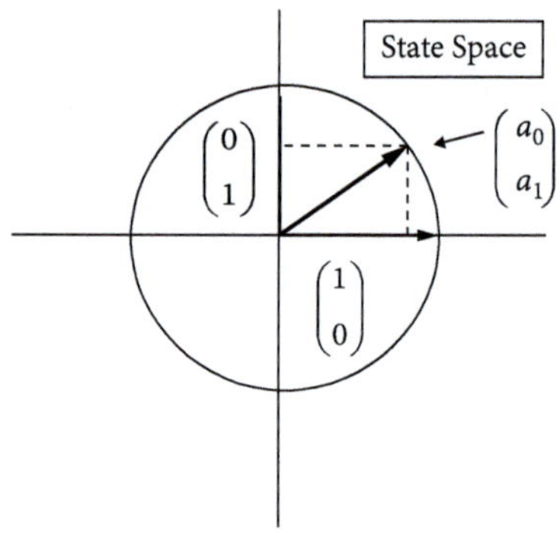

Fig. A.3 The state space vector tip has coordinates a_0 and a_1 ($|a_0|$ and $|a_1|$ if the amplitudes are complex numbers). The two basis states are shown in the column vector form. The circle has a radius equal to 1.

Figure A.3 shows how the state space coordinates of the tip of the state vector are combined to give a column vector representation of the state vector.

A.4 Products of State Vectors

There are several kinds of products of state vectors: inner, tensor, Kronecker, outer, and exterior.

A. **Inner Product**

Inner (*scalar* product or *dot* product) of two state vectors is given by the Dirac bracket $\langle s_0|\, s_1\rangle$, the projection of $|s_1\rangle$ along $|s_0\rangle$. If both state vectors are normalized, $\langle s_0|\, s_1\rangle = \cos\theta_{01}$, where θ_{01} is the state space angle between the two state vectors in state space. In terms of state amplitudes, we write

$$\begin{aligned} |A\rangle &= a_o|0\rangle + a_1|1\rangle \\ |B\rangle &= b_o|0\rangle + b_1|1\rangle \end{aligned} \tag{A.12}$$

and the inner product becomes

$$\langle B|A\rangle = b_0^* a_0 + b_1^* a_1. \tag{A.13}$$

The inner product of a vector with itself is the square of the length of the vector in state space:

$$\langle A\,|\,A\rangle = a_0^* a_0 + a_1^* a_1 = |a_0|^2 + |a_0|^2. \tag{A.14}$$

$\langle A|B\rangle = 0$ means the state vectors $|A\rangle$ and $|B\rangle$ are orthogonal. If $\langle A\,|\,A\rangle = 1 = \langle B\,|\,B\rangle$, the vectors are normalized. If all three relations hold, the vectors are said to be *orthonormal*.

B. **Tensor Product (Kronecker Product) of Two State Vectors**
The tensor product (Kronecker product) of state vectors is used to form basis vectors for the state space formed as a product of two subspaces. Consider two state vectors: $|\psi_A\rangle$ lives in the A subspace and $|\psi_B\rangle$ in the B subspace:

$$\begin{aligned} |\psi_A\rangle &= a_0|A_0\rangle + a_1|A_1\rangle \\ |\psi_B\rangle &= b_0\,|B_0\rangle + b_1\,|B_1\rangle\,. \end{aligned} \tag{A.15}$$

The tensor (Kronecker) product of two right vectors for a two-state system:

$$|\psi_A\rangle\,|\psi_B\rangle \doteq \begin{pmatrix} a_0 \\ a_1 \end{pmatrix}\begin{pmatrix} b_0 \\ b_1 \end{pmatrix} = \begin{pmatrix} a_0\begin{pmatrix} b_0 \\ b_1 \end{pmatrix} \\ a_1\begin{pmatrix} b_0 \\ b_1 \end{pmatrix} \end{pmatrix} = \begin{pmatrix} a_0b_0 \\ a_0b_1 \\ a_1b_0 \\ a_1b_1 \end{pmatrix}. \tag{A.16}$$

The tensor (Kronecker) product of two left vectors:

$$\begin{aligned} \langle\psi_A|\,\langle\psi_B| &\doteq \begin{pmatrix} a_0^* & a_1^* \end{pmatrix}\begin{pmatrix} b_0^* & b_1^* \end{pmatrix} = \begin{pmatrix} a_0^*\begin{pmatrix} b_0^* & b_1^* \end{pmatrix} & a_1^*\begin{pmatrix} b_0^* & b_1^* \end{pmatrix} \end{pmatrix} \\ &= \begin{pmatrix} a_0^*b_0^* & a_0^*b_1^* & a_1^*b_0^* & a_1^*b_1^* \end{pmatrix}. \end{aligned} \tag{A.17}$$

C. **Outer Product**
The outer product produces a matrix from two column vectors (not necessarily of the same dimension). Here is an example with each column vector having two rows:

$$|\psi_A\rangle \otimes_{\text{outer}} |\psi_B\rangle \Rightarrow \begin{pmatrix} a_0 \\ a_1 \end{pmatrix} \otimes_{\text{outer}} \begin{pmatrix} b_0 \\ b_1 \end{pmatrix} = \begin{pmatrix} a_0 \\ a_1 \end{pmatrix}\begin{pmatrix} b_0 & b_1 \end{pmatrix} = \begin{pmatrix} a_0b_0 & a_0b_1 \\ a_1b_0 & a_1b_1 \end{pmatrix}. \tag{A.18}$$

The outer product is most often written as

$$|\psi_A\rangle\langle\psi_B| \doteq \begin{pmatrix} a_0 \\ a_1 \end{pmatrix}\begin{pmatrix} b_0 & b_1 \end{pmatrix} = \begin{pmatrix} a_0b_0 & a_0b_1 \\ a_1b_0 & a_1b_1 \end{pmatrix}. \tag{A.19}$$

D. **Exterior Product**
The *exterior product* is a generalization of the vector cross product used in physics and engineering: $\vec{C} \times \vec{D} = \vec{E}$. The magnitude of $\vec{E}$ is $|C|\,|D|\sin\theta$, where θ is the angle between the two vectors. The exterior product of state vectors is not used in ordinary quantum mechanics.

E. **Changing Basis States—Two-Dimensional State Space**

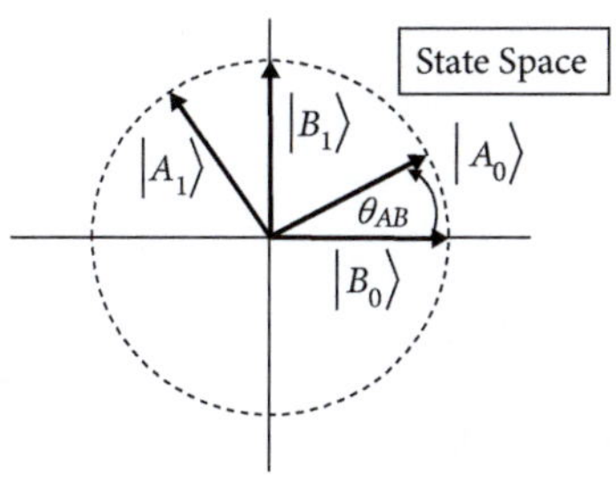

Fig. A.4 A state space diagram of two sets of basis states ($|A_0\rangle, |A_1\rangle$) and ($|B_0\rangle, |B_1\rangle$).

The general relationships among the basis states is (Fig. A.4)

$$|A_0\rangle = |B_0\rangle\langle B_0 \mid A_0\rangle + |B_1\rangle\langle B_1 \mid A_0\rangle$$
$$|A_1\rangle = |B_0\rangle\langle B_0 \mid A_1\rangle + |B_1\rangle\langle B_1 \mid A_1\rangle. \tag{A.20}$$

If all the Dirac brackets are real numbers, we can express the brackets in terms of the state space angle θ_{AB}:

$$|A_0\rangle = \cos\theta_{AB}\,|B_0\rangle + \sin\theta_{AB}\,|B_1\rangle$$
$$|A_1\rangle = -\sin\theta_{AB}\,|B_0\rangle + \cos\theta_{AB}\,|B_1\rangle, \tag{A.21}$$

where θ_{AB} is the *state space* angle between the $|A_0\rangle$ basis direction and the $|B_0\rangle$ basis direction.

In the matrix and column vector forms, Eq. (A.21) becomes

$$\begin{pmatrix} a_0 \\ a_1 \end{pmatrix} = \begin{pmatrix} \cos\theta_{AB} & \sin\theta_{AB} \\ -\sin\theta_{AB} & \cos\theta_{AB} \end{pmatrix} \begin{pmatrix} b_0 \\ b_1 \end{pmatrix}. \tag{A.22}$$

F. **Entangled States**
An entangled state is a quantum state for two (or more) particles that cannot be expressed as a simple product state $|\psi_A\rangle\,|\psi_B\rangle$ of individual particles. Example of a two-qubit system:

$$|AB\rangle = \frac{1}{\sqrt{2}}|0_A\rangle\,|0_B\rangle + \frac{1}{\sqrt{2}}|1_A\rangle\,|1_B\rangle. \tag{A.23}$$

The subscripts are the qubit labels.

A.5 Operators (Linear Transformations)

A. **Operators Act on Vectors**
An operator C acts on a state vector to produce a new state vector $C\,|\psi\rangle = |\phi\rangle$.
Right–left vector (outer product) form of an operator is $C = |\phi\rangle\langle\psi|$.

B. **Matrix Form of an Operator**
Most operators can be represented using both the left–right vector (outer product) form and matrices. For example, the Pauli X operator in outer product form in the computational basis is

$$\mathrm{X} = |0\rangle\langle 1| + |1\rangle\langle 0| \tag{A.24}$$

The *matrix form* in that basis is formed by "sandwiching" X between various basis states:

$$\mathrm{X}_{00} = \langle 0|\mathrm{X}|0\rangle = \langle 0|0\rangle\langle 1|0\rangle + \langle 0|1\rangle\langle 0|0\rangle = 0$$
$$\mathrm{X}_{01} = \langle 0|\mathrm{X}|1\rangle = \langle 0|0\rangle\langle 1|1\rangle + \langle 0|1\rangle\langle 0|1\rangle = 1$$
$$\mathrm{X}_{10} = \langle 1|\mathrm{X}|0\rangle = \langle 1|0\rangle\langle 1|0\rangle + \langle 1|1\rangle\langle 0|0\rangle = 1$$
$$\mathrm{X}_{11} = \langle 1|\mathrm{X}|1\rangle = \langle 1|0\rangle\langle 1|1\rangle + \langle 1|1\rangle\langle 0|1\rangle = 0. \tag{A.25}$$

Thus, the matrix form of X using these basis states is

$$\mathrm{X} \doteq \begin{pmatrix} 0 & 1 \\ 1 & 0 \end{pmatrix}. \tag{A.26}$$

This method works for any operator and any state basis set.

C. **Projection Operators**

$$P_{B_0} = |B_0\rangle\langle B_0| \tag{A.27}$$

projects a state $|\psi\rangle$ along the $|B_0\rangle$ direction in state space

$$P_{B_0}|\psi\rangle = |B_0\rangle\langle B_0 \mid \psi\rangle . \tag{A.28}$$

The resulting vector has length $|\langle B_0| \psi\rangle|$.

D. **Identity Operator I Is Defined as** $\mathrm{I}\,|\psi\rangle = |\psi\rangle$ **for All State Vectors**
The identity operator can be written as a sum of projection operators:

$$\mathrm{I} = |a_0\rangle\langle a_0| + |a_1\rangle\langle a_1| + \ldots , \tag{A.29}$$

where the sum includes all the projection operators that make up a "complete" set of states. (*Complete* means that any quantum state for the system can be expressed as a sum of the basis states, with appropriate amplitudes.) For a single-qubit state, there are only two terms corresponding to the two basis states: $\mathrm{I} = |a_0\rangle\langle a_0| + |a_1\rangle\langle a_1|$.

E. **Eigenvalues and Eigenvectors**
An operator A for a two-state system has associated with it a set of states that satisfy

$$\begin{aligned} A|\psi_1\rangle &= \lambda_1|\psi_1\rangle \\ A|\psi_2\rangle &= \lambda_2|\psi_2\rangle . \end{aligned} \tag{A.30}$$

λ_1 and λ_2 are numbers called the *eigenvalues* of the operator A. $|\psi_1\rangle$ and $|\psi_2\rangle$ are the operator's *eigenstates* (or *eigenvectors*). Often, the eigenstates are labeled by the corresponding eigenvalues: $|\psi_1\rangle \Rightarrow |\lambda_1\rangle$ and $|\psi_2\rangle \Rightarrow |\lambda_2\rangle$.

If the matrix is in diagonal the form $A \doteq \begin{pmatrix} a & 0 \\ 0 & b \end{pmatrix}$, you can read off the eigenvalues directly: $\lambda_1 = a$ and $\lambda_2 = b$. In general, the two eigenvalues satisfy the following two equations: $\lambda_1\lambda_2 = \det A$, where det is the determinant of A and $\lambda_1 + \lambda_2 = \mathrm{Tr}A$, where Tr A is the trace of the matrix and is given by the sum of the principal (upper-left and lower-right) matrix elements. It means that it is easy to find the eigenvalues of a 2×2 matrix because the determinant and the trace are the same for any matrix representation of the operator. These facts are proved in most textbooks on linear algebra. If the determinant of the matrix is zero, then one of the eigenvalues is 0. The other can be found from $\lambda_1 + \lambda_2 = \mathrm{Tr}A$.

Once we know the eigenvalues and eigenvectors associated with an operator, we can express the operator as a linear combination of the projection operators built from the right and left eigenvectors:

$$A = \sum_j \lambda_j \underbrace{|\lambda_j\rangle\langle\lambda_j|}_{\text{projection operator}} . \tag{A.31}$$

This form is called the spectral decomposition of the operator because the set of eigenvalues $\{\lambda_j\}$ is the "spectrum" of the eigenvalues.

F. **Functions of Operators**

We may define a mathematical function of an operator by using the spectral decomposition form in Eq. (A.31):

$$f(A) = \sum_j f(\lambda_j)|\lambda_j\rangle\langle\lambda_j| . \tag{A.32}$$

G. **Average Values (Expectation Values)**

If the system state is $|\psi\rangle = \sum_j c_j |\psi_j\rangle$, where $|\psi_j\rangle$ are the eigenstates of an operator M, the average value of the quantity represented by operator M is given by

$$\langle\psi| M |\psi\rangle = \sum_j \lambda_j |c_j|^2 . \tag{A.33}$$

If M represents a measurable property of the qubit, then M's eigenvectors are the corresponding *measurement basis states.*

A.6 Matrices and Tensors

In strict mathematical terms, a matrix is simply an array of entries, usually numbers, variables, or even other matrices. Tensors are also arrays of entries, but a tensor comes along with one or more transformation rules that show how the entries change, say, under rotations or translations of coordinates or the change of basis states in quantum mechanics, which, as shown before, can often be represented by rotations in state space. In this language, scalars, vectors, and "tensors" (two-dimensional arrays) are rank 0, rank 1, and rank 2 tensors, which transform differently under rotations of coordinate systems. In many cases, the transformation rules can be expressed as a matrix operating on the matrix representation of the tensor.

Matrices are often used to represent tensors. This leads to somewhat confusing language in which matrix and tensor terminology get mixed. In what follows, we restrict ourselves to operations on square $n \times n$ matrices.

A. **Matrix Elements**

The *entries in a matrix* (the so-called *matrix elements*) are labeled by subscripts indicating the row–column location of the entry within the matrix. For example, F_{23} indicates the entry in the second row and third column of the matrix F.

B. **Matrix Transpose**

Transpose F^T of a matrix F is the matrix obtained by swapping the corresponding elements across the upper-left to lower-right (main or primary or principal) diagonal:

$$\begin{pmatrix} F_{11} & F_{12} \\ F_{21} & F_{22} \end{pmatrix}^T = \begin{pmatrix} F_{11} & F_{21} \\ F_{12} & F_{22} \end{pmatrix}. \tag{A.34}$$

C. **Hermitian Conjugate**

Hermitian conjugate $F^\dagger$ means transpose (swap rows and columns, which is equivalent to swapping the corresponding elements across the upper-left to lower-right diagonal) and take the complex conjugate of the elements:

$$\begin{pmatrix} F_{11} & F_{12} \\ F_{21} & F_{22} \end{pmatrix}^\dagger = \begin{pmatrix} F_{11}^* & F_{21}^* \\ F_{12}^* & F_{22}^* \end{pmatrix}. \tag{A.35}$$

A matrix F is called *Hermitian* if $F = F^{\dagger}$. The eigenvalues of a Hermitian matrix are real numbers.

D. **Matrix Inverse**
Inverse: F^{-1}, which is defined such that $F^{-1}F = \mathrm{I}$.

E. **Orthogonal Matrix**
Orthogonal matrix: The matrix elements are real and $F^{-1} = F^{T}$, which is equivalent to $F^{T}F = \mathrm{I}$. Thus, it is easy to find the inverse of an orthogonal matrix: just calculate the transpose.
The rows of an orthogonal matrix constitute an orthonormal set of left (row) vectors. The columns contain an orthonormal set of right (column) vectors.

F. **Unitary Operators**
Unitary operators and matrices have $\mathrm{U}^{-1} = \mathrm{U}^{\dagger}$, that is, the inverse is equal to the Hermitian conjugate:

$$\mathrm{U}^{\dagger}\mathrm{U} = \mathrm{U}^{-1}\mathrm{U} = \mathrm{I}. \tag{A.36}$$

The eigenvalues of a unitary operator satisfy $|\lambda_u|^2 = 1$. The eigenvectors and eigenvalues for a unitary operator (matrix) can be written as

$$\mathrm{U}|\psi_u\rangle = \lambda_u|\psi_u\rangle = e^{ia_u}|\psi_u\rangle, \tag{A.37}$$

where λ_u is the eigenvalue associated with the state $|\psi_u\rangle$ and a_u is a real number.

G. **Involutory Matrices**
Involutory matrices are equal to their inverses. That is $A^{-1}A = AA = A^2 = \mathrm{I}$. We can think of an involutory matrix as the square root of the identity matrix. If the matrix is also symmetric, then it is an orthogonal matrix if the elements are real and a unitary matrix if the elements are complex.

A.7 Matrix Products

There are several types of products of matrices.

A. **Matrix Product**
Matrix product works only if the number of columns in one matrix is equal to the number of rows in the other. For 2×2 matrices,

$$\begin{pmatrix} a_{11} & a_{12} \\ a_{21} & a_{22} \end{pmatrix}\begin{pmatrix} b_{11} & b_{12} \\ b_{21} & b_{22} \end{pmatrix} = \begin{pmatrix} a_{11}b_{11} + a_{12}b_{21} & a_{11}b_{12} + a_{12}b_{22} \\ a_{21}b_{11} + a_{22}b_{21} & a_{21}b_{12} + a_{22}b_{22} \end{pmatrix}. \tag{A.38}$$

The matrix product of a $p \times m$ matrix with a $m \times q$ matrix is a $p \times q$ matrix.
In quantum mechanics, the matrix product gives the matrix representation of operators acting successively within a state space. Normally, the numbers of rows and columns match the dimensionality of the state space.

B. **Matrix Times a Vector**
A matrix multiplying a vector yields another vector (equivalent to an operator acting on a state vector):

$$\begin{pmatrix} F_{11} & F_{12} \\ F_{21} & F_{22} \end{pmatrix}\begin{pmatrix} a_0 \\ a_1 \end{pmatrix} = \begin{pmatrix} F_{11}a_0 + F_{12}a_1 \\ F_{21}a_0 + F_{22}a_0 \end{pmatrix}. \tag{A.39}$$

If the matrix is orthogonal (or more generally, unitary), the length of the newly produced vector is the same as that of the vector being multiplied. For quantum mechanics, we say that orthogonal (unitary) matrices maintain normalization.

C. **Outer Product**

The outer product of matrices F and G:

$$F \otimes G = \begin{pmatrix} F_{11} \otimes G & F_{12} \otimes G \\ F_{21} \otimes G & F_{22} \otimes G) \end{pmatrix} = \begin{pmatrix} F_{11}G_{11} & F_{11}G_{12} & F_{12}G_{11} & F_{12}G_{12} \\ F_{11}G_{21} & F_{11}G_{22} & F_{12}G_{21} & F_{12}G_{22} \\ F_{21}G_{11} & F_{21}G_{12} & F_{22}G_{11} & F_{22}G_{12} \\ F_{21}G_{21} & F_{21}G_{22} & F_{22}G_{21} & F_{22}G_{22} \end{pmatrix}. \tag{A.40}$$

For a multiqubit system in quantum mechanics, the outer product is used to build matrix representations of operators from those that act on the qubit subspaces. In most cases, the outer product is the same as the *Kronecker product*. However, the Kronecker product gives a product of matrices of any size, not necessarily the same, which is useful if the tensor product space is built from subspaces with different dimensions.

Index

E

F

G

H

I

J

K

L